International Conference on Manufacturing, Optimization, Industrial and Material Engineering

(MOIME 2013)

IOP Conference Series: Materials Science and Engineering Volume 46

Bandung, Indonesia
9 – 10 March 2013

ISBN: 978-1-63439-857-2
ISSN: 1757-8981

Printed from e-media with permission by:

Curran Associates, Inc.
57 Morehouse Lane
Red Hook, NY 12571

Some format issues inherent in the e-media version may also appear in this print version.

Printed by Curran Associates, Inc. (2015)

For permission requests, please contact the Institute of Physics
at the address below.

Institute of Physics
Dirac House, Temple Back
Bristol BS1 6BE UK

Phone: 44 1 17 929 7481
Fax: 44 1 17 920 0979

techtracking@iop.org

Additional copies of this publication are available from:

Curran Associates, Inc.
57 Morehouse Lane
Red Hook, NY 12571 USA
Phone: 845-758-0400
Fax: 845-758-2634
Email: curran@proceedings.com
Web: www.proceedings.com

International Conference on Manufacturing, Optimization, Industrial and Material Engineering

(MOIME 2013)

IOP Conference Series: Materials Science and Engineering Volume 46

Bandung, Indonesia
9 – 10 March 2013

ISBN: 978-1-63439-857-2
ISSN: 1757-8981

IOP Institute of Physics

TABLE OF CONTENTS

IOP Conf. Series: Materials Science and Engineering **46** (2013) 011001 doi:10.1088/1757-899X/46/1/011001

Preface

The 2013 International Conference on Manufacturing, Optimization, Industrial and Material Engineering (MOIME 2013), was held at the Grand Royal Panghegar Hotel, Bandung, Indonesia, from 9–10 March 2013.

The MOIME 2013 conference brought together researchers, engineers and scientists in the field from around the world. MOIME 2013 aimed to promote interaction between the theoretical, experimental, and applied communities, so that a high level exchange was achieved in new and emerging areas within Material Engineering, Industrial Engineering and all areas that related to Optimization.

We would like to express our sincere gratitude to all in the Technical Program Committee who reviewed the papers and developed a very interesting Conference Program as well as the invited and plenary speakers. This year, we received 103 papers and after rigorous review, 45 papers were accepted. The participants came from 16 countries. There were six Plenary and Invited Speakers.

It is an honour to present this volume of *IOP Conference Series: Materials Science and Engineering* (MSE) and we deeply thank the authors for their enthusiastic and high-grade contribution.

Finally, we would like to thank the conference chairmen, the members of the steering committee, the organizing committee, the organizing secretariat and the conference sponsors for the financial support that contributed to the success of MOIME 2013.

The Editors of the MOIME 2013

Dr Ford Lumban Gaol
Dr Raja Rizwan Hussain
Tumpal Pandiangan
Dr Amit Desai

Invited and Plenary Speakers

A Landau theory of ferroelectric nanoribbons

Professor Dr Jeff Webb
University of Nottingham Malaysia Camps

Abstract: Most of the work involving the Landau-Devonshire theory of ferroelectrics has been either for bulk materials or thin films. By extending such calculations to two dimensions it is possible to consider a ferroelectric strip. If the thickness and width of the strip are small compared to the surface are of the corresponding surfaces it is expected that the spontaneous polarization near the surfaces will be different from that at points further inside the strip. This talk will show how Landau-Devonshire theory can be applied to such a strip. This is of particular interest now with increasing focus on nanoscale dimensions. Thus the work would be relevant strips whose thickness and length are on the nanoscale but whose length is much greater. After showing how spontaneous polarization profiles can be calculated, an outline of how Maxwell's equations together with Landau-Khalatnikov dynamical equations can be brought in to study the interaction of the strip with incident electromagnetic waves.

Heuristic Optimization in Scheduling of Batch Processing Machines

Professor Dr M Mathirajan
Indian Institute of Science

Abstract: The research and development activities were essentially in the area of Heuristic Optimization for the problems related to Operations Management, Supply Chain/Logistics Management, particularly the emphasis on Scheduling Batch Processors, Scheduling Personnel, Vehicle Routing and Allocation problems, Inventory and Distribution problems. Globally there is greater preponderance towards heuristic optimization in Applied Operations Research/Management Science so that it is possible to deal within complex problems in the Management area. My current research work is focused on meta-heuristic and multi-criteria decision making approaches for Operations Management and Supply Chain/Logistics Management related problems.

Workshop

Optimization with Wolfram Mathematica

Ruben Garcia
Mathworks

Abstract: In this workshop, there are many *Mathematica* Functionalities to be learned.

Magnetic and nonmagnetic tunnel barriers in Sr_2FeMoO_6

Abhishek Nag[1], Somnath Jana[2], Srimanta Middey[2,3] and Sugata Ray[1,2]

[1]Department of Materials Science, Indian Association for the Cultivation of Science, Jadavpur, Kolkata 700 032, India
[2]Centre for Advanced Materials, Indian Association for the Cultivation of Science, Jadavpur, Kolkata 700 032, India

E-mail: mssr@iacs.res.in

Abstract. Tunnelling magnetoresistance (TMR) in polycrystalline Sr_2FeMoO_6 has been a matter of intense interest for more than a decade now, where the nature of the insulating tunnel barrier turned out to be the core issue of debate. Other than the nonmagnetic grain boundaries as conventional tunnel barriers, intragrain magnetic anti-phase boundaries (APB) as well as magnetically frustrated grain surfaces have also been proposed to act as tunnel barriers and influence the TMR response of Sr_2FeMoO_6 in different ways. In this paper, it is shown that depending upon the physical state of the sample, at least three different types of tunnelling mechanism may persist in polycrystalline Sr_2FeMoO_6 and a purely conventional TMR response, seldom reported in this important magnetoresistive material, is also possible.

1. Introduction

Sr_2FeMoO_6 had been identified as an important magnetoresistive (*MR*) material more than a decade ago [1]. The high ferromagnetic Curie temperature ($T_C \sim 420$K) and the consequent room temperature *MR* response at lower applied field [1] had enhanced curiosity considering the technological implication. The half-metallic ferromagnetic (HMFM) nature with large spin-polarization of conduction electrons [1, 2, 3], as well as the significant contribution of polycrystalline grain boundaries in the *MR* response [4] established Sr_2FeMoO_6 as a representative member of TMR class of materials. However, a serious debate about the nature of the tunnel barrier evolved from the fact that the double perovskite crystal structure as well as the involved magnetic ions in Sr_2FeMoO_6, could accommodate a special kind of order-disorder fluctuation [5, 6] which could also generate insulating but magnetic tunnel barriers that can intrinsically exist irrespective of the grain nature of the samples. The core idea was that the ideally alternating Fe/Mo occupancy at the perovskite *B*-site could be easily disrupted in Sr_2FeMoO_6 [5] giving rise to patches of Fe-O-Fe antiferromagnetic insulating regions (APBs) which could also act as tunnel barrier between two adjacent Fe/Mo ordered HMFM domains. Therefore, two alternative mechanisms were simultaneously discussed. In one, the intergrain tunnelling was considered important with physical grain boundaries acting as tunnel barriers, [7] while in the other, the intragrain APBs have been considered to be the main component of the observed TMR [8]. To end the controversy it was needed to independently vary the contributions of these two components and probe the effect on the *MR*, but it turned out to be a difficult task due to the following reasons. Any attempt to increase Fe/Mo ordering (*i.e.* reducing the density/thickness of the APBs) by

[3]Present address: Department of Physics-PHYS226, University of Arkansas, Fayetteville, Arkansas -72701, USA

thermal annealing also simultaneously changed the grain morphology, thereby affecting the intergrain TMR. Moreover, the degree of spin polarization of the conduction electrons was also found to get affected by increasing disorder [2] which further complicated the situation.

Few years ago, we had tried to vary the intergrain and intragrain barriers separately and concluded in favour of the intergrain tunnelling [9] as the dominating contribution to the observed TMR. In doing so, it was assumed that the intergrain tunnelling mechanism remains unaltered between a well annealed piece of sample and a cold-pressed pellet where grains are only physically connected to each other [9]. However, our study revealed unusual magnetic responses in intergrain TMR of cold-pressed samples indicating presence of distinct magnetic states at the grain surfaces. This particular behaviour was termed as spin-valve-type *MR* (SVMR) although the exact SVMR mechanism was revealed only later [10]. Now it is established that the surface of each ferromagnetic grain of Sr_2FeMoO_6 possesses a frustrated spin structure which interacts with the ferromagnetic core and consequently, strong exchange bias effect is seen in the field cooled *MR* measurements [10, 11]. This exchange biasing is intricately connected with the existence of the frustrated surface spin structure and SVMR. In any case, it was concluded that even the intergrain tunnelling in SFMO is far from conventional due to the magnetic nature of the grain surfaces, acting as tunnel barriers [9, 10], which influences the TMR response in a unique way.

In this paper, we show that all three possible TMR responses (intragrain tunnelling across APBs, intergrain tunnelling across magnetic barriers or SVMR and conventional intergrain tunnelling across nonmagnetic barriers) could be accessed in Sr_2FeMoO_6 by controlling the physical state of the sample under study. It is also established that the intragrain tunnelling contribution cannot be ignored and in fact, most of the available reports so far presented *MR* data having significant share from such magnetoresistance responses.

2. Experimental

Well annealed, highly ordered sample was prepared using conventional solid state route [1, 5] at 1250°C. One such pellet was thoroughly crushed in mortar and pestle and then a cold-pressed (CP) pellet was formed at room temperature. Different pieces from a CP pellet have been annealed at 300°C, 600°C, and 900°C. These low temperature annealing could not affect the degree of ordering as the samples were formed at much higher temperature (1250°C), while the grain connectivity was affected because the grains were loosely connected in CP samples. In another attempt, melt-grown pieces were grown in a vacuum arc furnace following the method described in Ref. 5. The as-grown sample possesses large degree of disorder, as reported earlier [5, 12]. Following this, pieces of such molten ingots were annealed at 900°C, 1100°C, 1250°C, 1350°C, and 1500° C. Here, the annealing could hardly affect the grain connectivity as large crystalline grains were already well connected in the molten samples, but could influence the Fe/Mo ion-ordering through high temperature ionic diffusion. The extent of ordering in all these samples was probed by x-ray diffraction studies, carried out in Bruker AXS: D8 Advance x-ray diffractometer with Cu K_α radiation. Magnetoresistance as a function of H ($MR(H)$) was measured on all these samples using a Cryogenics PPMS system and also a homemade resistivity measurement setup, equipped with a low-field electromagnet.

3. Results and discussion

The XRD studies (not shown here) confirmed that the degree of ordering did not improve by annealing the CP sample at 300°C, 600°C, and 900°C. Then, the SVMR behaviour was probed by measuring the exchange bias signal in $MR(H)$ measurements. In this measurement, the samples were at first cooled under an applied magnetic field of 2000 Oe down to 10 K. Then the field was switched off and the $MR(H)$ measurement was carried out at 5 K with the maximum measurement field being at least few times higher than the highest irreversibility field of the corresponding $M(H)$ data in order to avoid any minor loop effect [10]. The results of such measurements on the CP, and 300°C, 600°C, and 900°C annealed samples are shown in figure 1(a). Evidently, the cold-pressed sample exhibited large exchange biasing, reminiscent of the SVMR effect [10]. Interestingly, the exchange bias tends to decrease gradually with increasing annealing temperature and finally disappears in case of the 900ºC annealed sample. This clearly indicates that annealing triggers chemical bonding between the

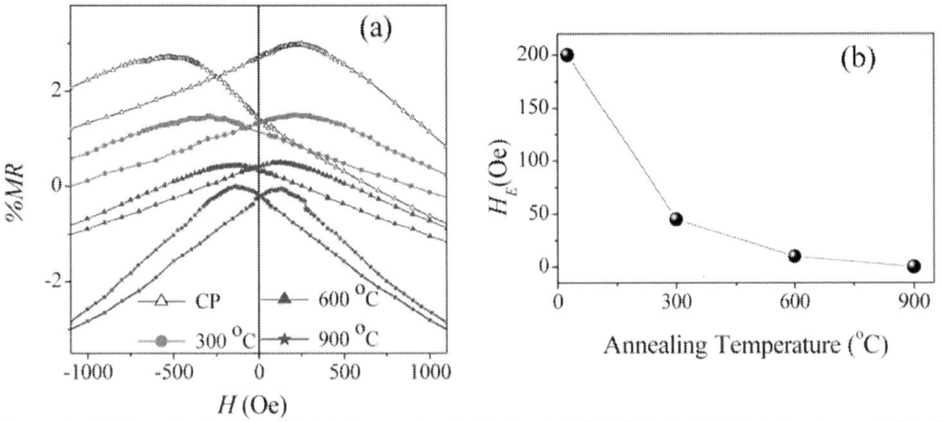

Figure 1. (Colour Online) Panel (a) shows the Field Cooled (2000 Oe) *MR*(*H*) behaviors at 10K for cold pressed (open triangle), 300ºC (full circle), 600ºC (full triangle) and 900ºC (asterisk) annealed samples. Panel (b) shows the exchange bias field (H_E) as a function of annealing temperature.

neighbouring grains which adversely affects the frustrated surface spin structure and as a result SVMR disappears. This is more clearly shown in figure 1(b), where the exchange bias field (H_E) has been plotted as a function of annealing temperature. It is observed that an exchange bias field of 200 Oe in case of the CP sample goes to zero in the sample annealed at 900ºC. One important conclusion that could be drawn from this observation is that SVMR operates only in polycrystalline samples with loosely connected grains while it becomes insignificant in case of well annealed samples. Therefore, as most of the literature discussed only well annealed samples, one can safely conclude that SVMR response could at the most be negligible in those data and therefore, would not qualify to participate in the running controversy about the nature of tunnel barrier in the TMR response of the highly annealed samples.

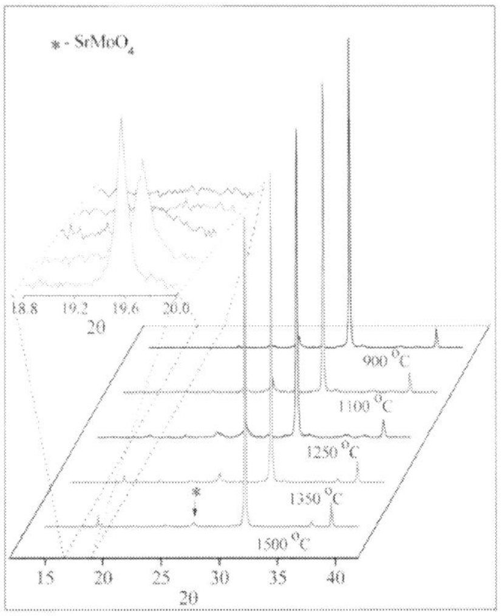

Figure 2. (Colour Online) Normalized x-ray diffraction data from molten samples annealed at different temperatures are shown. The expanded view shows the evolution of the superlattice peak with increasing annealing temperatures. Impurity peak has been marked with an asterisk.

The XRD data from the pieces of molten ingots, annealed at 900°C, 1100°C, 1250°C, 1350°C, and 1500°C are shown in figure 2. The regular increase in the intensity of the superlattice peak establishes that the degree of ordering could indeed be improved by higher temperature annealing.

In figure 3(a), the *MR*(*H*) data from the molten ingots, annealed at 900°C, 1100°C, 1250°C, 1350°C, and 1500◦ C are shown. Evidently, the nature of the *MR*(*H*) has undergone a large variation from 900°C annealed piece to the 1500°C. It is to be noted that all these pieces contain large physical grains and small density of physical grain boundaries while the degree of Fe/Mo disorder is expected to be different from sample to sample. The 900°C annealed sample shows no sharp low-field *MR* (LFMR) and varies almost linearly to reach a value of 11% at 7 Tesla. This behaviour is very similar to the

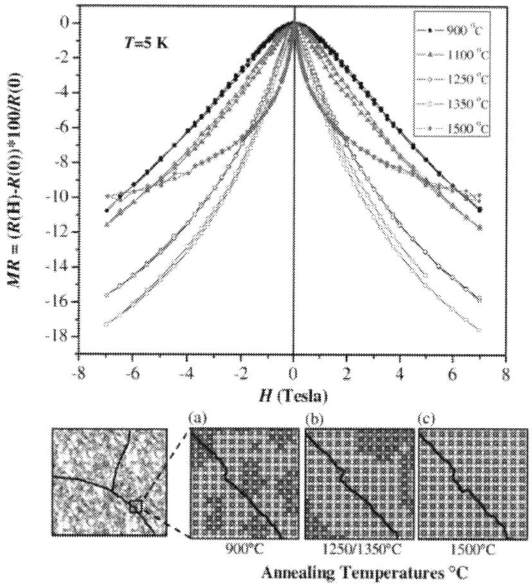

Figure 3. (Colour Online) The *MR*(*H*) behaviours at 5K for the molten ingots, annealed at 900°C (full circle), 1100°C (full triangle), 1250°C (open circle), 1350°C (open square), and 1500°C (asterisk) are shown in the upper part. The lower panels (a), (b), (c) are expanded schematic representations of *B*-site ordering about a grain boundary (dark lines) for different annealing temperatures of the molten ingots. Red and blue spheres represent ordered Fe and Mo atoms while gray and green spheres represent the same ordered atoms shifted *i.e.* red, blue and gray, green portions represent ordered Fe, Mo regions connected through antiphase relation.

behaviour of the as-grown ingot [5], which indicates that an annealing at 900°C hardly affects the degree of ordering. Interestingly, the nature of *MR*(*H*) curve starts to change substantially for samples annealed at or above 1100°C, where a TMR-like sharper low-field response starts to become visible. As the density or thickness of the physical grain boundaries do not change within this range and also there is no effect of SVMR, one can conclude that this observed TMR-like response must clearly be a manifestation of changing density or thickness of the APBs. Considering the fact that the *MR* goes above 17% for the sample annealed at 1350°C at 7 Tesla, it can be argued that the intragrain TMR across APB contributes significantly to the observed TMR of polycrystalline Sr_2FeMoO_6.

We had recently shown that the description of disorder in Sr_2FeMoO_6 is in reality very different in contrast to previous understanding [13]. The proposed scenario is schematically shown in panel (a) at the lower part of figure 3. It was observed that even in the *so-called* highly disordered sample, the local Fe/Mo ordering remains quite high [13] and the traditional disordering picture needs to be replaced by a description of nano-sized highly ordered domains connected via antiphase relationship (the sharp boundaries between the red-blue and gray-green regions). Contrary to previous prediction [2], the degree of spin-polarization is expected to remain significantly high in these *so-called* highly disordered samples. However, for extremely small-sized domains it becomes rather difficult to magnetize them along the direction of the magnetic field and therefore, the *MR*(*H*) remains linear as in the case with 900°C annealed sample. However, with increasing annealing temperature, the domains increase in size (panel (b)) and a nearly bulk-like magnetic behaviour is achieved. As a result, TMR-like response across APBs starts to dominate, as seen in *MR*(*H*) curves of figure 3. However, the most astonishing result is observed in case of the sample annealed at 1500ºC. Suddenly, both the nature of the *MR*(*H*) curve and the magnitude of *MR* changes. It is worth mentioning here that the same reversal was also observed earlier (see figure 1(b) of Ref. 9) although was not analyzed in detail. The most distinct features of this *MR*(*H*) curve is the presence of large low-field *MR* response as is normally observed in other conventional TMR materials like manganites [14] as well as a strong decrease in *MR* value (10% at 7 Tesla). It appears that the tunnelling mechanism transforms into the conventional one where only the nonmagnetic, insulating grain boundaries act as tunnel barriers and the TMR response is completely dominated by the low-field alignment of the magnetic grains, resulting in large LFMR signals. It can be argued that annealing at 1500ºC practically removes all the APBs (see panel (c)) and as a result only the conventional TMR retains. The low density of the physical grain boundaries in the ingots naturally reduces the TMR value. Overall, it appears that the intragrain TMR across APBs' contribute heavily in samples annealed between 1000ºC and 1500ºC and the nature of the *MR* (*H*) curves exhibit distinct differences (low LFMR and gradual changes) as compared with conventional TMR. It is important to note here that most of the results available in the literature fall into this category including Ref. 1 and it is expected that all the reported data had a mixed TMR response and the conventional TMR response in Sr_2FeMoO_6 is rarely observed.

4. Conclusion

In summary, we have shown that three different TMR mechanisms can operate in polycrystalline SFMO and the relative contributions could be easily controlled by carefully manipulating the grain connectivity and disorder density. SVMR type *MR* dominates in samples with poor or no grain connectivity while it disappears after annealing above 900ºC. TMR across APBs does have significant contribution till a perceptible amount of APBs exist. In this case, the *MR* (*H*) curves do not contain large LFMR signal which is a characteristic difference with the conventional TMR signal. Lastly, annealing at highly elevated temperatures (~1500ºC) get rid of all the APBs and only the conventional TMR across nonmagnetic physical grain boundaries remain, which is characterized by a sharp LFMR response.

References

[1] Kobayashi K I, Kimura T, Sawada H, Terakura K and Tokura Y 1998 *Nature* **395** 677.
[2] Saha-Dasgupta T and Sarma D D 2001 *Phys. Rev.* B **64** 064408.
[3] Sarma D D, Mahadevan P, Saha-Dasgupta T, Ray S and Kumar A 2000 *Phys. Rev. Lett.* **85** 2549; Meneghini C, Ray S, Liscio F, Bardelli F, Mobilio S and Sarma D D 2009 *Phys. Rev. Lett.* **103** 046403.
[4] Tomioka Y, Okuda T, Okimoto Y, Kumai R and Kobayashi K I 2000 *Phys. Rev.* B **61** 422.
[5] Sarma D D, Sampathkumaran E V, Ray S, Nagarajan R, Majumdar S, Kumar A, Nalini G and Guru Row T N 2000 *Solid State Commun.* **114** 465.
[6] Kobayashi M, Tanaka K, Fujimori A, Ray S and Sarma D D 2007 *Phys. Rev. Lett.* **98** 246401.
[7] Venkatesan M, Fitzgerald C B, Varadaraju U V and Coey J M D 2002 *IEEE Trans. Magn.* **38** 2901.
[8] Garca-Hernndez M, Martnez J L, Martnez-Lope M J, Casais M T and Alonso J A 2001 *Phys. Rev. Lett.* **86** 2443.
[9] Sarma D D, Ray S, Tanaka K, Kobayashi M, Fujimori A, Sanyal P, Krishnamurthy H R and Dasgupta C 2007 *Phys. Rev. Lett.* **98** 157205.
[10] Ray S, Middey S, Jana S, Banerjee A, Sanyal P,Rawat R, Gregoratti L and Sarma D D 2011 *Europhys. Lett.* **94** 47007.
[11] Middey S, Jana S, Ray S 2010 *J. Appl. Phys.* **108** 043918.
[12] Ray S, Kumar A, Sarma D D, Cimino R, Turchini S, Zennaro S and Zema N 2001 *Phys. Rev. Lett.* **87** 097204.
[13] Meneghini C, Ray S, Liscio F, Bardelli F, Mobilio S and Sarma D D 2009 *Phys. Rev. Lett.* **103**

046403.
[14] Li X W, Gupta A, Xiao G and Gong G Q 1997 *Appl. Phys. Lett.* **71** 1124.

Optimization studies on TFC membrane for Membrane Gas Absorption (MGA) application

A.R. Sunarti[1,*] and A.L. Ahmad[2]
[1]Faculty of Chemical and Natural Resources Engineering, Universiti Malaysia Pahang, Lebuhraya Tun Razak, 26300 Kuantan, Pahang, Malaysia.
[2]School of Chemical Engineering, Engineering Campus, Universiti Sains Malaysia, 14300, Nibong Tebal, Seberang Perai Selatan, Pulau Pinang. Malaysia.
Corresponding Author: Tel: +60 (9) 549-2822; Fax: +60 (9) 549-2889;
Email: sunartirahman@gmail.com

Abstract. A thin film composite (TFC) membrane has been developed by coating polydimethylsiloxane (PDMS) and glutaraldehyde (GH) on a surface porous polyvinylideneflouride (PVDF) membrane for membrane gas absorption (MGA) application. The optimum conditions for dip coating method were determined using response surface methodology (RSM). A central composite design (CCD) was used to investigate the effects of two independent factors, which PDMS concentrations (wt%) and dipping time (s) of GH on the four specific responses which are CO_2 and N_2 permeances, selectivity and contact angle (CA) value. The optimum conditions for PDMS concentration and GH dipping time are 10 wt% and 19 s, respectively where 354 GPU for CO_2 permeance, 66 GPU for N_2 permeance, 5.4 of selectivity and 132° of CA value were obtained. Through atomic force microscopy (AFM) analysis, the result shown the root mean square roughness (R_{ms}) of the TFC membrane was 381 nm and it was double from untreated membrane R_{ms} value. Therefore, the roughness of the surface membrane contributed to the performance of the separation in the process flow such as in MGA application. By coating PDMS, hydrophobicity of the surface membrane was improved as well.
Keywords: TFC Membrane; Dip Coating; PDMS; Response Surface Methodology

1. Introduction

Membrane gas absorption (MGA) is a gas–liquid contacting operation. The main element in the MGA process is a microporous hydrophobic polymeric membrane. The gas stream is fed to one side of the membrane, and at the same time, liquid absorbent flows on the other side of the membrane. The microporous hydrophobic membrane's serves as a wall to separate the gas phase from the liquid absorbent. The liquid absorbent should have a high attraction for components that must be removed from the gas stream. Meanwhile, these components will diffuse through the gas-filled pore of the membrane from the surface to the other side of the membrane, which is followed by absorption of the component in the liquid absorbent. Basically, absorption in the liquid occurred by physical absorption (e.g., water) [1] or by a chemical reaction (e.g., amine solution) [2]. Therefore, the selectivity of this process is determined by the strength of the liquid absorbents that absorb the component. Even the membrane does not contribute to the selectivity; however, the permeability of the component is dependent on the pores of the membrane. In addition, the membrane also plays an important role in ensuring that the two phases are separated and providing a large contact surface area [3,4]. MGA is also used for the removal of carbon dioxide (CO_2), sulphur dioxide (SO_2), nitrogen dioxide (NO_2) and other nitrogen oxides (NOx), hydrogen sulphide (H_2S), ammonia (NH_3), tropospheric ozone (O_3) and many other different compounds from gas streams, especially those that exist in flue gas, in natural gas, indoors and in off gases [5,6].

The selection of the type of membrane has a significant effect on the MGA process. Because the liquid absorbents are generally aqueous solutions that are used for absorption purposes, wetting problems seem like a major issue for MGA. Using a hydrophobic membrane leads to larger contact angles and thus is efficient in minimising the wetting problem. Polytetraflouroethylene (PTFE) membrane material has a very low surface energy (19.1 mN/m), and this material has a high resistance to wetting. However, PTFE membranes are very expensive. In contrast, polypropylene (PP) (30.0 mN/m) and polyvinylideneflouride (PVDF) (30.3 mN/m) membrane materials have a sufficiently low surface energy, are inexpensive and are commercially available with small pore diameters [7,8].

Thin film composite (TFC) technology has been developed for top surface coating supports in the fabrication of reverse osmosis membranes [9]. Many methods of fabricating TFC membranes have been reported in the literature. These include surface grafting, pore-filled grafting and dip coating. The dip-coating process was successfully implemented for membranes before the measurement of gas separation permeance because the silicone rubber was shown to be able to recover the intrinsic permselectivity of the skin layer of an asymmetric membrane [10]. Meanwhile, the separation and flux behaviour may be interpreted in terms of both chemical and structural changes occurring within the composite coating [11]. In the current study, the dip coating was adapted to coat the surface of a microporous PVDF membrane using silicone rubber. We propose an alternative method for decreasing water molecule interactions (increasing the hydrophobicity) using the dip-coating method. The aim of the present study was to determine the effect of the silicone rubber concentration and the dipping time on the cross-linking agent in a PVDF membrane based on CO_2 and N_2 single permeation, selectivity and contact angle (CA) measurements. To maximise the permeation of CO_2, an optimisation process was followed using response surface methodology (RSM) with a central composite design (CCD). The properties of the optimised surface of the TFC membrane were compared with an untreated PVDF membrane.

2. Methods

2.1 Materials

Flat sheets of 0.1 μm-thick PVDF membranes were purchased from Millipore Asia LTD. Polydimethylsiloxane (PDMS), silicone rubber and glutaraldehyde (GH, 25% aqueous solution, cross-linking agent) were procured from Sigma Aldrich (M) Sdn. Bhd. All solvents were used without further purification. The material properties are tabulated in Table 1 and Table 2.

Table 1: Properties of the membranes

Properties	0.1 μm PVDF membrane
Module:	Flat Sheet
Diameter:	470 mm
Thickness:	125 μm
Porosity:	70%
Wettability:	Hydrophobic

Table 2: Properties of the chemicals

Properties	PDMS	GH
Molecular formula:	$(C_2H_6OSi)_n$	$C_5H_8O_2$
Density:	965 kg /L	1.06 g/mL
Melting Point:	N/A	-14 °C
Boiling Point:	below approximately 200 °C	187 °C
Molar Mass:	N/A	100.12 g/mol
Chemical Structure:	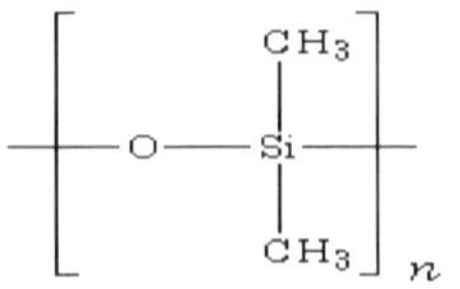	

2.2 TFC preparation

PDMS samples were mixed in various proportions of hexane. The solvent-exchanged membranes were dipped in the PDMS solution of various concentrations (in the range of 6-12%, w/v) for 30 s. This was followed by dipping the membrane pieces in 20% aqueous GH solution for varying amounts of time (in the range of 10-30 s). The membranes were then cured first by drying under ambient conditions for 5 min and then by curing in an oven at 60 °C for 30 min.

2.3 Pure gas permeation measurement

The permeability and selectivity of the TFC membrane were determined by the constant pressure variable volume method. The properties of CO_2 and N_2 are stated in Table 3.

Table 3: Properties of CO_2 and N_2

Gas	Kinetic Diameter, d (Å)	Molecular Weight, M (g/mol)	Viscosity, η (cm Hg.s)
CO_2	3.3	44	0.08×10^{-6}
N_2	3.64	28.01	1.81×10^{-9}

Gas permeation tests were performed with a rig setup, as shown schematically in Fig. 1. Circular membrane discs were cut and mounted on a porous support in a stainless steel, cylindrical membrane test cell and tightened by a rubber O-ring. The effective permeation area of each membrane was 7.38 cm^2. Prior to testing, care was taken to check for leaks and to fully flush and purge the system with a test gas. The gas permeation rates were measured by using pure gases, such a CO_2 and N_2, from compressed gas cylinders at 1 bar. The experiments were carried out at ambient temperature (30°C). Gas permeation rates were measured by a soap bubble flow meter. Each set of data was determined as an average of three replicates.

The permeance (P/l) for TFC membranes was calculated using eq. (1) and expressed in GPU (1 GPU = 1×10^{-6} cm^3(STP)/(cm^2 s cm Hg)).

$$\frac{P}{l}=\frac{V}{At\Delta p}\ (cm^3(STP))/(cm^2\ s\ cm\ Hg) \quad (1)$$

where l is the thickness of the membrane in cm, A the effective membrane area in cm^2, V is the volume (cm^3) displaced in time t (s) and Δp is the transmembrane pressure expressed in cm Hg. The selectivity (α) is expressed as the ratio of two pure gas permeabilities, as stated in eq. (2).

$$\alpha_{ij}=(P/l)_i/(P/l)_j \quad (2)$$

where i is CO_2 and j is N_2.

2.4 Contact angle (CA) measurement

In this work, the values of the CA of deionised water on different surface membranes were measured. The measurement was performed on sessile drops with a computerised Optical Contact Angle SCA 15 from Germany. Using the SCA 15, drop volume was controlled with a microsyringe followed by placing the surface near the needle; 6 μl of deionised water was squeezed out of the syringe. The surface was pulled out of the drop. Finally, the contact angles were measured by using the Laplace-Young equation, and the image was captured. The effects of evaporation of the liquids used were minimised by finishing each measurement within a period of less than 1 min. It was found experimentally that the contact angle remained constant within this period. The result of all measurements was the mean of at least 10 single measurements.

2.5 Experimental design and RSM

To design the experimental plan, response surface methodology (RSM) was utilised to optimise the TFC membrane performance, and a full fractional central composite design (CCD) was adopted to fit a second-order model. The design consisted of 13 sets of experiments. The basic theoretical aspects, the fundamentals assumptions and the experimental implications of RSM have been discussed elsewhere (Montgomery, 2001). The second-order model was selected for predicting the optimal point and is expressed as eq. (3).

$$Y=\beta_0+\beta_1X_1+\beta_2X_2+\beta_{11}X_1^2+\beta_{22}X_2^2+\beta_{12}X_1X_2 \quad (3)$$

where Y represents response variable (CO_2 and N_2 single permeation, selectivity and CA), β_0 is the interception coefficient, β_1 and β_2 are linear terms, β_{11} and β_{22} are quadratic terms and X_1 and X_2 are the independent variables studied (PDMS concentration and dipping time of GH). The ranges and the levels of the variables investigated in the study are given in Table 4.

Table 4: Experimental range and level of the independent variables

Variables	Range and level (coded)		
	-1	0	+1
PDMS concentration, *A* (%w/v)	6	9	12
Dipping time of GH, *B* (s)	10	20	30

A complete description of the experimental design is shown in Table 5. Panel I in Table 5 represents the level in terms of coded variables, while Panel II shows the conditions used with the original unit measurement.

Table 5: CCD for the study of two experimental variables in coded units

Run no.	Panel I		Panel II	
	PDMS concentration, *A*	Dipping time of GH, *B*	PDMS concentration, *A* (%w/v)	Dipping time of GH, *B* (s)
1	1	1	12	30
2	0	0	9	20
3	-1	1	6	30
4	-1.682	0	5	20
5	0	0	9	20
6	0	0	9	20
7	1	-1	12	10
8	0	-1	9	10
9	0	-1.682	9	6
10	1.682	0	13	20
11	-1	-1	6	10
12	0	0	9	20
13	0	1.682	9	34

Regression analysis was performed by Design Expert v.6.0.7 (Stat- Ease Inc. Minneapolis). Analysis of variances (ANOVA) was used for graphical analyses of the data to obtain the interaction between the process variables and the responses. The quality of the fit polynomial model was expressed by the coefficient of determination R^2, and its statistical significance was checked by the Fisher *F*-test in the same program. Model terms were selected or rejected based on the *P* value (probability) with a 95% confidence level. Three-dimensional plots and their respective contour plots were obtained based on the effects of the levels of two variables. From these three-dimensional plots, the simultaneous interaction of two factors on the responses was studied.

3. Results and discussion

3.1 Experimental results

Experiments were conducted based on the design in Table 5, and the relevant results are shown in Table 6, which lists the CO_2 and N_2 single permeation, the selectivity and the CA value.

Table 6: Experimental results

Run no.	CO_2 permeance $cm^3(STP)/cm^2s.cm\ Hg$	N_2 permeance $cm^3(STP)/cm^2s.cm\ Hg$	selectivity	CA (°)
1	343.11	132.42	3.96	122.38
2	337.07	62.02	5.03	132.53
3	157.67	83.41	1.35	128.87
4	235.33	82.91	0.91	126.92
5	339.01	61.21	5.31	132.42
6	337.28	61.73	5.00	132.45
7	390.01	91.03	5.31	134.75
8	337.30	60.88	4.69	132.36
9	396.19	112.54	7.09	143.21
10	405.32	123.97	3.88	122.72
11	327.44	142.76	4.00	124.02
12	337.01	61.06	5.23	132.40
13	216.93	133.21	2.79	123.05

3.2 ANOVA analysis

The results are further analysed using Design Expert Software. The relationship between two controllable variables (PDMS concentration and dipping time of GH) and four important responses (CO_2 and N_2 permeance, selectivity and CA value) for the TFC membrane performance is studied. Model terms that are significant are desired to obtain a good fit in a particular model. A CCFD shown in Table 4 allows the development of mathematical equations in which each response variable Y is assessed as a function of PDMS concentration (A) and dipping time of GH (B) and calculated as the sum of a constant, two first-order effects (terms in A and B), one interaction effect (AB) and two second-order effects (A^2 and B^2), according to Eq. (3). The results obtained are then analysed by ANOVA to assess the "goodness of fit". Only terms found to be statistically significant are included in the model. The non-significant terms can be reduced by reselecting only the significant terms to be included in the model. The model terms with "Prob > F > 0.5" will be eliminated from the model.

The A, B, A^2, B^2 and AB model terms are found to be significant for the CO_2 permeance. The A^2, B^2 and AB model terms are found to be significant for the N_2 permeance, and the A, A^2, B^2 and AB model terms are found to be significant for the selectivity. Meanwhile, the B, A^2 and AB model terms are found to be significant for the CA value. These results clearly show that interaction effect between the PDMS concentration and the dipping time of GH plays an important role in generating a high CO_2 permeance and producing a higher CA value. The quadratic model is well fitted to the observed data, and the following empirical models in terms of coded values are obtained for the CO_2 and N_2 permeance, selectivity and CA value, respectively:

1) CO_2 permeance
$Y_1 = +337.33 + 55.10A - 11.25B - 27.83A^2 - 54.23B^2 + 4.47AB$
2) N_2 permeance
$Y_2 = +61.49 + 22.42A^2 + 30.35B^2 + 25.18AB$
3) Selectivity
$Y_3 = +5.49 + 0.33A - 1.46A^2 - 1.77B^2 - 0.47AB$
4) CA value
$Y_4 = +132.46 - 4.51B - 4.59A^2 - 4.30AB$

The coefficients with one factor represent the effect of the particular factor, while the coefficients with second-order terms represent the quadratic effect. A positive sign in front of the term indicates a synergistic effect, while a negative sign indicates an antagonistic effect. The statistical parameters obtained from the ANOVA for the reduced model of the responses are given in Table 7.

Table 7: Statistical parameters obtained from the ANOVA for the reduced models

Variables	CO_2 permeance	N_2 permeance	Selectivity	Contact angle
Significant terms	A,B,A^2,B^2,AB	A^2,B^2,AB	A,A^2,B^2,AB	B,A^2AB
R^2	0.9997	0.9347	0.9868	0.8358
R^2 adjusted	0.9994	0.8879	0.9773	0.7185
Prob > F	<0.0001	0.0005	<0.0001	0.0013
Adequate precision	176.495	11.897	22.646	9.247
Standard deviation, SD	1.48	10.67	0.25	3.18
Coefficient of variance, CV	0.51	11.47	7.05	2.45
Probability of lack of fit	0.32	0.01	0.07	0.01

The ANOVA results of the quadratic models with Prob > F < 0.05 for all the responses presented in Table 7 indicate that the model equation adequately describes the response surfaces of CO_2 permeance, N_2 permeance, selectivity and CA value in the interval of the investigation. The effect of each variable on the response is a combination of the coefficients and the variable values, as well as a joint effect of the variables that cannot be observed by conventional experimental methods. The high R^2 value, close to 1, is desirable, and the predicted R^2 must be in reasonable agreement with the adjusted R^2 for a significant model (Nordin et al., 2004). The values of R^2 for CO_2 permeance, N_2 permeance and selectivity and CA value are 0.9997, 0.9347, 0.9868 and 0.8358, respectively. In this case, these values indicate that only 0.03–16.42% of the total variation is not explained by the model. The values of the adjusted R^2 of 0.9994, 0.8879, 0.9773 and 0.7185, respectively, for CO_2 permeance, N_2 permeance selectivity and CA value, are also high to support the high significance of the model [12]. The CV, the ratio of the standard error of the estimate to the mean value of the observed response (as a percentage), is a measure of the reproducibility of the model. As a general rule, a model is reasonably reproducible if its CV is not greater than 10%. The CV values obtained for all responses studied in Table 7 are below 10%, except the value for N_2 permeance. This result is reasonable because the goal of this research is to obtain a high permeability of CO_2 instead of N_2, which is more important for controlling the structure of the PVDF membrane for high selectivity and CA values.

It is important to check the fitted model to ensure that is provides an adequate approximation to the real system. Fig. 2 shows a comparison of the actual values of the respective responses that were obtained from the experiments and the estimation values from the regression model. As observed in Fig. 2, points above the diagonal line were those overestimated and vice versa. All four graphs shown in Fig. 2 generally indicate that all experiment design points were distributed along the diagonal line. This shows that the models are adequate for predicting the CO_2 permeance, N_2 permeance, selectivity and CA value within the range of the variables studied.

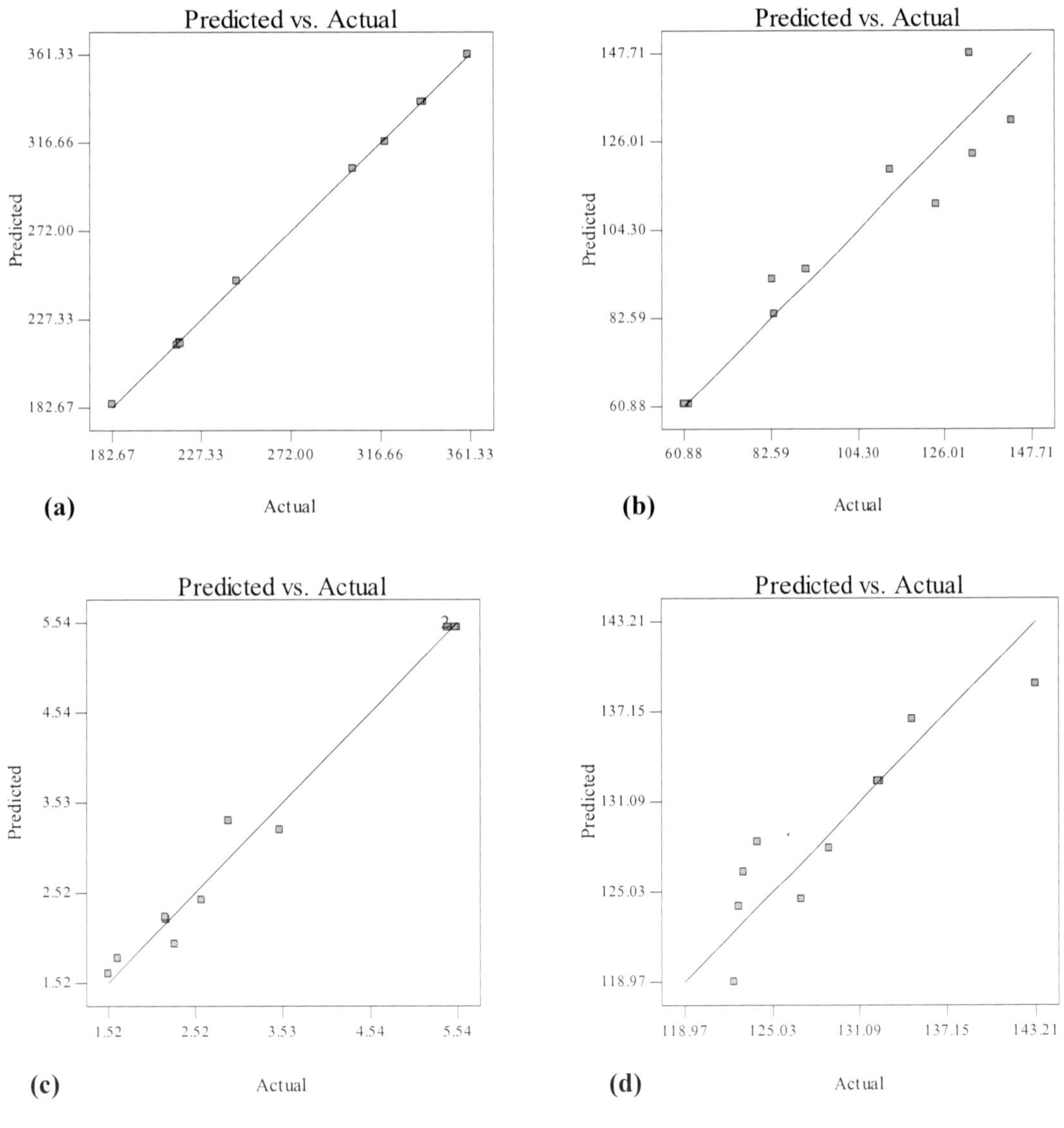

Fig. 2: Design-expert plot. Predicted vs. Actual plot for: (a) CO_2 permeance, (b) N_2 permeance, (c) selectivity and (d) contact angle

3.3 Analysis of the response surface

The response surface plot for CO_2 permeance is shown in Fig. 3(a). The figure shows the effect of the concentration of PDMS on the CO_2 permeance at the various dipping times for 20% wt of GH as a cross-linking agent. From the figure, a peak suggests that the optimum condition for maximum CO_2 permeance is well inside the design boundary. There is a clear elongated hill running along the time in glutaraldehyde axis on the plot of the three-dimensional response surface of the quadratic model for the CO_2 permeance. As can be observed in Fig. 3(a), the maximum CO_2 permeance is achieved at a concentration of PDMS of 10 -11 wt% and between 15- 20 s of GH dipping time. For further increases in the concentration of PDMS and the GH dipping time beyond the optimum condition, the CO_2 permeance will decrease.

Fig. 3(b) shows the response surface for N_2 permeance. The figure indicates that the N_2 permeance significantly decreases with the PDMS concentration and time of GH dipping. In the same range, 10 -11 wt% of PDMS concentration and between 15- 20 s of GH dipping time, the lowest peak of N_2 permeance is achieved. Fig. 3(c) shows the response surface for CO_2/N_2 selectivity. As can be observed in Fig. 3(c), the maximum peak of selectivity is very clear at the concentration of PDMS of 10 -11 wt% and between 15- 20 s of GH dipping time.

From Fig. 3(a-c), we also found that the defects on the surface of the skin layer of the membranes were completely sealed with PDMS silicone rubber film. Silicone rubber was proven to be able to recover the intrinsic permselectivity of the skin layer of an asymmetric membrane [13]. In addition, the CO_2 permeance decrease after the optimum region is due to a decrease in the surface porosity of the composite membrane by the PDMS layer. In contrast, the GH as a cross-linking agent may contribute to the existence of resistances in the substructure and the inner skin layer of the membranes. Therefore, the membrane pores became smaller and only allowed CO_2 to pass through rather than N_2. Therefore, the influence of these resistances is more significant to CO_2 permeance and CO_2/N_2 selectivity, respectively. However, according to Shieh et al., if multilayers of silicone rubber were applied on membrane surfaces, there would be a decrease in gas permeance because this leads to a thicker layer of coating. In contrast, the selectivity is significantly improved due to the sealing of defects by the silicone rubber [14]. Because the membrane does not contribute to the selectivity in the MGA process, it is important to produce a membrane with a high permeability towards CO_2. Meanwhile, Fig. 3(d) demonstrates the effect of the concentration of DPMS and the dipping time of GH on the CA value of the surface membrane. The hydrophobicity of membranes is represented in terms of the CA between the water and membrane. The CA value is an angle between the liquid and the solid. In the case of water, a material is deemed to be hydrophobic if the contact angle is higher than 90° (Wang et al., 2004). As the concentration of PDMS and the dipping time of GH are increased, the CA also increases. As we can see, the optimum value of CA is approximately 133° at the concentration of PDMS of 7.5- 11 wt% and between 15- 20 s of GH dipping time. However, Fig. 3(d) also shows that an increased concentration of DPMS and an increased dipping time of GH decrease the CA values. This might be due to the interaction between PDMS and glutaraldehyde past the optimum conditions, contributing to an incomplete seal on the surface of the membrane. The consequences would involve the creation of pores that were not hydrophobic enough, resulting in a decline of the CA values.

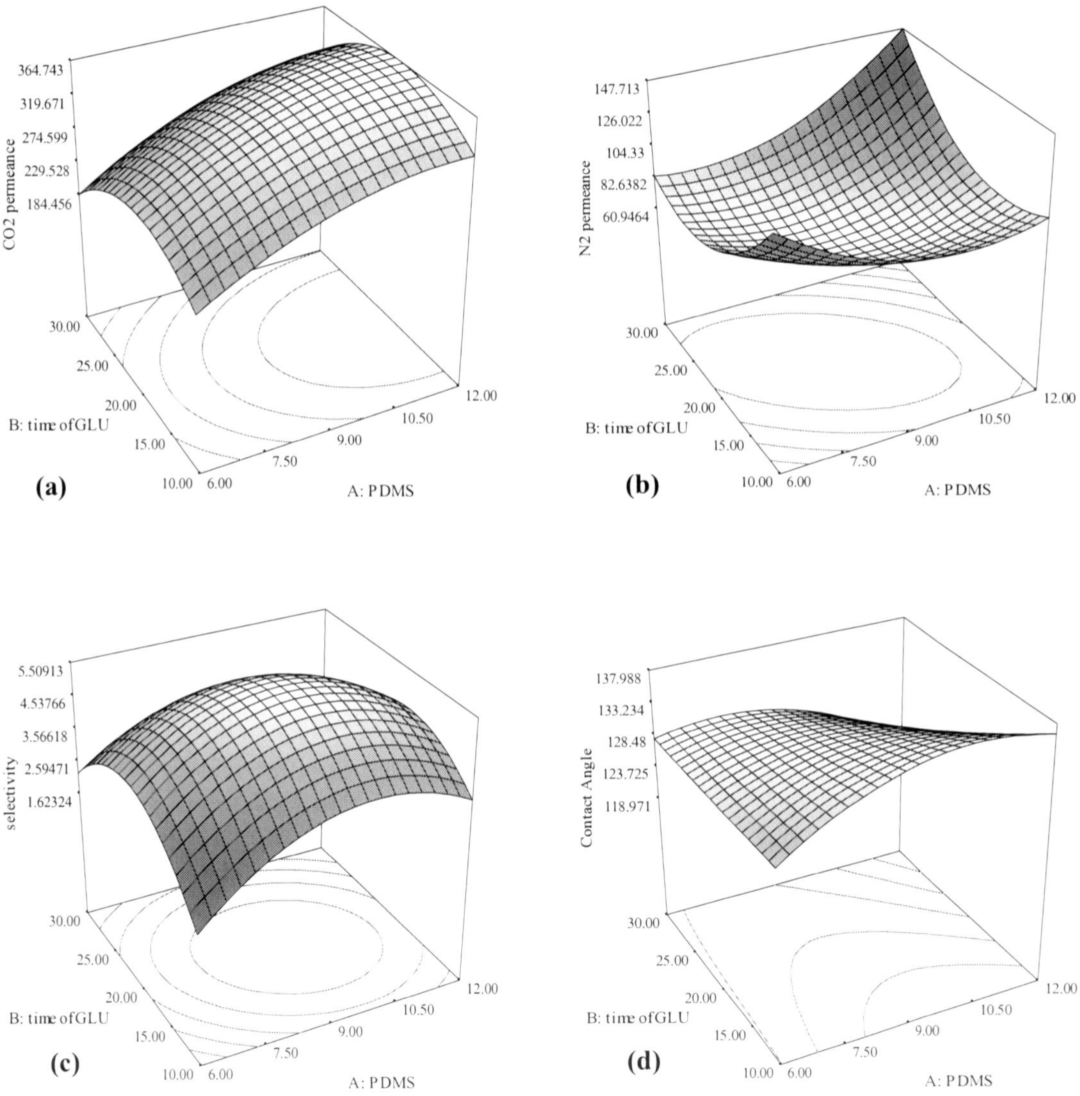

Fig. 3: Design-expert plot. Response surface plot for (a) CO_2 permeance, (b) N_2 permeance, (c) selectivity and (d) CA value

3.4 Optimisation analysis

Numerical optimisation was used to generate the optimal conditions for the targeted responses. Optimisation finds a set of operating conditions that either optimises all of the responses or at least keeps them within the desired range. In this model, the responses for CO_2 permeance, selectivity and CA value were required to be maximised. However, the N_2 permeance value was set in a range. All the criteria for the optimisation analysis are tabulated in Table 8. Design Expert software was used to obtain the optimal conditions for all factors.

Table 8: Model validation

Factors	Goal	Value
Variables		
PDMS concentration	is in range	10
Time for glutaraldehyde	is in range	19
Responses		
CO_2 permeance	maximise	360.32
Predicted		354.83
Experimental		342.76
N_2 permeance	is in range	
Predicted		66.69
Experimental		65.66
Selectivity	maximise	5.54
Predicted		5.41
Experimental		5.22
CA value	maximise	143.21
Predicted		132.29
Experimental		130.21

As mentioned in Table 8, the level of variables optimising the predicted response was 10 wt% of PDMS and 19 s of GH dipping time. Under these conditions, the predicted CO_2 and N_2 permeance was 354.83 GPU and 66.69 GPU, respectively. Meanwhile, the predicted selectivity and CA value was 5.41 and 132.29°, respectively. Using these conditions, a verification experiment was performed. The experimental values were close to the predicted values from the model. Small errors existed between the predicted and experimental values, which were 3.4%, 1.54%, 3.51% and 1.57%, respectively, for CO_2 permeance, N_2 permeance, selectivity and CA value.

After the optimisation was performed, the properties of the optimised surface of the TFC membrane were compared with those of an untreated PVDF membrane. The root mean square roughness (R_{ms}) and maximum and minimum elevations recorded for the optimised TFC membrane and an untreated PVDF membrane are shown in Table 9.

Table 9 shows that the TFC membrane surface has a roughness which is significantly higher than that of an untreated PVDF membrane surface.

Table 9: The comparison properties

Membrane	CA value	R_{ms} (nm)	Maximum elevation (nm)	Minimum elevation (nm)
Optimised TFC membrane	130.21	381	1277	-1141
Untreated PVDF membrane	122.17	197	746	-840

Atomic force microscopy (AFM) was used to show the roughness of both membranes, as shown in Fig. 4. The figure shows the lumpy aggregates were observed from the AFM image. The lumpy aggregates protrude out of the surface and can be observed as bright high peaks, while the pores are depressed into the surface and can be observed as dark regions. Despite this result, the optimised conditions of the TFC membrane provided the maximum CO_2 permeance, selectivity and CA value. The AFM analysis revealed that roughness contributes to the hydrophobicity of the TFC membrane. This membrane will be suitable for use in process flow separation for processes such as the MGA process.

4. Conclusions

Optimisation of the dip coating of PVDF membranes with respect to CO_2 and N_2 permeances, selectivity and CA values has been investigated. Response surface methodology using CCD was applied to determine the optimum conditions for obtaining the maximum CO_2 permeance, selectivity and CA value and the lowest N_2 permeance. The optimum conditions for PDMS concentration and glutaraldehyde dip coating time are 10 wt% and 19 s, respectively, where 354 GPU for CO_2 permeance, 66 GPU for N_2 permeance, a selectivity of 5.4 and a CA value of 132° can be obtained. The dip coating was also used to create a hydrophobic surface through surface roughness. This is proved by the R_{ms} values, which are 381 nm and 197 nm for the TFC membrane and an untreated PVDF membrane, respectively.

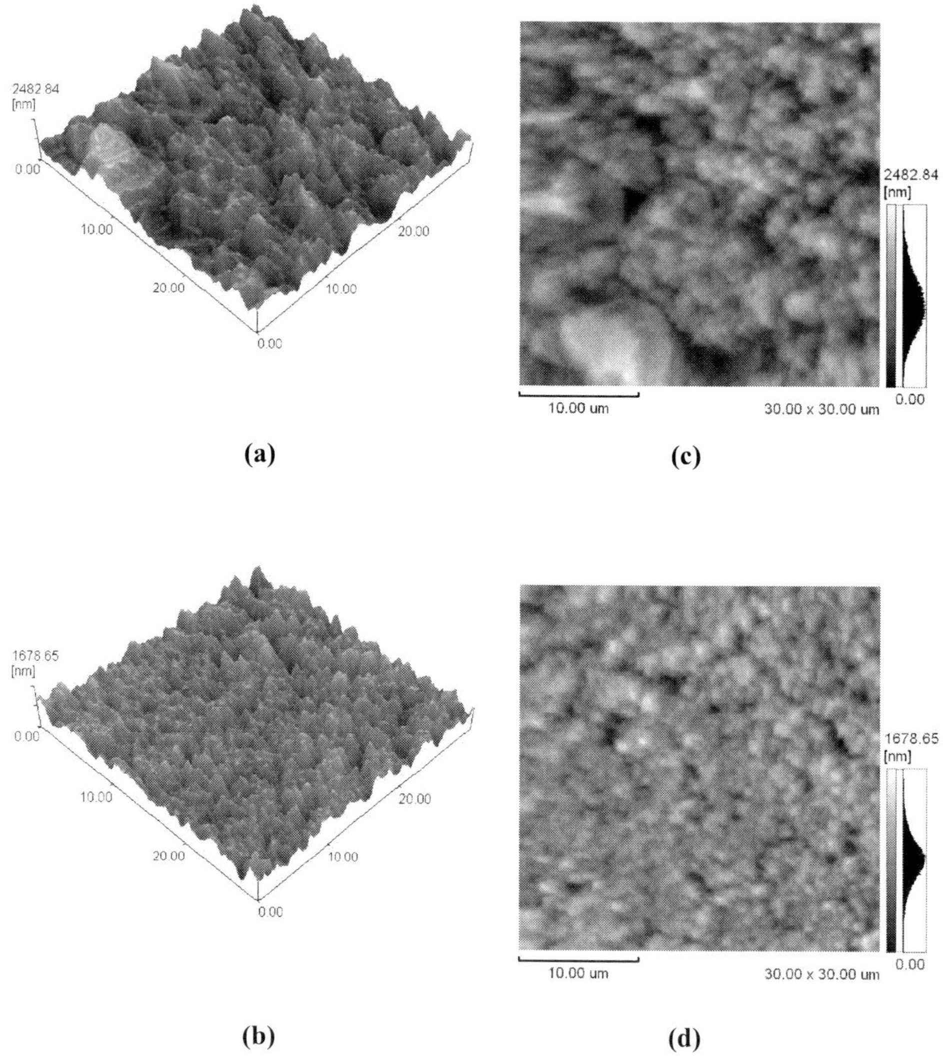

Fig. 4: AFM images: (a) three-dimensional optimised TFC membrane, (b) three-dimensional untreated PVDF membrane, (c) top surface of the optimised TFC membrane and (d) top surface of the untreated PVDF membrane

References

[1] Kreulen H *et al* 1993. *J. Membr. Sci.* **78** 197
[2] Matsumiya N *et al* 2005. *Sep & Purif Technol* **46** 33
[3] Klaassen R *et al* 2008. Desalination **224** 81
[4] Feron P H M and Jansen A E 2002. *Sep & Purif Technol* **27** 231

[5] Wang R *et al* 2004. *J. Membr. Sci.* **229** 147
[6] Todorovića J *et al* 2006. *Desalination* **193** 286
[7] Kumar P S *et al* 2002. *Chem. Eng. Sci.* **57** 1639
[8] Mulder M 1991. *Basic principle of membrane technology. 2nd ed.* (Dorddrecht: Kluwer Academic)
[9] Dickson J M *et al* 1998. *J. Membr. Sci.* **148** 25
[10] Qin J J and Chung T S 2004. *J. Membr. Sci.* **229** 1
[11] Trushinski B J *et al* 1998. *J. Membr. Sci.* **143** 181
[12] Khuri A I and Cornell J A 1996. *Response Surfaces, Design and Analyses, 2nd ed.* (Marcel Dekker Inc., New York)
[13] Henis J M S and Tripodi M K 1980. US Patent 4,230,463
[14] Shieh J J et al 1999. *Chem. Eng. Sci.* **54** 67

IOP Conf. Series: Materials Science and Engineering **46** (2013) 012003 doi:10.1088/1757-899X/46/1/012003

Electroless Ni-P plating and Heat Treatments on S45C Carbon Steel

Shih-Hsien Chang*, Chih-Chung Chang and Cheng Liang
Department of Materials and Mineral Resources Engineering, National Taipei University of Technology, Taipei 10608, Taiwan, R. O. China

E-mail: changsh@ntut.edu.tw

Abstract. This research focused on the effects of S45C carbon steel using different pH values of electroless Ni-P plating. The experimental results show that the specimens using acid and alkaline baths of pH5 and pH8 of Ni-P deposition can acquire a higher hardness; the hardness values were the $HV_{0.05}$ 701 and $HV_{0.05}$ 803, respectively. Moreover, for pH5 and pH8 of Ni-P deposition with heat treatment of 350°C and 300°C and soaking for 1 hr, the highest level of hardness appeared in $HV_{0.05}$ 1027 and $HV_{0.05}$ 1066, respectively. The specimens underwent the main precipitate phase of Ni_3P after heat treatment. Due to precipitation, strengthening of the Ni_3P phase increased the surface hardness of specimens after heat treatment; thus, the wear depth was decreased and wear resistance was enhanced. The optimal process appeared in Ni-P deposition by pH5 and pH8 after 350°C and 300°C heat treatment and for 1 hr of soaking, respectively.

1. Introduction

Electroless deposition is a variety of chemical deposition technology, involving the deposition of metals from solution onto surfaces without applying an external electric voltage and is thought to be the simplest and most economic method to finish steel, aluminum, copper, plastics and many other materials [1-3]. Electroless Ni-P (EN) plating technique, as an effective surface treatment method, has been widely used in a variety of environments due to its excellent corrosion resistance. In recent years, electroless Ni-P plating technology has been used in steel industry [4-6].S45C carbon steel (S45C) has been widely used in industrial applications, such as crank shafts, gears, main spindles of machine tools, connecting rods, etc., because of its distinguished mechanical property [7]. However, how to enhance the surface hardness and wear resistance for S45C is the object work in the industry. Several surface hardening for S45C and other materials have been discussed [8-12]. Most studies focus on the different pH values of electroless Ni-P plating and discuss various heat treatments for electroless Ni-P plating. However, no related paper has reported on the electroless Ni-P plating on S45C carbon steel by various heat treatments. The aim of this study is to evaluate surface microstructure, wear resistance and optimal parameters of heat treatment of electroless Ni-P plating in regard to S45C carbon steel.

2. Experimental Procedures

In this study, JIS S45C carbon steel was used as the substrate material for deposition of EN, the specimen size is Ø36 × D5 mm and the chemical compositions are listed in Table 1. In order to improve its microstructure and mechanical properties, first the sample of S45C was oil quenched at 850°C and tempered at 400°C, with the tempering process repeated twice to reach a hardness of $HV_{0.05}$

421. After these operations, the S45C was processed by grinding and polishing to finish the specimen surface; it was then put into the acetone solution and an ultrasonic vibrator used for 20 min, followed by cleaning and rinsing with de-ionized water. Secondly, it was immersed in the 15% hydrochloric acid for 20 min to remove the oxide film. The sample was cleaned and rinsed again with de-ionized water, so that the EN plating could be performed.

In addition, the plating bath was divided into an acid and alkaline bath (pH4, 5, 6, 8, 9 and 10); sulfuric acid and ammonia were used to adjust the pH value. The bath compositions, concentration and operational conditions are listed in Table 2. After these operations, all of the samples underwent five different heat treatment temperatures (300°C, 350°C, 400°C, 450°C, 500°C and soaked for 1 hr) to evaluate the effects on Ni-P deposition. Hardness of the specimens was measured by Vickers indenter (HV) with loading of 50 g, which complied with the ASTM E384-08a standard method. The wear resistance of specimens was evaluated in a ball-on-disk test. The wear parameters are: specimen size is Ø36 × D5 mm, diameter of Cr ball is 6 mm, axial load is 4.9 N, disc rotation is 100 rpm, sliding speed is 0.031 ms^{-1}, and total rotation is 3000 revolutions. The wear depth was measured by α-step equipment (ET- 4000A).

Table 1 Chemical composition of the S45C carbon steel (wt %)

C	Si	Mn	S	P	Fe
0.46	0.21	0.72	≦0.04	≦0.03	Bal.

Table 2 Bath composition and operating conditions of S45C carbon steel by electroless Ni-P deposition

Process	Bath composition	Concentration (g/L)	Condition
	$NiSO_4 \cdot 6H_2O$	30	84±1°C
Acid bath	$NaH_2PO_2 \cdot H_2O$	15	1 hour
	$Na_2C_4H_4O_4 \cdot 6H_2O$	12	pH 4, 5, 6
	$NiSO_4 \cdot 6H_2O$	30	84±1°C
Alkaline bath	$NaH_2PO_2 \cdot H_2O$	10	1 hour
	$NaC_6H_5O_7 \cdot 2H_2O$	40	pH 8, 9, 10

3. Results and Discussion

Table 3 shows the micro-hardness of non heat-treated and different heat treatment temperatures, of electroless Ni-P deposition. In this study, the hardness of substrate (S45C) was $HV_{0.05}$ 421. It was clearly shown that the hardness of the Ni-P deposition was higher than the substrate after electroless Ni-P deposition (non-heat treatment); both acid and alkaline plating of Ni-P deposition can enhance the hardness of the substrate. In the acid bath, the pH5 specimen had a higher level of hardness ($HV_{0.05}$ 701). Furthermore, compared to the alkaline bath, the pH8 specimen exhibited the highest degree of hardness; the hardness increased to $HV_{0.05}$ 803. In this study, heat treatment was performed since the author's major concerns were the microstructure and hardness of Ni-P deposition. The hardness of the Ni-P deposition was clearly enhanced after different heat treatments, as shown in Table 3. The highest levels of hardness ($HV_{0.05}$ 927) of pH4 and ($HV_{0.05}$ 1027) pH5 were under heat treatment of 400°C and 350°C, respectively. However, the increased temperature resulted in the hardness level decreasing; moreover, the pH6, pH8, pH9 and pH10 layers can elicit the highest level of hardness after 1 hr of heat treatment at 300°C. The above test results show that the pH5 and pH8 of Ni-P deposition exhibit an optimal hardness and microstructure; thus, we selected them to perform further analysis, heat treatment and wear tests.

Table 3 Compare the micro-hardness ($HV_{0.05}$) of non heat-treated and different heat treatment temperatures of electroless Ni-P deposition

pH value	4	5	6	8	9	10
non heat-treated	539	701	690	803	766	700
300°C	766	960	927	1066	986	816
350°C	841	1027	845	960	933	549
400°C	927	986	701	857	733	498
450°C	713	917	623	711	644	426
500°C	593	791	585	549	473	383

The deposition rate of Ni-P deposition by different pH values is shown in Figure 1; the highest deposition rate of acid and alkaline baths appeared in pH5 (16.5 μmh^{-1}) and pH8 (15.0 μmh^{-1}), respectively. In this study, it is reasonable to suggest that the higher pH values (pH9 and pH10) were adverse reactions as was the precipitation of Ni-P deposition.

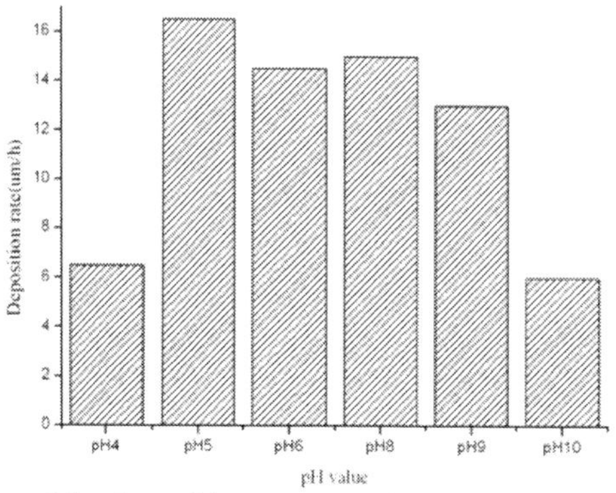

Figure 1 Comparison of the deposition rate on Ni-P deposition by different pH values

The properties of electroless deposits are attributed to their microstructural characteristics. The details of the structure of electroless deposits are not well understood, but plated EN films have been reported to be either crystalline, amorphous or a combination of both. Figure 2 shows the SEM observation of surface morphology on the Ni-P deposition by pH 5 after different heat treatments. The non-heat-treated specimen represents a smoother deposition, as shown in Fig. 2a. With increased heat treatment temperature, the continuity of a smooth morphology will gradually transform into one that is rougher, with slightly increased agglomeration. When the temperature was raised to 350°C, the particles contacted each other, forming large areas of uneven surface, as shown in Fig. 2b. When the temperature rose to 400°C, the agglomeration of particles contacted each other, forming large areas of uneven surface, as shown in Fig. 2c. Moreover, when the temperature was raised to 500°C, the precipitate particles on the surface showed more obvious agglomeration and also become denser, as shown in Fig. 2d.

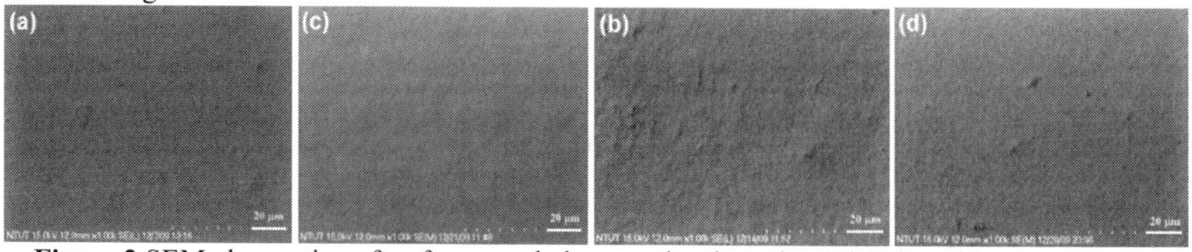

Figure 2 SEM observation of surface morphology on the electroless Ni-P deposition by pH5 after different heat treatments (a) non heat-treated, (b) 350°C, (c) 400°C, (d) 500°C.

Figure 3 shows the SEM image of surface morphology on the Ni-P deposition by pH8 after different heat treatments. Due to the lower phosphorus content, the surface deposition is rougher and more uneven. With increased heat treatment temperature, the surface morphology changed minimally; there was a slight difference after 1 hr of heat treatment at 300°C. As shown in Figs. 3a and 3b, the particle size of precipitation shows a trend towards a smaller and denser outcome. With an increase in the heat treatment temperature (400°C and above), the particle size of precipitation gradually became larger and more agglomerated, as shown in Figs 3c and 3d. As compared with Figs. 2 and 3, the precipitate particles of pH8 Ni-P deposition appeared significantly larger in the surface structure after heat treatment.

Figure 3 SEM observation of surface morphology on the electroless Ni-P deposition by pH8 after different heat treatments (a) non heat-treated, (b) 300°C, (c) 400°C, (d) 500°C.

Fig. 4a shows the XRD pattern of electroless Ni-P deposition by different pH values. The initial plating of Ni-P deposition is amorphous and coexists with the microcrystalline structure of the nickel phase. The diffraction peak of nickel tends to increase as the pH value increases. The result agrees with the relevant literature [13]. Acidic plating of Ni-P deposition has relatively broad diffraction peaks; the higher phosphorus content results in lower Ni diffraction peaks. However, the alkaline bath of Ni-P deposition has very clear diffraction peaks. As such, increasing the phosphorus content will produce a more disordered lattice. It is reasonable to suggest that the acidic plating of Ni-P deposition is closer to the amorphous structure, while the alkaline plating obtains better crystallization. In this study, the Ni (111) plane is the main diffraction peak of electroless Ni-P deposition by different pH values. Figs. 4b and 4c show the XRD patterns of electroless Ni-P deposition by pH 5 and pH 8 after different heat treatments. It can be observed that the Ni-P deposition clearly generated phase transformation (Ni → Ni_3P) after heat treatment. With increasing temperatures, the Ni_3P phase will start to precipitate and grow. When the heat treatment temperature increased, Ni_3P began to precipitate. Finally, the Ni-P deposition produced the Ni crystalline phase in coexistence with Ni_3P precipitation. Figs. 4b and 4c show that the pH value increases while the phosphorus content decreases; the crystallization of nickel was improved after heat treatment. In contrast, when the phosphorus content is high, the crystallization of the Ni_3P phase is better. XRD analysis indicated that two different kinds of structure were provided by the different pH values of electroless Ni-P deposition after heat treatment, namely, the crystalline structures of Ni and Ni_3P precipitations [14]. They belong to the Ni (111) plane, and (301), (231), (204) and (411) of the Ni_3P plane, respectively.

Figure 5 compares the coefficient of friction of electroless Ni-P deposition by pH5 and pH8 after different heat treatments. The substrate (S45C) has a low coefficient of friction at the beginning, but the coefficient of friction began to rise when the smooth surface of the sample was worn by the chrome ball. Conversely, the Ni-P depositions after heat treatments have a higher coefficient of friction in the initial test but the coefficient of friction then decreased; it began to change due to the static friction, and maintained a stable value after sliding 25 meters. Table 4 shows the average coefficient of friction and wear-depth of electroless Ni-P deposition by pH5 and pH8 at different heat treatment temperatures after the wear test. The result could be further compared with Table 3; the sample with the higher level of hardness is usually able to possess a lower coefficient of friction and wear depth. However, the hardness of Ni-P deposition by pH5 ($HV_{0.05}$ 701) was lower than pH8

($HV_{0.05}$ 803) after electroless treatment, while the pH5 had a lower coefficient of friction (0.357) instead.

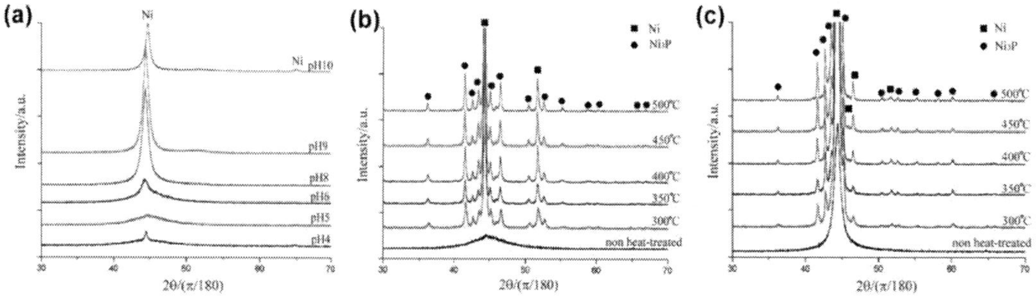

Figure 4 XRD patterns of the electroless Ni-P deposition by different pH values (a) non heat-treated, (b) pH 5 after different heat treatments, (c) pH 8 after different heat treatments.

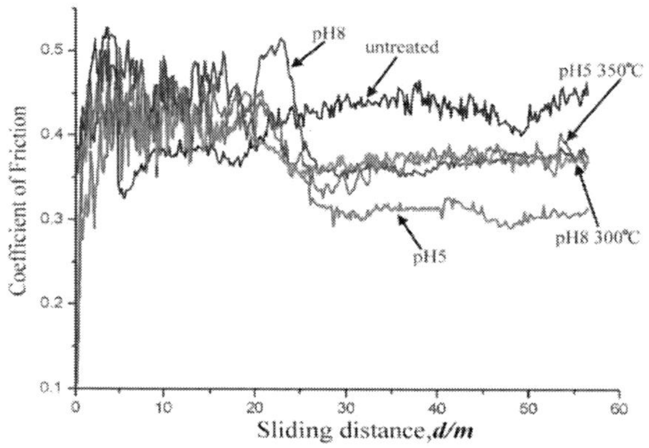

Figure 5 Comparison of the coefficient of friction on electroless Ni-P deposition by pH5 and pH8 and different heat treatments.

Table 4 Compare the average coefficient of friction and wear depth of electroless Ni-P deposition by pH5 and pH8, and different heat treatment temperatures after wear test

	Average coefficient of friction	Average wear depth (μm)
un-treated (S45C)	0.433	5.429
pH5	0.357	13.294
pH8	0.408	10.246
pH5 350°C	0.391	2.738
pH8 300°C	0.385	2.923

From the viewpoint of wear resistance, the pH8 of Ni-P deposition still had the lower wear depth (10.246 μm) because the oxygen atoms reacted with debris during the wear test and produced large amounts of oxide film. The oxide film has a lubricating effect, which causes a lower coefficient of friction [15]. As a result, the Ni-P deposition by pH5 plating has a better coefficient of friction than does the pH8, but lower hardness still results in weak wear resistance. The average wear depth of the substrate was 5.429 μm, pH5 was 13.294 μm and pH8 was 10.246 μm after the wear test, respectively. In addition, the heat treatment can improve the adhesion between the substrate and deposition. Regarding the specimens of Ni-P deposition by pH5 and pH8 after 1 hr of heat treatment at 350°C and 300°C and soaking, respectively, the hardness can be significantly improved so that it will result in the lower wear depth. The wear depths of pH5 and pH8 were 2.738 μm and 2.923 μm, respectively.

4. Conclusions

XRD analysis shows that the S45C surface in the acidic plating easily generates close to the amorphous structure of deposition, and that in the alkaline plating of Ni-P deposition good crystallization was obtained. Furthermore, regardless of acid or alkaline bath of the Ni-P deposition, the S45C surface obviously occurs in the phase transformation after heat treatment, resulting in the precipitation of the Ni_3P phase. In this study, the maximum hardness ($HV_{0.05}$ 1066) of the specimen appeared in the pH8 Ni-P deposition through a 1 hr of heat treatment at 300°C. The optimal wear depths (2.738 μm and 2.923 μm) appeared in Ni-P deposition by pH5 and pH8 after 1 hr of heat treatments at 350°C and 300°C and soaking, respectively. Significantly, the surface hardness and wear resistance could be effectively improved by the electroless Ni-P deposition via suitable pH values and heat treatment.

References

[1] Zhang W X, He J G, Jiang Z H, Jiang Q and Lia J S 2007 *Surf. Coat. Technol.* **201** 4594-4600
[2] Rahimi A R, Modarres H and Abdouss H 2009 *Sur. Eng.* **25** (2009) 367-371
[3] Wang L L, Chen H J and Chen Z L 2011 *Sur. Eng. 27* (2011) 57-60
[4] Zhanga H, Wangb S, Yaoa G and Huaa Z 2009 *J. Alloy. Compd.* **474** 306-310
[5] Zhang H and Liu R L 2011 *Sur. Eng.* **27** 5-10
[6] Abdeli M, Ahmadi, Parvini and Khosroshahi R A 2009 *Sur. Eng.* **25** 127-130
[7] Shin H J, Yoob Y T, Ahnc D G and Im K 2007 *J. Mater. Process. Technol.* **187-188** 467-470
[8] Chang S H, Lin Y K and Huang K T 2012 *Surf. Coat. Technol.* **207** 571-578
[9] Chang S H, Tang T P, Huang and Chen J K, 2012 *ISIJ Int.* **52** 499-504
[10] Matsubara H, Yonekawa T, Ishino Y, Saito N, Nishiyama H and Inoue Y 2006 *Electrochimica Acta.* **52** 402-407
[11] Chang S H, Tang T P and Tai F C 2011 *Sur. Eng.* **27** 581-586
[12] Chang S H, Huang K T and Wang Y H 2012 *Mater. Trans.* **53** 745-751
[13] Kumar P S and Nair P K 1996 *J. Mater. Process. Technol.* 56 511-520
[14] Wu F B, Li J J and Duh J G 2000 *Thin Solid Films.* **377-378** 354-359
[15] Chang S H, Tang T Pc and Huang K T 2010 *ISIJ Int.* **50** 569-573

Diameter Controlled of Carbon Nanotubes Synthesized on Nanoporous Silicon Support

N.A. Asli[a,b,*], M.S. Shamsudin[a,b], M.Maryam[a,b], S.F.M. Yusop[a,b], A.B.Suriani[d], M.Rusop[a,c], S.Abdullah[a,b]

[a] Centre of Nanoscience and Nanotechnology (NANO-SciTech Centre), Insitute of Science, Universiti Teknologi MARA, 40450 Shah Alam, Selangor D.E., Malaysia

[b] School of Physics and Material Studies, Faculty of Applied Sciences, Universiti Teknologi MARA, 40450 Shah Alam, Selangor D.E., Malaysia

[c] NANO-Electronic Centre, Faculty of Electrical Engineering, Universiti Teknologi MARA, 40450 Shah Alam, Selangor D.E., Malaysia

[d] Department of Physics, Faculty of Science and Mathematics, Universiti Pendidikan Sultan Idris, 35900 Tanjung Malim, Perak D.R., Malaysia

asnidasli@yahoo.com.

Abstract. Carbon nanotubes (CNTs) have been successfully synthesized on nanoporous silicon template (NPSiT) using botanical source, camphor oil. Diameter of CNTs synthesized was controlled by pore size of NPSiT prepared by photo-electrochemical anodization method. The diameter of CNTs grown on different NPSiT corresponded to the pore diameter of NPSiT. FESEM images showed self-organized bundles of fiber-like structures of CNTs with diameter of around 20nm which were successfully grown directly on nanoporous silicon while raman spectra obtained ratio of I_D/I_G at 0.67.

Keywords: Carbon nanotubes; Nanostructured porous silicon template; two-stage hot filament thermal chemical vapor deposition system; diameter of nanotubes

1. Introduction

Carbon nanotubes (CNTs) have attracted great attention in science and technology field since discovered by Iijima (1991) [1]. They are known due to their extraordinary mechanical, chemical, electronic and optoelectronic properties such as stronger than steel, harder and higher thermal conductivity compared to diamond and higher electrical conductivity than copper [1, 2]. Its various extraordinary properties are applicable in various applications especially in nanoelectronics and field emission devices which require the modification of carbon nanotubes with well orientation, internal structures and morphologies features [1, 3].

Recently, researchers have tried to synthesize carbon nanotube on various substrates in order to control their geometry and features. Earlier work focused on synthesis of CNTs on substrate in random orientation. [2]. Further approach had been verified in order to control the trend orientation of CNTs growth, as well as applying the external electric or magnetic field during the deposition of CNTs [3]. Additionally, diameters and lengths of tubes were considered as most controllable properties of CNT during fabrication process [4].

Study of growth condition such as diameter of nanotubes can clarify the size of catalytic nucleate particle, energy gap of the semiconducting nanotubes, as well as interaction between nanotubes [5]. As reported the diameter of nanotubes can be controlled by modulating growth temperature, growth time and pressure [6].

Previous studies reported that the diameter of carbon nanotubes can be easily controlled by nano support materials such as porous anodic aluminum oxide (AAO) template, where the diameter of nanotubes correspond to the pore size [7]. AAO was mostly used as template to fabricate nanomaterials including nanowires and nanotubes because of the well uniformity of pores [8-13]. Instead of AAO, porous silicon was used as template in this study because of the limitation of AAO in terms of its high cost. Silicon on the other hand, has its own advantages in device application because of its visible luminescence and lower cost [14].

This paper will report on the synthesis of carbon nanotubes on nanostructured porous silicon template using botanical hydrocarbon (camphor) by two-stage hot filament thermal chemical vapor deposition system. From observation, it can also be seen that changing the series of porous silicon template contributed to diameter distribution of carbon nanotubes synthesis.

2. Experimental

2.1. Preparation of Nanostructured Porous Silicon Template

NPSiT was prepared by photo-electrochemical anodization technique [15] on boron doped p-type Si wafer (1 0 0) orientation with resistivity of 1 Ωcm-1. The cleaning process of silicon wafer was carried out by immersion in acetone (CH3COCH3) and methanol (CH3OH) for at least 5 minutes at 40 °C in ultrasonic bath then followed by H2O: HF (40%) in ratio of 10:1 for at least 2 minute after rinsing with deionized water. The wafer then went through etching process where it was placed in Teflon cell with electrolyte consisting of ethanol: HF (48%) in 1:1 ratio with illumination of halogen lamp. Aluminum foil was attached behind the Si wafer during anodization process to increase the uniformity of anodic current flow. The etching process was carried out for 20, 30, 40 and 50 min at 20 mA/cm2 current density. The NPSiT was then rinsed in deionized water, dried under the dry nitrogen stream and stored at room temperature or in desiccators to avoid oxidation.

2.2. Preparation of Carbon Nanotube (CNTs) on Nanostructured Porous Silicon Template

A 100 cm long and 5 cm diameter of quartz tube serves as a thermal chemical vapour deposition (TCVD) reactor and placed horizontally inside a double stage tube furnace system. The camphor white oil as carbon source and ferrocene powder as catalyst were filled in separate alumina boats and placed at the middle of the first furnace while the NPSiT was placed in alumina boat at the center of the second furnace. The quartz tube was closed with end cap and connected to nitrogen gas supply through the inlet and exhaust through the outlet. Before the furnace is switched on, the nitrogen gas was flowed inside the furnace to remove the atmospheric air from tube reactor for 5 minutes and was reduced to 150 sccm of flow rate. The second furnace was first switched on at temperature of 800°C and stabilized for 15 minutes then followed by first furnace at temperature of 180 °C to allow camphor white oil and ferrocene powder to vaporized with flow of gas. Deposition time of CNTs on NPSiT was around 1 hour and when completed, the furnace was switched off and cooled down in ambient temperature.

3. Results and Discussion

2.2. Morphological features

The morphological study and cross-section of porous silicon template and carbon nanotube growth on template was visualized using Field Emission Scanning Electron Micrograph (FESEM). Figure 1 (a) showed FESEM micrograph of nanostructured porous silicon template (NPSiT) at 3 kV, x100 (plan view) and 5 kV, x5 magnification (cross-section) while figure 1 (b) showed carbon nanotubes deposited on porous silicon layer respectively at 5 kV, x1 and 5 kV, x100 magnification (high resolution).

Micrographs of NPSiT in figure 1(a) showed the pore size distribution of samples prepared by electrochemical anodization etching of p-type Si wafer for 20, 30, 40 and 50 minutes at a current density of 20 mA/cm^2 .The variation of average diameter pore and nanotubes of the NPSiT with different etching time is shown in figure 2.

Results showed that the sample prepared at 30 minutes and 20mA/cm^2 (1a,ii) have uniformed pore size which are about 20 nm, and the thickness of NPSi is about 17 μm while sample prepared at 20 minutes of etching time have very small pores and cannot be detected in the measurement but estimated to be around 1-10 nm size. The pore size increases as etching time increases up to 30 minutes. As the etching time was increased further, the average pore size decreased from 20 to 9 nm. The result agrees as reported [16], this is due to the Beale models which showed etching behaviors of silicon wafers in etchant and the resultant characteristic pore morphology of p-type NPSiT.

Micrographs of CNTs grown on NPSiT prepared by thermal chemical vapor deposition (double stage tube furnace system) for 1 hour of synthesis time at 800 °C are illustrated in figure 1(b). FESEM images showed that the carbon nanotubes was successfully synthesized on nanostructures porous silicon template respectively using botanical precursor, camphor white oil and presence of ferrocene as catalyst. The top view of FESEM images (1 b i, ii, iii) showed uniform growth of dense CNTs from entire template with different diameter of nanotubes respectively. Results indicate that the diameter of carbon nanotubes corresponds to the size of pores shows in figure 2. Carbon nanotubes have the biggest diameter and in random orientation as shown in figure 1(b,i) while figure 1 (b, ii) showed that carbon nanotubes were grown in self-organized bundles of fiber like structures and has lowest diameter of nanotubes of around 20 nm. As the etching time of template increases further, the diameter of carbon nanotubes growth gradually increases and most of them form spiral like structure.

These results agreed as reported by [7], the diameter of carbon nanotubes synthesis correlate with nanostructured pore size of template. Additionally, as referred from [7], CNTs growth depends on the catalytic behavior of template pores.

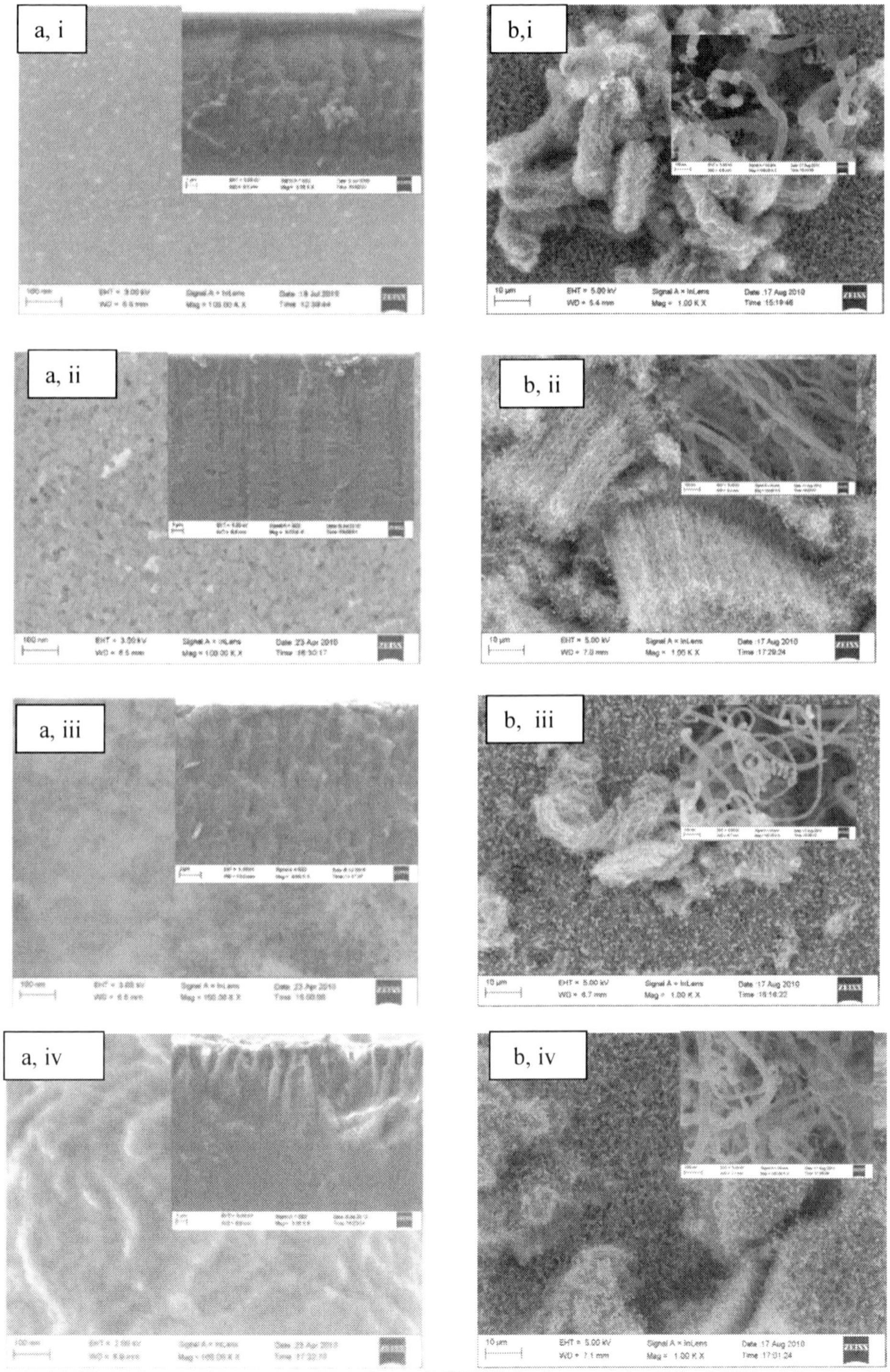

Figure 1. (a) Plan view and cross-section view of FESEM micrograph of the NPSi with magnification of ×200 and ×5 K at 20 mA/cm2 for i) 20 min, ii) 30 min, iii) 40 min, and iv) 50 min of etching time ; (b) FESEM micrograph of CNT grown on NPSiT respectively with magnification of x1 and x100 (high resolution).

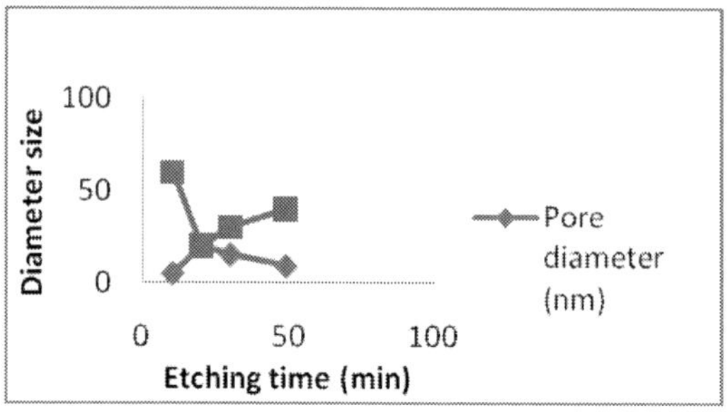

Figure 2. The variation curves of an average pore diameter and nanotubes diameter as a function of etching time.

2.2 ***Raman spectroscopy***

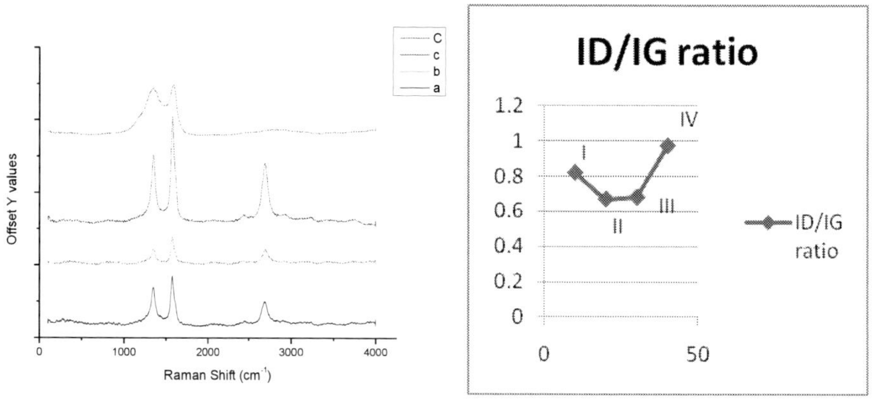

(a) (b)

Figure 3. (a) Raman spectra of CNTs grown on respective NPSiT for various etching time (20, 30, 40 and 50 minutes); (b) Variation curve I_D/I_G ratio of CNTs grow on respective NPSiT.

Figure 3 (a) showed the raman spectra of CNTs grown on NPSiT at different parameter of template from 20 to 50 minutes etching time and constant current density (20 mA/cm^2) in the range 100-4000 cm^{-1}. Raman characterization was carried at room temperature under 20 mW of power at 514 nm of wavelength with illumination of Ar laser.

Raman spectroscopy was used to determine the characteristic of overall graphitic structures growth. Raman spectra represent the G (~ 1573 cm^{-1}) and D (1348cm^{-1}) bands which were initiated from well-graphitized carbon and disorder-activated band respectively. It can be observed that the ratio of D and G band (I_D/I_G ratio) for CNTs growth on NPSiT prepared at 30 minutes and 20 mA/cm^2 (figure 3,a II) has the lowest value of 0.67. The variation trend of the I_D/I_G ratio is shown in figure 3 (b). The trend showed that the amount of I_D/I_G ratio is smaller in crystalline sample (<1). The I_D/I_G ratio for CNTs synthesized on NPSiT prepared at 20 minutes is about 0.8 and decreased sharply to 0.67 when template was change (30 minutes). The values increased gradually as the etching time of template was increased further, which is around 0.9 at preparation time 50 minutes. Therefore, it can be said that this experiment can only run for template prepared up to 50 minutes etching time at fix current density. As reported previously[17], smaller I_D/I_G ratio corresponds to fewer defects. Raman spectra showed no peak present at 0-500 cm-1 where refer to [18], radial breathing modes (RBM) is present at 186 cm^{-1} which means that there are no single wall CNTs obtained.

4. Conclusion

We have presented a simple method of growing multi-walled carbon nanotubes (MWCNTs) at smallest diameter corresponding to the diameter of nanostructured porous silicon template by using an economical, biomass and ecologically-friendly carbon source namely camphor white oil. Previous research focused more on TCVD parameters such as flow rate of gas, concentration of catalyst, temperature and synthesis time to control the diameter of nanotubes. This work is more on the controlled diameter of template which effects the diameter of nanotubes grown.

FESEM images showed the lowest diameter of nanotubes around 20nm deposited on NPSiT prepared at 30 minutes etching time at 20 mA/cm^2. Raman spectra revealed that the smallest defect of disorder crystallinity of carbon nanotubes was grown at 30 minutes etching time of NPSiT. Therefore, the best template to synthesize high quality, smallest diameter and aligned orientation is NPSiT prepared at 30 minutes of etching time and 20mA/cm^2 of current density.

Acknowledgement

I would like to thank all the staff of NANO-Inovation Centre and NANO-SciTech Centre, Universiti Teknologi MARA Malaysia for their support and cooperation. Research funding for this project was provided by Ministry of Higher Education (MOHE), (600-RMI/ST/FRGS 5/3/Fst (204/2010).

References

[1] Iijima, S.: Nature. **1991**, 354, 56-58.
[2] W.Y.Lee, C. H. W., Z.Y.Juang, C.H.Tsai: Diamond & Related materials. **2005**, 14, 1852-1856.
[3] Y. Ando, X. Z., T. Sugai , M. Kumar: Mater Today. **2004**, 7, 22-29.
[4] J.Li, W. L., X.Zhang, B.Wang, L.Ba: Solid-State Electronics. **2004**, 48, 2147-2151.
[5] H.S. Cheng, A. C. C., G.P. Pez,M.K. Kostov, P. Piotrowski, S.J. Stuart: J. Phys. Chem. B. **2005**, 109, 3780-3786.
[6] Liu, C. L. a. J.: J. Phys. Chem. B. **2006**, 110, 20254-20257.
[7] L.Kim, E. M. L., S.J.Cho, J.S.Suh: Carbon. **2005**, 43, 1453–1459.
[8] D. AlMawlawi D, N. C., M. Moskovits: J. Appl. Phys. **1991**, 70, 4421–4425.
[9] J. Choi, G. S., K. Nielsch, R.B. Wehrspohn, U. Gosele: Chem. Mater. **2003**, 15, 776–779.
[10] M.S. Sander, R. G., T. Sands, A.M.: Chem. Mater. **2003**, 15, 335-339.
[11] J.S. Suh, J. S. L.: Surface enhanced Raman scattering for CdS nanowires deposited in anodic aluminum oxide nanotemplate: Chem Phys Lett. **1997**, 281, 384–388.
[12] J. Li, C. P., J. Xu: Nature. **1999**, 402, 253–254.
[13] S.H. Jeong, H. Y. H., S.K. Hwang, K.H. Lee: Carbon. **2004**, 42, 2073–2080.
[14] L.T. Canham, M. R. H., W.Y. Leong, C. Pickering, J.M. Keen: J. Appl Phys. **1991**, 70, 422-431.
[15] Asli, N. A., Yusop, S. F. M., Rusop, M., and Abdullah, S.: Ionics. **2011**, 17, 653-657.
[16] M. I. J. Beale, N. G. C., M. J. Uren, A. G. Cullis, and J. D. Benjamin: Appl. Phys. Lett. **1985**, 46, p. 85.
[17] J. Robert, A., S.F. Catriona, J.A. Andrew: Carbon. **2006**, 44, 341-347.
[18] M.Kumar, Y. A.: **2003**, 12, 1845-1850.

IOP Conf. Series: Materials Science and Engineering **46** (2013) 012005 doi:10.1088/1757-899X/46/1/012005

Improvement sensitivity humidity sensor based on ZnO/SnO_2 cubic structure

N.D. Md Sin[1,*], S. Ahmad[1], M.F. Malek[1], M.H. Mamat[1], M. Rusop[1,2]
[1]NANO-ElecTronic Centre, Faculty of Electrical Engineering,
Universiti Teknologi MARA (UiTM), 40450 Shah Alam, Selangor, Malaysia
[2]NANO- SciTech Centre (NST), Institute of Science (IOS),
Universiti Teknologi MARA (UiTM), 40450 Shah Alam, Selangor, Malaysia

*E-mail: nordiyana86@yahoo.com

Abstract. ZnO nanorod, SnO_2 nanoparticle and ZnO/SnO_2 cubic structure were prepared using sol-gel immersion method. These three structures had been growth on top sputtered ZnO thin film. The surface morphology of all thin film were characterized using field emission scanning electron microscopy (FESEM) and energy dispersive X-ray spectroscopy (EDX). The prepared thin films were fabricated for humidity sensor application. The composite sensor ZnO/SnO_2 cubic structure gave excellent sensitivity compare to ZnO nanorod and SnO_2 nanoparticle. The improvement sensitivities of ZnO/SnO_2 cubic structure were 22.5 times from ratio of $R_{40RH\%}$ to $R_{90RH\%}$ at 25°C. ZnO/SnO_2 cubic structure had a great potential for humidity sensor application with good in repeatability and stability.

1. Introduction

Choosing material is an important factor to enhanced behaviour of humidity sensor. The material of sensor should be conductive during the reaction with gases or humidity especially at surface component of semiconductor [1-2]. N-type material with relatively little oxygen adsorption sites available is suitable for sensing application to create potential barrier such as zinc oxide (ZnO) and tin oxide (SnO_2). Besides that, insertion of additives to the semiconductor material also could improve the humidity sensor performance. Mixing process is adding or combining two or more materials such as metal mixed metal oxide, metal oxide mixed metal oxide and polymers mixed metal oxide[3]. The advantages composite sensors are thermally more stable, high electron mobility, have many hetero-contacts between phase and the catalytic activity of sensing matrix are be able to control [1, 4-5].

ZnO, as semiconducting II-VI metal oxide with wide band gap 3.37 eV. ZnO posses potential material for many devices applications such as gas sensor[6], solar cells[7], optoelectronic devices[8] and LED[9]. SnO_2 is n-type material with the band gap 3.5-3.6 eV. SnO_2 has been applied for various application such as solar cells[10], lithium ion batteries[11] and gas sensor[12-14]. SnO_2 have received attention from many researcher studied on gas sensor application for having high sensitivity and relatively low operating temperature [15]. ZnO and SnO_2 have unique properties due to their chemical and physical properties. Many type of deposition methods to prepare ZnO nanostructure and SnO_2 nanostructure such as electrochemical technique[16-17], sputtering[18-19], chemical vapour deposition[20-21] and hydrothermal[22-23]. Mixing ZnO and SnO_2 could form various kind of composite material such as ZnO/SnO_2 [24], $ZnSnO_3$ [25], Zn_2SnO_4 [26], Zn-doped SnO_2 [27] and Sn-doped ZnO [28]. Different type of ZnO/SnO_2 structured have been prepared such as nanofibers [29], nanorod [30], nanoflowers [31] and nanoparticles [32] using various of method including electrochemical [33], hydrothermal [34], thermal evaporation [35] and sputtering [36]. In this experiment, ZnO nanorod, SnO_2 nanoparticle and ZnO/SnO_2 cubic structure were deposited using sol-gel immersion method.

Some composites sensor that had been reported improved the humidity sensor performance. Leilei et. al reported on ZnO/TiO_2 composite nanorod array enhanced the sensitivity of humidity sensing [37]. Qing et. al found the preparation of composite ZnO-SiO_2 contribute a mesoporous that beneficial for humidity sensing properties [38].The aim of this project was to study the improvement of sensitivity towards humidity sensing for composite material ZnO/SnO_2 cubic structure compare to homogeneous material (ZnO nanorod and SnO_2 nanoparticle).To the best of my knowledge, there was no report on improvement sensitivity of humidity sensing performance for composite ZnO/SnO_2 cubic structure compare to single component material.

2. Methodology

2.1. Preparation of seeded catalyst

ZnO seeded catalyst were deposited on glass substrates by radio frequency (RF) magnetron sputtering. ZnO(99.999%) was used as target. The sputter was pump at $5x10^{-4}$ Pa using a molecular pump. The chamber was injected with gas flow $Ar:O_2$ (45:5) sccm. The deposition was set up for 60 min at substrate temperature 500°C with RF power 200 watt. Then as prepared sample were anneal for 1h at 500°C.

2.2. Preparation of sol-gel solution

The sol-gel solutions were prepared to deposit on ZnO seeded catalyst. Firstly, the ZnO nanorod was prepared using zinc chloride ($ZnCl_2$) and 20 mmol sodium hydroxide (NaOH). $ZnCl_2$ and NaOH were dissolved in 50 mL distilled water (DI) for 10 minutes at 50°C and stirred using magnetic stirrer on a hot plate stirrer. The $ZnCl_2$ solution was slowly dropped into NaOH solution under magnetic stirring. Then the solution was sonicated by ultrasonic for 10 minutes at 50°C. As prepared solution were transfer into a container with the seed catalyst were placed at the bottom of the container. The seeded were facing up to container caps. Next, the containers were immersed inside a water bath inside a water bath with 95°C DI water inside. The immersion time were carried out for 6h. Then, the samples were washed with DI water and dried for 15 min at 100°C. The samples were annealed at 500°C for 1h. The same procedure was repeated for SnO_2 nanoparticle preparation by replacing $ZnCl_2$ to 3 mmol tin tetrachloride ($SnCl_4.5H_2O$). Lastly, the ZnO/SnO_2 cubic structure was prepared using mixture of $ZnCl_2$ solution and $SnCl_4$ solution by following the same procedure for ZnO nanorod and SnO_2 nanoparticle.

2.3. Characterization

The structural property of the thin film was characterized using FESEM (JEOL JSM 7600F) and EDX. An Au contacts were deposited on the thin film using thermal evaporator to fabricate humidity sensor. The measurements of sensor were taken using two point probe I-V measurement (Keithley 2400). The characteristic of the sensor including sensitivity, stability and reproducibility were studied. The humidity condition had been controlled using a humidity chamber (ESPEC SH-261) by varying the relative humidity percentages (RH%) from 40 RH% to 90 RH% at constant temperature 25°C.

3. Results and discussions

3.1. Surface morphology

Figure 1 (a), (b), (c) and (d) shows the ZnO seeded catalyst, ZnO nanorod, SnO_2 nanoparticle and ZnO/SnO_2 cubic structure respectively. As been observed in Figure 1(a) show the ZnO seeded catalyst had columnar structured with average size 45nm. ZnO nanorod growth on ZnO seeded catalyst was shown in Figure 1(b). The average diameter of the ZnO nanorod was around 85nm. From the FESEM image, SnO_2 nanoparticle were form on seeded catalyst ZnO from the FESEM image as shown in Figure 1(c). As could been seen in Figure 1(d), the mixture of ZnO and SnO_2 resulting ZnO/SnO_2

IOP Conf. Series: Materials Science and Engineering **46** (2013) 012005 doi:10.1088/1757-899X/46/1/012005

cubic structure with average dimensions of cubic 120nm. The ZnO nanorod growth was based on seed layer[39]. However, SnO_2 nanoparticle was difficult to growth on ZnO seed layer due to lattice mismatch. There were aggregations of amorphous nanoparticles growth on ZnO/SnO_2 cubic structure and porosity as indicated in Figure 1(d). The formation of ZnO/SnO_2 cubic structure occurs when Zn^{2+} and Sn^{4+} particles were sonicated to let the both particles dispersed well and mix the $Zn^{2+}$and Sn^{4+} ions [39]. The combination of both particles formed cubic structure that favour from the thermal reaction during immersion process. The aggregations of amorphous nanoparticles that growth on ZnO/SnO_2 cubic structure might be due to Ostwald ripening law, which explained that the surface tension and surface to volume ratio [40]. Figure 2. (a), (b) and (c) shows the EDX spectrum of ZnO nanorod, SnO_2 nanoparticle and ZnO/SnO_2 cubic structure respectively. EDX used to analyze the element of the sample. EDX spectrum of ZnO nanorod was shows in Figure 2(a) contained element of Zn and oxygen.

Figure 1. (a) ZnO seeded catalyst, (b) ZnO nanorod, (c) SnO_2 nanoparticle and (d) ZnO/SnO_2 cubic structure with porosity as indicated in circle.

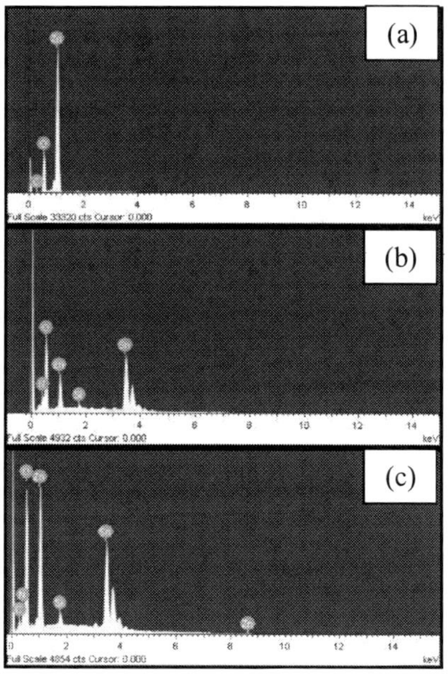

Figure 2. EDX spectrum of (a) ZnO nanorod, (b) SnO_2 nanoparticle and (c) ZnO/SnO_2 cubic structure.

Table 1. Atomic percentage of each element composition (%).

	Zn	Sn	Oxygen	Carbon	Silicon
ZnO nanorod	42.02		52.66	5.32	
SnO_2 nanoparticle	8.52	26.38	63.83		1.28
ZnO/SnO_2 cubic structure	16.04	16.55	61.32	6.09	

Figure 2(b) shows the EDX spectrum that the presence of Zn, Sn and oxygen in the SnO_2 nanoparticle. The Zn element inside the SnO_2 nanoparticle was due to the ZnO seeded catalyst. The Zn, Sn and oxygen element were detected in the ZnO/SnO_2 cubic structure as indicate in Figure 2(c). Table 1. show the list of atomic percentage of each element composition for ZnO nanorod, SnO_2 nanoparticle and ZnO/SnO_2 cubic structure.

3.2. Humidity sensor

Figure 3 shows the sensing characteristic of fabricated sensor of ZnO nanorod, SnO_2 nanoparticle and ZnO/SnO_2 cubic structure. Figure 3(a) shows the resistance measurement as a function of RH%. The value of resistance decreased as the RH% increased. The decrements of resistance occur due to presence of water vapour as the humidity increased. Figure 3(b) shows the sensitivity of ZnO/SnO_2 cubic structure show the highest with 22.5, ZnO nanorod sensitivity 7.5 and SnO_2 nanoparticle 2.3. The sensitivity data were calculated based from the ratio of resistance in air to resistance in humidity (at difference RH%). The highest sensitivity that produced by ZnO/SnO_2 cubic structure might be due to the contribution of porosity that enhanced the diffusion of water vapour [41]. Moreover, the shape of ZnO/SnO_2 cubic structure with three dimension (3D) favour large specific surface areas was another factor that influenced the sensitivity performance [42]. The stability of the sensor were measured at 90 RH% as indicated in Figure 3(c). The resistance were slightly the same when exposed to humidity every two hours. Figure 3(d) shows the reproducibility measurement of the ZnO/SnO_2 cubic structure when exposed to 40RH% and 90RH%. The current of the sensor increased steadily when the RH% increased and decreased rapidly when the RH% decreased. The sensor gave slightly same behaviour of current when it was tested for 4 cycles.

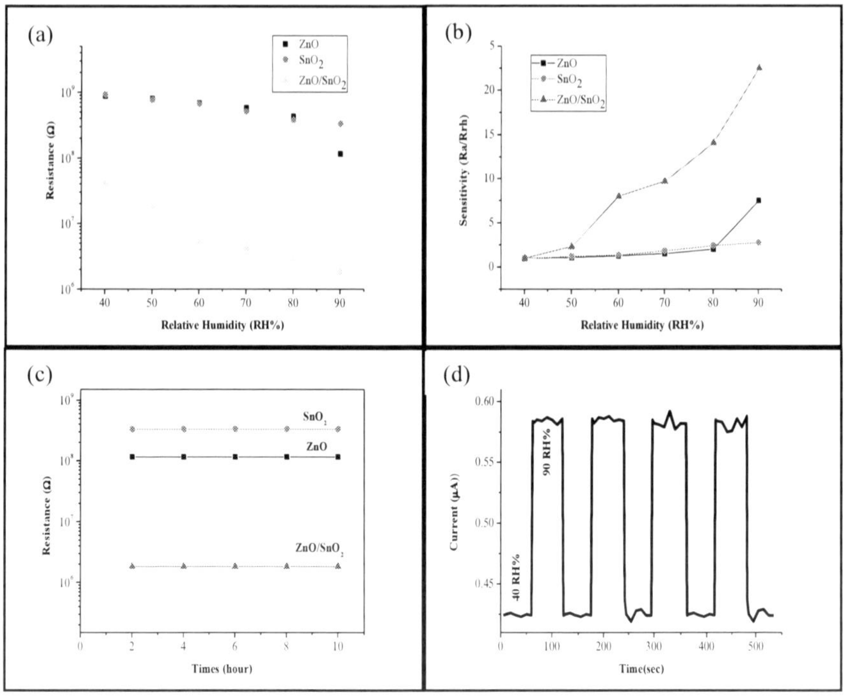

Figure 3. The humidity sensor characteristics of ZnO nanorod, SnO_2 nanoparticle and ZnO/SnO_2 cubic structure including (a) resistance measurement at various RH%, (b) sensitivity as function of RH% and (c) stability measurement at 90 RH%. **Figure 3.** (d) Repeatability measurement of ZnO/SnO_2 cubic structure with 1V voltage supplied tested for 4 cycles.

During high humidity, the water vapour increased and promotes the electron transport on the ZnO/SnO_2 cubic structure which controls the superconductivity of the thin film[43]. When the water vapour decreased, the water vapour dissociates appearance protons as charge carrier by hoping transports on the surface of the thin film[44].

4. Conclusions

ZnO nanorod, SnO_2 nanoparticle and ZnO/SnO_2 cubic structure were successfully prepared on ZnO seeded catalyst using sol-gel immersion method. The ZnO seeded catalyst was prepared on glass substrate by RF magnetron sputtering. The sensing characteristics towards humidity indicated that ZnO/SnO_2 cubic structure give highest sensitivity compare to ZnO nanorod and SnO_2 nanoparticle. It was due to distribution of porosity on surface of ZnO/SnO_2 cubic structure and the large surface specific area for adsorption and desorption process. Besides, the stability and reproducibility performance of ZnO/SnO_2 cubic structure carry out that it's was suitable for humidity sensor application.

5. Acknowledgement

The authors would like to express their thanks to Institute of Science and Faculty of Mechanical UiTM for providing the laboratory facilities. Thanks also to Ministry of Higher Education (MOHE) for financial support.

6. References

[1] Korotcenkov G 2007 Metal oxides for solid-state gas sensors: What determines our choice? *Materials Science and Engineering: B* **139** 1-23
[2] Wang C, Yin L, Zhang L, Xiang D and Gao R 2010 Metal oxide gas sensors: Sensitivity and influencing factors *Sensors* **10** 2088-106
[3] Singh N, Ponzoni A, Comini E and Lee P S 2012 Chemical sensing investigations on Zn–In_2O_3 nanowires *Sensors and Actuators B: Chemical* **171–172** 244-8
[4] Yu J H and Choi G M 1999 Electrical and CO gas-sensing properties of ZnO/SnO_2 hetero-contact *Sensors and Actuators B: Chemical* **61** 59-67
[5] Haeng Yu J and Man Choi G 1998 Electrical and CO gas sensing properties of ZnO–SnO_2 composites *Sensors and Actuators B: Chemical* **52** 251-6
[6] Singh G, Choudhary A, Haranath D, Joshi A G, Singh N, Singh S and Pasricha R 2012 ZnO decorated luminescent graphene as a potential gas sensor at room temperature *Carbon* **50** 385-94
[7] Chou C S, Chou F C, Ding Y G and Wu P 2012 The effect of ZnO-coating on the performance of a dye-sensitized solar cell *Solar Energy*
[8] Bu I Y Y and Yeh Y M 2012 Effects of sulfidation on the optoelectronic properties of hydrothermally synthesized ZnO nanowires *Ceramics International*
[9] Carotta M, Cervi A, Fioravanti A, Gherardi S, Giberti A, Vendemiati B, Vincenzi D and Sacerdoti M 2011 A novel ozone detection at room temperature through UV-LED-assisted ZnO thick film sensors *Thin Solid Films* **520** 939-46
[10] Chen Z, Tian Y, Li S, Zheng H and Zhang W 2011 Electrodeposition of arborous structure nanocrystalline SnO_2 and application in flexible dye-sensitized solar cells *Journal of Alloys and Compounds*
[11] Wang X, Zhou X, Yao K, Zhang J and Liu Z 2011 A SnO_2 graphene composite as a high stability electrode for lithium ion batteries *Carbon* **49** 133-9
[12] Ma X 2012 Towards Enhanced SnO2 Gas Sensor: Mini-Review *Journal of Nanoengineering and Nanomanufacturing* **2** 143-9
[13] Choi Y J, Hwang I S, Park J G, Choi K J, Park J H and Lee J H 2008 Novel fabrication of an SnO2 nanowire gas sensor with high sensitivity *Nanotechnology* **19** 095508
[14] Wang B, Zhu L, Yang Y, Xu N and Yang G 2008 Fabrication of a SnO_2 nanowire gas sensor and sensor performance for hydrogen *The Journal of Physical Chemistry C* **112** 6643-7

[15] Chen A, Huang X, Tong Z, Bai S, Luo R and Liu C C 2006 Preparation, characterization and gas-sensing properties of SnO_2–In_2O_3 nanocomposite oxides *Sensors and Actuators B: Chemical* **115** 316-21
[16] Elias J, Michler J, Philippe L, Lin M Y, Couteau C, Lerondel G and Lévy-Clément C 2011 ZnO Nanowires, Nanotubes, and Complex Hierarchical Structures Obtained by Electrochemical Deposition *Journal of electronic materials* **40** 728-32
[17] Yin Y X, Xin S, Wan L J, Li C J and Guo Y G 2012 SnO 2 hollow spheres: Polymer bead-templated hydrothermal synthesis and their electrochemical properties for lithium storage *SCIENCE CHINA Chemistry* 1-5
[18] Chen L, Li C, Yin W L, Liu J, Hei L and Lu F 2011 Effect of deposition temperature and quality of free-standing diamond substrates on the properties of RF sputtering ZnO films *Diamond and related materials* **20** 527-31
[19] Maleki M and Rozati S 2012 Structural, electrical and optical properties of transparent conducting SnO2 films: effect of the oxygen flow rate *Physica Scripta* **86** 015801
[20] Kang C G, Kang J W, Lee S K, Lee S Y, Cho C H, Hwang H J, Lee Y G, Heo J, Chung H J and Yang H 2011 Characteristics of CVD graphene nanoribbon formed by a ZnO nanowire hardmask *Nanotechnology* **22** 295201
[21] Kwoka M, Waczyńska N, Kościelniak P, Sitarz M and Szuber J 2011 X-ray photoelectron spectroscopy and thermal desorption spectroscopy comparative studies of L-CVD SnO_2 ultra thin films *Thin Solid Films* **520** 913-7
[22] Tian J H, Hu J, Li S S, Zhang F, Liu J, Shi J, Li X, Tian Z Q and Chen Y 2011 Improved seedless hydrothermal synthesis of dense and ultralong ZnO nanowires *Nanotechnology* **22** 245601
[23] Q.T. Khuc, X. Hien Vu, D. Vuong Dang, and D. Chien Nguyen 2010 The influence of hydrothermal temperature on SnO_2 nanorod formation *Advances in Natural Sciences: Nanoscience and Nanotechnology* **1** 025010
[24] Song X, Wang Z, Liu Y, Wang C and Li L 2009 A highly sensitive ethanol sensor based on mesoporous ZnO–SnO_2 nanofibers *Nanotechnology* **20** 075501
[25] ND Sin, N. Khadijah, MH Mamat, M.Z. Musa, and M. Rusop 2012 Influence of Cubic Structured-$ZnSnO_3$ Immersion Time to the Performance of Humidity Sensor *Nano Hybrids* **2** 1-11
[26] Parthibavarman M, Vallalperuman K, Sekar C, Rajarajan G and Logeswaran T 2012 Microwave synthesis, characterization and humidity sensing properties of single crystalline Zn2SnO4 nanorods *Vacuum* **86** 1488-93
[27] Wang W, Tian Y, Li X, Wang X, He H, Xu Y and He C 2012 Enhanced ethanol sensing properties of Zn-doped SnO_2 porous hollow microspheres *Applied Surface Science*
[28] Han N, Wu X, Zhang D, Shen G, Liu H and Chen Y 2011 CdO activated Sn-doped ZnO for highly sensitive, selective and stable formaldehyde sensor *Sensors and Actuators B: Chemical* **152** 324-9
[29] Zhang Z, Shao C, Li X, Zhang L, Xue H, Wang C and Liu Y 2010 Electrospun Nanofibers of ZnO−SnO_2 Heterojunction with High Photocatalytic Activity *The Journal of Physical Chemistry C* **114** 7920-5
[30] Tu Y F, Fu Q M, Sang J P and Zou X W 2012 Synthesis and photoluminescence properties of the ZnO@ SnO_2 core–shell nanorod arrays *Journal of materials science* **47** 1541-5
[31] Liu Z Q, Ding L X, Wang Z L, Mao Y C, Xie S L, Zhang Y M, Li G R and Tong Y X 2012 ZnO/SnO_2 hierarchical and flower-like nanostructures: facile synthesis, formation mechanism, and optical and magnetic properties *CrystEngComm* **14** 2289-95
[32] Ji H, Liu X, Wang X, Liang S, Ge X and Li Y 2011 Self-assembly of disk-like multiring ZnO–SnO_2 colloidal nanoparticles *Journal of Colloid and Interface Science* **356** 412-5
[33] Wu R, Chen X and Hu J 2012 Synthesis, characterization, and biosensing application of ZnO/SnO 2 heterostructured nanomaterials *Journal of Solid State Electrochemistry* 1-8
[34] Kowsari E and Ghezelbash M R 2012 Ionic liquid-assisted, facile synthesis of ZnO/SnO_2 nanocomposites, and investigation of their photocatalytic activity *Materials Letters* **68** 17-20

[35] Yu L M, Fan X H, Shui J Y, Cao L and Yan W 2012 Fabrication of SnO_2/Zn_2SnO_4/ZnO Nanocables through Thermal Oxidation of Zn and Sn Mixture Powders *Advanced Materials Research* **532** 70-3

[36] Choi S W, Park J Y and Kim S S 2009 Synthesis of SnO_2–ZnO core–shell nanofibers via a novel two-step process and their gas sensing properties *Nanotechnology* **20** 465603

[37] L. Gu, K. Zheng, Y. Zhou, J. Li, X. Mo, G.R. Patzke, and G. Chen 2011 Humidity sensors based on ZnO/TiO_2 core/shell nanorod arrays with enhanced sensitivity *Sensors and Actuators B: Chemical* **159** 1-7

[38] Yuan Q, Li N, Tu J, Li X, Wang R, Zhang T and Shao C 2010 Preparation and humidity sensitive property of mesoporous ZnO–SiO_2 composite *Sensors and Actuators B: Chemical* **149** 413-9

[39] Mamat M H, Khusaimi Z, Zahidi M M, Bakar S A, Siran Y M, Rejab S A M, Asis A J, Tahiruddin S, Abdullah S and Mahmood M R 2011 Controllable Growth of Vertically Aligned Aluminum-Doped Zinc Oxide Nanorod Arrays by Sonicated Sol--Gel Immersion Method depending on Precursor Solution Volumes *Japanese Journal of Applied Physics* **50**

[40] Tcholakova S, Mitrinova Z, Golemanov K, Denkov N D, Vethamuthu M and Ananthapadmanabhan K 2011 Control of Ostwald Ripening by Using Surfactants with High Surface Modulus *Langmuir* **27** 14807-19

[41] Huang J, Xu X, Gu C, Wang W, Geng B, Sun Y and Liu J 2012 Size-controlled synthesis of porous ZnSnO3 cubes and their gas-sensing and photocatalysis properties *Sensors and Actuators B: Chemical* **171–172** 572-9

[42] Huang J, Yu K, Gu C, Zhai M, Wu Y, Yang M and Liu J 2010 Preparation of porous flower-shaped SnO_2 nanostructures and their gas-sensing property *Sensors and Actuators B: Chemical* **147** 467-74

[43] Bauskar D, Kale B B and Patil P 2012 Synthesis and humidity sensing properties of $ZnSnO_3$ cubic crystallites *Sensors and Actuators B: Chemical* **161** 396-400

[44] Jamil H, Batool S S, Imran Z, Usman M, Rafiq M A, Willander M and Hassan M M 2012 Electrospun titanium dioxide nanofiber humidity sensors with high sensitivity *Ceramics International* **38** 2437-41

Sol-Gel Derived Nano-Magnesium Oxide: Influence of Drying Temperature to the Dielectric Layer Properties

Z Habibah[1,*], K A Yusof[1], L N Ismail[1], R A Bakar[1] and M Rusop[1,2]

[1]NANO-ElecTronic Centre (NET), Faculty of Electrical Engineering, Universiti Teknologi MARA, 40450 Shah Alam, Malaysia

[2]NANO-SciTech Centre (NST), Institute of Science, Universiti Teknologi MARA, 40450 Shah Alam, Malaysia

*E-mail: habibahzulkefle@yahoo.com

Abstract. Nano-magnesium oxide (Nano-MgO) had successfully deposited on glass substrate using sol-gel spin coating technique at different drying temperature (100, 200 and 300°C). The dielectric layer properties of deposited nano-MgO films were investigated in terms of its resistivity, leakage current density, relative permittivity, topology and morphology respectively. Changing in drying temperature resulted in variation of nano-MgO films properties and film dried at 200°C has compact, uniform, less porous and in nano size that lead to the enhancement in relative permittivity value.

1. Introduction

Over past decade, magnesium oxide (MgO) which is inorganic compound with NaCl structure had attracts great attention in many applications [1-3]. MgO also is a suitable candidate to replace existing dielectric layer material which is silicon dioxide (SiO_2) due to its unique properties such as high dielectric constant (9.8), wide band gap (7.8eV) and high breakdown field (12MV/cm) [4, 5]. MgO films can be prepared using several methods such as physical deposition and chemical deposition method and among other methods, chemical deposition via spin coating technique has its own advantages which are large area deposition, simple and also cost effective. There are quite number of researches had been done in order to investigate the effect of deposition parameters on the properties of MgO films [6, 7]. Heat treatment which includes drying temperature is the important deposition parameters that will influence the production of high quality MgO films. Since the effects of drying temperature on the properties of sol-gel derived MgO thin films is not well understood yet, it is therefore a need to carry out a details study on this matter. This paper reports on the effect of drying temperature on electrical, structural and dielectric properties of nano-MgO films deposited using sol-gel spin coating method.

2. Experimental Procedure

During MgO solution preparation, magnesium acetate tetrahydrate ((CH_3COO_2) Mg*$4H_2O$) was dissolved in ethanol (C_2H_6O) with small amount of nitric acid (HNO_3) which act as the stabilizer. The solution was then sonicated for 20 mins and stirred at 80°C for 3 hrs. Next, MgO solution was undergoes aging process at room temperature for 24 hrs. Afterward, the prepared MgO solution was deposited on cleaned glass substrate where the deposition parameter was set at 3200 rpm for 30 secs.

Next the deposited films were dried at different temperatures (100, 200 and 300°C) for 10 mins. The deposition and drying processes were repeated for several times to achieve desired film thickness and finally the deposited nano-MgO films were annealed at 500°C to restructure the MgO particles.

There are three main characterizations have been done in this research work which are electrical, dielectric and structural characterizations. Electrical characterization was conducted using two-point probes *I-V* measurement (Bukoh Keiki) while dielectric properties were determined by impedance analyzer (Solartron S1 1260A-1296). Surface topography and morphology of the deposited nano-MgO films were observed using atomic force microscopy, AFM (Park System, XE-100) and field emission scanning electron microscope, FESEM (JEOL JSM 7600F) respectively. The thickness of deposited films was determined using surface profiler (Veeco).

3. Results and Discussion

3.1. Electrical Properties

Fig. 1a and 1b shows the electrical behaviour of the deposited MgO films in terms of resistivity, conductivity and leakage current density respectively. As shown in Fig. 1a, the resistivity values were increased as the drying temperature increased. Rise in resistivity value was due to the small grain and high porosity [8] film that lead to the increase of carrier scattering and low carrier mobility [9]. Besides, it was also due to thicker film produced as the drying temperature increased. The resistivity values obtained are 10.8×10^4, 12.8×10^4 and 17.4×10^4 Ω.cm for films dried at 100, 200 and 300°C respectively which are in same magnitude with work done by S. Lee et.al which their resistivity was around 10^4Ω.cm [10].

(a)

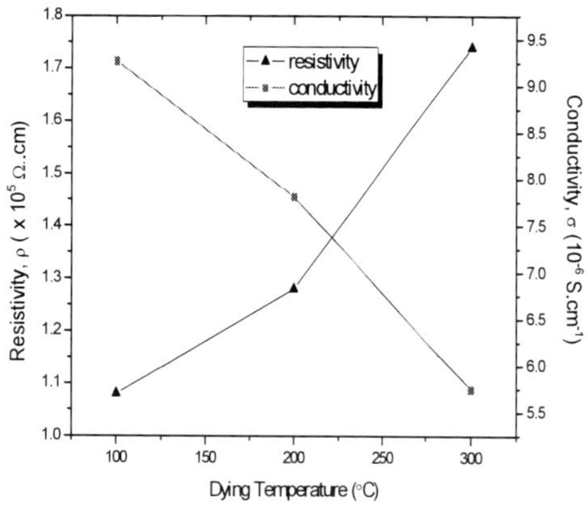

(b)

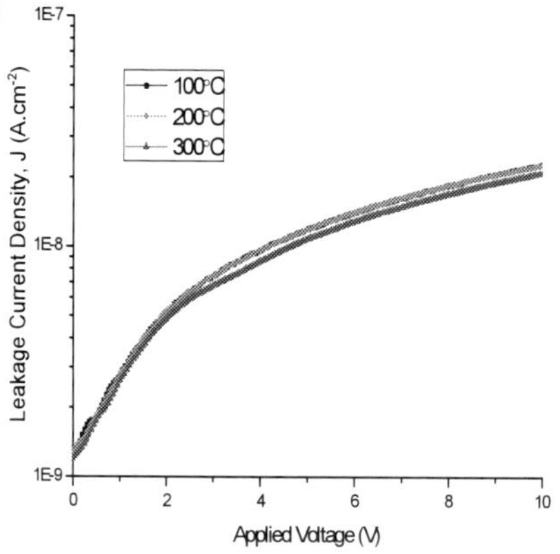

Figure 1. Electrical properties in terms of (a) resistivity and conductivity, (b) leakage current density of MgO films dried at different temperatures.

From Equation 1, it clearly shows that the conductivity is inversely proportional to the resistivity value and it is proven by conductivity graph shown in fig. 1a.

$$\sigma = \frac{1}{\rho} \tag{1}$$

where σ is the conductivity and ρ is the resistivity of deposited MgO films. The leakage current density, *J* was plot in Fig. 1b where the value was below than 10^{-9} at applied voltage of 2V which was five order of magnitude lower than common dielectric material ($SiO_2 \sim 4 \times 10^{-4}$ A/cm) . Moreover, it also reveals that there were slightly decreased in the *J* values as the drying temperature increased from

100 to 300°C. This was due to the reduction in grain boundary area produced in the film that reduced a direct path for the carrier to pass through the film.

3.2 Relative Permittivity Characteristic

The dielectric properties of MgO films with the influence of drying temperature were measured at frequency range from 0 to 40 kHz and their relative permittivity, ε_r value was shown in Fig. 2. High ε_r values obtained for all samples at low frequency region (below 10 kHz) and its value drastically decreased as the frequency increased. As stated by Shaikh et. al, decrease in ε_r value with frequency is due to inhomogeneous film that cause the charge trapped at the interface structure [11]. At high frequency region (above 40 kHz), ε_r value of each film decreased to a constant value which due to non-polarized charges [12].

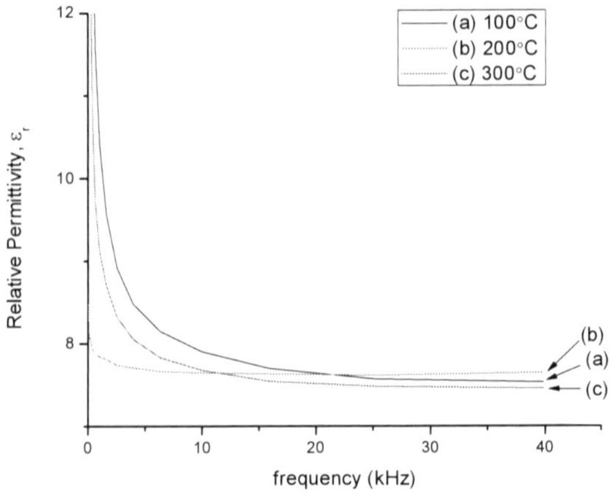

Figure 2. Relative permittivity, ε_r as the function of frequency at various drying temperature.

In addition, Fig. 2 also revealed that ε_r value increases as the drying temperature increased from 100 to 200°C. However, as the temperature increase to 300°C, leads to the reduction of ε_r value. The variation in ε_r value for various drying temperature is related to the structural changes such as grain size and surface roughness of the film [13]. From the FESEM images (Fig. 3), it clearly shows that the MgO film dried at 200°C has a compact, uniform, small roughness and less porosity which contributes to the high ε_r value produced.

3.3 Structural Properties

Deposited MgO thin films growth was characterized in term of their thickness using surface profiler and their values were tabulated in Table 1. As can be seen, rise in the drying temperature leads to the formation of thicker film.

Table 1: Nano-MgO films average thickness and roughness.

Drying Temperature (°C)	Average Film thickness (nm)	Root Mean Square, RMS Surface Roughness (nm)
100	137.49	44.360
200	192.97	20.660
300	224.72	21.543

As stated by G. Hu et. al, film surface roughness is define as the measure of the surface texture of the film in terms of their RMS value of the surface height profile of the film [14]. The surface roughness of the deposited MgO films was determine using atomic force microscopy, AFM with scan area of (10 x 10) m^2. From the tabulated data, the RMS surface roughness value for MgO films shows an improvement in film uniformity as the drying temperature increased from 100 to 200°C, however it value slightly increase for film dried at 300°C.

Fig. 3 shows the FESEM images of deposited MgO films with magnification and voltage of 30k and 5kV respectively. From the images (Fig. 3), it indicates that the drying temperature affect the MgO film morphology. Increased in drying temperature from 100 to 200°C resulted to the formation of uniform structure with few defects. Nevertheless increase in drying temperature to 300°C produced a non-uniform and cracks film. This was due to the magnesium evaporation process [15].

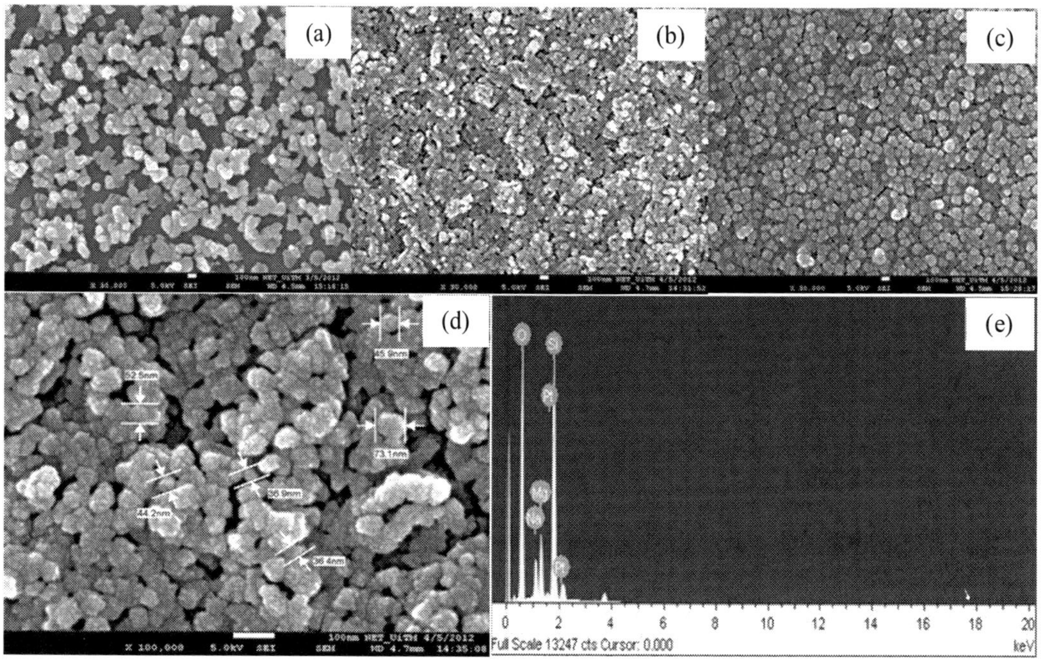

Figure 3. Surface morphology at 30k magnification (a) 100°C, (b) 200°C, (c) 300°C and 200°C film (d) at 100k magnification with (e) EDS analysis of nano-MgO films.

MgO film dried at 200°C gives better electrical and dielectric properties compared to others which are related to it uniform and less porosity on the film. Besides particle size of MgO film produced was in nano-meter size (Fig. 3d) which in a range of 36.4nm to 73.1nm. P. K. Ghosh et.al stated that, the enhancement in polarization that lead to the increased in the ε_r value can be obtained with nano-meter particle size of the film due to its high surface area per unit volume [16]. Since the film dried at this temperature has good film properties, the EDS analysis (Fig. 3e) was then performed to confirm magnesium, Mg and oxygen, O_2 element are exist in the film. Other than that, silicon, sodium and platinum element detected are due to the glass substrate and coating respectively. Thus, the film dried at this temperature is suitable to be used as a dielectric layer.

Conclusion

Sol-gel spin coating method was used as deposition technique of prepared nano-MgO films. The drying temperature was varied at 100, 200 and 300°C in order to investigate the effect of drying temperature towards dielectric layer properties of nano-MgO films. Dielectrics properties of deposited nano-MgO films were investigated in terms of its electrical, relative permittivity and structural properties. Changing in drying temperature applied resulted in the variation of nano-MgO film properties. Nano-MgO film dried at 200°C was suggested suitable to be used as dielectric layer due to its high resistivity (12.8x10^4 Ω.cm) and low leakage current density (~10^{-9} A.cm^{-2}). Besides, it also has merit in surface properties such as dense, homogeneous, less porous and nanometer particle size which resulted increment in relative permittivity.

Acknowledgement
The authors were grateful to the financial support from Excellent Grant UiTM (600-RMI/ST/DANA 5/3 Dst (401/2011) from Research Management Institute (RMI), UiTM Shah Alam and Ministry of Higher Education (MOHE) Malaysia.

References

[1] N. Yasui, H. Nomura, and A. Ide-Ektessabi, "Characteristics of ion beam modified magnesium oxide films," *Thin Solid Films,* vol. 447-448, pp. 377-382, 2004.
[2] H. S. Jung, J.-K. Lee, J.-Y. Kim, and K. S. Hong, "Crystallization behaviors of nanosized MgO particles from magnesium alkoxides," *Journal of Colloid and Interface Science,* vol. 259, pp. 127-132, 2003.
[3] S. K. Shukla, G. K. Parashar, A. P. Mishra, P. Misra, B. C. Yadav, R. K. Shukla, L. M. Bali, and G. C. Dubey, "Nano-like magnesium oxide films and its significance in optical fiber humidity sensor," *Sensors and Actuators B: Chemical,* vol. 98, pp. 5-11, 2004.
[4] V. P. Shital Patil, "Properties of magnesium oxide and magnesium oxide-bismuth oxide mixed thick film," *Microelectronics International,* vol. 29, pp. 3-9, 2012.
[5] H.-Y. Tsao, Y.-J. Lin, Y.-H. Chen, and H.-C. Chang, "Leakage currents through In/MgO/n-type Si/In structures," *Solid State Communications,* vol. 151, pp. 693-696, 2011.
[6] M. A. Shah and A. Qurashi, "Novel surfactant-free synthesis of MgO nanoflakes," *Journal of Alloys and Compounds,* vol. 482, pp. 548-551, 2009.
[7] J. Cheng, X. Meng, P. Yang, and J. Chu, "Preparation and characterization of Pt/MgO/Si multilayer thin films," *Journal of Crystal Growth,* vol. 194, pp. 89-93, 1998.
[8] S. Patil and V. Puri, "Oxidation temperature dependent properties of MgO thin film on alumina," *Applied Surface Science,* vol. 258, pp. 1535-1540, 2011.
[9] J.-H. Lee, K.-H. Ko, and B.-O. Park, "Electrical and optical properties of ZnO transparent conducting films by the sol-gel method," *Journal of Crystal Growth,* vol. 247, pp. 119-125, 2003.
[10] S. Lee and T. Ito, "Oxygen treatment effects on properties of MgO thin films grown on single-crystalline diamond (100)," *Materials Research Bulletin,* vol. 40, pp. 951-961, 2005.
[11] P. A. Shaikh and Y. D. Kolekar, "Study of microstructural, electrical and dielectric properties of perovskite $_{(0.7)}$ PMN-$_{(0.3)}$ PT ferroelectric at different sintering temperatures," *Journal of Analytical and Applied Pyrolysis,* vol. 93, pp. 41-46, 2012.
[12] S. F. Khor, Z. A. Talib, W. M. Daud, H. A. A. Sidek, and B. H. Ng, "Effects of MgO on dielectric properties and electrical conductivity of ternary zinc magnesium phosphate glasses," *Journal of Non-Crystalline Solids,* vol. 355, pp. 2533-2539, 2009.
[13] K.-T. Kim and C.-I. Kim, "The effects of drying temperature on the crystallization of YMnO3 thin films prepared by sol-gel method using alkoxides," *Journal of the European Ceramic Society,* vol. 24, pp. 2613-2617, 2004.
[14] G. Hu, G. Orkoulas, and P. D. Christofides, "Regulation of film thickness, surface roughness and porosity in thin film growth using deposition rate," *Chemical Engineering Science,* vol. 64, pp. 3903-3913, 2009.
[15] S. H. Tamboli, V. Puri, and R. K. Puri, "Adhesion and stress of magnesium oxide thin films: Effect of thickness, oxidation temperature and duration," *Applied Surface Science,* vol. 256, pp. 4582-4585, 2010.
[16] P. K. Ghosh, S. Jana, S. Nandy, and K. K. Chattopadhyay, "Size-dependent optical and dielectric properties of nanocrystalline ZnS thin films synthesized via rf-magnetron sputtering technique," *Materials Research Bulletin,* vol. 42, pp. 505-514, 2007.

Nanozeolite Produced by Wet Milling at Different Milling Time

N.Z.F. Mukhtar[1,2,*], M.Z. Borhan[1,2], S.Abdullah[2] , M. Rusop[1]

[1]NANO-SciTech Centre, Institute of Science, Universiti Teknologi MARA (UiTM), 40450 Shah Alam, Selangor, Malaysia

[2]Faculty of Applied Sciences, Universiti Teknologi MARA (UiTM), 40450 Shah Alam, Selangor, Malaysia

*E-mail: n.zahidatul@gmail.com

Abstract. Recently, there has been a considerable growing interest in utilizing nanozeolites due to their advantages over conventional micron sized materials. Zeolite particle may be reduced by mechanical treatment such as ball milling or grinding in order to get smaller particles. In this paper, effect of milling time on particle size and surface morphology of Zeolite were investigated by a few designed of experiments in aqueous environment. Zeta-sizer Nano series of particle sizer and Field Emission Scanning Electron Microscopy (FESEM) have been used to characterize this nanozeolite. Results shown that, there were changes of particle size and also the surface morphology of Zeolite.

1. Introduction

Zeolite is a large group of crystalline materials and have a framework of linked tetrahedral that consisting of four oxygen atoms around their cation. Zeolites are also known as molecular sieves and aluminosilicate materials that are characterized by their open structure of interconnected cavities that can be accessed by molecules, atoms and ions species [1]. The properties and uses of zeolites are being explored in many scientific disciplines: modern inorganic chemistry, physical chemistry, colloid chemistry, biochemistry, mineralogy, geology, surface chemistry, oceanography, crystallography, catalysis and in all types of chemical engineering process technology [2].

Ball milling treatment, as a cost-lowering and environmental friendly method has been used in this project at aqueous environment. Ball milling refers to the use of friction, collision, impingement, shear or other mechanical actions to modify the structures and properties of the Zeolite [3]. The reduction of the particle size of Zeolite causes larger external surface areas available for interaction, shorter diffusion path lengths reducing mass and heat transfer resistances in catalytic and sorption applications, decreasing of side reactions, enhancing selectivity as well as lowering tendencies to coke formation in some catalytic reactions [4].

Therefore, the objective of this study is to investigate the effect of milling time on particle size and surface morphology of Zeolite by wet ball milling treatment through Zeta-sizer Nano series of particle sizer and Field Emission Scanning Electron Microscopy (FESEM).

2. Methodology

2.1. Materials
The commercial powders of Zeolite was purchased from Fluka with means size about ≥45μm.

2.2. Ball Milling Treatment
Wet grinding has been selected in this study because it has some advantages over dry grinding such as the higher energy efficiency, lower magnitude of excess enthalpy and the elimination of dust formation [5]. The milling process was performed by mean of a planetary ball mill with some parameters that have been highlighted with respect to size reduction. Grinding time was varied for different experiment between 2 to 8 h while the rotational speeds were constant at 550 rpm for each experiment. All of the experiments were carried out in a 50 mL stainless steel jar with protective jacket of zirconium oxide and 2 mm of zirconium oxide balls were used for this milling process respectively. The grinding jar was set up and arranged eccentrically on the sun wheel of the planetary ball mill. The direction of movement of the sun wheel being opposite to that of grinding jars was selected with the ratio of 1: 1. A certain amount of Zeolite: 0.5 g, water: 10 ml and balls: 45 g were weighted accurately and placed in the jar at room temperature also at atmospheric pressure. The jar then sealed and imposed to milling. After milling, all sample were carried out then was allowed to be cooled down at 40°C for 24 hours. Then all samples were continuing milling for 15 minutes and then were characterized.

2.3. Characterization
Particles size measurements were carried out using Zeta Nano Sizer S at 25 °C and the reading was taken at 173 ° scattering angle. Prior the measurement, 1 ml of nanosuspensions for each samples were taken out to introduced in the disposable cuvette to be measured. The samples were measured in triplicate. The dried powders were taken for FESEM analysis.

3. Results and discussion

3.1. Particle Sizer
The effects of milling time on particle size of Zeolite were investigated by a few designed experiments in aqueous environment and different time. The particle size distributions of all the Zeolite samples prepared were obtained by Zeta Nano Sizer S, as mentioned before. Figure 1 showed the influences of milling time on zeolite particle. The particle size without milling, the Z-average (mean value in nm) was 1889 nm then after milling treatment at 2 h the Z-average was 211.6 nm while for 4, 6 and 8 h were 185.2, 171.4 and 164.9 nm.

From original Zeolite it reduced about 88.80 % of particle size when imposed to milling time at 2 hours. Then, the particle size of Zeolite continuous to reduce when continuous imposed to milling process until 8 hours. 8 hours of milling time shown the smallest average of Zeolite particle size compared to others with reduction about 91.27% from the original size. From these sizes produced, we observed that the particles size is inversely proportional with grinding time. This closely related to the generation of energy over time during grinding process.

As we can see from figure 1, the shorter grinding time will produce larger mean particles compare to the longer grinding time. This result is consistent with the finding by Takatsuka et.al. [6] which found that the particles size reduce with increasing of running time of grinding. This can be understood from the interaction of particles and the grinding ball.

Interaction between frictional and impact forces due to the collision of ball and inner wall of jar which move at the difference speeds which will unleashes a high dynamic of energies. For a 'nano' process to take place, energy in the jar should be sufficient enough. Fadda et al [7] proposed that the particles breakage happen at initial stage of milling then equilibrium between re-agglomeration and de-agglomeration occur.

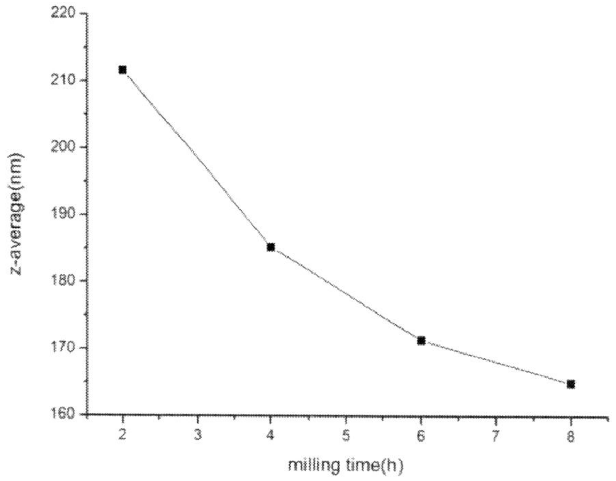

Figure 1. Average particle size of Zeolite at different milling time.

3.2. Surface Morphology

To study probable changes of morphology for ground powders of Zeolite after milling as well as to confirm the particle size results, all samples were subjected to Field Emission Scanning Electron Microscopy (FESEM). Some of the FESEM images of samples were shown in Figure 2. FESEM images indicated that almost in all samples, Zeolite powder with particles size less than 100nm may be recognise as a separated particle or in the form of larger agglomerates.

Amir Charkhi et al. [4] also reported the similar phenomenon when grinding Clinoptilolite by a wet ball milling treatment. Moreover most particles have lost their initial cubic shape(Fig.2a) and converted into spherical and also irregular shapes after imposed to the milling process with varied in milling times(Fig 2b,2c,2d,2e).

Moreover, Chunlong et al [8, 9] hypothesized that fractures or de-agglomerations easily happen at the defects, impurities and at the interface of the intergrowth within large Zeolite particles. Thus the mechanical stress that was generated by the jar ball mill randomly fractured or de-agglomerated the Zeolite particles into smaller fragment with varying size distribution and different morphology. The larger fragments were typically irregular shape whereas the small fragments tended to be spherical. From this observation of Zeolite morphology, we could summarize that FESEM results were in accordance with the characterization results by the particle size analyser that have been explained before.

IOP Conf. Series: Materials Science and Engineering **46** (2013) 012007 doi:10.1088/1757-899X/46/1/012007

Figure 3. The surface morphology of nanozeolite as a function of milling time: a) 0 h b) 2 h c) 4 h d) 6 h and e) 8 h

4. Conclusion

As a conclusion for this study we can conclude that at milling time 8 hours, the particle size is smaller than others while for surface morphology, we can observed that the nanozeolite tended to be spherical shape and also irregular shape after milling treatment. Therefore, ball milling is a powerful mechanical method to reduce size of Zeolite particles. We proposed that, the condition of milling should be carefully selected to get the nanozeolite powder.

Acknowledgement

The authors would like to thank Universiti Teknologi MARA for the financial support. We are grateful to the all staffs of NANO SciTech Centre (NST) and also NANO ElecTronic Centre (NET) for technical support of this study.

References

[1] Vladimir V M, Mary A W 2002 Thermal properties of zeolites: effective thermal conductivity of dehydrated powdered zeolite 4A *Mater Chem and Phys.* **75** 178-180
[2] Kanokorn H *et al* 2011 Synthesis of zeolite A from by- product of aluminium etching process: Effects of reaction temperature and reaction time on pore volume *American Journal of Environmental Science* **7** (1): 35-42
[3] Tian Y L *et al* 2011 The effect of ball milling treatment on structure and porosity of maize starch granule *Innovative Food Science and Emerging Technologies* **12** 586–593
[4] Amir C *et al* 2010 Optimized experimental design for natural clinoptilolite zeolite ball milling to produce nano powders *Powder Technology* **203** 389-396
[5] Kani A *et al* 2004 Wet ball milling of zeolite HY *Powder Technology* **142** 121 – 128
[6] Takatsuka *et al* 2010 *Chemical and Pharmaceutical Bulletin* **57** (10): 1061-1067
[7] Fadda *et al* 2009 *Powder Technology* **194** (3): 207-216
[8] Chunlong K *et al* 2010 Zeolite nanocrystas prepared from zeolite microparticles by a centrifugation – assisted grinding method *Chemical Engineering and Processing* **49** 809 – 814
[9] Pankaj S *et al* 2008 Synthesis and morphological studies of nanocrystalline MOR type zeolite material *Journal of Colloid and Interface Science* **325** 547–557

Comparative Study of Performance and Combustion Characteristics of Conventional and Low Heat Rejection (Mullite Coated) Diesel Engines

S B Patond[1], S A Chaple[2], P N Shrirao[3], P I Shaikh[4]

[1,2,3,4] Jawaharlal Darda Institute of Engineering and Technology, Near M.I.D.C., Lohara, Amravati Road, Yavatmal[445001], Maharashtra, India

E-mail: swaps.patond@gmail.com, schaple@ymail.com

Abstract. Tests were performed on a single cylinder, four stroke, direct injection, diesel engine whose piston crown, cylinder head and valves were coated with a 0.5 mm thickness of $3Al_2O_3$ $.2SiO_2$ (mullite) (Al_2O_3= 60%, SiO_2= 40%) over a 150 μm thickness of NiCrAlY bond coat. The working conditions for the conventional engine (without coating) and LHR (mullite coated) engine were kept exactly same to ensure a comparison between the two configurations of the engine. This paper is intended to emphasis on performance and combustion characteristics of conventional and LHR (Mullite coated) diesel engines under identical conditions. Tests were carried out at same operational constraints i.e. air-fuel ratio and engine speed conditions for both conventional engine (without coating) and LHR (mullite coated) engines. The results showed that, there was as much as 1.8 % increasing on brake power for LHR (mullite coated) engine compared to conventional engine (without coating) at full load The average decrease in brake specific fuel consumption in the LHR engine compared with the conventional engine was 1.76 % for full engine load. However, there was increasing on cylinder gas pressure and net heat release rate for LHR engine compared to conventional engine. Also the results revealed that, there was as much as 22% increasing on exhaust gas temperature for LHR engine compared to conventional engine at full engine load.

Keywords: Performance characteristic; ceramic coating; Mullite; LHR; SE

1. Introduction:

It is well known fact that about 30% of the energy supplied is lost through the coolant and the 30% is wasted through friction and other losses, thus leaving only 30% of energy utilization for useful purposes. In view of the above, the major thrust in engine research during the two decades has been on development of low heat rejection engines. Several methods adopted for achieving low heat rejection to the coolant were using ceramic coatings [1] on piston, liner and cylinder head and creating air gap in the piston [2] and other components with low-thermal conductivity material like superni, mild steel etc. However, this method involved the complication of joining two different metals. [3] Jabez Dhinagar et al. used different crown materials with different thickness of air gap in between the crown and the body of the piston. Ceramics have a higher thermal durability than metals; therefore it is usually not necessary to cool them as fast as metals. Low thermal conductivity ceramics can be used to control temperature distribution and heat flow in a structure [4-5].

Thermal barrier coatings (TBC) provide the potential for higher thermal efficiencies of the engine, improved combustion and reduced emissions. In addition, ceramics show better wear characteristics than conventional materials. Lower heat rejection from combustion chamber through thermally insulated components causes an increase in available energy that would increase the in-cylinder work and the amount of energy carried by the exhaust gases, which could be also utilized [6-7]. A major breakthrough in diesel engine technology has been achieved by the pioneering work done by Kamo and Bryzik [8-9]. Kamo and Bryzik used thermally insulating materials such as silicon nitride for insulting different surfaces of combustion chamber. An improvement of 7% in the performance was observed [10]. Sekar and Kamo [11] developed an adiabatic engine for passenger cars and reported an improvement in performance to the maximum extent of 12%. The experimental results of [12] indicate that the higher temperatures of the insulated engine cause reduction in the in-cylinder heat rejection, which is in accordance with the conventional knowledge of convective heat transfer. Woschni [13] state that 5% of the input fuel energy cannot be accounted for which is of the order of the expected improvements. Havstad [14] developed a semi-adiabatic diesel engine and reported an improvement ranging from 5 to 9% in ISFC, about 30% reduction in the in-cylinder heat rejection. Prasad [15] used thermally insulating material, namely partially stabilized zirconia (PSZ), on the piston crown face and reported a 19% reduction in heat loss through the piston.

Among possible alternative materials, one of the most promising is mullite. Mullite has low density, high thermal stability, stability in severe chemical environments, low thermal conductivity and favorable strength and creep behavior. It is a compound of SiO_2 and Al_2O_3 with composition $3Al_2O_3.2SiO_2$. Compared with Yttria-stabilized zirconia (YSZ), mullite has a much lower thermal expansion coefficient and higher thermal conductivity, and is much more oxygen-resistant than YSZ. For the applications such as diesel engines where the surface temperatures are lower than those encountered in gas turbines and where the temperature variations across the coating are large, mullite is an excellent alternative to zirconia as a TBC material. Engine tests performed with both materials show that the life of the mullite coating in the engine is significantly longer than that of zirconia [16-17]. Above 1273 K, the thermal cycling life of mullite coating is much shorter than that of YSZ [18]. Mullite coating crystallizes at 1023–1273 K, accompanied by a volume contraction, causing cracking and de-bonding. Mullite has excellent thermo-mechanical behavior; however its low thermal expansion coefficient creates a large mismatch with the substrate [19]. To avoid the mismatch with the substrate, a 150 μm thickness of NiCrAlY bond coat was used.

Table 1: Properties of Bond coat material and mullite material was used as thermal barrier coating (TBC) for diesel engine

Material Properties	NiCrAlY	Mullite
Modulus of elasticity (GPa)	64.5	19
Poisson's ratio	0.30	0.25
Thermal conductivity (W/ mK) @ RT	3.88	1.29
Coefficient of thermal expansion ($\times 10^{-6} K^{-1}$) @ RT	10.3	5.1
Density (Kg /m^3)	6290	2710
Specific heat (J/kg K) @ RT	460	760

2. Experimental Setup

A four stroke, direct injected, water-cooled, single cylinder, naturally aspirated diesel engine was used for investigation. Details of the engine specifications are given in Table 2.

Table 2: Engine specifications

Engine type	**Kirloskar AV1, DI**
Stroke number	4
Cylinder number	1
Bore (mm)	80
Stroke (mm)	110
Compression ratio	16.5:1
Maximum engine power (KW)	3.7
Maximum engine speed (rpm)	1500
Specific fuel consumption (g/Kwh)	245
Injection timing	20^0 Before Top Dead Centre(BTDC) static

Note: T1, T3-Inlet Water Temperature , T2- Outlet Engine Jacket Water Temperature ,T4-Outlet Calorimeter Water Temperature, T5-Exhaust Gas Temperature before Calorimeter, T6- Exhaust Gas Temperature after Calorimeter,F1-Fuel Flow DP (Differential Pressure) unit, F2- Air Intake DP unit, PT- Pressure Transducer, N- RPM Decoder, EGA- AVL Di-Gas Analyzer (5 gas), SM - AVL Smoke meter.

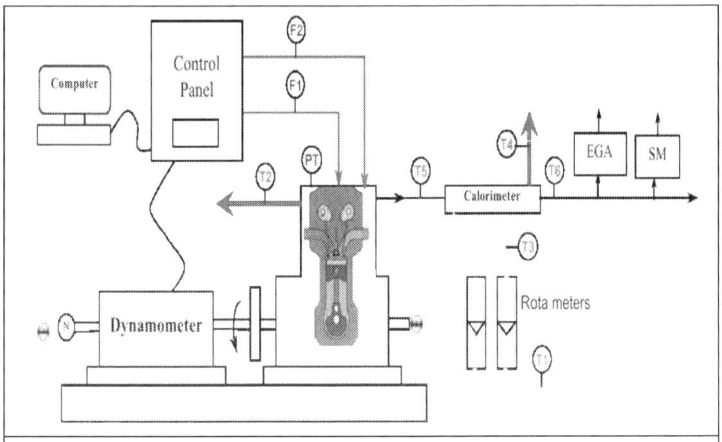

Figure 1: Experimental Set up

The first stage tests were performed at different engine loads for conventional engine. The experiments were conducted at five load levels, viz. 0, 25, 50, 75% of full load and full load. The required engine load percentage was adjusted by using the eddy current dynamometer. At each of these loads, engine performance and combustion characteristics such as brake power, brake thermal efficiency, brake specific fuel consumption, exhaust gas temperature, cylinder gas temperature and net heat release rate were recorded. The second stage tests were conducted on engine when combustion chamber insulation was applied. A piston crown, cylinder head and valves were coated with ceramic material over super alloy bond coating (NiCrAlY). The bond coat was first applied to these engine components to avoid mismatch in thermal expansion between substrate and ceramic material. A piston crown, cylinder head and valves were coated with 0.5 mm coating of Mullite is commonly denoted as $3Al_2O_3$ $.2SiO_2$ (i.e. 60 mol% Al_2O_3). The ceramic material was coated by using plasma-spray technique. The engine was insulated and tested at baseline conditions to see the effect of insulated surfaces on engine performance and combustion characteristics. In present investigation, a centrifugal blower was used for LHR (mullite coated) diesel engine to maintain air-fuel ratio nearly same as in conventionally cooled (without coating) diesel engine. Thus the working conditions for the conventional engine (without coating) and LHR (mullite coated) diesel engine were kept exactly same to ensure a comparison between the two configurations of the engine.

The results of performance and combustion characteristics of LHR engine were compared with the conventional engine.

3. Plasma Spray Technique:

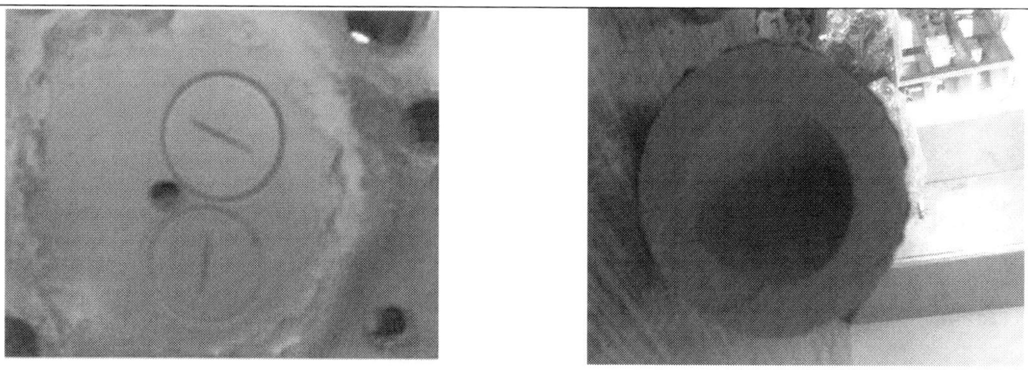

Figure 2: Photographic view of Cylinder head, Cylinder valves and Piston crown after ceramic coating.

The gas tunnel type plasma spraying torch was used. The experimental method to produce ceramic coating by means of the gas tunnel type plasma spraying is as follows. After igniting plasma gun, the main vortex plasma jet is produced in the low pressure gas tunnel. The spraying powder is fed from central inlet of plasma gun. The coating was formed on the substrate traversed at the spraying distance L. The power input to the plasma torch was about P= 25 KW. The current and voltage applied was about 837 amp and 37.3 volts respectively. The inputs were given by Miller Thermal, Inc. Model3702. The power input to the pilot plasma torch, which was supplied by power supply PS1, was turned off after starting of the gas tunnel type plasma jet. The spraying distance was short distance of L=40 mm. The working gas was Argon gas, and the flow rate for gas tunnel type plasma spraying torch was Q= 180 l/min, and gas flow rate of carrier gas was 10 l/min [20].

4. Results and Discussions:

After conducting long-term experimental investigations on a single cylinder, four stroke, direct injection, conventional (without coating) and LHR (mullite coated) diesel engines, the engine performance and combustion characteristics such as brake power, brake thermal efficiency, brake specific fuel consumption, exhaust gas temperature, cylinder gas temperature and net heat release rate for both the Conventional and LHR engines are evaluated. The engine performance and combustion characteristics are evaluated for25, 50, 75% of full engine load and full engine load condition for both conventional and LHR diesel engines.

Figure 3 Shows the comparison of brake power as a function of engine load for conventional and LHR (mullite coated) diesel engines. It is observed that, the values of brake power are slightly higher for LHR (mullite coated) engine as compared to conventional engine. This is due to effect of insulation; the heat free flow is restricted, which leads to reduction in heat transfer in case of LHR engine. The reduction in heat transfer leads to increase in combustion temperature, which leads to better combustion. The higher combustion temperature will lead to more expansion work. The increase of combustion temperature causes the brake power to increase up to 1.8 % with LHR engine at full engine load condition compared to conventional engine.

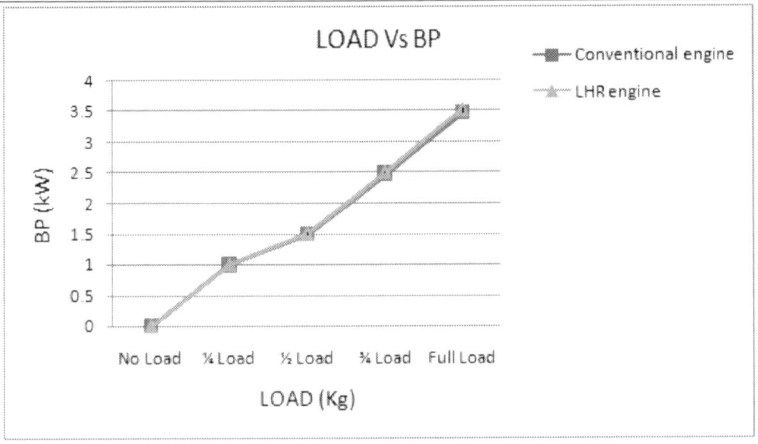

Figure 3: Engine Load Vs Brake Power

A comparison of BSFC for conventional and LHR engine for all loads is as shown in figure 4. Because of higher surface temperatures of combustion chamber of LHR engine, the BSFC values of LHR engine are lower than those of conventional engine. The improvement in fuel economy achieved by LHR engine may be attributed to: higher premixed combustion, lower diffused combustion, reduced heat transfer loss and higher rate of heat release in the main portion of combustion. It is observed that BSFC value is decreased by 1.76 % for LHR (mullite coated) engine as compared to conventional engine at full engine load.

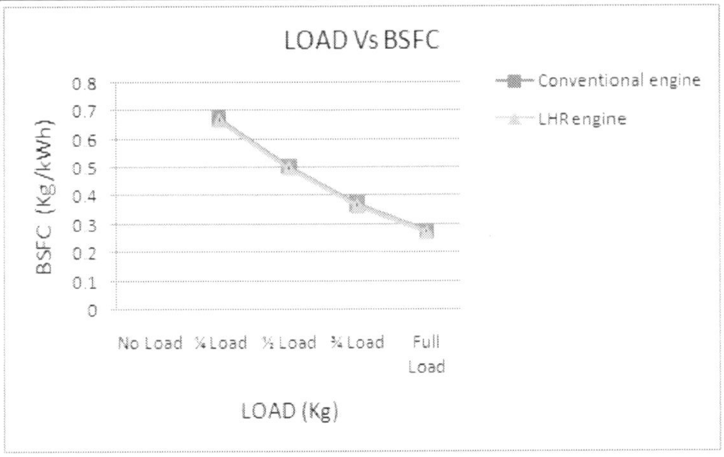

Figure 4: Engine Load Vs Brake Specific Fuel Consumption

It is also evident from figure 5 that, the amount of increase in thermal efficiency for LHR engine with 0.5 mm thickness of mullite insulation is 1.8 % compared to conventional engine at full engine load while at low and medium loads thermal efficiency shows slight variation for LHR engine when compared to the conventional engine. The reason is heat recovered by the insulation, which is normally lost to the cooling, is converted into indicated work. But all the heat recovered by the insulation perhaps may not be able to get converted into useful work. Hence, the rate of increasing thermal efficiency for LHR engine is marginal compared to conventional engine.

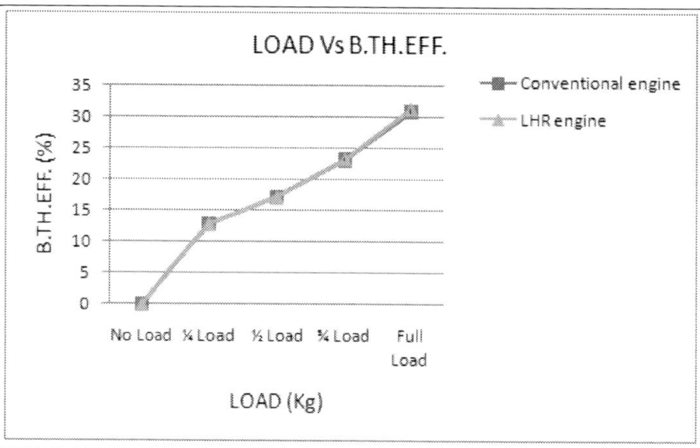

Figure 5: Engine Load Vs Brake Thermal Efficiency

Figure 6 shows the comparison of cylinder gas pressure as a function of crank angle for conventional and LHR engines with 0.5 mm coating of mullite thermal insulation on combustion chamber. It is observed that, the peak of gas pressure curve occurs 58.0 bars for conventional engine while for LHR (mullite coated) engine the peak of gas pressure curve occurs 58.88 bars. The increase in peak of gas pressure for LHR engine is due to elevated temperature of the insulated engine and better combustion of fuel.

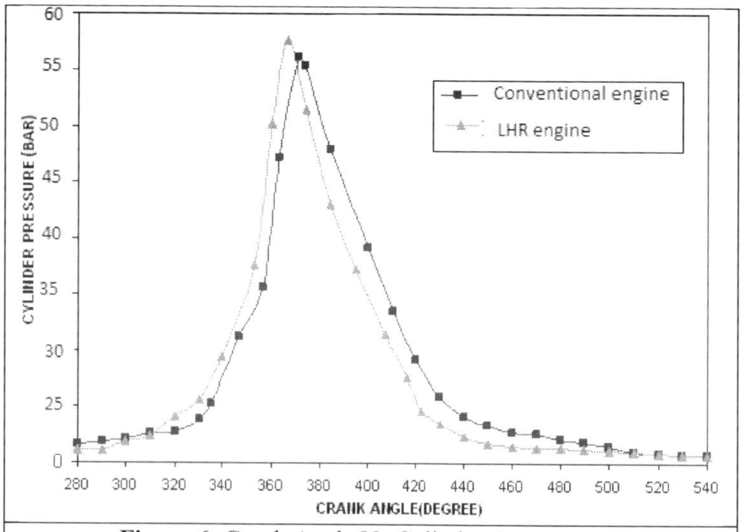

Figure 6: Crank Angle Vs Cylinder Gas Pressure

Figure 7 shows the comparison of instantaneous heat release rate as a function of crank angle for conventional and LHR engines with 0.5 mm insulation coating of mullite on combustion chamber. The trend shows that, the LHR engines are exhibiting a higher rate of heat release. The reason for high rate of heat release is due to insulation, the higher heat retainment inside the combustion chamber is exhibited. This leads to evaporate the fuel at faster rate, which helps possibly to better premixing, reduced diffused combustion. Further, it results in complete combustion of fuel. Hence, it releases more amount of heat.

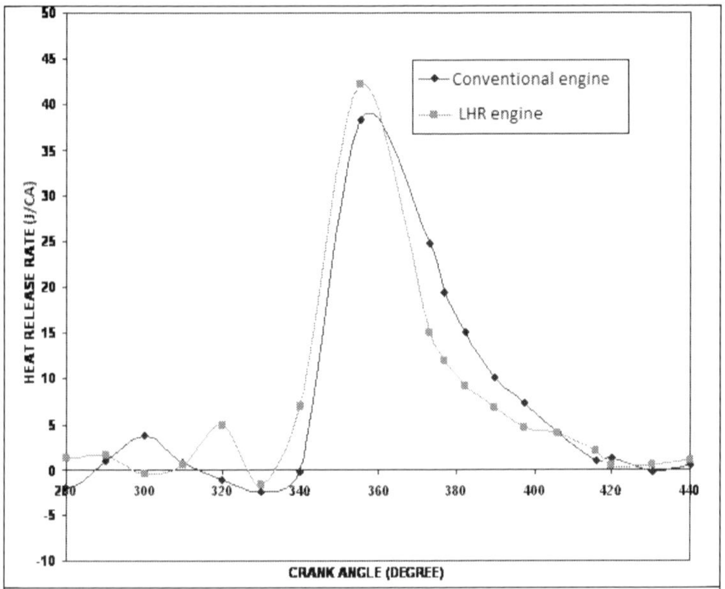

Figure 7: Crank Angle Vs Heat Release Rate

Figure 8 shows variations of exhaust gas temperature depending on the load of the conventional and LHR (mullite coated) engines. As seen in figure exhaust gas temperature increases as the engine load increases for both engines. This is due to the amount of fuel per unit time increases as the engine load increases, and consequently, more heat energy is produced. As a result, exhaust gas temperature increases. The increase in exhaust gas temperature in the LHR engine, compared with the conventional engine is 22% at full engine load. The increase in exhaust gas temperature for the LHR engine compared with the conventional engine may be due to the decrease in heat losses going into the cooling system and outside due to the coating and the transfer of this heat to the exhaust gas.

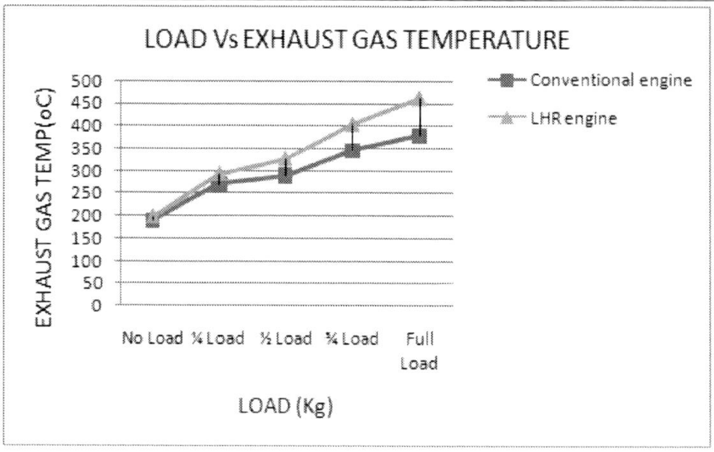

Figure 8: Engine Load Vs Exhaust Gas Temperature

5. Conclusions:

The main conclusions drawn from present experimental investigation on LHR (mullite coated) and conventional diesel engines are as follows

- LHR engine with 0.5 mm of mullite ($3Al_2O_3.2SiO_2$) insulation coating on piston crown, cylinder head and valves of diesel engine exhibits lower brake specific consumption than the conventional diesel engine. This insulation coating exhibits the brake specific consumption very close to conventional engine with deviation by about 1.76% higher at full engine load. This is due to effect of insulation; the heat free flow is restricted, which leads to reduction in heat transfer in case of LHR engine. The reduction in heat transfer leads to increase in combustion temperature, which leads to better combustion. The higher combustion temperature will lead to more expansion work.
- LHR engine with 0.5 mm of mullite ($3Al_2O_3.2SiO_2$) insulation coating on piston crown, cylinder head and valves of diesel engine gives marginal rise in brake thermal efficiency when compared with conventional diesel engine. The brake thermal efficiency for LHR engine is higher by about 1.8 % than the conventional diesel engine at full engine load level. The insulation coating reduces the heat loss through combustion chamber resulting in increase in the charge temperature. This higher charge temperature leads to better combustion. However, this increased heat release is not converted into useful work in direct proportion but leaves with exhaust as seen from the rise in the exhaust temperature. It is observed that, the peak of gas pressure curve occurs 58.0 bars for conventional engine while for LHR (mullite coated) engine the peak of gas pressure curve occurs 58.88 bars. The increase in peak of gas pressure for LHR engine is due to elevated temperature of the insulated engine and better combustion of fuel.
- The LHR engines are exhibiting a higher rate of heat release compared with conventional engine. The reason for high rate of heat release is due to insulation, the higher heat retainment inside the combustion chamber is exhibited. This leads to evaporate the fuel at faster rate, which helps possibly to better premixing, reduced diffused combustion. Further, it results in complete combustion of fuel. Hence, it releases more amount of heat.

6. Acknowledgement:

We are very thankful to Mr. M. NAGESWARA RAO, Managing Director, Sai Surface Coating Technologies Pvt. Ltd., Hyderabad, for coating the Diesel engine components using Plasma Spray Technique. We are also thankful to J.D. college of Engg. Yavatmal for providing Research recognized I. C. Engine Lab. for testing.

7. References:

[1] D. B. Krishnan et.al. Performance of an Al-Si graphite Particle composite piston in a diesel engine, Transactions of Wear, 60(2): 205-215, 1980.

[2] K. Rama Mohan, Performance evaluation of an air gap insulated piston engine. Ph.D.Thesis, Kakatiya University, Warangal, 1995.

[3] S. Jabez Dhinagar, B. Nagalingam, and K. V. A. Gopala Krishna, comparative study of the performance of a low heat rejection engine with four different levels of insulation. International Conference on Small Engines and Fuels. Chang Mai, Thailand. Proceedings, pp121-126, 1993.

[4] A.C. Alkidas, 1989.Performance and emissions achievements with an uncooled heavy duty, single cylinder diesel engine. SAE paper 890141.

[5] A.Uzun, I. Cevik, and M.Akcil, 1999.Effects of thermal barrier coating material on a turbocharged diesel engine performance. Surf. Coat. Technol. pp.116–119.

[6] T.Hejwowski, and A. Weronski, 2002.The effect of thermal barrier coatings on diesel engine performance, Vacuum 65.

[7] K.Toyama, T. Yoshimitsu, and T.Nishiyama, 1983.Heat insulated turbo compound engine.SAE Transactions, vol. 92, pp. 3.1086.

[8] R.Kamo, and W. Bryzik, 1978. Adiabatic turbo compound engine performance prediction. SAE Paper 780068.

[9] R. Kamo, and W. Bryzik, 1979. Ceramics in heat engines. SAE Paper 790645.

[10] R. Kamo, and W.Bryzik, 1978.Adiabatic turbocompound engine performance prediction. SAE Paper 780068.

[11] R R Sekar, and R.Kamo, 1984.Advanced adiabatic diesel engine for passenger cars. SAE Paper 840434.

[12] T. Morel, EF.Fort, and PN.Bulumberg, 1985.Effect of insulation strategy and design parameters on diesel engine heat rejection and performance. SAE Paper 850506.

[13] G.Woschni, W. Spindler, and K.Kolesa, 1987.Heat insulation of combustion chamber Walls—A measure to decrease the fuel consumption of I.C. Engines. SAE Paper 870339.

[14] PH.Havstad, IJ. Gervin, and WR.Wade, 1986.A ceramic insert uncooled diesel engine. SAE Paper 860447.

[15] R. Prasad, and NK.Samria, 1990. Heat transfer and stress fields in the inlet and exhaust valves of a semi-adiabatic diesel engine.Comput. Struct.,34(5), pp.765–77.

[16] K. Kokini, Y.R. Takeuchi, and B.D.Choules, 1996. Surface thermal cracking of thermal barrier coatings owing to stress relaxation: zirconia vs mullite. Surf. Coat. Technol., 82, pp.77–82.

[17] T. M. Yonushonis, 1997. Overview of thermal barrier coatings for diesel engines, J. Therm. Spray Technol., 6(1), pp. 50–56.

[18] P. Ramaswamy, S. Seetharamu, K. B. R. Varma, and K. J. Rao,1999. Thermal shock characteristics of plasma sprayed mullite coatings. J. Therm. Spray Techno., 7(4), pp. 497–504.

[19] H. Samadi, and T. W. Coyle, 2005.Alternative Thermal Barrier Coatings for Diesel engines. X.Q. Cao, R. Vassen, and D. Stoever, 2004. Journal of the European Society 24, pp.1-10.

[20] Y.Arata, A. Kobayashi, Y. Habara, and S. Jing, 1986. Gas Tunnel Type Plasma Spraying Trans. of JWRI., vol.15-2, pp.227-231.

Synthesization of Nanostructured Titanium Dioxide at Low Molarity of Sol-Gel Process

S.K.M. Maarof[1,2,*], S.Abdullah[1], M. Rusop[2]

[1]NANO-SciTech Centre, Institute of Science, UniversitiTeknologi MARA (UiTM), 40450 Shah Alam, Selangor, Malaysia

[2]Faculty of Applied Sciences, Universiti Teknologi MARA (UiTM), 40450 Shah Alam, Selangor, Malaysia

*E-mail: khadijahmaarof@gmail.com

Abstract. Titanium dioxide, TiO_2 is one of the semiconductor materials. The aim of this paper is to determine the production of TiO_2 by sol-gel method. The sol-gel method used because this method is quite simple compare to other methods such as dip-coated and refractive sputtering. The parameter will be used in this paper is the concentration. The sol-gel TiO_2 solution then spin coated on the glass substrate to form homogenous and transparent thin film. The thin film was coated at 8 layers to adjust the refractive index characteristics and to get the clear images during characterize by spin coating technique. Other than that, this work also to get the band gap energy similar to the standard bandgap for TiO_2. The optical properties such as absorption and transmittance of TiO_2 can be done by Ultraviolet-visible Spectroscopy (UV-Vis). The changes on the surface morphology were observed using Atomic Force Microscopy (AFM), Field Emission Scanning Electron Microscopy (FESEM). Based on the result, higher the molarity of TiO_2, the uniformity of the surface morphology, and the energy bandgap is much better. Higher the molarity, the bandgap will be lower with 0.1M (3.78 eV) and its too large compare to the standard value (3.2 eV).

1. Introduction

TiO_2 is a white powder with maximum whiteness and great opacity. TiO_2 can be crystallizing in two forms which is anatase and rutile [1] while the natural polymorph of TiO_2 includes anatase, rutile and brookite[2]. Great opacity of TiO_2 is because of the refractive index. Higher the refractive index, the opacity will be greater. The refractive index of anatase is 2.55 while rutile is 2.73[3]. TiO_2 is also one of the semiconductor materials with a larger bandgap, approximately 3.2 eV and it can absorb UV-light [4].

TiO_2 nanoparticles has a wide range of application such as antireflection coatings for photovoltaic cells and passive solar collectors[5], photocatalytic purification of polluted air or wastewater because it is nontoxicity, high chemical stability, and excellent degradation for organic pollutants[6]. Akira Fujishima in 1967 discovered the properties of TiO_2 as a photocatalyst, including self-cleaning surfaces, and most recently into the area of photoinduced hydrophilicity, which involves not only self-cleaning surfaces, but also antifogging ones[7].

As mentioned before, one of the application of TiO_2 as a catalyst and can absorb only UV light. When photocatalyst TiO_2 absorbs Ultraviolet (UV)* radiation from sunlight, it will produce pairs of electrons and holes. The electron of the valence band of titanium dioxide becomes excited when illuminated by light. The excess energy of this excited electron promoted the electron to the conduction band of TiO_2 therefore creating the negative-electron (e-) and positive-hole (h+) pair. The positive-hole of titanium dioxide breaks apart the water molecule to form hydrogen gas and hydroxyl radical. The negative-electron reacts with oxygen molecule to form super oxide anion. This cycle continues when light is available.

TiO_2 thin films can be prepared in several ways such as dip-coating, reactive sputtering, chemical vapor deposition (CVD), sol–gel technique[8]. TiO_2 also can be coat by different thermal spray techniques: atmospheric plasma spraying (APS), suspension plasma spraying (SPS), and high-velocity oxygen fuel (HVOF) spray process and liquid phase deposition[9]. The sol–gel process involves the formation of inorganic glass from solution and yields high-purity homogenous ceramic material. The nanostructured materials on the basis of a sol–gel approach have exhibited new optical properties of interest in many areas of application such as electroluminescent devices, optical and chemical sensors. For example, the thin films of anatase from nano-sized TiO_2 show a very high photocatalytic activity due to large internal surfaces. Nanocrystalline TiO_2 can be prepared by a sol–gel route assisted by a complexing agent-assisted sol–gel route. The advantage of nanostructured particles is an enhancement of the surface area, and related properties. Recently, it was revealed that titanium can be bonded to living bone by treatment with alkali followed by heat [10].

2. Methodology

2.1Materials

All the solution was prepared using spectroscopic grade solvents. Absolute ethanol (100%) was obtained from Aldrich acts as a solvent. Tetra (IV) Isopropoxide, TTIP as a precursor, Acetic Acid Glacial (GAA) as a stabilizer, Triton-X-100 as a surfactant may act as wetting agent detergents, distilled water. All the reagents were used as received.

2.2 Preparation of glass substrate

In this experiment the substrate will be use is microscope glass. The glass substrate is cutting in dimensions of 2 cm x 2 cm. Then, it is treating with acetone two times, methanol two times and distilled water in ultrasonic bath for 10 minutes each. Lastly, the substrate was blow with N_2 gas for drying purpose [11].

2.3 Synthesis of pure TiO_2 solution by sol-gel method.

Using a sol-gel technique, TiO_2 thin film was prepared by mixing tetra (IV) isopropoxide, glacial acetic acid, ethanol, deionized water and triton x-100. In order to investigate the influence of amount of precursor, the amount of tetra (IV) isopropoxide was varies in the molarity for 0.01M, 0.05M and 0.10M. In this technique, tetra (IV) isopropoxide was used as precursor, glacial acetic acid as a stabilizer or chelating agent, ethanol as a solvent and deionised water as a function of adding the oxygen (O_2). Triton X-100 used acts as a surfactant to avoid precipitation in solution and at the same time used to increase the conductivity of films. The hydrolysis and the polycondensation of titanium alkoxides proceeds according to the following scheme:

$$Ti\,(OR)_4 + H_2O \rightarrow Ti(OR)_3(OH) + ROH$$

$$Ti\,(OR)_4 + Ti(OR)_3(OH) \rightarrow Ti_2O(OR)_6 + ROH.$$

The reaction stops with the inclusion of two water molecules: $Ti\,(OR)_4 + 2H_2O \rightarrow TiO_2 + 4ROH$.
0.01 Molar equal to 10 ml divided by 1000 ml. To find the value of molar mass of TTiP in 1 ml is by the division of molar mass of TTiP by 1000 ml. To prepare 0.01 Molar in 50 ml TiO_2 solution, the amount of TTiP must be multiply by 50ml and then divided by 1000. Since the amount of TTiP was

found, the value of solvent then can be calculated. The amount of solvent comes from the equation below:

Amount of solvent = 50 ml (TiO_2) – TTiP – GAA – DI water

Since the first chemical is put into the Schott Blue Cap Bottle, the solution will be starting to stir and heat. The time for heating is 1 hour, since the times up heating must be switch off and solution are left in continuous stirring for another 23 hours. The purpose for heating is to break the carbon chain, while continuous stirring is to avoid from agglomeration.

2.4 TiO_2 Thin Films

The TiO_2 solution was coated onto a glass substrate by spin coating method. The spin coating was done in 6000 rpm at 30 seconds. The spin coated was done in 8 layers. Every layer will be heated at 100 °C for 5 minutes. The purpose of heated the thin film is to vaporize the solvent or volatile material and to remove the contamination. The process repeated represents the number of layer of TiO_2 thin film. As prepared, sample was carried to annealing process at 450 °C in 2 hour to restructure the lattice through deformation of atoms.

2.5 Characterization

UV-Vis spectroscopy was measured with spectrophotometer in the range of 300 nm – 800 nm. Particle morphology and composite structure were examined by field-emission scanning electron microscope (FESEM) equipped with an energy–dispersive spectroscopy (EDS). Atomic force microscope (AFM) was used to observe the morphology and measure the roughness of the thin films on glass substrate.

3. Results and discussion

a. Optical properties by UV-Vis.

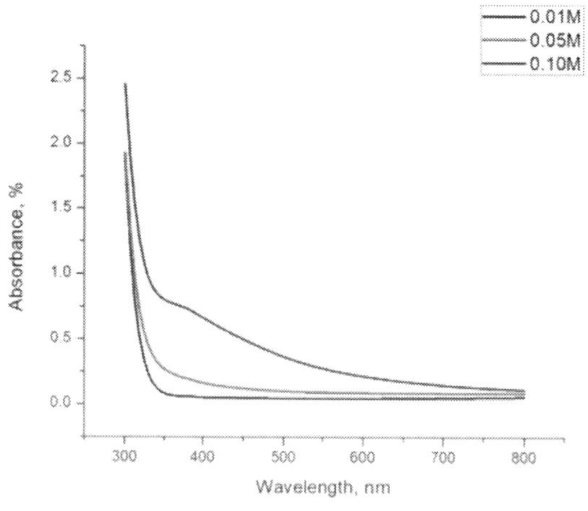

Figure 1. UV-Vis spectroscopy for different molarity of TiO_2

UV–Vis absorption study was carried out in order to characterize the optical absorbance of pure TiO2 for different molarity. The absorption spectrums of the synthesized pure TiO2 with different molarity are shown in figure 1. The optical band gap can be estimated by plotting (α*hv)2 versus photon energy (hv) based on the relation αhv = A (hv - Eg)n/2 where a is the absorption coefficient, A is a constant, Eg is the band gap and n is the exponent depending on quantum selection rule for a particular material, n = 1 for a direct and n = 4 for indirect bandgap. According to the above relation, the intercept of the tangent on the photon energy axis corresponds to optical band gap [12]. From the (α*hv)2 versus photon energy (hv) plots, the optical band gaps Eg for all the three synthesized samples were estimated and tabulated in Table 1.

b. Surface Morphology

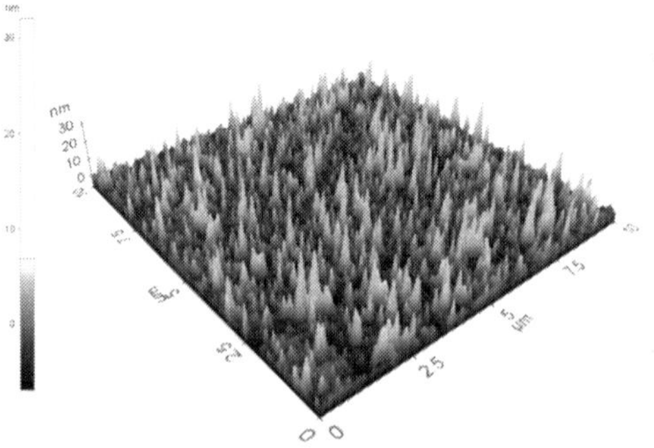

a. 0.01 M

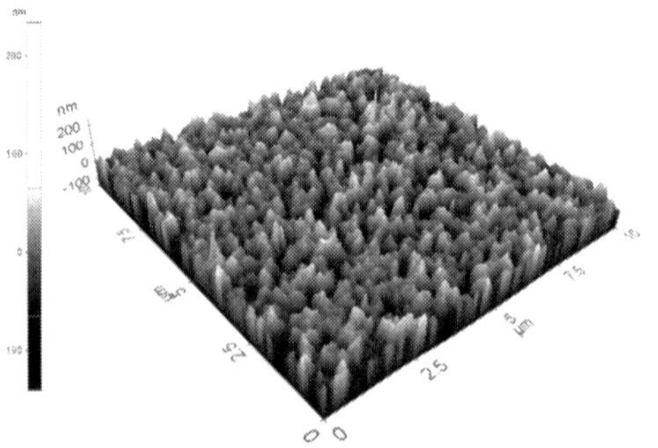

b. 0.05 M

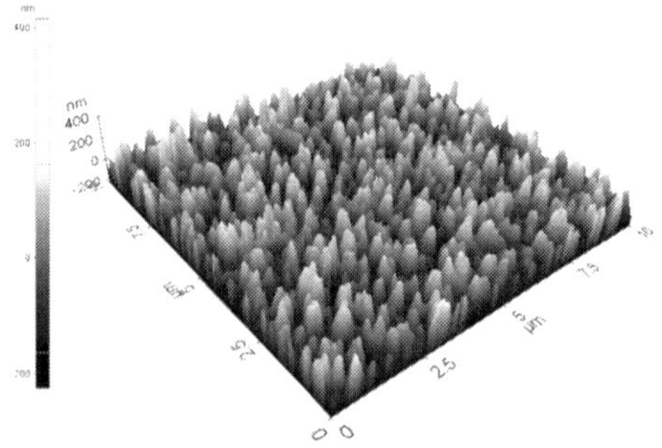

c. 0.10 M

Figure 2. AFM result for a. 0.10M, b. 0.05M and c. 0.10M.

Figure 2 shows the AFM results of TiO_2 for different molarity. At low molarity, which is 0.01M the thickness of TiO_2 is only 8.483 nm. The thickness for 0.05M and 0.10M are 58.139 nm and 178.755 nm respectively. The roughness also shows the difference for every molarity. At low molarity, roughness is 1.331 nm, followed by 9.292 nm for 0.05M. At the higher molarity, roughness is 32.264 nm.

The top view for low molarity (0.01M) in figure 2a is not uniform. The gap is very obvious between higher peak and the lower peak. For figure 2b, the top view of the surface is uniform where the gap between higher and lower peak is almost same. For figure 2c, it's quit uniform because the gap is low between higher and lower peak.

Table 1. Summarize for AFM and optical bandgap result.

Molarity	Thickness, nm	Roughness, nm	Band gap, eV
0.01M	8.483	1.331	3.90
0.05M	58.139	9.292	3.84
0.10M	178.755	32.264	3.78

IOP Conf. Series: Materials Science and Engineering **46** (2013) 012009 doi:10.1088/1757-899X/46/1/012009

c. Surface morphology by FESEM

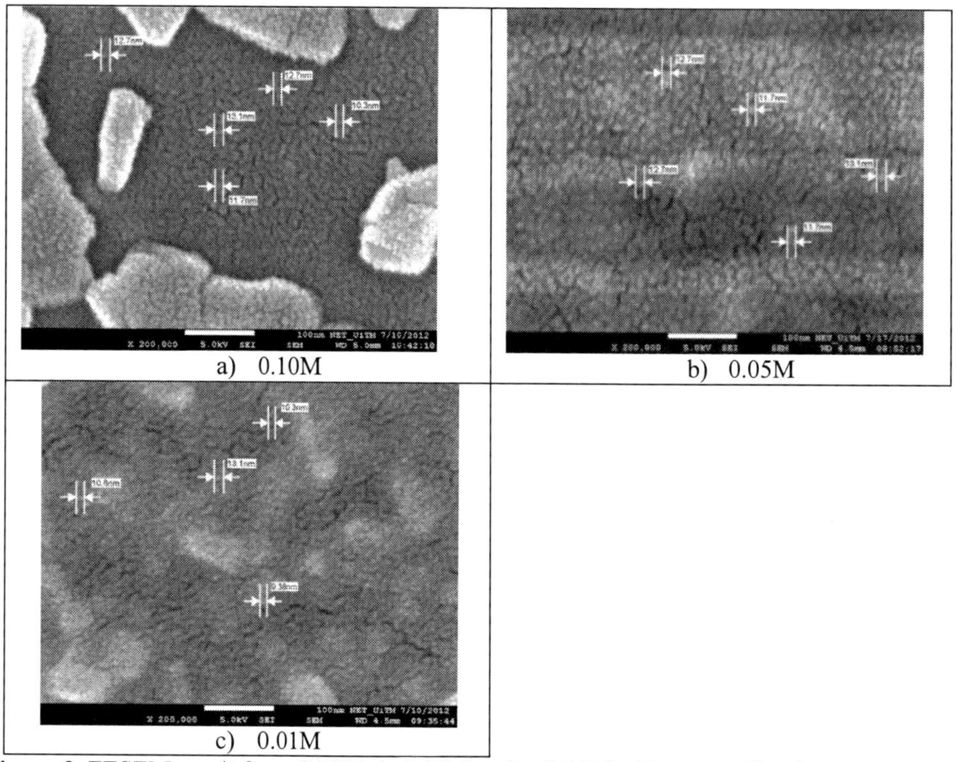

Figure 3. FESEM result for a. 0.10M, b. 0.05M and c. 0.01M with a magnification x 200,000 at 5 kV power.

FESEM is one of the characterizations to found the particle morphology and the diameter of the particles. Figure 3 shows the detailed of FESEM studies on the material indicated that the mean diameter of the nanoparticles, their diameter distribution and tendency to agglomerate are strongly correlated to the experimental parameters, due to the mixing technique by using magnetic. When the solution was not continuously stirred the nanoparticles were irregular in shape and easily formed agglomerates. The temperature of the process strongly influenced the mean diameter of the TiO_2 nanoparticles. Higher temperature of the process resulted in the preparation of TiO_2 nanoparticles with lower mean diameter. It was already investigated that more energy delivered through increase of the temperature influences the reaction rate and more nuclei particles are formed[11]. As a consequence this leads to the preparation of higher number of TiO_2 nanoparticles with smaller diameter.

Figure 3a shows some agglomeration of the particles because the molarity is higher than figure 3b and 3c. The morphology for figure 3b and 3c doesn't give much comparison because the molarity for both figures is only slightly different. In this study used TiO_2 nanoparticles produced at 60°C with continuous magnetic stirring for 2h. Statistical analysis of the nanostructures revealed a mean diameter of 11nm with the diameter distribution ranging between 10 and 12 nm [12].

4. Conclusion

Sol-gel is the simple method to produce pure TiO_2 nanostructured at nano-scale. This solution then coated onto a glass substrate by do the spin coating at 6000 rpm in 30 seconds. For UV-Vis result, it can be concluded that higher the concentration of the solution, lower the bandgap as shown in Table 1. For AFM, higher the molarity of the solution, the roughness and the thickness will be higher. Lastly, based on the result of FESEM, higher the molarity of the solution, the mean diameter of the particle will be lower. This is might be initial result for future research direction and it's a part for optimization TiO_2 to hybrid TiO_2 with metal, to modified the bandgap become wider and can absorb other than UV light.

Acknowledgement

The authors would like to thank UniversitiTeknologi MARA for the financial support. We are grateful to the all staffs of NANO SciTech Centre (NST) and also NANO ElecTronic Centre (NET) for technical support of this study.

References

[1] M. P. F. Graça*, C. C. Silva, L. C. Costa, and M.A. Valente, Study of the structural, morphological and electric characteristics of TiO2 nano-powders., Nanoelectronics and Materials 3 (2010) 3.
[2] Thilagam, A., D.J. Simpson, and A.R. Gerson, A first-principles study of the dielectric properties of TiO(2) polymorphs, J Phys Condens Matter 232 (2011) 025901.
[3] Chang Kwon, H., P. Yong Jun, and S. Kma, Optical Properties of TiO2 Zigzag Films Prepared by Oblique Angle Deposition, Journal of the Korean Physical Society 5642 (2010) 1378.
[4] Kannaiyan, D., M.-A. Cha, Y.H. Jang, B.-H. Sohn, J. Huh, C. Park, and D.H. Kim, Efficient photocatalytic hybrid Ag/TiO2 nanodot arrays integrated into nanopatterned block copolymer thin films, New Journal of Chemistry 3312 (2009) 2431.
[5] A. Shanaghi, A. Sabour Rouhaghdam, T. Shahrabi, and M. Aliofkhazraei, Study of Ti O2 nanoparticles coatings by the sol-gel method for corrosion protection, Journal of Materials Science 442 (2008) 233-247.
[6] Wei Wu, Xiangheng Xiao, Shaofeng Zhang, Feng Ren, and C. Jiang, Facile method to synthesize magnetic iron oxides/TiO2 hybrid nanoparticles and their photodegradation application of methylene blue, Nanoscale Research Letters 6 (2011) 1-15.
[7] Akira Fujishima, Tata N. Rao, and D.A. Tryk, Titanium dioxide photocatalysis, Journal of Photochemistry and Photobiology 1 (2000) 1-21.
[8] Park, S. and J. Heo, Characteristics of N-doped titanium oxide prepared by the large scaled DC reactive magnetron sputtering technique, Separation and Purification Technology 581 (2007) 200-205.
[9] Khakpash, N., A. Simchi, and T. Jafari, Adsorption and solar light activity of transition-metal doped TiO2 nanoparticles as semiconductor photocatalyst, Journal of Materials Science: Materials in Electronics 233 (2011) 659-667.
[10]Liang Zhou, Hendrik Heinz, Mark D. Soucek, Elvin A. Alemán, and D.A. Modarelli, UV-Absorption and Silica/Titania Colloids Using a Core–Shell Approach, Silicon 2 (2010) 10.
[11]Isrihetty Senain, Nafarizal Nayan, and H. Saim, Structural and Electrical Properties of TiO2 Thin.
[12]A.Y. Orala, E. Men¸sur, M.H. Aslan, and E.Ba¸saran, The preparation of copper(II) oxide thin films and the study of their microstructures and optical properties, Materials Chemistry and Physics 831 (2004) 140-144.

IOP Conf. Series: Materials Science and Engineering **46** (2013) 012010 doi:10.1088/1757-899X/46/1/012010

Application of Viral Systems for Single-Machine Total Weighted Tardiness Problem

Budi Santosa and Umar Affandi[1]
Industrial Engineering, Institute Technology of Sepuluh Nopember (ITS),
Kampus ITS, Surabaya 60111, Indonesia

E-mail: budi_s@ie.its.ac.id , umar12345678910@gmail.com

Abstract. In this paper, a relatively new algorithm inspired by the viral replication system called Viral Systems is used to solve the Single-Machine Total Weighted Tardiness (SMTWTP). SMTWTP is a job scheduling problem which is one of classical combinatorial problems known as np-hard problems. This algorithm makes the process of finding solutions through neighborhood and mutation mechanism. The experiment was conducted to evaluate its performance. There are seven parameters which are required to tune in to find best solution. The experiment was implemented on data sets of 40 jobs, 50 jobs, and 100 jobs. The results show that the algorithm can solve 235 optimally out of 275 problems.

Keywords : ***Viral Systems, Single-Machine Total Weighted Tardiness Problem, Metaheuristics, combinatorial problem***

1. Introduction

Scheduling and sequencing is one form in decision making that has an important role in the manufacturing and services industries. In the existing competitive environment, sequencing and scheduling becomes an important issue to maintain market share. Companies must determine the exact shipping time which is already determined based on agreement with the consumer. The failure of the fulfillment of this could lead to the loss of loyalty. Therefore, they need to schedule an activity based on the available resources in an efficient manner. [2].

Machine Scheduling Problems (MSPs) is one of the classical combinatorial optimisation problems [3]. Total weighted tardiness scheduling problem is one of MSPs that is not only a NP-hard problem [4], but also difficult to solve pratically. Potts and Van Wassenhove [5] have found problem in solving problem of 50 jobs with state of the art branch and bound algorithm [6]. According to the difficulty of this problem, metaheuristics method are recommended to find a good solution for computing time reason. Metaheuristics have been successfully applied to many problems including Genetic Algorithm (GA), Tabu Search (TS), and Ant Colony Optimization (ACO) [7]. In this study, Viral Systems is proposed to solve Single Machine Total Weighted Total Tardiness Problem (SMTWTP).

Viral system is an optimization algorithm based metaheuristics introduced by Cortes et. al. [8]. This algorithm is relatively a new method that is inspired by analogy of virus behavior. The virus continues to reproduce by means of infectious cells. Then, the cells also strive to provide resistance in the form of antibodies. The paper is organized as follows. Section 2 explains about Viral Systems in brief .

Next, section 3 discusses the SMTWTP and section 4 elaborates the implementation of Viral Systems on SMTWTP. The numerical experiment is presented in section 5 and the results are conluded in Section 6.

2. Viral Systems

A transition creature between living things and inanimate objects named viruses are unique in terms of reproduction. The virus uses the cells of organisms as a container for replication. When the virus injects DNA into the host cell in the body, there are two types of replication: lytic cycle and lysogenic cycle. Lytic cycle allows the virus to replicate in the host cell. At a certain amount of viral replication will break the wall and out of the host cell. While the lysogenic cycle allows the virus to remain in the host cell is switched off until there is something to activate it. Then the cell of organism that has been injected was broken or mutated, it depends on the health of the cell itself.
In the system of the virus, the virus always seek and infect susceptible cells. Each cell was trying to produce antibodies to fight the viral infection. The cell that fail to produce antibodies will be grouped into the clinical picture. VS define the clinical picture of the infected population as a description of all the cells infected by viruses. In terms of computing, Clinical Picture containing solutions are being explored (the genome of infected cells, in biological terms) and total nucleus-capsids were replicated in the form of NR (for lytic replication) or IT (for lysogenic replication) [9].
In the infected cells, there are virus replication in the lytic or lysogenic. Both replication is done based on probability of lytic (p_{lt}) and the probability of lysogenic (p_{lg}) and $p_{lt} + p_{lg} = 1$. Initially, the number of nucleus capsid is zero (NR = 0 and IT = 0).

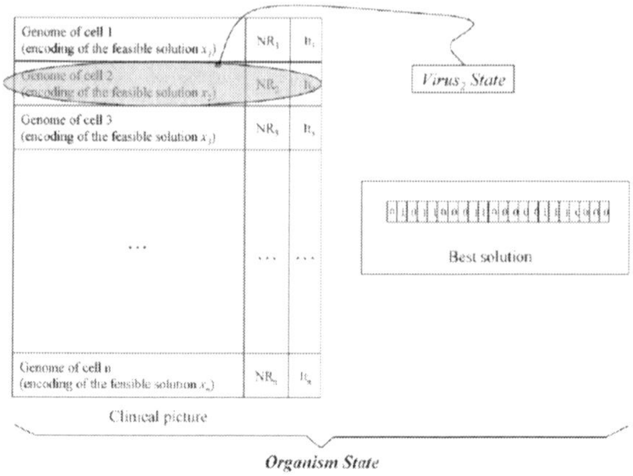

Figure 1 *Clinical picture* (Source: [8])

2.1. Lisogenic replication

In the lysogenic replication, the activation process of mutation occurs after iteration reaches the maximum limit (LIT). LIT value depends on the health condition of the cells, so the cells are healthy (minimization function (f (x)) of high value) would have a lower probability of infection so the value of LIT becomes higher. In contrast, unhealthy cells will have a lower value of LIT [9]. The value of LIT can be calculated from the value of initial LIT (LIT^0) based objective function value relative to the best objective function has been found so far, which can be calculated by the following equation.

$$LIT_{sel-x} = LIT^0 \left(\frac{f(x)-f(\hat{x})}{f(\hat{x})}\right) \qquad (1)$$

While :
LIT_{sel-x} = LIT value of cell-x
LIT^0 = initial LIT value
$f(x)$ = Objective function value of cell-x
$f(\hat{x})$ = The best value of objective function has been found so far.

2.2. Lytic replication

In lytic replication, first we need to obtain the number of replication nicleus capsid (NR). NR is calculated for each iteration as a function of the binomial variable (Z), its value is added to the existing NR on the clinical picture. Z is calculated using the binomial distribution is given by the maximum value of the nucleus replication capsid (LNR) and the probability of the single replication (p_r).

$$z = \text{bin}(LNR, p_r) \qquad (2)$$
$$NR_{iter} = NR_{iter-1} + z \qquad (3)$$

LNR represents the limit cell wall rupture and viruses that inhabit it out. LNR value depends on the value of the objective function (f (x)). Cells with the value f (x) which has a high probability of infection is lower, then the value will be higher LNR [9]. LNR Value can be calculated from the value of initial LNR (LNR^0) based objective function value relative to the best objective function has been found, which can be calculated by the following equation.

$$LNR_{sel-x} = LNR^0\left(\frac{f(x)-f(\hat{x})}{f(\hat{x})}\right) \qquad (4)$$

where:
$LNR_{sel\text{-}x}$: LNR value of cell-x
LNR^0 : initial LNR value
$f(x)$: Objective function value of cell-x
$f(\hat{x})$: The best value of objective function has been found so far.

After that, the cells that have a value exceeding NR LNR be broken so that the virus released. Each virus liberated would have probability (p_i) to infect new cells other being around. If the cardinality of x is defined as | V (x) |, the number of cells infected by the virus in the environment can be obtained by using the binomial random value | V (x) | and the probability p_i.

$$y = \text{bin}(|V(x)|, p_i) \qquad (5)$$

On the other hand, to maintain from the growth of a viral infection, the organism (combined cell) responds by producing antigen. In the clinical picture, individual infected cells produce antibodies with Bernoulli probability distribution A (x) = Ber (p_{an}), which is the probability p_{an} antibodies released by cells in the clinical x-picture. Therefore, the population of infected cells produce antibodies based on a binomial distribution with parameters characteristic clinical picture size (n) and the probability of producing antibody (p_{an}) [9].

$$A(\text{populasi}) = \text{Bin}(n, p_{an}) \qquad (6)$$

2.3. Mutation

Lysogenic replication allows the algorithms to perform searching by mutation. Mutation has a goal to bring new solutions that are totally different from the previous solution in order to get out of the local optimal solution [10]. A feasible solution can be mutated and transferred to obtain other feasible solutions. Given n jobs, each jon is represnted by a gene, there are several types of mutation, for examples [10]:

1. *Flip*

This mutation is done by reversing the mutated segment left-to-right to be right-to-left. Firstly, we choose two different random-integer uniform number at range 1-n, let them a and b, and a < b. Then we flip the segment between job a and b, $[j_a, j_{a+1}, \ldots j_b]$, reverse to be $[j_b, j_{b-1}, \ldots j_a]$.

2. *Swap*

This mutation is done by exchanging a gene with the other gene. Firstly, we choose two different random-integer uniform number at range 1-*n*, *a* and *b*, and *a* < *b*. Then, we interchange the position of *a* and *b*.

3. *Slide*

Firstly, we choose two different random-integer uniform number at range 1-*n*, *a* and *b*, and *a* < *b*. *a* is called the "head" of segment and *b* is called the "tail". Then, place the head of backward and the other genes of mutated segment are stepped forward, so the original segment $[j_a, j_{a+1}, \ldots j_b]$ will become $[j_{a+1}, j_{a+2}, \ldots j_b, j_a]$.

2.4. Neighborhood

The lytic replication allows the algorithms to perform searching by neighborhood. Neighborhood tracking is a method of solution adjacent to a good solution that has been found before. Therefore, this method makes the algorithm to do local search faster. Neighboring cells can be defined by exchanging a pair of genes in the genome of the cell before. Starting from the left gene first, interchange with a second gene that produced the first neighboring cells. For cells whose genome consists of n genes, the procedure is repeated until the (n-1) neighboring cells are obtained [11].
For example, the feasible solution 1-2-3-4-5 has following several neighborhood solutions:
The first job is excanged with the second one. 2-1-3-4-5
The second job is excanged with the third one. 1-3-2-4-5
The third job is excanged with the fourth one. 1-2-4-3-5

3. Single machine total weighted tardiness problem

Single machine total weighted tardiness problem (SMTWTP) is known as a planning operation problem. In SMTWTP, a set of jobs $j = (j_1, ..., j_n)$ need to be processed on a single machine. Each job (j_j) consists of only a single operation, which involves the processing time $p_j > 0\ \forall\ j = 1, ..., n$. The penalty of tardiness, $w_j > 0\ \forall\ j = 1, ..., n$. Machine can process only one job at a time, there was no parallel work processes. Each job j should be completed before the deadline d_j. If the job is done over time d_j, then the delay (lateness) is calculated as a function of the following:

$$T_j = \max(s_j + p_j - d_j, 0) \qquad (7)$$

The problem is also formulated to the following integer programming model [12]:

$$\text{Minimize } TWT = \sum_{j=1}^{n} (w_j . \max(0, s_j + p_j - d_j)) \qquad (8)$$

Subject to: $s_j - s_{j'} >= p_j - M\, Z_{j,j'}$ (9)

where

s_j = starting processing time of job j
p_j = processing time of job j
d_j = due date of job j
T_j = lateness of job j
w_j = weight of lateness of job j

with M is a large positive number and $z_{j,j'}$ is a binary variable, equal to 1 if job j followed by job j', and 0 otherwise.

4. Viral Systems for SMTWTP

The following *pseudocode* describes the application of Viral Systems for SMTWTP.

```
Procedure VS_SMTWTP(N_max, CPS, p_lt, p_i , p_an , LNR, LIT,data)
   CP=∅
   iteration = 0
         for i = 1 to CPS
            /* Get randomly feasible solutions and their replication type
            CP(i) = Random permutation ()
            CP(i).Replication_Type = Get_Randomly_Replication_Type (p_lt)
            F(i) = SMTWTP (CP(i), data)
         Next
         Fbest = min (F)
   Do
```

```
        iteration = iteration+1
        For c =1 to clinical_size
            If Replication_Type (CP (c)) = 'Lytic' Then
                Litic (c , CP, plt ,pi, pan, pr, LNR )
            Else
                Lisogenic (c, CP, plt, LIT)
            End If
              Fbest = min (F)
        Next
    Loop Until iterations = Nmax or ceck_gap(CP ) = True
End VS_SMTWTP
```

```
procedure Lytic_Replication (c , CP, plt ,pi, pan, pr, LNR)
    CS = CP(c)  /* Current_Solution
    /* Get the value of  nucleus capsid
    z =  Get_Problability_of_Random_Binomial (LNR, pr)
CP(c).NR = CP(c).NR + z
    CS.LNR=LNR(F(c)-Fbest)/Fbest
    /* Ceck infection
    if CS.NR > CS.LNR then
      /* Arrange descendly solutions of neighbourhood based on health of cells
      Vs = Neighbourhood (CS)
      VAS = Get_Arranged_Neighbourhood(Vs)
      /* Arrange CP ascendingly based on health of cells
      CPA = Get_Clinical_Picture_Arranged (CP)
      i = 1
      for each S'∈VAS
          if i <= |CPA|  and i <= |VAS| then
            a = random_value
            b = random_value
            if a < pi and b > pan then
            /* Replace CPA(i) with a new solution CS'
                CPA(i) = CS'
                CPA(i). Replication_type= Get_Randomly_Replication_Type (plt)
                FA(i) = SMTWTP(CPA(i), data)
                replace = true
    end-if
            i = i + 1
      end-for
    end-if
end  Lytic_Replication
```

```
procedure  Lysogenik_Replication(c, CP, plt, LIT)
    CS = CP(c)
    CS.IT = CS.IT + 1
    CS.LIT=LIT(F(c)-Fbest)/Fbest
    if CS.IT > CS.LIT' then
       s = Get_Random_Gen ()
       /* Do Mutatuin CS
       CS_NEW = Mutation(CS, s)
       CS_NEW.Replication_Type = Get_Replication_Type (plt)
       F(c) = SMTWTP (CS_NEW, data)
End-if
end Lysogenic_Replication
```

IOP Conf. Series: Materials Science and Engineering **46** (2013) 012010 doi:10.1088/1757-899X/46/1/012010

5. Numerical Experiment

The experiment was conducted using some data sets obtained from OR Library. The data sets consists of 125 instances of 40 jobs, 12 instances of 50 jobs and 25 instances of 100jobs.The experiment was done by varying the values of parameters using in the algorithmon the case of 40 jobs. Best parameters combination obtained from case of 40 jobs then was applied on other cases.

5.1. Parameters of the algorithm

The experiment was conducted by alternating the values of the parameters. Table 1 shows the parameters values used in the experiment.

Table 1 Parameters values

	Scenario					
Parameter	1	2	3	4	5	6
CPS	10	20	30	40	50	60
Plt	0	0.2	0.4	0.6	0.8	1
Pi	0	0.2	0.4	0.6	0.8	1
Pan	0	0.2	0.4	0.6	0.8	1
Pr	0.2	0.4	0.6	0.8	1	-
LNR	5	10	15	20	25	30
LIT	5	10	15	20	25	30
Mutation	swap	Flip	fslide	bslide	random	combination

The basic parameters that are used to test the performance are :
CPS = 20, plt = 0.6, pi = 0.7, pan = 0.3, pr = 0.7, LNR = 5, and LIT = 5. To get best parameters we ran the algorithm on the case of 40 jobs. The best parameters obtained in this case then applied for all other problems. Each parameter was tested by running the algorithm 10 times in order to evaluate the performance of the algorithm. Then, we varied the parameters in order to get better results in terms of minimum total weighted tardiness.

5.1.1. Clinical Picture Size (CPS)

CPS indicates the size of the solutions contained in the Clinical Picture (CP). In the initialization phase, CP is obtained randomly based on CPS value. The influence of CPS on the performance is shown in Fig 1.

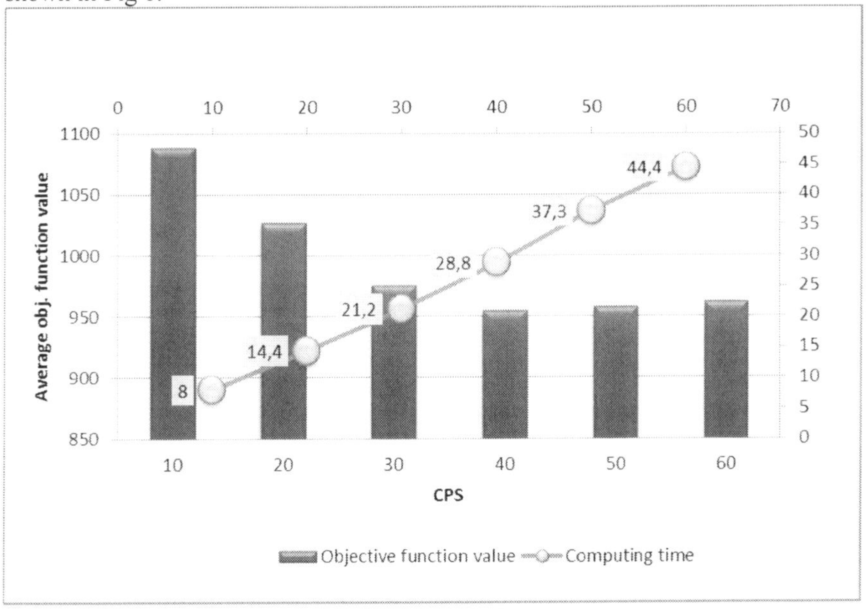

Figure 1 Influence of CPS on objective function

Based onthe results shown in Figure 1, the greater the value of CPS, the better the searching process in achieving optimum solution. In terms of computing time, the higher CPS value require longer time to do searching proce ss. Thus we can not make this CPS as high as possible, otherwise we require very high computing time. The value of CPS of 40 is reasonably good.

5.1.2. Lytic Probability (plt)

Probability of lytic replication (plt) shows the proportion of lytic and lysogenic replication in the sample, which is analogous to the neighborhood and mutation . If the value of plt is greater than 0.5, then the searching process with neighborhood is more dominant than mutation. This triggers obtaining the optimul values, but may be local solutions. If the value of p_{lt} is less than 0.5, then the searching process is using more mutation than neighborhood. In this way. the search process is more slowly but can potensially reach the global optimum. It might get out of the local optimum trap.

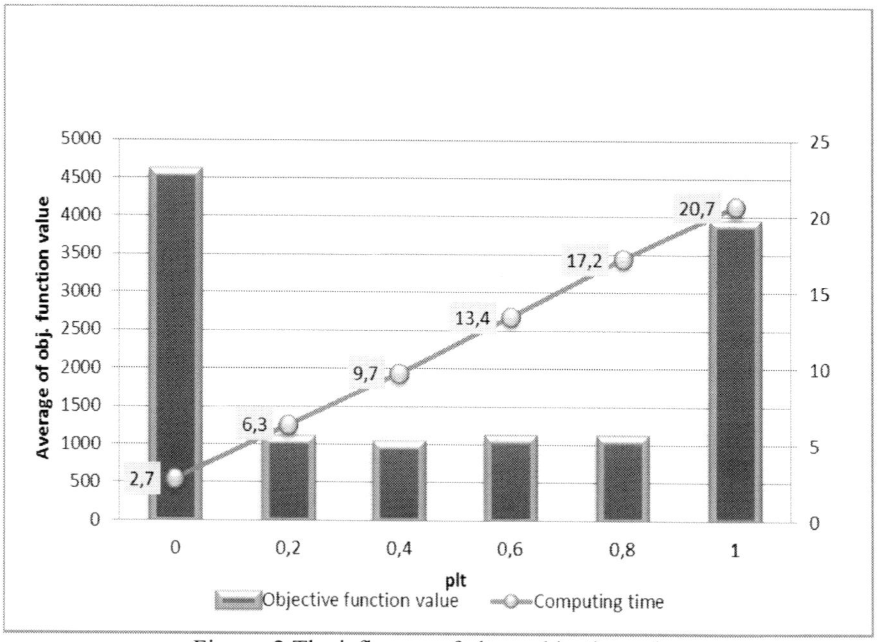

Figure 2 The influence of plt on objective value

Based on the experimental results shown in Figure 2, we see that for the data set with 40 jobs, the best plt is about 0.4. Therefore, a good value of the plt make search faster and can reach a global optimum. Based on this experiment, the combination of two ypes of search (neighborhood and mutation) provides a good result.

5.1.3. Infection probability (Pi) and antigen (Pan)

The probability of infection (Pi) indicates the probability of infection of the virus to the cell that was around cells previously seized. While, the probability of antigen (Pan) indicate the probability of cells secrete antibodies to against virus infection. With the same value of Pan (Pan = 5), the experiment that has been done for Pi values between 0.2 to 1, did not indicate significant influence on objective value. However, if the value of Pi = 0, the results obtained are very bad. The value of $P_i = 0$ indicates the analogy that no single virus that infects neighboring cells, so there is no neighborhood. In terms of computing time, the greater the value of Pi, the longer the computing time. With the same value of Pi (Pi =0.5), the experiment has been done with Pan values between 0 and 0.8. The change of Pan does not indicate significant influence on objective value. However, if the value of Pan = 1, the results obtained are very bad. Pan = 0 indicates the value of the analogy that all the neighboring cells can prevent infection, so the neighborhood-was not the case. In terms of computing time, the greater the value of Pi computation time becomes faster. The value of Pi and Pan has an effect on the size of the neighborhood around solutions that are previously been broken. The bigger Pi_i, the more the number of neighborhood.

5.1.4. Single probability of one replication (pr)

The Single probability of one replication is based on an analogy to the probability that nucleus capsid inside the cell to replicate. The greater value of *pr* make number of nucleus capsid be more developed. This triggers the virus to quickly solve the resulting cell neighborhood. At each iteration, increasing value of NR for random values obtained from pr and value of LNR. Therefore, this parameter affect how often neighborhood occurs.

5.1.5. LNR

LNR value is the limit value of NR. If the NR value exceeds LNR, then nucleus capsid within the cell will break the cell wall and out of the cell to carry out an infection to another cells that was around. This is analogized to the new solution approach to the previous solutions. LNR value of a cell is also analogous to the health of the cell. In this algorithm, the value of LNR cells is obtained from the value of the function relative to the best function value multiplied by the basic LNR. Parameter LNR essentially does not make a significant effect on the algorithm, since the development of the NR value is influenced by the amount of pr which is also based on the value of LNR.

5.1.6. LIT

LIT value indicates the limit value of IT. If the IT value exceeds LIT value, then the nucleus capsid within the cell can be mutated. This is analogized to a mutation of the previous solution to the new solution. IT value is incremented by one on each iteration, while the value derived from the value of LIT-cell function relative to the value of the best functions of the base multiplied by the LIT. LIT This parameter determining the frequency of mutations.

5.1.7. Mutation

In the case of combinatorial, mutation can be done in several ways that include swap, flip, and slide. In this research, the experiment perform six scenarios of mutations, swap, flip, slide forward, slide backward, random mutation, and combination mutation. Based on preliminary experiment on alternating mutation type, swap mutation type provides the best solution. Using a combination of the mutation types also produce a good result but taking a longer time.

Based on parameter setting experiment, the best parameters combination is presented in Table 2.

Table 2 Selected Parameters values

No	Data	Nmax	CPS	Plt	pi	Pan	Pr	LNR	LIT
2	40 jobs	1000	20	0.6	0.7	0.3	0.7	5	5
4	50 jobs	1000	20	0.6	0.7	0.3	0.7	5	5
5	100 jobs	2000	20	0.4	0.7	0.3	0.7	5	5

5.2. Results

Based on the experiment results , Viral System can solve optimally 125 out of 125 instances with 40 jobs. A total of 123 istances were solved optimally using the maximum iteration 500, while for 2 istances, (instances 40_58 and 40_85) the optimal solutions were found the maximum iteration 1000. Computing time required on average is 23.7 sec. For those problems with 50 jobs, 101 out of 125 instances can be solved optimally. Computing time required on average is 41,6 sec. For problem with 100 jobs, 9 out of 25 instances can be solved optimally. Computing time required on average is 344 sec. We notify that the greater the number of jobs, the harder to find optimal solutions. Parameters used in the algorithm should be tuned correctly in order to have the best peformance. The results is compared with some algorithms applying on the same problems [1].

Table 3, Comparison with other algorithms

Algorithm	40 job	50 job	100 job
SAmulti-s	121	115	59
TS	115	111	96
TSmulti-s	118	113	103
CE_ST	124	118	80
ILS-dyna	123	125	82
VS	125	101	9 *

*For case with 100 jobs, we only ran experiment with 25 intances

The results show that VS is promising in solvin scheduling problem. But for big problem the computing time is still high.

6. Conclusion

Viral Systems has been successfully applied for solving Single Machine Total Weighted Total Tardiness Problem (SMTWTP). The experimental results show that Viral System solve optimally 125 of 125 instances of 40 jobs, 101 of the 125 instances of 50 jobs, and 9 of the 25 instances of 100 jobs. The increasing of number of jobs affect decreasing of the number of instances solved. Viral systems algorithms have 8 parameters that have an influence on the search process. For the problem with large number of job, intensive experiment is needed and modification on the algorithm might improve the performance.

REFEREENCES

[1] Borgulya, I. (2002). "A Cluster-based Evolutionary Algorithm for the Single Machine Total Weighted Tardiness-scheduling Problem", Journal of Computing and Information Technology - CIT 10, 2002, 3, 211–217.

[2] M. L. Pinedo, Scheduling Theory, Algorithms, and Systems. New York: Springer, 2008.

[3] H. Nazif, *A Genetic Algorithm on Single Machine Scheduling Problem to Minimise Total Weighted Completion Time*, *European Journal of Scientific Research,* vol. 35, pp. 444-452, 2009.

[4] J. K. Lenstra, A. H. G. R. Kan, and P. Brucker, "Complexity of machine scheduling problems," *Annals of Discrete Mathematics,* vol . 1 pp. 343–362, 1997.

[5] C. N. Potts and L. N. V. Wassenhove., *A branch and bound algorithm for the total weighted tardiness problem, Operations Research,* vol. 33 pp. 363–377, 1985.

[6] R. K. Congram, C. N. Potts, and S. L. v. d. Velde, *An Iterated Dynasearch Algorithm for the Single-Machine Total Weighted Tardiness Scheduling Problem, INFORMS J. on Computing,* vol. 14, pp. 52-67, 2002.

[7] X. Wang and L. Tang, *A Population-Based Variable Neighborhood Search for The Single Machine Total Weighted Tardiness Problem, Computers and Operations Research,* vol. 36, pp. 2105-2110, 2009.

[8] P. Cortés, J. M. García, J. Muñuzuri, and L. Onieva, *Viral systems : A new bio-inspired optimisation approach, Computers & Operations Research,* vol. 35, pp. 2840 - 2860, 2008.

[9] P. Cortes, J. M. Garcia, J. Munuzuri, and J. Guadix, *A viral system massive infection algorithm to solve the Steiner tree problem in graphs with medium terminal density, Industrial Engineering,* vol. 2, pp. 71-77, 2010.

[10] Santosa, B and P. Willy, *Metode Metaheuristik, Konsep dan Implementasi.* Surabaya: Guna Widya, 2011.

[11] Suryadi, D and E. K. Kartika, Viral Systems Application for Knapsack Problem, Proceedings of the 2011 Third International Conference on Computational Intelligence, Communication Systems and Networks, 2011.

[12] Santosa, B. and M. A. Budiman, *Simple Algorithm for Single Machine Total Weighted Tardiness Problem,*" in *Proceeding of Industrial Engineering and Service Science, 2011*

Preparation of nanostructures $LaPO_4$ films using sol-gel reaction with different annealing temperature

A Z Zainurul[1,2*],F.S.Husairi[1,2], M Rusop[2,3], S Abdullah[1,2]
[1]Faculty of Applied Sciences,
[2]NANO-SciTech Centre (NST), Institute of Science;
[3]NANO-ElecTronic Centre (NET), Faculty of Electrical Engineering;
Universiti Teknologi MARA,
40450 Shah Alam, Selangor, Malaysia.

E-mail: *z.nurulatikah@gmail.com

Abstract. Thin films of lanthanum phosphate ($LaPO_4$) nanostructures were fabricated on quartz substrate using sol-gel spin coating method subsequent with drying and annealing process. Increase of annealing temperature will increase the intensity of diffraction peaks; reveal that the particle size and crystallinity increase with temperature. A monoclinic structure was shown to be present in $LaPO_4$ thin films after first anneal at 200°C. The $LaPO_4$ nanostructures of average size 30 – 50 nm were observed with field emission scanning electron microscopy. While the crystallize size were measured with x-ray diffraction (XRD). The images of $LaPO_4$ nanostructure films show uniformity of samples with homogenously close-packed morphology.

1. Introduction

Lanthanide compounds have been applied in many applications of luminescence and display such as lighting, cathode ray tubes, and field emission display [1]. One of the lanthanide compounds are Lanthanum phosphate ($LaPO_4$) also known as monazite had been used widely as a phosphor and proton conductor due to properties such as very low solubility in water, high thermal stability, high index of refraction [2] high chemical stability and high light yields of the doped materials. $LaPO_4$ has a monoclinic phase of monazite structure where La^{3+} ion is nine-coordinated to oxygen atoms, four oxygen forming a distorted tetrahedron interpenetrating a quasiplanar pentagon formed by other five [3]. It expected smaller size with nanosize lanthanum phosphate can increase the luminescent quantum efficiency and display resolution as a result of both their marked shape-specific and quantum confinement effect [4]. At nanometer phase, the material will consist of high surface to volume ratio and enhanced structural, electronic and optical properties compare to the other phases. Many methods had been used to synthesis $LaPO_4$ nanostructure such as solid-state reaction, sol-gel, precipitation, micro emulsion, hydrothermal and mechano chemical method [5],[6],[7]. The fabrication of $LaPO_4$ nanostructured films could be achieve such as by several coating techniques, layer-by-layer (LbL) assembly method , spin coating , electrospining, ink-jet printing and sol-gel process [8].

Generally, $LaPO_4$ is one of the oxide material that has an excellent host to fabricate the luminescent material. Sol-gel process is one of the promising candidate for fabricating optical devices [9] due to many advantages, such as low processing temperature, easy coating of large surface, homogenous multicomponent oxide films and simple with cheap experimental setup. In this study, we adopted the sol-gel process to prepare $LaPO_4$ with good morphologies and fine crystal structures. The present paper reports the phase of obtained thin films with different annealing temperature through X-ray diffraction (XRD). Their morphology was acquired by field emission scanning electron microscope (FESEM).

2. Experimental

2.1. Preparation of $LaPO_4$ thin film by sol-gel method

Preparation of the $LaPO_4$ nanostructure was followed by the method similar to Gao et al [10]. All chemical reagents are of analytic pure without further purification. In a typical procedure, 0.01 mol of ethylenediamine tetraacetic acid (EDTA) solution was dissolved in 1 M ammonia solution under vigorous stirring. Then, 0.45 M $La(NO)_3$ solution was added dropwise to the 1 M Citric Acid solution with strong stirring. This mixed solution with 0.5 M of $NH_4H_2PO_4$ solution were added to the already EDTA and ammonia solution. The solution was heated with 90°C in water bath with strong stirring until transparent mix solution obtained. The solution was leave at room temperature for ageing time. The films were prepared from this solution on glass quartz by spin coating with three step shown below.

Table 1. Three step of coating solution.

Step	**Duration**	**Spin Rate**
1	10 s	1500 rpm
2	30 s	3000 rpm
3	30 s	1500 rpm

The following step was used in other to ensure the solution will coat smoothly on the quartz glass substrate. The films were pre-baked (dry) at 100°C for 1 hour and post-baked (anneal) at different temperature for 2 hours.

2.2. Characterization

The formation phase of films was confirmed by X-ray diffraction measurement in other to investigate the purity of samples and also it crystallite size. X-ray diffraction was done using high resolution Philips X'Pert MRD diffractometer with Cu Kα radiation (λ= 0.15406 nm) in scan range of 20° – 70°. The morphology of the film was studied by field-emission scanning electron microscope (FESEM). But before that, the $LaPO_4$ films were sputtered with platinum in other to increase the conductivity of the films.

3. Results and discussion

3.1. Phase of $LaPO_4$ films

The composition and phase purity of sample were firstly characterized using Philips X'pert Pro MPD X-ray Diffractometer (XRD). The Cu anode X-ray was operated at 40 kV and 30 mA in combination with a Ni filter to give monochromatic Cu Kα radiation. The XRD data were processed using PANanalycal X'pert PRO software to examine the peak position and its corresponding intensity. Particle size can be obtained using X-ray diffraction (XRD) technique as the particle size is correlated to the diffraction peak broadening.

Diffraction can occur when electromagnetic radiation interacts with a periodic structure whose repeat distance is about the same as the wavelength of the radiation. Figure 1 shows the typical X-ray diffraction (XRD) patterns of pure $LaPO_4$ figure 1(a) and as-prepared $LaPO_4$ figure 1(b), 1(c), 1(d), at different annealing temperature. $LaPO_4$ nanostructure consist of monoclinic phases (space group P21/n) which refer to the high-temperature phases [8]. The XRD patterns were found to be match well with monoclinic phase of $LaPO_4$ JCPDS card files no. 01-084-0600 data. The high and shape peaks indicate that the solutions are well crystallized on the film. The main peak was found around 27°–30° corresponding to the $LaPO_4$ reported by Y.S.Patil [11]. The grain sizes of nanostructure were calculated using Scherrer equation (1) below from value FWHM (full width half maximum). The crystalline sizes are around 5 - 30 nm for nanostrured $LaPO_4$ thin films respectively.

$$D = 0.9\,\lambda / \beta\cos\theta \qquad (1)$$

Where β represent full width at half maximum (FWHM) of XRD lines
λ = Wavelength of the X-rays
θ = Braggs angle of the XRD peak

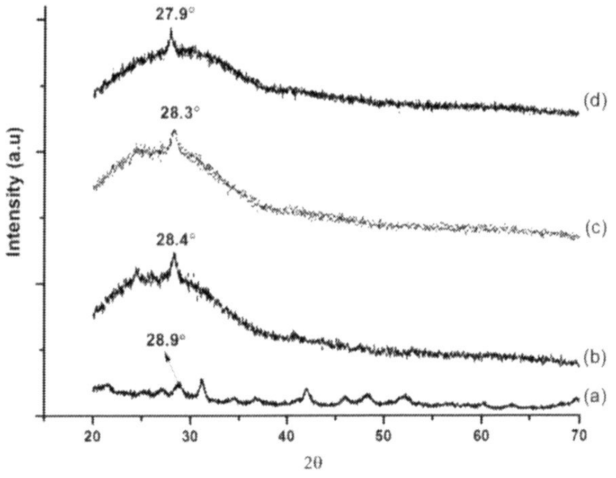

Figure 1. X-ray diffraction patterns of (a) pure $LaPO_4$ without annealing and as-prepared (b) 200°C (C) 400°C and (d) 600°C.

Table 2. Crystallite size and FWHM of nanostructure $LaPO_4$ films anneal at different temperature

Sample	**FWHM**	**Crystallize size (nm)**
(a)	0.08055	27.68
(b)	0.08940	11.45
(c)	0.09804	10.45
(d)	0.19532	5.24

Table 2 show the crystallite size of as-prepared $LaPO_4$ nanostructure films. From crystallize size above it show the intensity of sample were increase with increase of anneal temperature.

3.2. Morphology of $LaPO_4$ films

Figure 2 show Field Emission Scanning Electron Microscope (FESEM) images of rod-like nanostructures $LaPO_4$ for different annealing temperature on quartz substrate. The top view of film illustrates a homogenous flat surface with covering glass quartz well. The magnification (50K X) was constant for all samples. The deposit solution on substrate typically composed of uniform nanorods with length 30 - 50 nm. In this study, Ethelynediamine tetraacetic acid (EDTA) was used as it will form constant coordination compound when respond with metal ions in neutral aqueous solution. Thus, lanthanide (Ln^{3+}) which divided into three (La, Ce, Tb) were dissolve in $^{-}$EDTA solution. Therefore, La^{3+} possibly will co-exist with PO_4^{3-} without forming any precipitation as shown in equation below [2].

$$Ln^{3+} + PO_4^{3-} = LnPO_4 \quad (2)$$

The formation of rod like nanostructure is due to the amount of EDTA solution [12]. The pH value of 10 was prepared with amount 0.01 mol of EDTA. Increasing concentration of EDTA have great effect on morphologies and sizes of product due to the result that not shown here. It can be seen that long nano-wire with 30 - 40 nm in width and about 100nm in length are observed.

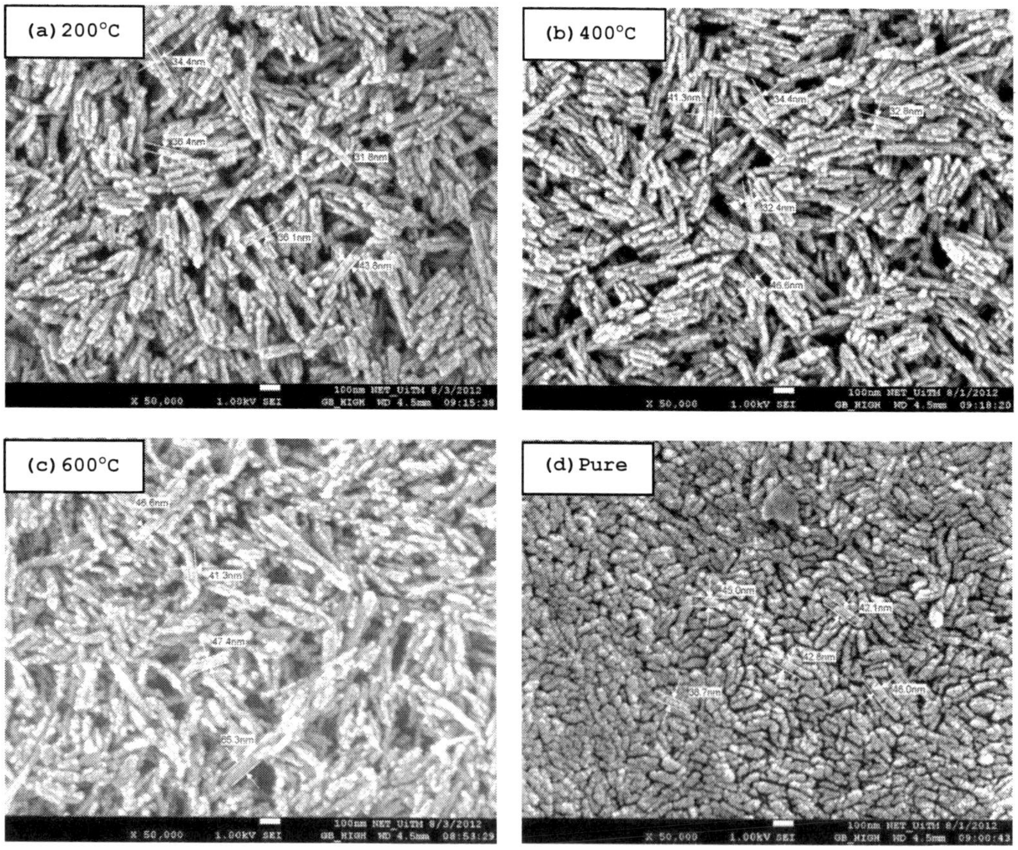

Figure 2. FESEM images of $LaPO_4$ anneal at (a) 200°C (b) 400°C (c) 600°C and without anneal (d) Pure $LaPO_4$

The increasing of annealing temperature do promote the formation of $LaPO_4$. Figure 3 (b) show an enlarged FESEM image rod like nanostructure which assembled by abundant of nanoparticels fig 3(a). The thick rod was constructed by the smaller rod as shown in Figure 3. Each of samples were annealed for at least 2 hours and above [13]. No obvious precipitation were found when solution only react for 1 hour . But when the time of reaction is increase, the yield of phosphate will gradually increased. This is understand-able. When the reaction was short, only small amount of PO_4^{3-} were produced, thus led to the low yield of phosphate and vice versa. From the FESEM observation, the product were similar to rod like nanostructure in all cases (different annealing temperature) unless for the pure $LaPO_4$ [13].

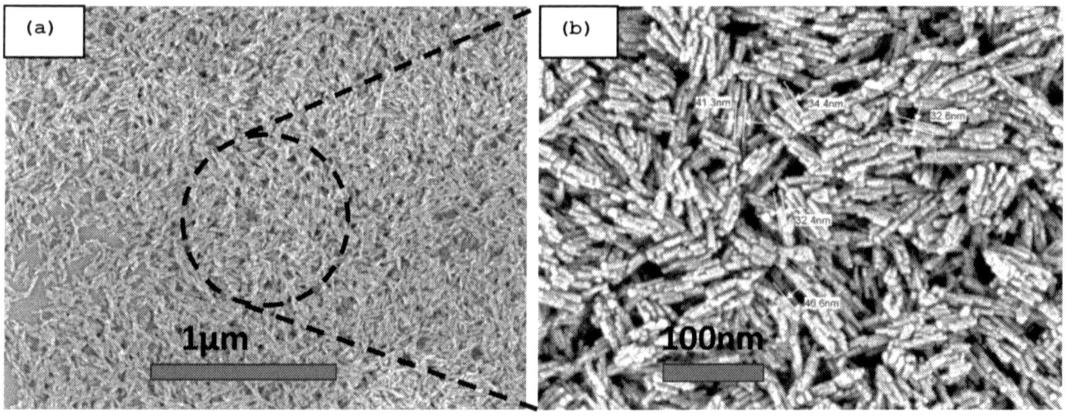

Figure 3. FESEM images of $LaPO_4$ films : top view obtained from annealed at 400°C with (a) low magnification, 10K x (b) high magnification, 50K x.

4. Conclusion

$LaPO_4$ films were successfully synthesis using sol-gel spin coating method. $La(NO)_3$ was used as a precursors to fabricate $LaPO_4$ films while concentration of EDTA play a main role on the morphology of films. The main peak in XRD pattern was found between 27° - 30° corresponding to the previous research [11]. Based on XRD results, monoclinic phase of $LaPO_4$ was identified. FESEM images also revealed homogenously quite long rod like nanostructures with length of around 30 – 50 nm. From the experiment, the reaction could not be initiated until the temperature is in a proper range. In this study, the annealing temperature started with 200°C till 600°C.

5. Acknowledgement

The authirs would like to thank the UNIVERSITI Teknologi MARA and the Ministry of Higher Education for their support and funding.

6. References

[1] Hui Wang, Run Liu, Liying Liu, Xiaofang Shi, Z. Xu, Electrodeposition of high aspect ratio LaPO4 and LaPO4:Ln3+ (Ln3+ = Ce3+, Tb3+) nanostructures, Electrochemistry Communications, 17 (2012) 79-81.

[2] V. Pankratov, A.I. Popov, L. Shirmane, A. Kotlov, C. Feldmann, LaPO4:Ce,Tb and YVO4:Eu nanophosphors: Luminescence studies in the vacuum ultraviolet spectral range, Journal of Applied Physics, 110 (2011) 053522-053527.

[3] H.-K. Jung, J.-S. Oh, S.-I. Seok, T.-H. Lee, Preparation and luminescence properties of LaPO4:Er,Yb nanoparticles, Journal of Luminescence, 114 (2005) 307-313.

[4] K. Kömpe, H. Borchert, J. Storz, A. Lobo, S. Adam, T. Möller, M. Haase, Green-Emitting CePO4:Tb/LaPO4 Core–Shell Nanoparticles with 70 % Photoluminescence Quantum Yield, Angewandte Chemie International Edition, 42 (2003) 5513-5516.

[5] M.T. Colomer, S. Gallini, J.R. Jurado, Synthesis and characterisation of a green NiO/La(Sr)PO4−δ cermet anode for phosphate based solid oxide fuel cells, Journal of the European Ceramic Society, 27 (2007) 4237-4240.
[6] K. Rajesh, P. Shajesh, O. Seidel, P. Mukundan, K.G.K. Warrier, A Facile Sol–Gel Strategy for the Synthesis of Rod-Shaped Nanocrystalline High-Surface-Area Lanthanum Phosphate Powders and Nanocoatings, Advanced Functional Materials, 17 (2007) 1682-1690.
[7] S. Gallini, J.R. Jurado, M.T. Colomer, Synthesis and characterization of monazite-type Sr:LaPO4 prepared through coprecipitation, Journal of the European Ceramic Society, 25 (2005) 2003-2007.
[8] Sangmoon Park, Zhao Zhen, D.H. Park, Preparation of Eu-doped LaPO4 films using successive-ionic-layer-adsorption-and-reaction, Materials Letters, 64 (2010) 1861-1864.
[9] B.Y. Ahn, S.I. Seok, S.-I. Hong, J.-S. Oh, H.-K. Jung, W.J. Chung, Optical properties of organic/inorganic nanocomposite sol-gel films containing LaPO4:Er,Yb nanocrystals, Optical Materials, 28 (2006) 374-379.
[10] R. Gao, D. Qian, W. Li, Sol-gel synthesis and photoluminescence of LaPO4:Eu3+ nanorods, Transactions of Nonferrous Metals Society of China, 20 (2010) 432-436.
[11] Y.S.Patil, K.G.Chaudhari, N.V.Poornachandra Rao, K.V.R.Murthy, Effect of Erbium Doping on Structural and Photoluminescence Properties of LaPO4:Eu Phosphor, Advances in Applied Science Research, 2 (2011) 303-309.
[12] H. Dong, Y. Liu, P. Yang, W. Wang, J. Lin, ChemInform Abstract: Controlled Synthesis and Characterization of LaPO4, LaPO4:Ce3+ and LaPO4:Ce3+, Tb3+ by EDTA Assisted Hydrothermal Method, ChemInform, 41 (2010) no-no.
[13] K. Mi, Y. Ni, Y. Xu, X. Ma, J. Hong, A simple mixed-solvothermal route for LaPO4 nanorods: Synthesis, characterization, affecting factors and PL properties of LaPO4:Ce3+, Journal of Colloid and Interface Science, 356 (2011) 490-495.

The Photovoltaic Performance of Doped-CuI Hole Conductors for Solid State Dye-Sensitized Solar Cells

M.N. Amalina[1], A.A.E. Najwa[1], M. H. Abdullah[1], M.Z. Musa[1] and M. Rusop[1,2]

[1]NANO-ElecTronic Centre, Faculty of Electrical Engineering,
Universiti Teknologi MARA, 40450 Shah Alam, Selangor, Malaysia

[2]NANO SciTech Centre, Institute of Sciences;
Universiti Teknologi MARA, 40450 Shah Alam, Selangor, Malaysia

Abstract. The iodine doped copper (I) iodide (I_2: CuI) at different weight of iodine dopant have been prepared to investigate its thin films properties and photovoltaic performance. A novel method of mist atomization technique has been used for the deposition of CuI materials. The structural and electrical properties of CuI thin films deposited on the glass substrates were studied. The thin films morphology examined by FESEM shows a variation of crystal size and structure. Brick-like structure with smooth faces and sharp edges were seen for the doped thin films. The CuI thin films at 30 mg of iodine doping shows the highest resistivity of 4.56 x 101 Ω cm which caused by the surface traps create by iodine doping. The photovoltaic performance of ss-DSSC on the effect of variation iodine doping was investigated. The ss-DSSC fabricated with undoped CuI materials shows the highest efficiency of 1.05% while the 40 mg I_2 content shows the lowest conversion efficiency of 0.45%. The crystals size of CuI and its degree of crystallization are greatly contributed to the high filling fraction of the porous TiO_2 layer and hence the cells performance.

Keywords: iodine doping; dye sensitized solar cells; hole transport material; pore filling; copper (I) iodide

1. Introduction

Photovoltaic is a promising alternative renewable energy that utilizes the sunlight energy to obtained electricity which has pronounced potential to solve the future energy problem faces worldwide. Dye sensitized solar cells (DSSC) is one of the auspicious technology in the photovoltaic fields. It offers two separate function of light harvesting and charge transport which only uses low cost materials with less stringent procedure in device fabrications [1]. Generally, there are two types of DSSC which are liquid based and solid state DSSC. The efficiency of over 11% has been achieved for DSSC using the redox couple (I^-/I^{3-}) liquid electrolytes [2]. This organic liquid electrolyte is however, highly corrosive, volatile and photo reactive [3]. Besides that, it is also facing a leakage problem due to the interaction between the electrolytes and the sealing materials [3]. Therefore, in the quest of replacing the liquid electrolytes, the solid-state DSSC (ss-DSSC) is the promising option by exploiting the hole conductor to perform the dye regeneration. The recent progress of ss-DSSC has reached the power conversion efficiency of about 4-4.8% [4-6]. The solid state hole transport material (HTM) includes inorganic and organic material such as CuBr, CuSCN, CuI, poly (3,4-ethylenedioxythiophene) (PEDOT). The inorganic material such as copper based material is likely to be used since it is able to be deposited with low temperature which helps to preserve the dye materials [7]. Furthermore, the CuI has good conductivity of 10^2 Scm^1, which expedites the hole conducting ability [8]. Solid state DSSC employing the inorganic hole transport material (HTM) of copper (I) Iodide have first fabricated by K.Tennakone and had achieved efficiency of 2.4% [9]. Despite the advantages of using HTM to fabricate ss-DSSC, the low efficiency of ss-DSSC compared to the liquid based DSSC is mainly contributed by two factors which are (1) poor light harvesting efficiency due to the thinner electrode use (~2-3 *u*m) and (b) rapid recombination dynamics between the hetero-junction of ss-DSSC [10].

Some approaches have been proposed to improve the overall efficiency of ss-DSSC. These includes a study done to reduce a recombination rate between the electron, oxidized dye and the hole conductor. The recombination rate can be reduced by improved the pore filling fraction between the nano porous layer of TiO_2 and the HTM and also by increasing the hole mobility of the HTM. Various methods had been done to improve the pore filling penetration, including solution casting, soaking in precursor solution, spin coating and many more [9, 11, 12].

Therefore, in this research, we are targeting to study the effect of doped hole conductor which in this case is copper (I) iodide (CuI) to the photovoltaic performance of TiO_2/dye/CuI solar cells. In this paper, the iodine is chosen as the dopant that act as an electron acceptor that creates holes in the valence band [13]. The anionic nonmetal dopants are favorable than cationic dopants in this case, because the related impurity states are near the valence band edge and do not act as charge carriers [14]. Furthermore, their function as recombination centers might be minimized as compared to cation doping [14]. Other than that, this research also proposed a new method to deposit hole conductor which is by mister atomizer technique that will hope to improve the pore filling problem between the porous layer of dyed TiO_2 and HTM. The I_2 dopant concentrations were varied and the ss-DSSCs were fabricated at different dopant concentration. The thin film characteristic which includes surface morphology and electrical properties at different doping concentration were studied. Then, further investigation was done to see the doping effect to the photovoltaic performance of TiO_2/dye/CuI solar cells.

2. Experimental Details

2.1 Preparation of CuI thin film

Glass substrates of 2.5cm x 2.5 cm were used as the substrate for the CuI deposition. The glass substrates were cleaned by using sonication in acetone, methanol and de-ionized water respectively for 10 min. The deposition of CuI films were taken at room temperature by atomization technique. Mist-atomization is a process similarly known as spray pyrolysis which involves the process of spraying precursor solution onto heated surface and the constituents react to form the intended products [15]. The CuI solution was formed by mixing the precursor of copper (I) iodide powder (ALDRICH, 98%) with acetonitrile as a solvent. All chemicals were used without any further purification. The iodine which acts as dopant was varied at 10, 20, 30 and 40 mg. The prepared solutions were stirred for 3 hours at room ambient condition. Before the deposition, the substrates were preheated at 50℃ in the furnace before the deposition process. This solution was then deposited by mister atomizer to form fine droplets using argon as a carrier gas. The solution volume sprayed was 50 ml for all samples. Then, after the completion of the deposition process, the films were heated for about 5 min at 50°C to evaporate the solvent. All other parameters, including, the flow rate, solution molarity, height between substrate and nozzle were kept constant during the deposition to obtain a uniform thin films throughout the experiment process.

2.2 Fabrication of TiO_2/dye/CuI solar cells
The indium tin oxide (ITO) coated glass (10 Ω /□) was used as the conducting substrate. The mesoporous layer of TiO_2 was coated on the ITO coated glass by the following method. The TiO_2 paste was made by adding 0.6g TiO_2 nanopowder (<25nm, Sigma Aldrich) to the 5 ml absolute ethanol (99%). Then, 200 μl of titanium isopropoxide (TTIP, 99.99%) and 0.24 g of polyethylene glycol (PEG) 20000 were added and the mixture was stirred for 1 day to obtain a well dispersed solution. The TiO_2 deposition was carried out by spin coating technique at 500 rpm and 2000 rpm for 15s and 45s respectively. The TiO_2 film was then annealed at 450 °C for 60 min. The TiO_2 thin film was immersed into an ethanolic solution of 0.5mM N719 dye and was kept at room temperature for 24 hours. The CuI was deposited onto the dyed TiO_2 thin film at varied doping concentrations of 10, 20, 30 and 40 mg. The Au sputtered ITO coated glass substrate was used as the counter electrode the device active area is 0.28cm^2.

2.3 Characterization of CuI thin film and TiO_2/dye/CuI solar cells
The characterizations of CuI thin films were done for structural and electrical properties. The surface morphology of the films was observed with a field emission scanning electron microscope (JEOL JSM- J600F. The electrical properties of CuI thin films have been characterized by 2 point probe current-voltage (I-V) measurement by solar simulator CEP 2000 Spectral Sensitivity Analyzing System. Au was used as metal contacts for I-V measurement and deposited using sputter coater (EMITECH K550X). The DSSCs performance was measured using solar simulator device (CEP 2000). The current density-voltage (*J-V*) characteristics of the solar cells were measured under illumination at 100 mW/cm^2 by using an AM 1.5 solar simulator CEP 2000 Spectral Sensitivity Analyzing System. All of the measurements were done in room ambient.

3. Results and Discussion

3.1 Surface Morphology and Electrical Properties of CuI Thin Film
Fig. 1 shows surface morphology of SEM images of CuI thin films at various doping concentration for undoped and doped thin films. From the figures, it could be seen that the surface morphology of CuI films are crystalline in nature. The undoped thin films in fig. 1 (a) shows a good uniformity with crystal grains of about 80 nm – 150 nm. When the CuI solution was doped from 10 mg to 40 mg, there morphology of all the doped thin films was very much affected by the doping process. The SEM images of doped films revealed that there are existence of brick – like crystals with smooth facets and sharp corners. For 10 mg and 20 mg, the crystal grains are much larger than the undoped films. The 20 mg films show poor uniformity of films by having large and small CuI brick-like crystals in the films. When the I_2 dopant increases to 40 mg, the films composed of small brick-like crystals with sharp edges with sizes of about 75-90 nm compared to other doping concentration. By comparing between the undoped and doped thin films, the undoped thin films had a grain crystal structure of CuI while the doped thin films generally had brick – like crystals with smooth facets and sharp corners structures. The reasons of this to happened is due to the CuI nuclei formed simultaneously when the iodine concentration was higher, which preferred the growth of the integrated polyhedral CuI crystals on the surface [16] . The CuI crystal size obtained in this research was in the good agreement by Y. Yan et.al which also used spraying method for CuI deposition [17]. For application as a hole conductor in ss-DSSC, the pore penetration of HTM into TiO_2 which will limit the efficiency of the device. There are several ways studied by other researchers to overcome the pore penetration issues, including by using crystal growth inhibitor and melt infiltration [18-21]. The used of crystal growth inhibitor to the precursor solution might be tedious since it may change the properties of the deposited materials. While, the melt infiltration technique is not suitable for CuI since it has very high melting temperature of 878K [22]. Therefore, we proposed a new deposition method of CuI by using mister atomizer that hope to lessen the pore filling problem in ss-DSSC.

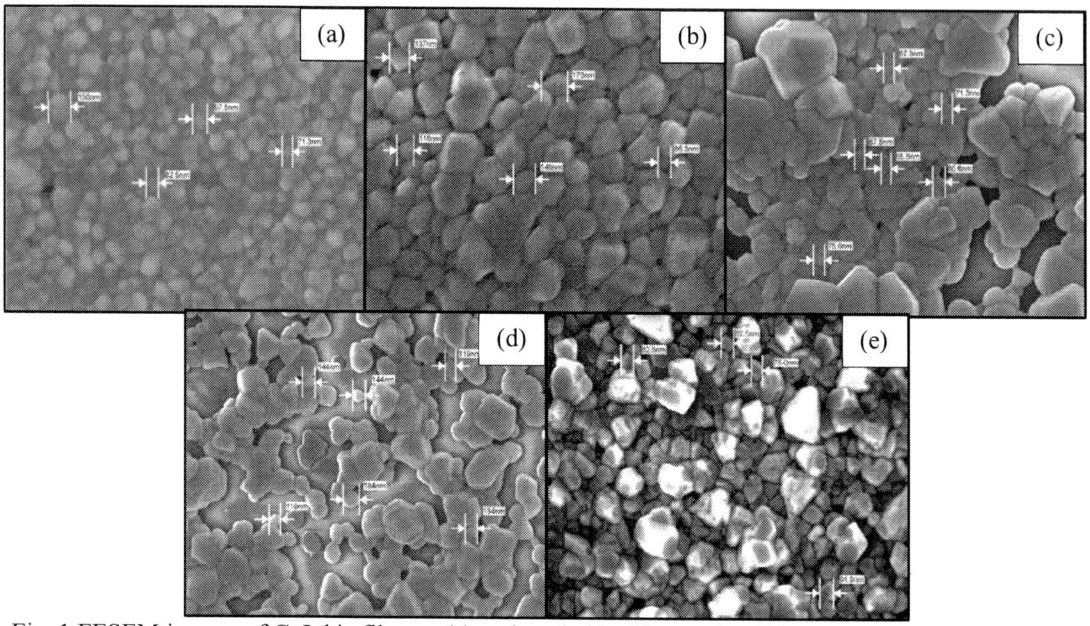

Fig. 1 FESEM images of CuI thin films at (a) undoped thin films and at different weight of I_2 dopant of (a) 10 mg, (b) 20 mg, (30) mg, (c) 40 mg

The effects of iodine doping of CuI thin films to the electrical properties are presented in fig. 2. The electrical measurement of CuI films were carried out by using 2 point probe I-V measurement with a constant voltage in the range of -5V and 5V. The metal contact used for the measurement is aurum (gold). From the I-V curves the resistivity, ρ of CuI thin film was calculated using following equation (1) :

$$\rho = \left(\frac{V}{I}\right)\frac{wt}{l} \qquad (1)$$

where V is supplied voltage, I is measured current, t is the film thickness, w is the electrode width and l is the length between electrodes.

Figure 2 shows the plot of resistivity of CuI films deposited at various doping concentration. It is observed that the resistivity first increased with increased of doping concentration up to 30 mg and then decreased at doping concentration of 40 mg. The resistivity of CuI thin films at various doping concentration showed resistivity ranging from 1.8×10^{-1} Ωcm to 4.56×10^{1} Ωcm. The p-type conductivity of CuI is depends on the excess of iodine in the stoichiometric excess which due to the acceptor-type copper vacancies (V_{cu}). In this investigation, the doped thin films have much higher resistivity compared to the doped thin films. This is because as the I^- atoms increases consequently the increasing of hole concentration, the number of free electron carrier is reduced due to the attraction between the electron and hole. Hence, the reduction of free charge carrier of electron will significantly increase the resistivity value of the doped CuI thin films. Other than that, the increment of resistivity for doped thin films may be due to surface traps created above the valence band upon the introduction of iodine atoms. The excess of I_2 atoms in the lattice will create layer of electrons charge trapping and thus will reduce its mobility. Perera et. al had described that the surface trapping sites is 0.2 eV above the valence band of CuI [23]. However, the low resistivity of CuI thin films at 40 mg of doping concentration may be due to the morphology of the thin film itself which was attributed to the improvement of grain boundaries that lead to better less carrier scattering and enhances the carrier mobility of CuI atoms.

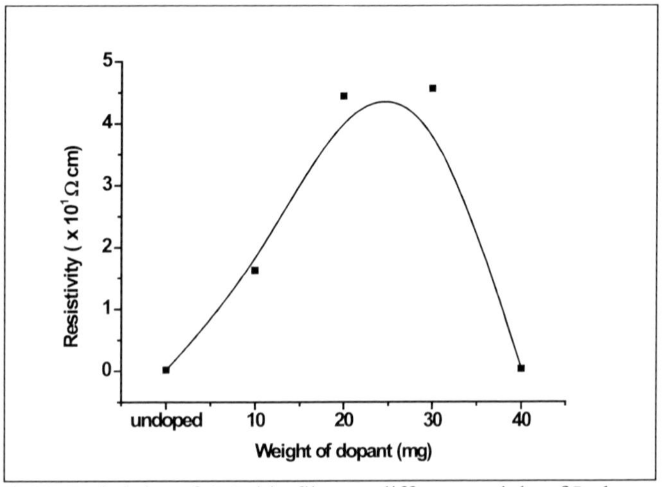

Fig. 2 Resistivity of CuI thin films at different weight of I_2 dopant

3.2 *Photovoltaic Performance of Solid State DSSC*

The *J-V* curves of TiO_2 based DSSC at different weight of I_2 dopant was shown in fig. 3. The cells were illuminated under AM 1.5 (100mW/cm^2) with cells area of 0.28 cm^2. Good rectification behaviour is observed for all samples which are a key function required in solar cells. The summarized photovoltaic efficiency for TiO_2/dye/ CuI cells fabricated at weight of I_2 content has been tabulated in table 1. The best device efficiency was achieved for undoped CuI thin films with a current density of J_{SC} = 3.51 mA/cm^2, open circuit voltage of V_{oc} = 0.65V, a fill factor of 0.46 and a power conversion efficiency of η= 1.05%. While the lowest conversion efficiency was obtained for solar cells device with the I_2 content of 40 mg with efficiency of 0.45%. The J-V curve for 40 mg of I_2 content for the fabricated DSSC is contradictory to the finding of I-V characterization of 40 mg of I_2 content in CuI thin film.

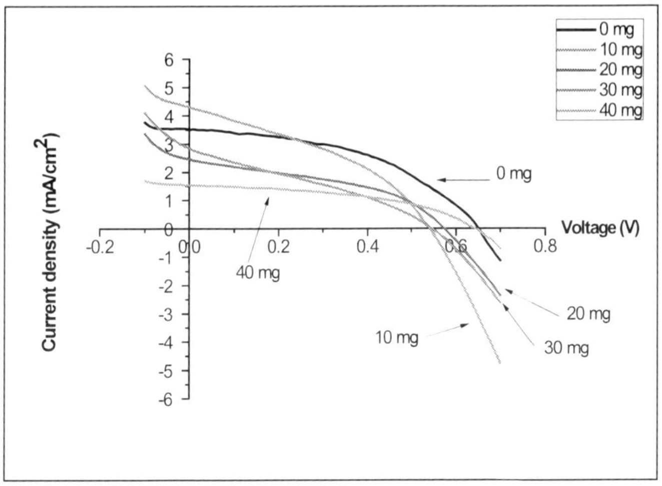

Fig. 3 The *J-V* curves of the TiO_2/dye/CuI solar cell at different weight of I_2 dopant

Table 1. Photovoltaic performance of TiO_2/dye/CuI solar cells at different weight of I_2 dopant

Weight of I_2 dopant (mg)	J_{SC} (mA/cm^2)	V_{oc} (V)	FF (%)	η (%)
0	3.51	0.65	0.46	1.05
10	4.30	0.54	0.38	0.87
20	2.46	0.57	0.42	0.59
30	2.85	0.55	0.31	0.48
40	1.54	0.64	0.46	0.45

The resistivity curves as shown in fig. 2 shows a low resistivity value of undoped and 40 mf I_2 dopant content CuI thin films. However, the same trend was not observed for the DSSC fabricated at different I_2 weight. The reasons of this to happen are maybe due to the high crystallization of CuI doped at 40 mg weight of I_2. Although, the low resistivity of CuI thin films was observed at this condition, which means high hole transfer yield, these 40 mg of I_2 content showed strong tendency to crystallize compared to undoped thin films. The high crystallization of CuI atoms attributed to the formation of crystal boundaries which will act as barriers of hole transport [24]. Even though the 40 mg fabricated DSSC shows low conversion efficiency, the fill factor of this device is same as the DSSC fabricated for undoped CuI which is 0.46 and is the highest fill factor among other samples. Basically, the fill factor is defined by the ratio of the current and the voltage at the maximum power point to the short circuit current and the open circuit voltage. The squareness of the J-V curves indicates low recombination reaction caused by the high conductivity observed for these (undoped and 40 mg of I_2 content) CuI thin films. From table 1, it could be seen that, as we increased the amount of dopant, the overall performance of the device was slightly decreased. The main factor contributes to the decrease of cell performance is attributed to the pore penetration of CuI onto the dyed TiO_2 layer. The bigger crystallite size of CuI at 10, 20 and 30 mg of I_2 content will results poor pore filling of porous TiO_2 layer and hence increase the recombination reaction thus less device efficiency was perceived.

4. Conclusion

In summary, we have demonstrated the study on I_2 doped CuI thin films deposited by mist atomization technique at different weight of I_2 dopant. The surface morphology and electrical properties shows an effect towards the doping process. The FESEM images reveal the structure of brick-like crystal with smooth faces and sharp edges as the I_2 amount increases. The undoped and 40 mg doped films shows smaller grain sizes compared to others. This morphology affects the electrical resistivity of the doped films. The smallest resistivity of 10^{-1} Ωcm was shown for undoped thin films and sample doped at 40 mg of I_2 content. For the electrical device performance, the results shows that, as the I_2 dopant increases from 0 mg- 40 mg, the cells efficiency decreases with the highest efficiency obtained for undoped CuI of 1.05%. The strong tendency of CuI crystallization for sample doped at 40 mg I_2 weight attributed to the low device performance.

Acknowledgement

The authors would like to thank Universiti Tekonologi Mara (UiTM) and Ministry of Higher Education (MOHE) and UiTM for the financial support.

References

[1] M. G. Brian O'regan, "Low cost and highly efficient solar cells based on the sensitization of colloidal titanium dioxide " *Nature,* vol. 353, pp. 737-740, 1991.

[2] M. K. Nazeeruddin, A. Kay, I. Rodicio, R. Humphry-Baker, E. MÃller, P. Liska, N. Vlachopoulos, and M. GrÃtzel, "Conversion of light to electricity by cis-2bis(2,2-bipyridyl-4,4-dicarboxylate)ruthenium(II) charge-transfer sensitizers (X = Cl -, Br -, I -, CN -, and SCN

-) on nanocrystalline TiO 2 electrodes," *Journal of the American Chemical Society,* vol. 115, pp. 6382-6390, 1993.
[3] I. Chung, B. Lee, J. He, R. P. H. Chang, and M. G. Kanatzidis, "All-solid-state dye-sensitized solar cells with high efficiency," *Nature,* vol. 485, pp. 486-489.
[4] L. Schmidt-Mende, U. Bach, R. Humphry-Baker, T. Horiuchi, H. Miura, S. Ito, S. Uchida, and M. Grätzel, "Organic Dye for Highly Efficient Solid-State Dye-Sensitized Solar Cells," *Advanced Materials,* vol. 17, pp. 813-815, 2005.
[5] H. J. Snaith, A. J. Moule, C. d. Klein, K. Meerholz, R. H. Friend, and M. GrÃtzel, "Efficiency Enhancements in Solid-State Hybrid Solar Cells via Reduced Charge Recombination and Increased Light Capture," *Nano Letters,* vol. 7, pp. 3372-3376, 2012/11/11 2007.
[6] M. Wang, S.-J. Moon, M. Xu, K. Chittibabu, P. Wang, N.-L. Cevey-Ha, R. Humphry-Baker, S. M. Zakeeruddin, and M. Grätzel, "Efficient and Stable Solid-State Dye-Sensitized Solar Cells Based on a High-Molar-Extinction-Coefficient Sensitizer," *Small,* vol. 6, pp. 319-324.
[7] B. Li, L. Wang, B. Kang, P. Wang, and Y. Qiu, "Review of recent progress in solid-state dye-sensitized solar cells," *Solar Energy Materials and Solar Cells,* vol. 90, pp. 549-573, 2006.
[8] G. P. Smestad, S. Spiekermann, J. Kowalik, C. D. Grant, A. M. Schwartzberg, J. Zhang, L. M. Tolbert, and E. Moons, "A technique to compare polythiophene solid-state dye sensitized TiO2 solar cells to liquid junction devices," *Solar Energy Materials and Solar Cells,* vol. 76, pp. 85-105, 2003.
[9] G. R. R. A. K. K Tennakone, A R Kumarasinghe, K G U Wijayantha and P M Sirimanne, "A dye-sensitized nano-porous solid-state photovoltaic cell " *Semiconductor Science and Technology,* vol. 10, p. 1689, 1995.
[10] M. Wang, J. Liu, N.-L. Cevey-Ha, S.-J. Moon, P. Liska, R. Humphry-Baker, J.-E. Moser, C. GrÃ¤tzel, P. Wang, S. M. Zakeeruddin, and M. GrÃtzel, "High efficiency solid-state sensitized heterojunction photovoltaic device," *Nano Today,* vol. 5, pp. 169-174.
[11] Y. Zhenzhen, C. P. Katherine, L. Di-Jia, R. Yang, and X. Tao, "Solid dye-sensitized solar cells prepared through a counter strategy for filling of solid hole transporter," *Journal of Renewable and Sustainable Energy,* vol. 3, p. 063101.
[12] I. K. Ding, N. Tétreault, J. Brillet, B. E. Hardin, E. H. Smith, S. J. Rosenthal, F. Sauvage, M. Grätzel, and M. D. McGehee, "Pore-Filling of Spiro-OMeTAD in Solid-State Dye Sensitized Solar Cells: Quantification, Mechanism, and Consequences for Device Performance," *Advanced Functional Materials,* vol. 19, pp. 2431-2436, 2009.
[13] M. Rusop, T. Shirata, S. Prasad Manjusri, T. Soga, and T. Jimbo, "Properties of CuI films in the power output of TiO/sub 2/|Dye|CuI cells," in *Photovoltaic Energy Conversion, 2003. Proceedings of 3rd World Conference on*, 2003, pp. 140-143 Vol.1.
[14] J. Ding, Y. Yuan, J. Xu, J. Deng, and J. Guo, "TiO_2 nanopowder Co-doped with iodine and Boron to enhance visible-light photocatalytic activity," *Journal of Biomedical Nanotechnology,* vol. 5, pp. 1-7, 2009.
[15] J. B. Mooney and S. B. Radding, "Spray Pyrolysis Processing," *Annual Review of Materials Science,* vol. 12, pp. 81-101, 1982.
[16] B. Huang, Z. Zheng, F. Yang, Y. Zhang, D. Pu, H. Zhao, and D. Li, "A rapid sonochemical approach to semiconductor thin films: The case of metal iodides," *Solid State Ionics,* vol. 179, pp. 2006-2010, 2008.
[17] Y.-h. Yan, Y.-c. Liu, L. Fang, Z.-c. Lu, Z.-b. Li, and S.-x. Zhou, "Growth of CuI buffer layer prepared by spraying method," *Transactions of Nonferrous Metals Society of China,* vol. 21, pp. 359-363.
[18] G. R. A. Kumara, A. Konno, K. Shiratsuchi, J. Tsukahara, and K. Tennakone, "Dye-Sensitized Solid-State Solar Cells: Use of Crystal Growth Inhibitors for Deposition of the Hole Collector," *Chemistry of Materials,* vol. 14, pp. 954-955, 2012/02/24 2002.

[19] G. R. A. Kumara, S. Kaneko, M. Okuya, and K. Tennakone, "Fabrication of dye-sensitized solar cells using triethylamine hydrothiocyanate as a CuI crystal growth inhibitor," *Langmuir,* vol. 18, pp. 10493-10495, 2002.

[20] M. H. M. A.R. Zainun, U.M. Noor and M. Rusop, "Particles Size and Conductivity Study of P-Type Copper (I) Iodide (CuI) Thin Film for Solid State Dye-Sensitized Solar Cells," *IOP Conference Series: Materials Science and Engineering,* vol. 17, p. 012009, 2011.

[21] K. Fredin, E. M. J. Johansson, T. Blom, M. Hedlund, S. Plogmaker, H. Siegbahn, K. Leifer, and H. Rensmo, "Using a molten organic conducting material to infiltrate a nanoporous semiconductor film and its use in solid-state dye-sensitized solar cells," *Synthetic Metals,* vol. 159, pp. 166-170, 2009.

[22] R. R. Khalid Siraj, "Electrical Conductivity Behavior of CdHgI4 – CuI Mixed System," *International Journal of Chemistry,* vol. 3, pp. 174-179, 2011.

[23] V. P. S. Perera and K. Tennakone, "Recombination processes in dye-sensitized solid-state solar cells with CuI as the hole collector," *Solar Energy Materials and Solar Cells,* vol. 79, pp. 249-255, 2003.

[24] J. E. Kroeze, N. Hirata, L. Schmidt-Mende, C. Orizu, S. D. Ogier, K. Carr, M. Grätzel, and J. R. Durrant, "Parameters Influencing Charge Separation in Solid-State Dye-Sensitized Solar Cells Using Novel Hole Conductors," *Advanced Functional Materials,* vol. 16, pp. 1832-1838, 2006.

IOP Conf. Series: Materials Science and Engineering 46 (2013) 012013 doi:10.1088/1757-899X/46/1/012013

A Mathematical Model for Scheduling a Batch Processing Machine with Multiple Incompatible Job Families, Non-identical Job dimensions, Non-identical Job sizes, Non-agreeable release times and due dates

M. Ramasubramaniam[1], M. Mathirajan[2]

[1]Scool of Maritime Management, Indian Maritime University, Chennai, India.

[2]Principal Research Scientist, Department of Management Studies, Indian Institute of Science, Bangalore, India.

E-mail: rams@liba.edu

Abstract. The paper addresses the problem scheduling a batch processing machine with multiple incompatible job families, non-identical job dimensions, non-identical job sizes and non-agreeable release dates to minimize makespan. The research problem is solved by proposing a mixed integer programming model that appropriately takes into account the parameters considered in the problem. The proposed is validated using a numerical example. The experiment conducted show that the model can pose significant difficulties in solving the large scale instances. The paper concludes by giving the scope for future work and some alternative approaches one can use for solving these class of problems.

1. Introduction

Scheduling of parallel batch processing machines has recently attracted a great deal of interest among the researchers. Parallel batch processing machines, by definition, can simultaneously process a number of jobs as a batch where all the jobs have the same beginning and ending times. One could find these problems classified as bounded batch processing machines (BPM) scheduling problem because the size of batch is limited based on the capacity of the BPM.

This paper tries to address one such problem namely scheduling of a Heat-treatment furnace(HTF) that finds application in steel casting industry. A general characteristic of the steel casting industry is the product variety. Castings (or jobs) of any size, shape, quantities can be produced. Also, the materials used in this process are widely different that are selected based on the type of environment and application of the jobs which makes them incompatible from the point of view of processing. Thus, these job families are called multiple incompatible job families. Steel casting industry is a capital intensive industry and competition in this industry is high. So, there is a pressure on the steel casting industries to reduce the lead time to satisfy the customers in meeting the deadlines. One way of reducing the lead time for production is to schedule the bottleneck machines effectively.

From the viewpoint of throughput and utilization of the important and costly resources in the foundry manufacturing, it was felt that the process-controlled furnace operations for the melting and pouring operations as well as the heat-treatment furnace operations are critical for meeting the overall production schedules. The two furnace operations are batch processes that have distinctive constraints on job-mixes in addition to the usual capacity and technical constraints associated with any industrial processes. The benefits of effective scheduling of these batch processes include higher machine utilization, lower work-in-process (WIP) inventory, shorter cycle time and greater customer satisfaction. Although the melting and pouring operations are of central importance, the scheduling of heat-treatment furnaces is also of considerable importance. This is primarily due to the fact that the processing times of heat treatment furnaces are often longer compared to any other operations in the process. This paper attempts to propose methods to reduce the lead time of this process by addressing this bottleneck machine.

Accordingly, in this paper, we address a new research problem of scheduling a BP with Incompatible Job Families (MIJF), NIJS, NIJD and NARD to minimize makespan which is a surrogate measure for increasing throughput and reducing the lead time.

2. Research Problem

The scheduling problem defined in this paper can be denoted in three-field notation: $\alpha \mid \beta \mid \gamma$ as 1 | p-batch, multiple incompatible job family, non-identical job sizes, non-identical job-dimensions, non-agreeable release times and due dates | Cmax problem. We propose a mathematical model for the research problem with the following assumptions:

- There are n jobs (n single castings) to be processed and these jobs are from a single job-family. Each job has a size s_i (in terms of weight) with dimensions l_i - length, w_i – width and h_i - height
- All jobs must pass through the BP
- There is a single BP, which has capacity limit in terms of size S (in terms of weight) and dimensions L - Length, W - Width, and H - Height
- The BP is available continuously
- Orientation of jobs is not allowed
- All data of S, L, W, H, s_i , l_i, w_i, and h_i are deterministic and known a priori
- It is assumed that $s_i \leq S$; $l_i \leq L$; $w_i \leq W$; and $h_i \leq H$ for all jobs i
- Jobs are assumed to have non-agreeable release times and due dates. i.e .$r_i \leq r_j$ does not imply $d_i \leq d_j$
- Processing time of job family is assumed to be independent of number of jobs in the batch and constant
- Preparation time, if any, is included in the processing time, and there is no setup time in switching from one batch to another
- Once processing of a batch is initiated, the BP cannot be interrupted and other jobs cannot be introduced into the BP until the current processing is completed.
- Machine breakdowns are not considered

The paper is organized as follows: In section 3, we review the related work. In section 4, we propose a mathematical model. In section 5, we validate the mathematical model with a numerical example. We conclude the paper with a summary and possible future research direction in Section 6.

3. Review of Related Works:

Semiconductor industries have attracted the interest of researchers the most [10] whereas, there are only a few studies close to the research problem under consideration. This could be primarily due to the economic & technological importance associated with semiconductor industries.

[5] addressed the problem of scheduling a BPM with incompatible job families with the objectives of minimizing makespan, total completion time and total weighed completion time. The authors propose a mathematical model and a set of heuristic algorithms and a meta heuristic for the different objectives.

Other than this study, other studies focus on meta-heuristic approaches for addressing the scheduling of BPMs [4], [11], [2].

Our studies differ from other studies in that we address an important aspect namely non-identical job dimension which may not be relevant in the semiconductor industry but important in steel casting industry. Also, our study considers non-agreeable release times which are not addressed in the literature. Finally, unlike studies which propose meta-heuristics, our study attempts to develop a mathematical model which would yield optimal solution for the research problem.

Apart from the semiconductor industry, there are only a few studies which focus on addressing the problem of scheduling heat treatment furnaces. Thus we only focus on these few studies which are very close to the research problem.

[8] examine the problem of scheduling multiple heat-treatment furnaces (HTF), of steel casting industry with non-identical job size and assuming identical job dimensions to maximize the utilization of the batch processors under dynamic job arrivals with agreeable release times and due dates. They take a three-stage approach to schedule the heat-treatment furnaces namely BPM selection, job family selection. They conduct a series of computational experiments in which the solution quality of heuristic algorithms is compared with estimated optimal solutions for each problem instances. They conclude that heuristic algorithms based on job family selection criteria which use: (a) the simple average priority of job, (b) the simple average size of job and (c) the weighted average size of job perform better with lesser computational effort.

[9] study the scheduling problem addressed earlier [8] with the objective of minimizing total weighted tardiness. They proposed four greedy heuristic algorithms which differ based on the job family selection. The detailed computational experiments conducted by them reveal that on an average, the heuristic algorithms which use job family selection criteria based on (a) cumulative due date and (b) weighted cumulative due date with job size as weight, perform better when compared with statistical estimation procedure.

Our study differs from these studies in that we consider non-identical job dimensions which take our study closer to reality.

4. Development of mathematical model for scheduling a BP with MIJF, NIJS and NIJD

The problem of scheduling a BPM with MIJF, NIJD, NIJS and NARD could be viewed as an extension of the three-dimensional bin-packing problem.

There are different approaches for solving this type of research problem like mathematical models, heuristic methods or metaheuristic methods. Since out of all this three, mathematical models provide optimal solutions for the problem this paper approaches the problem by proposing a mathematical model.

Mathematical models proposed by various researchers ([3], [12], [1]) for bin-packing problem consider NIJD characteristics explicitly and there is no restriction on the capacity of the bins in terms of job sizes considered in these studies.

Scheduling a BPM, however, is more complex. First, there are multiple incompatible job families which mean jobs from one family cannot be combined with jobs from another family. Also, even within a single job family, job (casting) sizes vary widely (from 10 g to 1000 kg), which puts a tremendous restriction on the number of jobs that can be accommodated in a batch to satisfy the capacity restrictions of BP in terms of size, in addition to dimensions of job. Accordingly, among the mathematical models available in the literature for bin-packing problem, the MILP model proposed by [1] is appropriately modified to address the problem of scheduling a BP with incompatibility limitation and capacity restrictions on both job dimensions and sizes. The proposed MILP model for scheduling a BP with a MIJF, NIJD, NIJS and NARD is as follows:

Indices

i, k	-	$1, 2, 3, \ldots, n$ Jobs
j	-	$1, 2, 3, \ldots, m$ Batches
p	-	$1, 2, 3, \ldots, q$ Families

Parameters

M	-	A large arbitrary number
(l_i, w_i, h_i)	-	Parameters indicating length , width and height of job i
s_i	-	Size of job i
(L, W, H)	-	Length, Width and Height of BP
S	-	Size capacity of BP
t_i	-	Processing time of job i (= processing time of job family)
r_i	-	Release time of job i

Continuous Decision Variables

(a_i, b_i, c_i)	-	Coordinates of the front left bottom corner of job i. It is assumed that the origin is fixed at Front Left Bottom corner of the BP
(a_k, b_k, c_k)	-	Coordinates of the front left bottom corner of job k. It is assumed that the origin is fixed at Front Left Bottom corner of the BP
C_{max}	-	Completion time of last batch
c_i	-	Completion time of job i
R^j	-	Release time of batch j
P^j	-	Processing time of batch j
C^j	-	Completion time of batch j

Binary Decision Variables

$$x_{ipj} - \begin{cases} 1 \text{ if job } i \text{ of family } p \text{ is assigned to batch } j \\ 0 \text{ otherwise} \end{cases}$$

$$y_{pj} - \begin{cases} 1 \text{ if family } p \text{ is assigned to batch } j \\ 0\text{otherwise} \end{cases}$$

$$left_{ik} = \begin{cases} 1 \text{ if job } i \text{ is on left side of job } k \; \forall \; i < k; \\ 0 \text{ otherwise} \end{cases}$$

$$right_{ik} = \begin{cases} 1 \text{ if job } i \text{ is on right side of job } k \; \forall \; i < k; \\ 0 \text{ otherwise} \end{cases}$$

$$behind_{ik} = \begin{cases} 1 \text{ if job } i \text{ is behind job } k \; \forall \; i < k; \\ 0 \text{ otherwise} \end{cases}$$

$$front_{ik} = \begin{cases} 1 \text{ if job } i \text{ is in front of job } k \; \forall \; i < k; \\ 0 \text{ otherwise} \end{cases}$$

$$below_{ik} = \begin{cases} 1 \text{ if job } i \text{ is below job } k \; \forall \; i < k; \\ 0 \text{ otherwise} \end{cases}$$

$$above_{ik} = \begin{cases} 1 \text{ if job } i \text{ is above job } k \; \forall \; i < k; \\ 0 \text{ otherwise} \end{cases}$$

Min C_{max}

Subject to:

$$\sum_{j=1}^{m}\sum_{p=1}^{q} x_{ipj} = 1 \qquad \forall i \qquad (1)$$

$$\sum_{p=1}^{q} y_{pj} \leq 1 \qquad \forall j \qquad (2)$$

$$a_i + l_i \leq L \qquad \forall i \qquad (3)$$

$$b_i + w_i \leq W \qquad \forall i \qquad (4)$$

$$c_i + h_i \leq H \qquad \forall i \qquad (5)$$

$$x_{ipj} \leq y_{pj} \qquad \forall i, p, j \qquad (6)$$

$$\sum_{i=1}^{n} s_i x_{ipj} \leq S \qquad \forall p, j \qquad (7)$$

$$a_i + l_i \leq a_k + (1 - left_{ik})M \qquad \forall i, k, i < k \qquad (8)$$

$$a_k + l_k \leq a_i + (1 - right_{ik})M \qquad \forall i, k, i < k \qquad (9)$$

$$b_i + w_i \leq b_k + (1 - behind_{ik})M \qquad \forall i, k, i < k \qquad (10)$$

$$b_k + w_k \leq b_i + (1 - front_{ik})M \qquad \forall i, k, i < k \qquad (11)$$

$$c_i + h_i \leq c_k + (1 - below_{ik})M \qquad \forall i, k, i < k \qquad (12)$$

$$c_k + h_k \leq c_i + (1 - above_{ik})M \qquad \forall i, k, i < k \qquad (13)$$

$$left_{ik} + right_{ik} + behind_{ik} + front_{ik} +$$
$$below_{ik} + above_{ik} \geq x_{ipj} + x_{kpj} - 1 \qquad \forall i, k, p, j,\ i < k \qquad (14)$$

$$left_{ik} + right_{ik} + behind_{ik} + front_{ik} +$$

$$below_{ik} + above_{ik} \leq 1 + x_{ipj} - x_{kpj} \qquad \forall i, k, p, j,\ i < k \qquad (15)$$

$$P^{j} \geq t_i x_{ipj} \qquad \forall i, p, j \qquad (16)$$

$$R^{j} \geq r_i x_{ipj} \qquad \forall i, p, j \qquad (17)$$

$$C^{j} \geq R^{j} + P^{j} \qquad \forall j \qquad (18)$$

$$R^{j+1} \geq C^{j} \qquad \forall j \text{ and } j < n \qquad (19)$$

$$c_i \geq C^{j} - M(1 - x_{ipj}) \qquad \forall i, p, j \qquad (20)$$

$$C_{max} = C^{n} \qquad (21)$$

Binary restriction:
$left_{ik}, right_{ik}, behind_{ik}, front_{ik}, below_{ik}, above_{ik}, x_{ipj}, y_{ipj} = 0$ or 1

Non-negativity constraints:
$c_i, P^j, R^j, C^j \geq 0$

The objective of the model is to minimize the completion time of the last batch. Constraint set (1) ensures that each job of a job family is assigned only once to the batch. Constraint set (2) stipulates that at most only one family can be assigned to a batch. Constraints (3)- (5) ensure that all the jobs assigned to a batch fit within the physical dimensions of the BP.Constraint (6) ensures that a job from a job family is assigned to a batch if and only if that job family is assigned to that batch.Constraint (7) ensures that the weight capacity of BP is not exceeded. Constraints (8–13) ensure that jobs do not overlap with one another. This check for overlap is necessary only if a pair of jobs is placed in the same BP. This is taken care of by Constraints (14–15).

Constraint (16) defines the processing time of a batch as greater than or equal to the processing time of each job (= processing time of the job family) in that batch. Constraint (17) ensures that release time of a batch is the maximum of the release times of all jobs in the batch. Calculation of Completion time of the jobs is taken care by (18). Ensuring those jobs have their release time atleast equal to the completion of time of previous batch is by (19). Constraint (20) ensures that completion time of job is atleast equal to the completion time of batch, if and only if the job is in that batch. Constraint (21) is completion time of the last batch.

The number of binary variables, in the proposed MILP model is $nm + fm + 3n(n-1)$ and the model has $3n(n-1)+2n(n-1)m+4nfm+5n+3m+fm+1$ number of constraints, where f is the number of job families, n number of jobs, m number of constraints.

5. Validation of the Proposed MILP model

To validate the proposed MILP model, we develop a numerical example assuming one BP with two job families having n = 10 jobs with the processing time of the first job family as 15 h and second as 12 h. The Capacity of the BP in terms of size of the BP = 1000 kg; and in terms of dimensions: L = 1500 mm; W = 950 mm; H = 900 mm. In addition, the size, dimensions, and processing time of the 10 jobs are given in Table 1.

The optimal solution for the proposed MILP model is obtained using LINGO. The appropriate LINGO set code to generate is given in Appendix 1. Using this set code, an ILP model is generated for the numerical example presented in Table 1 and solved.

Table 1: A numerical example for research problem

Parameter	Job = j =									
	1	2	3	4	5	6	7	8	9	10
$Size_j(kg^*)$	275	115	189	250	133	160	157	222	298	291
$length_j(mm^{\#})$	466	223	321	164	165	251	359	474	253	155
$width_j(mm)$	105	188	142	258	292	149	301	190	202	197
$height_j(mm)$	245	256	204	213	103	213	281	150	236	290
Release time$_j$(h)	3	3	4	6	10	13	15	15	19	19
Due date$_j$(h)	20	35	34	50	36	61	47	45	56	70
Job Family	1	1	1	1	1	2	2	2	2	2

* Kilogram, # Millimetre

Table 2: Optimal solution obtained using the proposed MILP model

Batch	Optimal Solution				
	Jobs in the batch	Family	Release time of the batch (h)	Processing time of the batch (h)	Completion time of the Batch (h)
1	1,2,3,4,5	1	10	15	25
2	6, 7, 8	2	25	12	37
3	9, 10	2	37	12	49
C_{max} = **49 h**; Computational Time = 00: 19:01 in P4 with 2.40 GHz, 1 GB of RAM; Number of Iterations = 2161072					

The optimal solution obtained for the numerical example is presented in Table 2 which gives the batch wise job details, job family details, release time, processing time and the completion time of each batch.

6. Summary and Future Work

Based on our observation from steel casting industry, we addressed an important research problem of scheduling a HTF in steel casting industry. This problem has not been given due attention in the literature. This led us to development of a mathematical model for scheduling the HTF with MIJF, NIJD, NIJS and NARD with the objective of minimizing makespan. The proposed mathematical model is validated using a numerical example by generating and solving the model in LINGO.

It could be noted that even for a small scale instance such as the one used for solving the mathematical model in this paper, the computational time is reasonably high. In general, it is widely accepted that integer programming models are difficult to solve to optimality, especially for large instances. Thus, it is possible that larger instances can pose significant difficulties in solving the model proposed in this paper. Considering these things in mind, one could first test the model extensively on large scale instances to estimate the computational time for this class of models. Secondly, one could think of reformulating this model to make it solvable in reasonable time limit.

As an alternative, the problem can be approached by using heuristic methods that could result in near optimal solutions very quickly. Also, meta-heuristic methods like genetic algorithms, tabu search etc. could be developed for solving the problem which can provide better solutions than a simple heuristic method at the expense of more computational effort.

7. References

[1] Chen C S Lee s and Shen Q S 1995 An Analytical Model for the Container Loading Problem *European Journal of Operational Research* 80(1) 68-76

[2] Damodaran P and Velez-Gallego M.C 2012 A simulated annealing algorithm to minimize makespan of parallel batch processing machines with unequal job ready times *Expert Systems with Applications* 39 1451-1458

[3] Fasano G 2008 MIP based heuristic for non standard 3D-bin packing problems *4OR: A Quarterly Journal of Operations Research* 6(3) 291–310

[4] KashanA Karimi B 2007Scheduling a single batch-processing machine with arbitrary job sizes and incompatible job families: an ant colony framework*Journal of Operational Research Society* 59 1269-1280

[5] Koh S G Koo P H Kim D C and Hur W S 2005 Scheduling a single batch processing machine with arbitrary job sizes and incompatible job families *International Journal Of Production Economics*98(1) 81-96

[6] Martello S Pisinger D and Vigo D2000 The Three dimensional Bin Packing Problem *Operations Research*48(2) 256-267

[7] Mathirajan M (2002) Heuristic scheduling algorithms for parallel heterogenous batch processors *PhD Thesis* Department of Management Studies Indian Institute of Science

[8] Mathirajan M Sivakumar A I and Chandru V (2004a) Scheduling algorithms for heterogeneous batch processors with incompatible job-families *Journal Of Intelligent Manufacturing* 15(6) 787-803

[9] Mathirajan M and Sivakumar A I (2006a) Minimizing total weighted tardiness on heterogeneous batch processing machines with incompatible job families *International Journal Of Advanced Manufacturing Technology*28(9) 1038-1047

[10] Mathirajan M and Sivakumar A I (2006b) A literature review classification and simple meta-analysis on scheduling of batch processors in semiconductor *International Journal Of Advanced Manufacturing Technology*29(9-10) 990-1001

[11] Nong Q Ng C T and Cheng T C E (2008) The bounded single-machine parallel-batching scheduling problem with family jobs and release dates to minimize makespan *Operations*

*Research Letters*36(1) 61-66
[12] Padberg M (2000) Packing small boxes in a big box *Mathematical Methods of Operations Research* Vol 52 pp1–21
[13] Pinedo M L (1995) Scheduling: Theory Algorithms and Systems Prentice Hall
[14] Ramasubramaniam M (2008) 'Batch Processor Scheduling – A Class of Problems in Steel Casting Industries' *PhD thesis* Department of Management Studies Indian Institute of Science Bangalore India

Appendix 1

```
SETS :
  FAMILIES:STARTIME;
  JOBS:P,Q,HT,WEIGHT,X,Y,Z,RELTIME,DUEDATE,PROCTIME,COMPTIMEJOB;

FNCS:CTIME,FNCHEIGHT,FNCCAPACITY,FNCLENGTH,FNCWIDTH,PROCTIMEBAT,RELTIMEBAT,
COMPTIMEBAT;
ORIENTATION(JOBS,JOBS)|&1 #LT# &2:LEFT,RIGHT,BEHIND,INFRONT,BELOW,ABOVE;
FAMTOFNCS(FAMILIES,FNCS):FTOB;
JOBTOFAMTOFNC(JOBS,FAMILIES,FNCS):JTOFTOB;


 ENDSETS
DATA :

 FAMILIES = F1,F2 ;

JOBS,FNCS,P,Q,HT,WEIGHT,FNCLENGTH,FNCWIDTH,PROCTIME,RELTIME,DUEDATE,FNCHEIGHT
,FNCCAPACITY = @OLE ('C:\\3DINCOMPCH2\\12jobs.XLS');
 ENDDATA

 BIGM=1500;

 MAXJOBS = @SIZE (FNCS);

!OBJECTIVE TO MINIMIZE MAKESPAN;
MIN = CMAX;
!SUBJECT TO CONSTRAINTS:
!ASSIGN EACH JOB TO ONE BATCH;
 @FOR (JOBS(I):
   @SUM (FAMILIES(FAM):
 @SUM (FNCS(B): JTOFTOB(I,FAM,B)))=1);

!AT MOST ONE JOB FAMILY CAN BE ASSINGED TO A BATCH;
 @FOR (FNCS(B):
  @SUM (FAMILIES(FAM): FTOB(FAM,B))<=1);

 ! A JOB FROM JOB FAMILY IS ASSINGED TO A BTATCH ONLY IF
 THE JOB FAMILY IS ASSIGNED TO A BATCH;
 @FOR (JOBS(I):
   @FOR (FAMILIES(FAM):
    @FOR (FNCS(B): JTOFTOB(I,FAM,B) <= FTOB(FAM,B))));

!SIZE CONSTRAINT;
 @FOR (FAMILIES(FAM):
 @FOR (FNCS(J):
 @SUM (JOBS(I): WEIGHT(I)*JTOFTOB(I,FAM,J)) <= 1000));


!ORIENTATION OF JOBS CONSTRAINT;
 @FOR (JOBS(K):
  @FOR (JOBS(I)|I #LT# K: X(I)+ P(I) <= X(K)+(1-LEFT(I,K))*BIGM) );


 @FOR (JOBS(K):
  @FOR (JOBS(I)|I #LT# K: X(K)+ P(K) <= X(I)+(1-RIGHT(I,K))*BIGM ));
```

```
@FOR (JOBS(K):
 @FOR (JOBS(I)|I #LT# K: Y(I)+ Q(I)<= Y(K)+(1-BEHIND(I,K))*BIGM ));

 @FOR (JOBS(K):
 @FOR (JOBS(I)|I #LT# K: Y(K)+Q(K) <= Y(I)+(1-INFRONT(I,K))*BIGM ));

@FOR (JOBS(K):
 @FOR (JOBS(I)|I #LT# K: Z(I)+ HT(I) <= Z(K)+(1-BELOW(I,K))*BIGM ));

@FOR (JOBS(K):
 @FOR (JOBS(I)|I #LT# K: Z(K)+ HT(K) <= Z(I)+(1-ABOVE(I,K))*BIGM ));

@FOR (FAMILIES(FAM):
@FOR (FNCS(J):
@FOR (JOBS(K):
  @FOR (JOBS(I)|I #LT# K:
LEFT(I,K)+RIGHT(I,K)+BEHIND(I,K)+INFRONT(I,K)+BELOW(I,K)+ABOVE(I,K)
>= JTOFTOB(I,FAM,J)+JTOFTOB(K,FAM,J)-1))));
@FOR (FAMILIES(FAM):
@FOR (FNCS(J):
@FOR (JOBS(K):
  @FOR (JOBS(I)|I #LT# K:
LEFT(I,K)+RIGHT(I,K)+BEHIND(I,K)+INFRONT(I,K)+BELOW(I,K)+ABOVE(I,K)
<= 1+JTOFTOB(I,FAM,J)-JTOFTOB(K,FAM,J)))));

!LENGTH CONSTRAINT;
 @FOR (JOBS(I): X(I)+ P(I)<= 1500);

!WIDTH CONSTRAINT;
 @FOR (JOBS(I): Y(I)+ Q(I)<= 950);

!HEIGHT CONSTRAINT;
 @FOR (JOBS(I): Z(I)+ HT(I)<= 900);

!CALCULATE PROCESSING TIME OF A BATCH;
 @FOR (FAMILIES(FAM):
 @FOR (FNCS(B):
  @FOR (JOBS(J):   PROCTIMEBAT(B) >= PROCTIME(J)*  JTOFTOB(J,FAM,B))));


!RELEASE TIME OF A BATCH IS MAXIMUM OF ALL RELEASE TIMES OF JOBS IN A BATCH;
 @FOR (FAMILIES(FAM):
 @FOR (FNCS(B):
  @FOR (JOBS(J):   RELTIMEBAT(B) >= RELTIME(J)*  JTOFTOB(J,FAM,B))));


!CALCULATE COMPLETION TIME OF A BATCH;
 @FOR (FNCS(B):  COMPTIMEBAT(B) = RELTIMEBAT(B)+PROCTIMEBAT(B)   );


!RELEASE TIME OF A BATCH SHOULD BE LESS THAN THE COMPLETION TIME OF
 THE PREVIOUS BATCH;
 @FOR (FNCS(B)|B #LT#  @SIZE (FNCS):  RELTIMEBAT(B+1) >= COMPTIMEBAT(B) );

!CALCULATE COMPLETION TIME OF A JOB;
```

```
 @FOR (FAMILIES(FAM):
 @FOR (FNCS(B):
  @FOR (JOBS(J):   COMPTIMEJOB(J) >= COMPTIMEBAT(B)-1000*(1-JTOFTOB(J,FAM,B)))));

!COMPLETION TIME OF A JOB IS LESS THAN ITS DUE DATE;
 !@FOR(JOBS(J): COMPTIMEJOB(J) <= DUEDATE(J));

!CALCULATE CMAX;
 @FOR (FNCS(B)|B #EQ# @SIZE (FNCS): CMAX = COMPTIMEBAT(B)   );


!DECISION VARIABLES;
 @FOR (JOBS(I)|I  #GE# 1 #AND# I #LE# 5:
   @FOR (FAMILIES(FAM)|FAM #EQ# 1:
 @for (FNCS(B):  @bin (JTOFTOB(I,FAM,B)))));

 @FOR (JOBS(I)|I  #GE# 6 :
   @FOR (FAMILIES(FAM)|FAM #EQ# SETS :
  FAMILIES:STARTIME;
  JOBS:P,Q,HT,WEIGHT,X,Y,Z,RELTIME,DUEDATE,PROCTIME,COMPTIMEJOB;

FNCS:CTIME,FNCHEIGHT,FNCCAPACITY,FNCLENGTH,FNCWIDTH,PROCTIMEBAT,RELTIMEBAT,
COMPTIMEBAT;
ORIENTATION(JOBS,JOBS)|&1 #LT# &2:LEFT,RIGHT,BEHIND,INFRONT,BELOW,ABOVE;
FAMTOFNCS(FAMILIES,FNCS):FTOB;
JOBTOFAMTOFNC(JOBS,FAMILIES,FNCS):JTOFTOB;

 ENDSETS
DATA :

 FAMILIES = F1,F2  ;


JOBS,FNCS,P,Q,HT,WEIGHT,FNCLENGTH,FNCWIDTH,PROCTIME,RELTIME,DUEDATE,FNCHEIGHT
,FNCCAPACITY =  @OLE ('C:\\3DINCOMPCH2\\12jobs.XLS');
 ENDDATA

 BIGM=1500;

 MAXJOBS =  @SIZE (FNCS);

!OBJECTIVE TO MINIMIZE MAKESPAN;
MIN  = CMAX;
!SUBJECT TO CONSTRAINTS:
!ASSIGN EACH JOB TO ONE BATCH;
 @FOR (JOBS(I):
   @SUM (FAMILIES(FAM):
 @SUM (FNCS(B): JTOFTOB(I,FAM,B)))=1);

!AT MOST ONE JOB FAMILY CAN BE ASSINGED TO A BATCH;
 @FOR (FNCS(B):
  @SUM (FAMILIES(FAM): FTOB(FAM,B))<=1);

 ! A JOB FROM JOB FAMILY IS ASSINGED TO A BTATCH ONLY IF
 THE JOB FAMILY IS ASSIGNED TO A BATCH;
 @FOR (JOBS(I):
```

```
    @FOR (FAMILIES(FAM):
     @FOR (FNCS(B): JTOFTOB(I,FAM,B) <= FTOB(FAM,B))));

 !SIZE CONSTRAINT;
  @FOR (FAMILIES(FAM):
  @FOR (FNCS(J):
  @SUM (JOBS(I): WEIGHT(I)*JTOFTOB(I,FAM,J)) <= 1000));


 !ORIENTATION OF JOBS CONSTRAINT;
  @FOR (JOBS(K):
   @FOR (JOBS(I)|I #LT# K: X(I)+ P(I) <= X(K)+(1-LEFT(I,K))*BIGM) );


  @FOR (JOBS(K):
   @FOR (JOBS(I)|I #LT# K: X(K)+ P(K) <= X(I)+(1-RIGHT(I,K))*BIGM ));

  @FOR (JOBS(K):
   @FOR (JOBS(I)|I #LT# K: Y(I)+ Q(I)<= Y(K)+(1-BEHIND(I,K))*BIGM ));


  @FOR (JOBS(K):
   @FOR (JOBS(I)|I #LT# K: Y(K)+Q(K) <= Y(I)+(1-INFRONT(I,K))*BIGM ));

  @FOR (JOBS(K):
   @FOR (JOBS(I)|I #LT# K: Z(I)+ HT(I) <= Z(K)+(1-BELOW(I,K))*BIGM ));

  @FOR (JOBS(K):
   @FOR (JOBS(I)|I #LT# K: Z(K)+ HT(K) <= Z(I)+(1-ABOVE(I,K))*BIGM ));


  @FOR (FAMILIES(FAM):
  @FOR (FNCS(J):
  @FOR (JOBS(K):
    @FOR (JOBS(I)|I #LT# K:
LEFT(I,K)+RIGHT(I,K)+BEHIND(I,K)+INFRONT(I,K)+BELOW(I,K)+ABOVE(I,K)
>= JTOFTOB(I,FAM,J)+JTOFTOB(K,FAM,J)-1))));
  @FOR (FAMILIES(FAM):
  @FOR (FNCS(J):
  @FOR (JOBS(K):
    @FOR (JOBS(I)|I #LT# K:
LEFT(I,K)+RIGHT(I,K)+BEHIND(I,K)+INFRONT(I,K)+BELOW(I,K)+ABOVE(I,K)
<= 1+JTOFTOB(I,FAM,J)-JTOFTOB(K,FAM,J)))));

 !LENGTH CONSTRAINT;
  @FOR (JOBS(I): X(I)+ P(I)<= 1500);

 !WIDTH CONSTRAINT;
  @FOR (JOBS(I): Y(I)+ Q(I)<= 950);

 !HEIGHT CONSTRAINT;
  @FOR (JOBS(I): Z(I)+ HT(I)<= 900);


 !CALCULATE PROCESSING TIME OF A BATCH;
  @FOR (FAMILIES(FAM):
```

```
 @FOR (FNCS(B):
  @FOR (JOBS(J):   PROCTIMEBAT(B) >= PROCTIME(J)*  JTOFTOB(J,FAM,B))));


!RELEASE TIME OF A BATCH IS MAXIMUM OF ALL RELEASE TIMES OF JOBS IN A BATCH;
 @FOR (FAMILIES(FAM):
 @FOR (FNCS(B):
  @FOR (JOBS(J):   RELTIMEBAT(B) >= RELTIME(J)*  JTOFTOB(J,FAM,B))));


!CALCULATE COMPLETION TIME OF A BATCH;
 @FOR (FNCS(B):  COMPTIMEBAT(B) = RELTIMEBAT(B)+PROCTIMEBAT(B)   );


!RELEASE TIME OF A BATCH SHOULD BE LESS THAN THE COMPLETION TIME OF
 THE PREVIOUS BATCH;
 @FOR (FNCS(B)|B #LT#  @SIZE (FNCS):  RELTIMEBAT(B+1) >= COMPTIMEBAT(B) );

!CALCULATE COMPLETION TIME OF A JOB;
 @FOR (FAMILIES(FAM):
 @FOR (FNCS(B):
  @FOR (JOBS(J):   COMPTIMEJOB(J) >= COMPTIMEBAT(B)-1000*(1-JTOFTOB(J,FAM,B)))));

!COMPLETION TIME OF A JOB IS LESS THAN ITS DUE DATE;
 @FOR (JOBS(J): COMPTIMEJOB(J) <= DUEDATE(J));

!CALCULATE CMAX;
 @FOR (FNCS(B)|B #EQ#  @SIZE (FNCS): CMAX = COMPTIMEBAT(B)   );


!DECISION VARIABLES;
 @FOR (JOBS(I)|I  #GE# 1 #AND# I #LE# 5:
   @FOR (FAMILIES(FAM)|FAM #EQ# 1:
 @for (FNCS(B):  @bin (JTOFTOB(I,FAM,B)))));

 @FOR (JOBS(I)|I  #GE# 6 :
   @FOR (FAMILIES(FAM)|FAM #EQ# 1:
 @for (FNCS(B): JTOFTOB(I,FAM,B)=0)));

 @FOR (JOBS(I)|I  #GE# 6 :
   @FOR (FAMILIES(FAM)|FAM #EQ# 2:
 @for (FNCS(B):  @bin (JTOFTOB(I,FAM,B)))));

 @FOR (JOBS(I)|I  #LE# 5 :
   @FOR (FAMILIES(FAM)|FAM #EQ# 2:
 @for (FNCS(B): JTOFTOB(I,FAM,B)=0)));

 @FOR (FAMTOFNCS: @BIN (FTOB));
 @FOR (ORIENTATION: @BIN (LEFT));
 @FOR (ORIENTATION: @BIN (RIGHT));
 @FOR (ORIENTATION: @BIN (BEHIND));
 @FOR (ORIENTATION: @BIN (INFRONT));
 @FOR (ORIENTATION: @BIN (BELOW));
 @FOR (ORIENTATION: @BIN (ABOVE));

# 1:
```

```
@for (FNCS(B): JTOFTOB(I,FAM,B)=0)));

@FOR (JOBS(I)|I #GE# 6 :
  @FOR (FAMILIES(FAM)|FAM #EQ# 2:
@for (FNCS(B):  @bin (JTOFTOB(I,FAM,B)))));

@FOR (JOBS(I)|I #LE# 5 :
  @FOR (FAMILIES(FAM)|FAM #EQ# 2:
@for (FNCS(B): JTOFTOB(I,FAM,B)=0)));

@FOR (FAMTOFNCS: @BIN (FTOB));
@FOR (ORIENTATION: @BIN (LEFT));
@FOR (ORIENTATION: @BIN (RIGHT));
@FOR (ORIENTATION: @BIN (BEHIND));
@FOR (ORIENTATION: @BIN (INFRONT));
@FOR (ORIENTATION: @BIN (BELOW));
@FOR (ORIENTATION: @BIN (ABOVE));
```

Effect of Iron and Cobalt Catalysts on The Growth of Carbon Nanotubes from Palm Oil Precursor

A.B.Suriani[a], N.A. Asli[b], M. Salina[c], M.H. Mamat[d], A.A. Aziz[e], A.N. Falina[a], M. Maryam[b], M.S. Shamsudin[b], Roslan Md Nor[f], S. Abdullah[b], M. Rusop[b,d]

[a]Department of Physics, Faculty of Science and Mathematics, Universiti Pendidikan Sultan Idris, 35900 Tanjung Malim, Perak D.R., Malaysia.

[b] Centre of Nanoscience and Nanotechnology (NANO-SciTech Centre), Insitute of Science, Universiti Teknologi MARA, 40450 Shah Alam, Selangor D.E., Malaysia.

[c]Faculty of Engineering, Universiti Industri Selangor, Bestari Jaya Campus, Jalan Timur Tambahan, 45600 Bestari Jaya, Selangor D.E., Malaysia.

[d] NANO-Electronic Centre, Faculty of Electrical Engineering, Universiti Teknologi MARA, 40450 Shah Alam, Selangor D.E., Malaysia.

[e]Advanced Rubber Technology Unit, Technology & Engineering Division, Malaysian Rubber Board, RRIM Research Station Sg. Buloh, 47000, Selangor, Malaysia.

[f]Department of Physics, Faculty of Science, University of Malaya, 50603 Kuala Lumpur, Malaysia.

absuriani@yahoo.com.

Abstract. Catalysts which are typically a transition metal is mandatory and plays an important role in the production of CNT. In this work, the effect of iron (Fe) and cobalt (Co) nitrate catalyst on the growth of carbon nanotubes (CNT) were systematically studied. Green bio-hydrocarbon precursor namely palm oil was used as a precursor. The synthesis was done using thermal chemical vapour deposition method at temperature of 750°C for 15 min synthesis time. The Fe and Co solution were spin-coated separately on silicon substrate at speed of 3000 rev.min-1. The CNT characteristics were analyzed using field emission scanning electron microscopy and micro-Raman spectroscopy. The experimental results revealed that CNT properties were strongly affected by the catalyst type. CNT catalyzed by Co yields large diameter, crooked tube and lower quality, whereas CNT produced by Fe catalyst results in the smallest diameter and reasonably good graphitization. As a conclusion, Fe was considered as the optimum catalyst for better CNT structure and crystallinity. This was due to efficient, uniform and stable Fe catalytic activity as compared to Co catalyst in producing CNT.

1. Introduction

Carbon nanotubes (CNT) were discovered in 1991 [1], since then many studies have been performed on their synthesis [2-6]. The synthesis of CNT can be classified into catalytic and non-catalytic methods. As for the catalytic method, nickel (Ni), iron (Fe) and cobalt (Co) are common transition metal utilized as pure-metal to catalyze the CNT growth [7-12]. The catalyst particles essentially serve as seeds for the growth of CNT. The main challenge in CNT production is to mass produce at low cost. In this regard, the catalytic method has been claimed to be the best approach due to its capability in lowering the reaction temperatures and cost. Between Ni, Fe and Co as a catalyst, Fe was reported to be the most efficient catalyst [13]. CNT produced from Fe catalyst were found to have higher degree crystallinity character, high density and small diameter tubes. However in term of growth rate Ni and Co catalysts have demonstrated better growth rate which outperformed Fe catalyst [13]. The ability of those catalyst were related to its catalytic activity for the decomposition and diffusion of hydrocarbon precursor on the active site, the formation of C-hexagon as well as the catalyst deactivation and the formation of meta-stable carbides [14, 15]. The catalyst itself can be used either separately or combined together to form bi or tri-catalyst combination [4]. Each single catalyst has its own role, a synergistic effect was expected when the catalyst were combined. Although the effect of catalyst on the growth of CNT have been extensively study but the decomposition of palm oil precursor over Fe and Co surface for the production of CNT have not been clearly discussed.

Here we report a systematic study of the effect of Fe and Co-nitrate on the growth of CNT using palm oil precursor. The synthesis was done in double furnace thermal chemical vapor deposition (TCVD) system. The results clearly revealed that the catalysts were not only affecting the diameter and growth rate but also the morphology and microstructure of CNT. The structural properties of CNT were analyzed using field emission scanning electron microscope (FESEM) and micro-Raman spectroscopy.

2. Experimental

First, the Fe and Co-nitrate catalyst were diluted separately with ethanol and stirred for 20 min. The catalyst films *were* prepared by dropping 3 drops of catalyst on Si substrates using a spin coater at speed of 3000 rev.min^{-1}. The catalyst coated substrate was dried at 90 °C for 24 hours before loaded into deposition furnace of TCVD system. 3 mL of palm oil was used as carbon source and loaded into *precursor* furnace. The deposition furnace was then heated to 750°C with argon gas supplied continuously. When the temperature was stable at 750°C, precursor furnace was turned-on at temperature of 450°C. The CNT synthesis was carried out for 15 min followed by 10 min of annealing time. When the growth completed, a Zeiss Supra 40VP FESEM and Jobin-Yvon, Horiba miro-Raman spectrometer were used to characterize the samples.

3. Results and Discussion

3.1. Morphological features

The CNT synthesis was done using two different catalysts namely Fe and Co-nitrate at synthesis temperature of 750°C. Figure 1 shows the morphologies of CNT synthesized using (a) Fe and (c) Co catalyst. The magnified images of CNT are shown in Figure 1 (b) and (c) for Fe and Co catalyst respectively. For comparison purposes, the FESEM images for 700°C samples are also shown in Figure 2 (a)-(d). Both catalysts gave a dense distribution of nanotubes growth with the presence number of un-reacted catalyst particles. In term of diameter, CNT catalyzed by Fe produced smaller and uniform diameter ranged at 31.2 nm. The tubes were longer as compared to Co catalyst. The diameter of CNT produced by Co catalyst were in range of 80-100 nm. The nanotubes were also severely crooked, obviously mixed with impurities namely amorphous carbon (a-C) and non-tubular carbon structures. A synthesis temperature of 700°C gave less CNT population particularly for Fe catalyst where large catalyst particles that remained un-reacted amidst the nanotubes were seen. For Co catalyst, it was clearly seen that the nanotubes diameter were getting smaller (42.2 nm) as compared to 750°C synthesis temperature. In contrast to CNT catalyzed by Fe at 750 °C the tubes diameter greatly increases when it was synthesized at 700°C to 83.3 nm. From FESEM images in Figure 2 (e) and (f) it showed that better and smooth morphology CNT have been repeatedly demonstrated by CNT catalyzed by Fe catalyst which was regardless of the temperature utilized during the synthesis process.

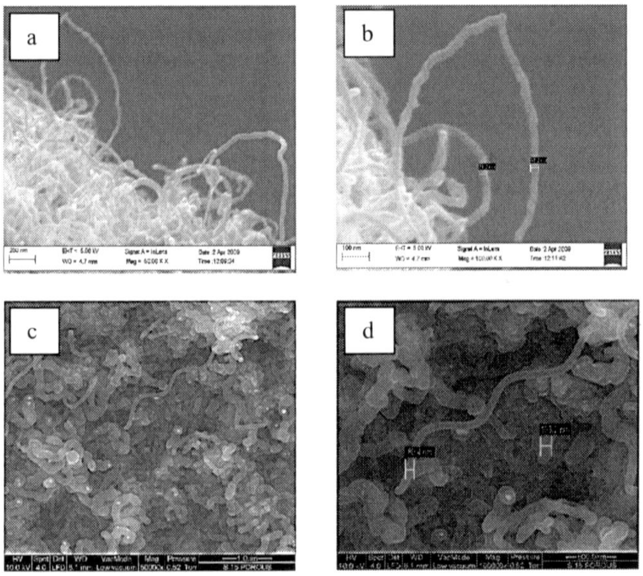

Figure 1. FESEM images of CNT synthesized at 750°C using (a)-(b) Fe and (c)-(d) Co-nitrate catalyst.

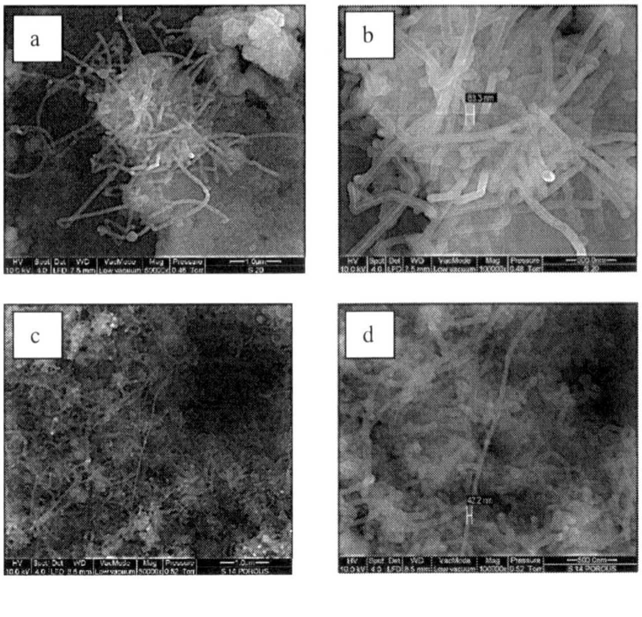

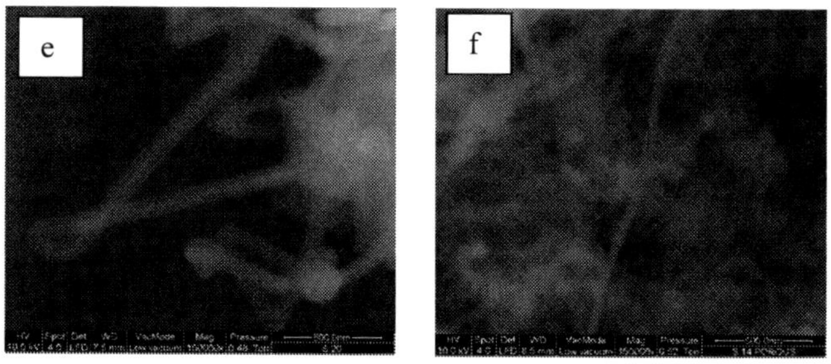

Figure 2. FESEM images of CNT synthesized at 700°C using (a),(b),(e) Fe-nitrate catalyst and (c),(d),(f) Co-nitrate.

3.2. Raman spectroscopy

The grown CNT were then characterized by Raman spectroscopy as shown in Figure 3. The well separated two Raman peaks at ~1592.67-1600.12 cm^{-1} for G peaks and at ~1346.30-1353.23 cm^{-1} for D peaks was observed for all samples. The I_D/I_G ratios were calculated to estimate the variation of CNT crystallinity with Fe and Co catalyst. The I_D/I_G value was found to be lower at Fe-750°C sample which was 0.65 and slightly higher using Co-nitrate (0.78). Irrespective of the type of catalyst used, the ratio values increases at 700°C synthesis temperature (Table 1). This reveals that the trend of CNT crystallinity varies with catalyst and synthesis temperature. Our results indicated that the best crystallinity was shown by CNT grown on Fe catalyst at synthesis temperature of 750°C. There was no lower frequency radial breathing mode (RBM) peaks detected for all samples which indicate no single-walled CNT (SWCNT) was produced [16].

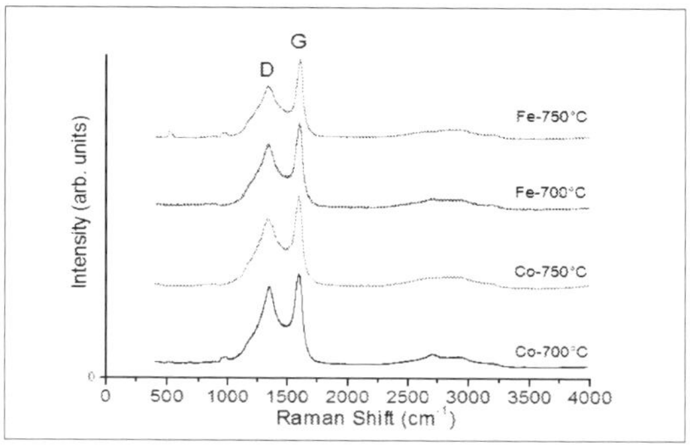

Figure 3. Typical Raman spectra of palm oil based CNT catalyzed by Fe and Co-nitrate at 750 and 700°C synthesis temperatures

Table 1. The Summary of Raman Scattering Data of CNT Catalyzed by Fe and Co-Nitrate at 750 and 700°C Synthesis Temperatures.

Samples	G peak (cm^{-1})	G-Width (cm^{-1})	D peak (cm^{-1})	D-Width (cm^{-1})	I_D/I_G ratio
Fe-750°C	1600.12	59.76	1353.23	236.93	0.65
Fe-700°C	1596.50	64.60	1346.30	237.54	0.72
Co-750°C	1592.67	69.91	1350.32	243.21	0.78
Co-700°C	1594.79	72.71	1352.42	232.69	0.81

Analyzing the efficiencies of Fe and Co-nitrate on the morphology and microstructure quality of CNT synthesized at 700 and 750°C, it was observed that good graphitization nanotubes (from ID/IG ratio) was best catalyzed by Fe and followed by Co catalyst. Co catalyst results in crooked, biggest diameter and impurities-rich tubes. This deduced that Co catalyst exhibit inefficient, less uniform and instable catalytic activity as compared to Fe which was much more efficient, uniform and stable across catalyst surface [4]. Figure 4 (a)-(b) shows FESEM images of Fe and Co-nitrate catalyst. From the images it can be seen that the Co particles have much bigger particles size distribution as compared to Fe. The catalyst particles size distribution were greatly affecting the efficacy of the catalytic activity and the size of the tubes itself. High surface area to volume ratio of small particles size of Fe catalyst was a crucial factor for faster and better catalytic reaction as compared to Co catalyst. This was because more surface area were involved in the reaction. As a result, the CNT catalyzed by Fe show better microstructure and less crooked. Therefore, the growth optimization with low I_D/I_G ratio and low nanotubes diameter were obtained when Fe-nitrate was used as catalyst at synthesis temperature of 750°C. However, the common features shown either by Fe or Co-nitrate catalyst was that there was no vertically aligned carbon nanotubes (VACNT) successfully produced and overall entangle, bigger diameter with moderate quality and quantity CNT were synthesized. In addition, this catalyst preparation method (through spin coating) produces non-continuous nanotube growth due to catalyst driven growth which would be a drawback, for some characterization [17] particularly for the application in nanoelectronic devices. For this reason, this creates the need to seek for better catalyst preparation method which able to produce densely packed CNT-bundle that cover the entire sample area.

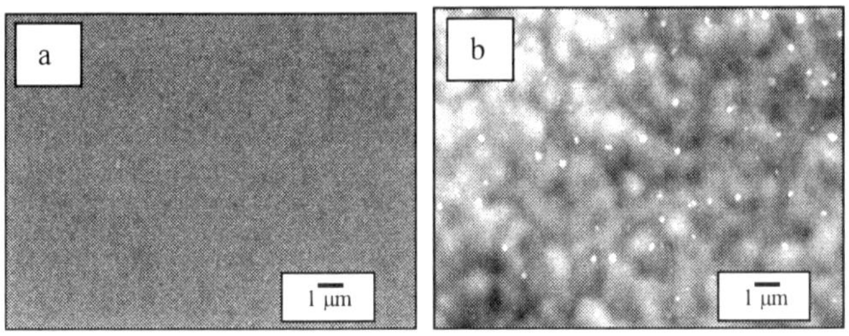

Figure 4. FESEM images of (a) Fe and (b) Co-nitrate catalyst.

4. Conclusion

In this study, the effect of Fe and Co-nitrate on the morphology and microstructure of CNT were investigated. The finding here suggest that Fe catalyst was an effective catalyst in obtaining the optimal CNT graphitization with lower diameter. Crooked, bigger diameter and impurities tubes were produced by Co catalyst. Due to low surface area to volume ratio of Co catalyst yielded inefficient, less uniform and instable catalytic activity in comparison to Fe catalyst. Through this catalyst preparation method there was no VACNT and SWCNT produced and the CNT population did not cover the entire substrate surface.

Acknowledgement

This work is supported by Fundamental Research Grant Scheme (FRGS) (Project Code: 2010-0064-102-02) funded by the Malaysia Government.

References

[1] Iijima S 1992 Nature 354, 56
[2] Suriani A B, Nor R M and Rusop M 2010 J. Ceram. Soc. Jpn. 118, 963
[3] Suriani A B, Azira A A, Nik S F, Nor R M and Rusop M 2009 Mater. Lett. 63, 2704
[4] Osorio A G and Bergmann C P 2013 Appl. Surf. Sci. 264, 794
[5] Yan X and Lui C 2013 Diamond Relat. Mater., 31, 50
[6] Suriani A B, Salina M, Sarah M S P, Nor R M, Anuar A, Siran Y M, Rejab S A M, Asis A J, Tahiruddin S, Abdullah S and Rusop M 2011 Defect. Diffus. Forum, 312, 900
[7] Tran K Y, Heinrichs B, Colomer J F, Pirard J P and Lambert S 2007 Appl. Catal. A-Gen. 318, 63
[8] Avdeeva L B, Kochubey D I and Shaikhutdinov S K 1999 Appl. Catal. A-Gen. 177, 43
[9] Bulusheva L G, Okotrub A V, Kudashov A G, Shubin Y V, Shlyakhova E V, Yudanov N F, Pazhetnov E M, Boronin A I and Vyalikh D V 2008 Carbon 46, 864
[10] Emmenegger C, Bonard J M, Mauron P, Sudan P, Lepora A, Grobety B, Züttel A and Schlapbach L 2003 Carbon, 41, 539
[11] Guczi L, Stefler G, Geszti O, Koppány Z, Kónya Z, Molnár E, Urbán M and Kiricsi I 2006 J. Catal., 244, 24
[12] Suriani A B, Salina M, Sarah M S P, Zobir S A M, R. M. Nor, Anuar A, Siran Y M, Rejab S A M, Asis A J, Tahiruddin S, Abdullah S and Rusop M 2011 Defect. Diffus. Forum, 312, 906
[13] Lee C J, Park J, Yu Y A 2002 Chem. Phys. Lett., 360, 250
[14] Kiang C H 2000 J. Chem. Phys., 113, 4763
[15] Andriotis A N, Menon M, Froudakis G 2000 Phys. Rev. Lett., 85, 3193
[16] Saito R, Dresselhaus G and Dresselhaus M S 2000 Phys. Rev. B. 61, 2981
[17] Deck C P, Ph.D. Thesis San Diego, California, USA: University of California, 2009

A deterministic aggregate production planning model considering quality of products

Najmeh Madadi, Kuan Yew Wong

Department of Manufacturing and Industrial Engineering, Faculty of Mechanical Engineering, Universiti Teknologi Malaysia, Skudai, Malaysia

E-mail: najmeh.madadi@gmail.com, wongky@fkm.utm.my

Abstract. Aggregate Production Planning (APP) is a medium-term planning which is concerned with the lowest-cost method of production planning to meet customers' requirements and to satisfy fluctuating demand over a planning time horizon. APP problem has been studied widely since it was introduced and formulated in 1950s. However, in several conducted studies in the APP area, most of the researchers have concentrated on some common objectives such as minimization of cost, fluctuation in the number of workers, and inventory level. Specifically, maintaining quality at the desirable level as an objective while minimizing cost has not been considered in previous studies. In this study, an attempt has been made to develop a multi-objective mixed integer linear programming model that serves those companies aiming to incur the minimum level of operational cost while maintaining quality at an acceptable level. In order to obtain the solution to the multi-objective model, the Fuzzy Goal Programming approach and max-min operator of Bellman-Zadeh were applied to the model. At the final step, IBM ILOG CPLEX Optimization Studio software was used to obtain the experimental results based on the data collected from an automotive parts manufacturing company. The results show that incorporating quality in the model imposes some costs, however a trade-off should be done between the cost resulting from producing products with higher quality and the cost that the firm may incur due to customer dissatisfaction and sale losses.

1. Introduction

Nowadays, due to huge improvement of operation management methods and emerging new methodologies, companies are faced with intensive competitive circumstances. Under these conditions, companies try to apply appropriate techniques and methods to have a better utilization of their available resources. Facing with expanding competition, companies try to move from disconnected decision making processes towards a more aggregated planning and control of their activities in order to reduce total cost, improve performance, and increase service level [1]. Among all those techniques, the role of aggregate production planning is getting more and more highlighted.

"Aggregate Production Planning" (APP) is a process by which a company determines ideal levels of capacity, production, subcontracting, inventory, stock-out, and even pricing over a specified time horizon [2]. It can also be defined as a medium-term capacity planning often from 3 to 18 months ahead which is concerned with the lowest-cost method of production planning to meet customer's requirements and to satisfy fluctuating demand over the planning horizon [3].

APP problem has been studied widely, since it was introduced and formulated in1950s. The interest in the problem stems from the ability that such a model provides for effective control of production and inventory costs, which are a substantial portion of the overall cost of the manufacturers [4]. In several conducted studies in the APP area, most of the researchers have concentrated on some common objectives such as minimization of cost, fluctuation in the number of workers, and inventory level. However, due to intensive competitive circumstances, just focusing on some specific objectives can no longer fulfill the requirements of demanding customers. Specifically, to the best of the researchers' knowledge, maintaining quality at the optimal level as an objective while minimizing cost has not been considered in previous studies. In this study, an attempt has been made to respond to this issue by developing a multi-objective mixed integer linear programming model in which both objectives of minimizing cost and maximizing quality have been considered simultaneously. As the aim of this study is to develop a more practical model, it tries to incorporate both quantitative and qualitative objectives in developing the APP model. In order to test the applicability of the model, it has been applied to an APP problem of an automotive parts manufacturing company in Iran. This paper consists of seven parts. In the next section, a review on some previous studies is presented. The model formulation and the procedure for solving it will be presented in sections 3 and 4 respectively. In Section 5, the model is applied to the company under study. Section 6 presents and discusses the experimental results and finally, the conclusion will be presented in section 7.

2. Literature review

Following the work done by Foote and Ravindran [5], various models with different levels of sophistication have been introduced and many researchers tried to present a classification of them. In 1992 Nam and Ogendar [6] conducted a survey of APP techniques and categorized them into these classes: trial and error method, graphical technique, mathematical technique, linear decision rule, search decision rule, management coefficients method, parametric production planning, production switching heuristic, linear programming, goal programming, mixed integer programming, transportation method and simulation model. Three years later, Pan and Kleiner [7], categorized available models into seven classes including informal approach, mathematical model, linear programming model, linear decision rule [8], [9], heuristic technique, the management coefficients model and search procedure using computer simulation. However, another classification of recently used techniques was given in 2004 by Reay-Chen and Tien-Fu [10]. They categorized conventional models of aggregate planning into six divisions: linear programming model, decision rule method, transportation model, management coefficient approach, search decision rule and simulation method. In the case of dealing with uncertain environments, some approaches including stochastic programming, fuzzy programming, stochastic dynamic programming, and robust optimization have been followed as the main approaches to incorporate uncertainty into the production planning process [11].

In cases where we are sure about the linearity of objective functions, linear programming techniques can be used [12]. Kanyalkar and Adil [13] developed an APP model using linear programming for a multi-plant, multi-selling-location problem. The solution serves both detailed and aggregated problems at the same time and in a single step. This scheme claims to eliminate the disadvantages of monotheistic plans (huge computational demand and imprecise forecast) as well as hierarchical plans (having infeasibility and sub-optimality). The model formulation consists of some factors such as storage space limitation, availability of raw materials and production capacity of each plant as well as inventory buffer level. Solution to the sample problem was obtained using GLPK optimizer software.

In newer study, Sillekensa et al. [14] applied a Mixed Integer Linear Programming (MILP) model in order to respond to certain challenges in the production planning process of the automotive industry. The challenges came from the possible flexibility in workforces and corresponding working times in APP. Other researchers applied MILP for other purposes. Christoua et al. [15] presented a MILP model and a new approach of hierarchical decomposition of production planning to generate an aggregate production plan that serves the multi-commodity production and distribution scheduling problem. To improve the hierarchical approach of production planning, Stefansson et al. [16] suggested another new approach based on a "hierarchically structured moving horizon" that can address the problem of considering fixed time horizon and the hypothesis of availability of all data at the time of implementation. Jayaraman [17] proposed a linear programming model called Remanufacturing Aggregate Production Planning (RAPP) to consider the return flow of recoverable materials for manufacturers.

Some researchers incorporated the optimization of operational functions while developing APP. For instance, Vassiliadis et al. [18] studied the interaction between production and maintenance plans and established a MILP model to integrate APP with continuous Markov chain maintenance for a multi-purpose batch plant. However, Suryadiy and Papageorgiouy [19] expanded their work by incorporating preventive maintenance and some aspects of design into the model. Both studies provided the solution to the problem using the branch and bound method.

In many cases where a decision maker is faced with a multi-objective mathematical model, "Goal Programming" techniques have been applied in order to obtain the optimal solution that satisfies all objectives simultaneously [20], [21], [22]. Rifai [23] introduced goal programming as an extension of linear programming, which is commonly applied to deal with multi-objective problems. Based on the idea of Leung et al. [20], this approach can manipulate various scenarios with a high degree of flexibility by means of setting either targets or weights. This study also follows a branch of the goal programming approach in order to obtain the solution to the developed multi-objective APP model. This study provides a way of balancing operational resources with fluctuation of customer demand in a deterministic environment while maintaining the cost and quality at the optimal levels. In addition, the developed APP model can provide a better interaction between procurement and production departments in a company.

3. Mathematical Model

This section presents a multi-objective MILP model, was constructed based on a multi-product, multi-period APP problem of an automotive parts manufacturing company in Iran. The company under study produces three different parts in three separate production lines.

3.1. Operational conditions and assumptions of the model

The operational conditions related to the case study and the assumptions of the model are as follows:

- All input data are assumed to be deterministic during the considered time horizon.
- Production lines are balanced and the required number of workers for each production line has been determined and hired at the beginning of the first period. Therefore, the cost of hiring is not included in the variable costs.
- Regular time production, overtime production, and warehouse space cannot exceed their maximum levels determined by the decision maker.
- Demand can be either satisfied or backordered, but backorders must be fulfilled in the next time period.
- No backorder level at the end of the last period is allowed.
- No subcontracting for all types of products is allowed during the time horizon.
- As the company hires permanent workers, firing hired workers because of fluctuation in demand is not allowable. However, based on the required skill level of workers, they will be trained to obtain the required higher level of qualification in different time periods.
- The number of workers with certain skill level in a specified time period is equal to or greater than the number of workers with the same skill level in the previous time period.
- The cost of Work In Process (WIP) inventory is negligible.
- A time horizon of T, consisting of six monthly planning periods has been considered.

- Two separate warehouses are available; the first one is for storing finished products of types one and two, and the second one is used to store the products of type three.
- Inventory level at the beginning of the first period is assumed to be zero.
- Each production line is assigned to just one type of product.
- All raw material components can be purchased from all suppliers.
- Reject rates and costs of raw materials, purchased from different suppliers are different.
- The company does not pay for rejected raw materials and the associated cost is incurred by the corresponding supplier.
- Salary of workers is not included in the cost of producing one unit of product and is considered separately. Workers with higher skill levels are paid more.

3.2. Parameters definition

The related parameters have been defined as follows:

- T Planning time horizon consisting of six time periods
- t Time period
- m Type of product to be produced in a given time horizon T ($m \in M$)
- k Supplier index ($k \in K$)
- i Components (items) of raw materials ($i \in I$)
- n_{im} Coefficient of using component i in product type m
- h Type of production time (regular (h=1) or overtime (h=2)) ($h \in H$)
- sk Level of workers' skill (ordinary (sk=1), good (sk=2) and excellent (sk=3))
- S_{sk} Salary of workers with skill level of sk
- C_{ik} Cost of component I purchased from supplier k (cost of placing orders and receiving materials)
- θ_{ik} Reject rate of component i purchased from supplier k
- D_{mt} Forecasted demand of product type m in period t
- τ_m Ideal cycle time for producing one unit of product type m
- C^{P}_{mh} Cost of producing one unit of product type m in production time h
- C^{I}_{mt} Inventory carrying cost per unit of product m in period t
- C^{B}_{m} Backorder cost per unit of product m
- C_{ht} Cost to train one worker in period t
- L^{W}_{mt} Number of workers to be assigned for producing product type m in time period t
- Max^{wm1} Maximum warehouse space for storage of product types 1 and 2
- Max^{wm2} Maximum warehouse space for storage of product type 3
- Max^{ht} Maximum allowable regular time (h=1) or overtime (h=2) production in period t
- W_m Required warehouse space per unit of product m

All input data related to each defined parameter will be made available upon request.

3.3. Decision variables

The outputs of the model are:

- P_{mht} Unit of product type m to be produced in production time h (regular or overtime) in time period t
- L^{h}_{mt} Number of workers to be trained in period t for product type m
- B_{mt} Backorder level at the end of period t for product type m

- I_{mt} Available inventory level of product type m at the end of period t
- Q_{ikt} Quantity of component i to be purchased from supplier k in time period t
- L_{skmt} Number of workers with skill level of sk to produce product type m in time period t

3.4. Objective functions

For the process of production planning in the company under study, qualitative objective function has been considered in the model as well. The objective functions of the model are as follows:

Quantitative objective functions:

- Minimization of total cost

$$Min\, Z_1 = \sum_{t=1}^{T}\sum_{m=1}^{M}\sum_{h=1}^{H} C^{p}_{mh} P_{mht} + \sum_{t=1}^{T}\sum_{m=1}^{M} (C^{B}_{m} B_{mt} + C^{I}_{mt} I_{mt}) + \sum_{t=1}^{T}\sum_{m=1}^{M} C_{ht} L^{h}_{mt} + \sum_{t=1}^{T}\sum_{m=1}^{M}\sum_{sk=1}^{SK} S_{sk} L_{skmt} + \sum_{t=1}^{T}\sum_{k=1}^{K}\sum_{i=1}^{I} C_{ik} Q_{ikt} \quad (1)$$

The first element is cost of producing products in regular and overtime production hours, the second element demonstrates backorder cost and inventory cost, the third element is related to cost of training, and the last two elements are costs that the company should incur for salary of workers and purchasing raw materials respectively.

- Maximization of products' quality

$$Min\, Z_2 = \frac{\sum_{i=1}^{I}\sum_{k=1}^{K}\sum_{t=1}^{T} Q_{ikt}\theta_{ik}}{\sum_{i=1}^{I}\sum_{m=1}^{M}\sum_{t=1}^{T} D_{mt} n_{im}} + \frac{\sum_{t=1}^{T}\sum_{sk=1}^{1}\sum_{m=1}^{M} L_{skmt}}{\sum_{m=1}^{M}\sum_{t=1}^{T} L^{W}_{mt}} \quad (2)$$

This objective function consists of two elements. The first element tries to maximize the quality of raw materials by minimizing the quantity of raw materials with higher reject rates to be purchased from different suppliers and the second element focuses on applying higher skilled workers by minimizing the number of lower skilled workers. These two elements have a tremendous effect on the quality of finished products. One can add other elements that are effective on the quality of finished products, depending on the situation of the company under study.

Qualitative objective function:

- Customer service level should be "rather high".

The third objective function is a linguistic term defined by the decision maker. For this qualitative objective function, the linguistic term "customer service level should be rather high" is represented by a membership function [25] which has been defined based on the expectation of the decision maker of the company under study as shown in Figure 1.

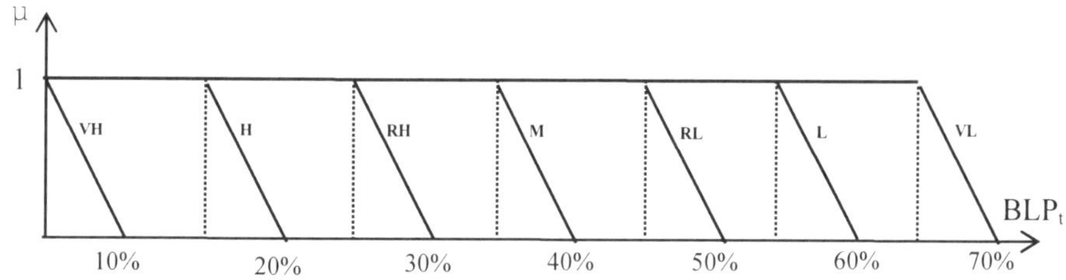

Figure 1. Membership functions for different "Backorder Level Percentages" (BLP).

As depicted in Figure 1, the linguistic set has been defined as: A= {VL (Very Low), L (Low), RL (Rather Low), M (Medium), RH (Rather High), H (High), VH (Very High)}.

So, the membership function corresponding to the defined qualitative objective function can be formulated as:

$$\mu_{BLP_t} = \begin{cases} 1 & \mathrm{BLP}_t \leq 25 \\ \dfrac{30 - BLP_t}{5} & 25 < \mathrm{BLP}_t < 30 \\ 0 & 30 \leq \mathrm{BLP}_t \end{cases} \tag{3}$$

in which, BLP_t is defined as backorder level percentage at the end of period t and is formulated as:

$$\mathrm{BLP}_t = \frac{\text{Backorder level at the end of period t}}{\text{Demand in period t}} \times 100 \tag{4}$$

Therefore, for the third objective function, the following expression can be considered:

$$Z_3 = Max \sum_{t=1}^{T} \frac{30 - BLP_t}{5} \tag{5}$$

3.5. Constraints

Based on the operating conditions and assumptions of the model described earlier, the following constraints have been considered:

$$\sum_{h=1}^{2} P_{mht} = D_{mt} - I_{m(t-1)} - B_{mt} + I_{mt} + B_{m(t-1)};\ \forall t > 1,\ \forall m \tag{6}$$

$$\sum_{h=1}^{2} P_{mht} = D_{mt} - B_{mt} + I_{mt};\ \forall m,\ t = 1 \tag{7}$$

$$\sum_{sk=1}^{3} (L_{skmt} - L_{skm(t-1)}) = L^{h}_{mt};\ \forall t > 1,\ \forall m \tag{8}$$

$$L_{skmt} \geq L_{skm(t-1)};\ \forall t > 1,\ \forall m,\ \forall sk \tag{9}$$

$$\sum_{sk=1}^{3} L_{skmt} = L^{W}_{mt};\ \forall m,\ \forall t \tag{10}$$

$$P_{mht} \tau_m \leq Max^{ht};\ \forall m,\ \forall t,\ \forall h \tag{11}$$

$$B_{mt} \leq D_{mt};\ \forall t,\ \forall m \tag{12}$$

$$B_{mt} = 0;\ \forall m,\ t = 6 \tag{13}$$

$$\sum_{m=1}^{2} W_m I_{mt} \leq Max^{wm1};\ \forall t \tag{14}$$

$$W_m I_{mt} \leq Max^{wm2};\ \forall t,\ m=3 \tag{15}$$

$$\sum_{k=1}^{K} Q_{ikt} = \sum_{m=1}^{M}\sum_{h=1}^{H}(P_{mht} \times n_{im});\ \forall t,\ \forall i \tag{16}$$

Constraints 6 and 7 describe the total quantity of products to be produced in both regular and overtime production hours. Constraint 8 determines the number of workers that should be trained to obtain the required higher level of skill in different periods. Constraint 9 is constructed based on the operational condition of the company that forbids the firing of workers during the time horizon under study. Therefore, once an operator is trained in a certain period, she/he has the same or higher skill level (due to training) in the following periods. Constraint 10 states that the total number of workers with different skill levels assigned to a specific production line should be equal to the dedicated number of workers of that production line in each period. Constraint 11 puts a limitation on the quantities of products to be produced based on the capacity of the corresponding production line. Constraint 12 is a reasonable constraint, which limits the level of backorder quantities in each period; whereas constraint 13 puts a barrier on having any backorder level at the end of the specified time horizon (the time horizon consists of 6 time periods). Constraints 14 and 15 specify the space limitations of the available warehouses for storing all three types of finished products (inventory) at the end of each period. Constraint 16 determines the total quantity of each component (raw material) to be purchased from different suppliers based on the usage coefficients of that component in different products and the quantity of products to be produced in each period.

4. Providing solution to the model

To solve the resulted multi-objective MILP model, the Fuzzy Goal Programming approach has been applied. This approach involves determining a goal value for each objective function, defining a membership function for each of them and finally transforming them to an equivalent single objective using an aggregation operator. Consider an objective function Z_i (with minimization objective), the corresponding membership function is presented in Figure 2.

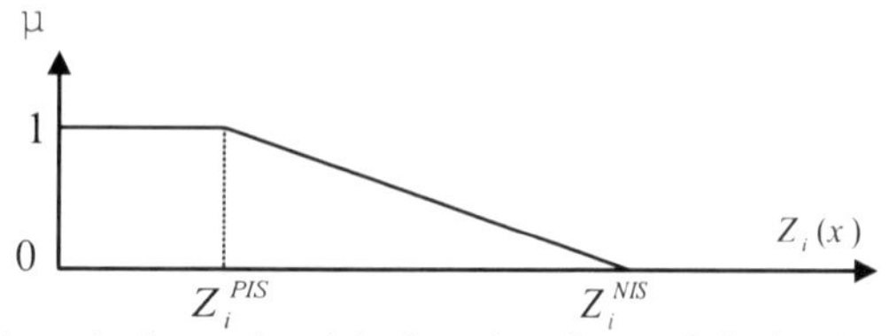

Figure 2. A typical membership function for a minimization objective.

Z_i^{PIS} and Z_i^{NIS} are positive and negative ideal solutions of objective function Z_i respectively. The main problem in defining a membership function is to determine positive and negative ideal solutions with respect to each objective function Z_i. The approach followed by Waiel and Sang [24] can best answer this issue. Based on this approach, the maximum aspiration level of an objective function Z_i (PIS) is the solution obtained by solving the model for each Z_i separately. However, the negative ideal solution of an objective function is obtained by one of the following equations:

$$Z_i^{NIS} = \max\left\{Z_i\left(v_j^*\right);\ i \neq j\right\} \text{ in case of having a minimization objective} \tag{17}$$

$$Z_i^{NIS} = \min\left\{Z_i\left(v_j^*\right);\ i \neq j\right\} \text{ in case of having a maximization objective} \tag{18}$$

where v_j^* is the positive ideal solution of objective function Z_j. Once all membership functions are constructed, the FGP model can be formulated.

5. Applying the model to the company under study

Applying the data gathered from the company under study and following the approach described in section 4, the positive and negative ideal solutions of each objective function were obtained using IBM ILOG CPLEX Optimization Studio (version 12.4) software as shown in Table 1.

Table 1. Pay off table for obtaining positive and negative ideal solutions.

	v_1^*	v_2^*	v_3^*
Z_1	5457721944	9859252719	6604138849
Z_2	6.05	0.25	3.727
Z_3	5.808	4.791	6

Therefore from the table above and equations 17 and 18, it is concluded that:
$Z_1^{NIS}=9859252719,\ Z_2^{NIS}=6.05,\ Z_3^{NIS}=4.791,\ Z_1^{PIS}=5457721944,\ Z_2^{PIS}=0.25,\ Z_3^{PIS}=6.$

The constructed membership functions along with their formulations are as depicted in Figures 3 to 5 and Equations 19 to 21.

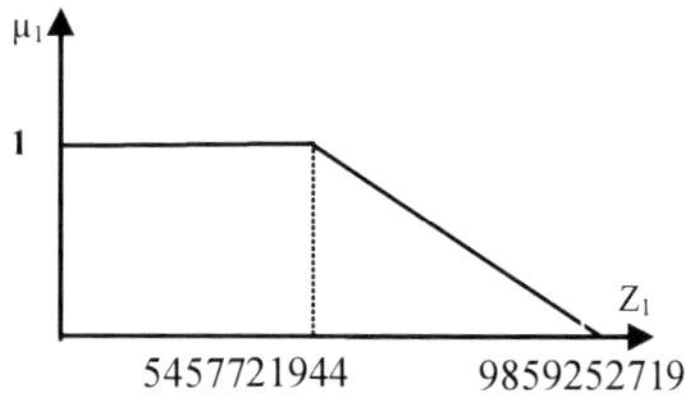

Figure 3. Membership function related to objective function Z_1.

Figure 4. Membership function related to objective function Z_2.

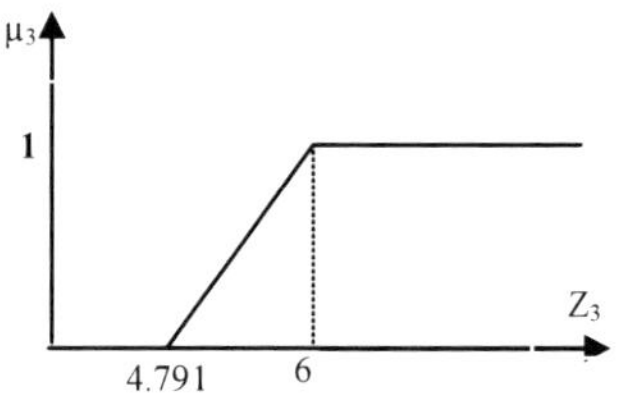

Figure 5. Membership function related to objective function Z_3.

$$\mu_1=\begin{cases}1 & Z_1\le 5457721944\\ \dfrac{9859252719-Z_1}{9859252719-5457721944} & 5457721944<Z_1<9859252719\\ 0 & 9859252719\le Z_1\end{cases}\tag{19}$$

$$\mu_2=\begin{cases}1 & Z_2\le 0.25\\ \dfrac{6.05-Z_2}{6.05-0.25} & 0.25<Z_2<6.05\\ 0 & 6.05\le Z_2\end{cases}\tag{20}$$

$$\mu_2=\begin{cases}1 & 6\le Z_3\\ \dfrac{Z_3-4.791}{6-4.791} & 4.791<Z_3<6\\ 0 & Z_3\le 4.971\end{cases}\tag{21}$$

As the final step for solving the fuzzy goal programming model, the max-min operator of Bellman-Zadeh [26] was applied as an aggregation operator to convert the multi-objective linear model to an equivalent single objective one. Therefore the following model can be constructed as the final model.

$$
\begin{aligned}
&Max\ \Phi \\
&S.t. \\
&\Phi \le \mu_1; \Phi \le \mu_2; \Phi \le \mu_3 \qquad (22)\\
&\mu_1 = \frac{9859252719 - Z_1(x)}{9859252719 - 5457721944};\ \mu_2 = \frac{6.05 - Z_2(x)}{6.05 - 0.25};\ \mu_3 = \frac{Z_3(x) - 4.791}{6 - 4.791}
\end{aligned}
$$

Constraints 6 to 16;
$x \ge 0$
$0 \le \Phi \le 1.$

Using IBM ILOG CPLEX Optimization Studio software (version 12.4), the solutions to the above MILP model were obtained as presented in the following section.

6. Experimental Results

As shown in Table 2, incorporating quality as an objective function in the model imposes an increase of 36510000 units in the total cost. The increase stems from purchasing some raw materials from suppliers with lower reject rates and higher prices and also applying higher skilled workers with higher salaries when considering the objective of quality in the model. Due to space limitation, we just present a general comparison related to the skill levels of workers in Figures 6 and 7. From the figures, it can be found that, in the presence of quality as an objective function, higher qualified workers are required to produce the products.

Although including quality in the APP process imposes additional cost, it has tremendous effect on enhancing customer satisfaction due to the higher reliability of products. Therefore, a firm should perform a trade-off between the additional cost resulted from enhancing the reliability of products and the cost incurred because of customer dissatisfaction and sale losses.

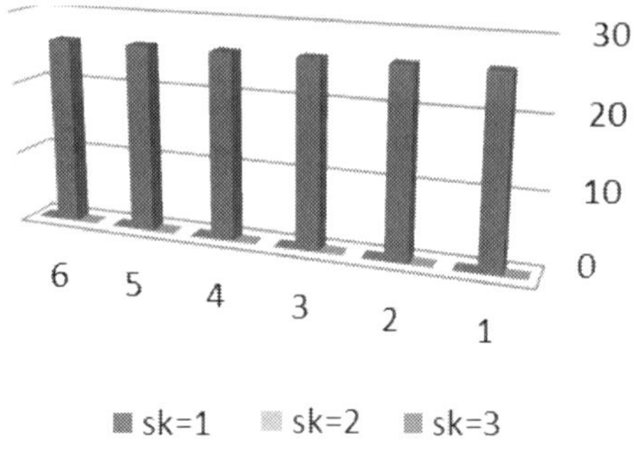

Figure 6. Required number of workers with different skills in different time periods without considering the objective of quality.

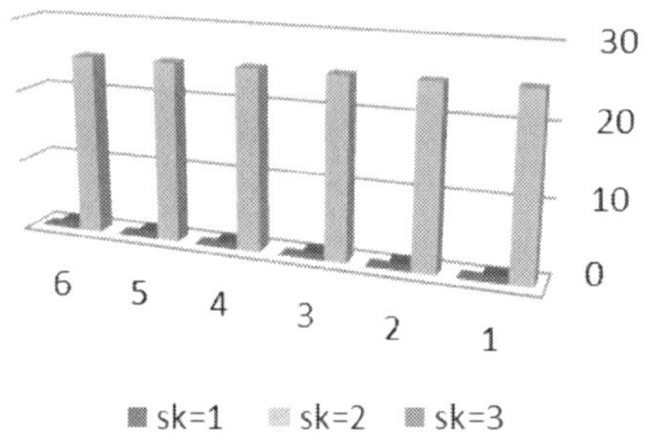

Figure 7. Required number of workers with different skills in different time periods considering the objective of quality.

Table 3 presents the obtained results for other decision variables including inventory, backorder and production levels in different time periods. The results shown in this table obviously indicate that bearing some inventories or/and having overtime production hours are more preferred than having some backorders. The reason lies in the fact that the cost of having backorder is far higher than keeping inventory and/or producing during overtime production hours in the case under study. In addition, the objective of optimizing customer service level has its own effect on obtaining such results.

Table 2. Comparison of obtained results with and without considering quality as an objective.

Considering the objective of quality			Without considering the objective of quality		
Z_1	Z_2	Z_3	Z_1	Z_2	Z_3
5532400000	0.28	6	5495890000	-----	6

Table 3. Obtained results for decision variables of backorder, inventory and production levels.

	Backorder unit			Inventory unit			Production unit					
							$h=1$			$h=2$		
t	m_1	m_2	m_3	m_1	m_2	m_3	m_1	m_2	m_3	m_1	m_2	m_3
1	0	0	0	6606	5440	0	10800	6780	10472	702	0	0
2	0	0	0	5148	0	1950	10800	0	19260	6750	0	0
3	0	0	0	6858	0	0	10800	2340	23040	6750	0	14400
4	0	0	0	0	4580	2938	10800	9700	19994	6750	0	0
5	0	0	0	0	0	9882	10800	0	23040	6480	0	14400
6	0	0	0	0	0	0	3000	6483	23040	0	0	14400

In addition to the results discussed above, the outputs related to the decision variable "required number of workers to be trained in different time periods" recommend having no training (the values are zero in all time periods) for workers during the specified time horizon. In other words, it is more profitable for the firm to hire workers with the required qualification level instead of training them to acquire higher skills.

7. Conclusion

In this study, an attempt has been made to present a multi-objective MILP model that best serves those planners who aim at optimizing cost and quality simultaneously. The application of qualitative objective functions in the model makes it more valuable and practical for those who deal with linguistic terms in the process of production planning. To solve the constructed model, the fuzzy goal programming approach along with the aggregation operator of Bellman-Zadeh was used. The numerical results were obtained using IBM ILOG CPLEX Optimization Studio (version 12.4) software based on the data gathered from an automotive parts manufacturing company. The results show that incorporating quality as an objective into the model can impose additional cost. However the firm should perform a trade-off between the additional cost resulted from producing products with higher quality and the cost incurred because of customer dissatisfaction and sale losses.

For the company under study, the results also show that it is more profitable to have some level of inventories and/or overtime production rather than having backorders. Furthermore, the numerical outputs of the model emphasize on hiring workers with the required qualification level instead of training them to acquire demanded skills.

References

[1] Yan F., D'Amours S. and Beauregard R. 2008 The value of sales and operation in oriented strand board industry with make to order manufacturing system: cross functional integration under deterministic demand and spot market resource *Int. J. Prod. Econ* **115** 189-209.

[2] Chopra S. and Meindl P. 2010 *Supply Chain Management, Strategy, Planning, and Operation* (New Jersey: Pearson education)

[3] Ramezanian R, Rahmani D and Barzinpour F 2011 An aggregate production planning model for two phase production systems: Solving with genetic algorithm and tabu search *Expert Syst. Appl.* **39** 1256–63

[4] Akhil J and Udatta S P 2005 Aggregate production planning for a continuous reconfigurable manufacturing process *Computer & Operations Res.* **32** 1213–36

[5] Foote B. L. and Ravindran A. 1988 Computational feasibility of multi-criteria models of production planning and scheduling *Comput. Ind. Eng.* **15** 129-138

[6] Nam. S. J. and Ogendar N.R. 1992 Aggregate production planning- a survey of models and methodologies. *Eur. J. Oper. Res.* **61** 255-272

[7] Pan L. and Kleiner B.H. 1995 Aggregate planning today. *Work Study* **44** 4- 7

[8] Silva J. Lisboa J. and Huang Ph. 2000 A labor-constrained model for aggregate production planning *Int. J. Prod. Res.* **38 (9)** 2143- 52

[9] Holt CC, Modigliani F and Simmon HA. A. 1955 Linear decision rule for production and employment scheduling *Manage. Sci.* **2 (1)** 1–30

[10] Reay-Chen W. and Tien-Fu L. 2004 Application of fuzzy multi-objective linear programming to aggregate production planning *Comput. Ind. Eng.* **46** 17–41

[11] Mirzapour S.M.J., Al-e-hasheman, Malekly H. and Aryanezhad M.B. 2011 A multi- objective robust optimization model for multi- product, multi-site aggregate production planning in a supply chain under uncertainty *Int. J. Prod. Econ.* **134** 28-42

[12] Kogan K., Khmelnitsky E. and Maimon O. 1998 Balancing facilities in aggregate production planning: make-to-order and make-to-stock environments *Int. J. Prod. Res.* **36(9)** 2585-96

[13] Kanyalkar A. P. and Adil G. K. 2005 An integrated aggregate and detailed planning in a multi-site production environment using linear programming *Int. J. Prod. Res* **43 (20)** 4431–54

[14] Sillekensa Th., Kobersteinb A. and Suhlb L. 2010 Aggregate production planning in the automotive industry with special consideration of workforce flexibility *Int. J. Prod. Res.* **49 (17)** 5055–78

[15] Christoua I.T., Lagodimos A.G. and Lycopoulou D. 2007 Hierarchical production planning for multi-product lines in the beverage industry *Prod. Plann. Control* **18 (5)** 367–376

[16] Stefansson H., Jensson P. and Shah N. 2006 Integration of multi scale planning and scheduling *Comput.-Aided chem. Eng.* **21** 2111-16

[17] Jayaraman V. 2006 Production planning for closed-loop supply chains with product recovery and reuse: an analytical approach. *Int. J. Prod. Res* **44 (5)** 981–998

[18] Vassiliadis C. G., Vassiliadou M. G., Papageorgiou L. G. and Pistikopoulos E. N. 2000 *Simultaneous Maintenance Considerations and Production Planning in Multi-purpose Plants: Proc. Annual Reliability and Maintainability Symposium* (Los Angeles, California, USA, 24-27 January 2000) pp 228–233

[19] Suryadiy H. and Papageorgiouy L. G. 2004. Optimal maintenance planning and crew allocation for multipurpose batch plants *Int. J. Prod. Res.* **42 (2)** 355–377

[20] Leung S.C.H, Shirley S.W.Ch. 2008 A goal programming model for aggregate production planning with resource utilization constraint *Comput. Ind. Eng* **56** 1053–64

[21] Romero C. 2004 A general structure of achievement function for a goal programming model. *Eur. J. Ope*. Res. **153** 675–686

[22] Baykasoglu A. 2001 Aggregate production planning using the multiple objective tabu search *Int. J. Prod. Res.* **39 (16)** 3685-3702

[23] Rifai A. K.1994 A note on the structure of the goal programming model: assessment and evaluation *Int. J. Oper. Prod. Manag.* **16** 40–49

[24] Waiel F.A., Sang M. L., 2006 Interactive fuzzy goal programming for multi-objective transportation problems *Omega* **34** 158 – 166

[25] Ghasemy Yaghin R., Torabi S.A., Fatemi Ghomi S.M.T. 2012 Integrated markdown pricing and aggregate production planning in a two echelon supply chain: A hybrid fuzzy multiple objective approach *Appl. Math. Modell.* 36 (12) 6011-30

[26] Bellman R. Zadeh L. A. 1970 Decision-making in a fuzzy environment *Manage. Sci.* **17** 141–161

Optimization algorithms intended for self-tuning feedwater heater model

P Czop[1] T Barszcz and J Bednarz
AGH University of Science and Technology, Department of Robotics and Mechatronics

E-mail: pczop@agh.edu.pl.

Abstract. This work presents a self-tuning feedwater heater model. This work continues the work on first-principle gray-box methodology applied to diagnostics and condition assessment of power plant components. The objective of this work is to review and benchmark the optimization algorithms regarding the time required to achieve the best model fit to operational power plant data. The paper recommends the most effective algorithm to be used in the model adjustment process.

Introduction

Great advances in control and monitoring systems applied in thermal power plants have created a possibility to develop new efficient methodologies for diagnostics and condition assessment of power unit components. The state-of-the-art gray-box model diagnostic approach, proposed and developed by Barszcz and Czop in [1-2], requires efficient methods to adjust the first-principle models to operational data. One of the most demanding tasks is tuning of first-principle models which involve three categories of parameters: geometrical, physical and phenomenological (e.g. heat exchange coefficients). Geometrical and physical parameters are deduced from construction or operational documentation, while phenomenological ones are adjusted using optimisation routines. A feasibility study of optimisation routines applied in a tuning process of the first-principle is presented in this paper based on a feedwater heater installed in a 225MW coal-fired power block. Heaters installations are typically affected by fouling and corrosion phenomena which may have effect on heat transfer and fluid transportation process [1-2]. Feedwater heaters are typically designed as three-zone heat exchangers with a condensing section, desuperheater and integrated subcooler (Figure 1).

[1] Corresponding author, Al. Mickiewicza 30, 30-059 Cracow, Poland.

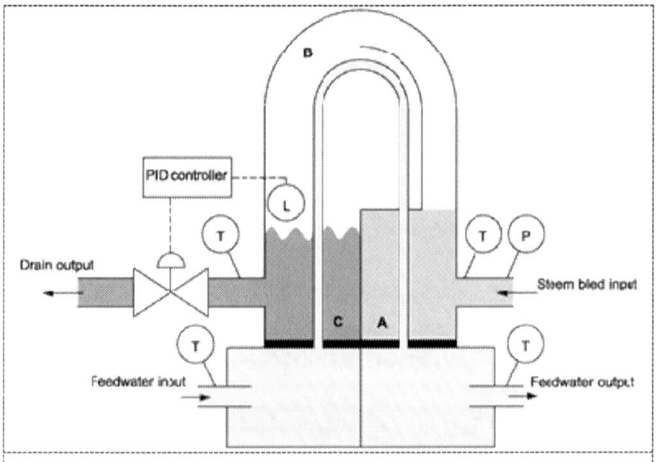

Figure 1. Schematic representation of a three-zone feedwater heater; A: desuperheating area; B: condensing area; C: subcooling area; T: temperature sensor; L: level sensor; P: pressure sensor

The drainage system of the feedwater heater consists of a drain removal path from each heater. The normal drain flow path is cascaded to the next lower stage heater, and the alternate path is diverted to the condenser. When the turbine is loaded at a given rate, steam is allowed to enter the bank of heaters through extraction outlets [1-2].

2. Review the methods to update the model parameters

The process of heater model calibration consists of (i) selection of representative data sets, including a sufficiently broad operating range that the model proved could work correctly using a data fit measure and (ii) adjustment of model parameters to fit a model response to data. The model is represented as a set of non-linear state-space equations formulated in the continuous-time domain. The objective of the estimation is to minimize the error function between the measured signals and model responses by means of an iterative numerical technique. The function describing the error has to be a positive and decreasing function. The procedure for updating model parameters consists of two in-a-loop phases: (i) simulation of a model by solving differential equations numerically in Simulink [4], and (ii) numerical minimization in the parameter space with respect to an error-related criterion function using the Matlab Optimization Toolbox [4]. After each simulation of the model for fixed-length input signals, the simulated output data are sampled and the criterion function is re-evaluated to calculate a new set of model parameters. Detail discussion on model tuning algorithms and validation techniques is given in [3]. A general form of the first-principle model which represents power-block thermodynamic behaviour in the continuous domain is as follows:

$$\begin{aligned} &\frac{d}{dt}x(t) = f\big(t, x(t), u(t), w(t); \theta\big) \\ &y(t) = h\big(t, x(t), u(t), v(t); \theta\big) \\ &x(0) = x_0 \end{aligned} \tag{1}$$

where vector f(.) is a nonlinear, time-varying function of the state vector x(t) and the control vector u(t), while vector h(.) is a nonlinear measurement function; w(t) and v(t) are sequences of independent random variables and θ denotes a vector of unknown parameters. In nonlinear systems, the state vectors and the measurements vectors are not Gaussian distributed. The predictor resulting from the model (1) takes the form

$$\hat{y}(t \mid \theta) = g(t, Z^{t-1}; \theta) \tag{2}$$

while the prediction error equation has the form

$$\varepsilon(t, \theta) = y(t) - g(t, Z^{t-1}; \theta) \tag{3}$$

The sum of squared errors is used as an error criterion. This problem is known in numerical analysis as "the nonlinear least-square problem" [5].

Three methods of minimizing the error function are feasible as presented in Table 1. These are (i) direct search, (ii) first-order and (iii) second-order methods.

Table 1. Summary of algorithms minimizing the error function.

Method	Algorithm
Direct search methods	Simplex
	Pattern search
First-order (gradient)	Gradient (steepest-descent)
Second-order (gradient and Hessian)	Gauss-Newton
	Quasi-Newton
	Trust-regions
	Damped Gauss-Newton (Adaptive Gauss-Newton)
	Levenberg-Marquardt

Direct search methods use only the value of the function to find the minimum. The first-order method uses the information provided by the first derivative (gradient) of the error function, while the second-order method uses both, information regarding the first and the second order derivatives (gradient and Hessian form) of the error function. First- and second-order methods update the estimates of the error function iteratively according to the general scheme

$$\hat{\theta}^{(k+1)} = \hat{\theta}^{(k)} + \alpha_k \cdot f^{(k)} \tag{4}$$

where $\hat{\theta}^{(k)}$ denotes the k-th iteration point in the search, f(k) is a search direction based on the information about $V(\theta)$ acquired at previous iterations, and the sequence of positive scalars αi determine velocity in which the value of $V(\theta)$ is decreased. In general, the function f(k) has the form

$$f^{(i)} = -\left[V''(\hat{\theta}^{(i)})\right]^{-1} V'(\hat{\theta}^{(i)})^T \tag{5}$$

The first and second derivatives are obtained as follows

$$V_N'(\theta) = -\frac{1}{N}\sum_{t=1}^{N}\varepsilon(t,\theta)\psi(t,\theta) \tag{6}$$

$$V_N''(\theta) = \frac{1}{N}\sum_{t=1}^{N}\psi(t,\theta)\psi^T(t,\theta) - \frac{1}{N}\sum_{t=1}^{N}\psi'(t,\theta)\varepsilon(t,\theta) \tag{7}$$

where $\psi'(t,\theta)$ denotes the d-by-d Hessian matrix of $\varepsilon(t,\theta)$, d = dim(θ) [5]. A general family of iterative search routines is described by the scheme

$$\hat{\theta}_N^{(k+1)} = \hat{\theta}_N^{(k)} + \mu_N^{(k+1)}\left[R_N^{(k)}\right]^{-1} V_N'(\hat{\theta}_N^{(k)}, Z^N) \tag{8}$$

where $R_N^{(k)}$ is a d-by-d matrix, which modifies the search direction (changes the search direction according to user preferences), while the step size $\mu_N^{(k+1)}$, is chosen so that it is satisfied in the inequality

$$V_N(\hat{\theta}_N^{(k+1)}, Z^N) < V_N(\hat{\theta}_N^{(k)}, Z^N) \tag{9}$$

It is often necessary to use variable step length to improve the convergence of the algorithm; a sequence of scalars $\mu_N^{(k+1)}$ is used to control the step length. If the term $\mu_N^{(k+1)} \neq 1$, equation (8) describes the so-called damped Gauss-Newton algorithm.

3. The feedwater heater model

The family of feedwater heater models was extensively discussed in [1]. It has four phenomenological parameters, which are the heat exchange coefficients [1]. Two parameters describe the heat exchange process between steam-condensate volumes and feedwater volumes, while the remaining two parameters describe the heat exchange process between steam-condensate volumes and the metal housing [1]. The two latter parameters have to be neglected due to lack of measurements of the temperature of the metal housing. This simplification neglects a bypass energy flow to/through the metal housing. This simplification is allowed since the heat exchange process through the heater wall is significantly smaller compared to heat exchange through metal tubes with feedwater. This allows the model simplifications to be verified experimentally. The control system could not be directly reconstructed in the simulation, due to its complexity and limited relevance to the functionality required in the model (e.g. trip logic). Hence, the module maintaining a constant level of the condensate inside the heater was simplified using a PID controller model. The parameters of the first-principle model are updated based on operational data. The total length of each sequence corresponds to 12 hours and includes daily fluctuation in the produced electrical energy form 225MW to 160MW.

Performance of the procedure of adjusting model parameters is evaluated by analysing the value of Pearson's product-moment correlation coefficient between the measured and simulated values of the N-sample-long output signals (Figure 2 bottom graph). For each output signal, the correlation coefficient is computed separately; in the particular case of the model discussed in this paper, the output signals were the temperature of the condensate and the temperature of the feedwater. The cconvergence trajectory plots were also investigated to ensure a stable trend towards constant values of the parameters, which correspond to convergence towards the minimum of the criterion function, within less than 10 iterations (Figure 2 top and mid graph).

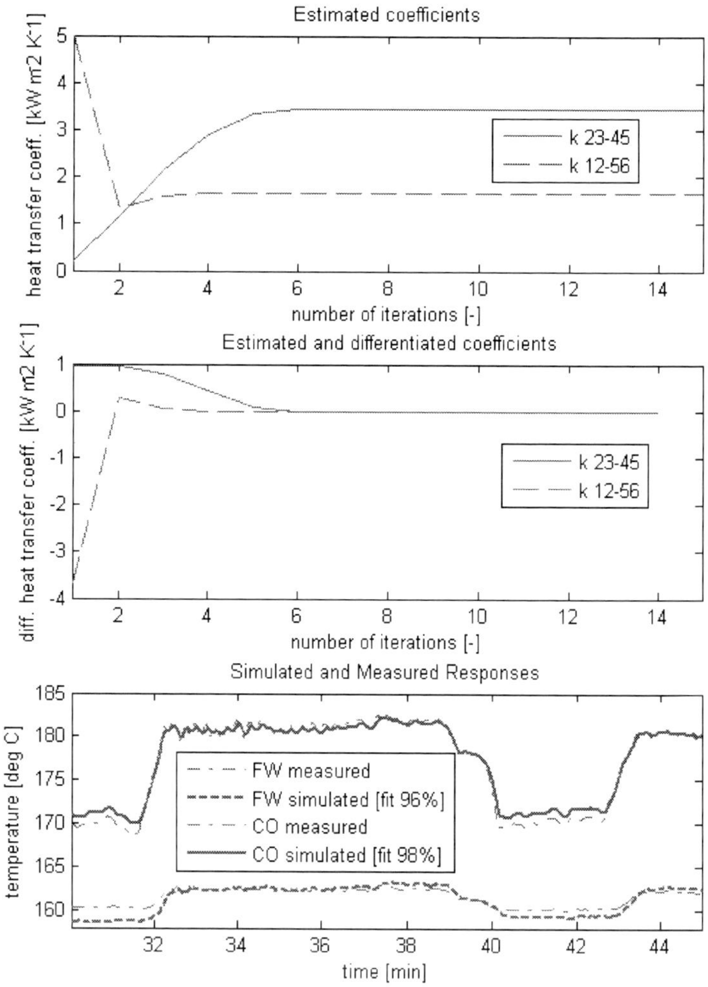

Figure 2. Exemplary fitting accuracy of models to operational data as a function of time (legend: FW – feedwater, CO – condensate).

4. The benchmark of the model tuning methods

The validation tests were conducted using a few minimization algorithms representing three categories of methods as summarized in Table 1, i.e. direct search, first-order, and second-order. The sum of squared errors was used as the error criterion to evaluate the fit of the model to operational data. A direct search approach [5], which does not use numerical or analytic gradients, is represented by two algorithms: (i) pattern search involving different sub-routines to perform a search and the simplex search method. Both methods are recommended for optimization tasks that contain discontinuities. The pattern search method has the following choices: no search step, Latin hypercube, a pattern search

(GPS or MADS methods), or genetic algorithm [6]. The simplex search method uses the Nelder-Mead algorithm [7]. The first-order and the second-order gradient methods are represented by the gradient descent and nonlinear least squares optimization algorithms from the Optimization Matlab Toolbox [4]. Second-order methods are more effective compared to the first-order ones when the second-order information is obtainable in an analytical way. The calculation of second-order information, using numerical differentiation, is computationally expensive. The second-order method is available in two versions: (i) a subspace trust-region [8-9], and (ii) Levenberg-Marquardt method. A subspace trust-region method approximates only a certain region (the so-called trust region) of the objective function with a model function (often a quadratic) as opposed to the entire function as with the Newton–Raphson optimization algorithm. When an adequate model of the objective function is found within the trust region, then the region is expanded. If the approximation is poor then the region is contracted. Table 2 shows the evaluation results of different algorithms, for the stopping criterion chosen as the exceeded number of iterations or approached stable value for the parameters, i.e. the difference between the current and the previous value is zero.

Table 2. Benchmark results.

Method	Algorithm	Matlab function	Condensate temperature fit [-]	Feedwater temperature fit [-]	T [min]
direct search	Nelder-Mead	fminsearch	0. 865	0.603	35.1
first-order	Steepest-descent	fmincon	0. 874	0.514	235.8
second-order	Levenberg-Marquardt	lsqnonlin	0. 897	0.615	48.2
second-order	Trust-region	lsqnonlin	0. 894	0.632	27.9

The results that provide the best trade-off between data fit to operational data and calculation time were achieved for direct methods, while the worst results were for the first-order method. Nevertheless, if iteration limits for algorithms are replaced by the parameter tolerance value bounds, the convergence is faster for second-order methods. The pattern search algorithms provide step-like convergence showing a few significant variations in the values of parameters before they approach the stable value. This variation requires many additional iterations compared to gradient methods, thus the stable value is not approached as fast as in the gradient methods. Moreover, the pattern search function in Matlab ignores the initial values of model parameters (Latin hypercube) by replacing the guess values with pre-calculated values which differ significantly. In turn, the iteration process took much more time than was required for second-order algorithms.

5. Conclusions

The objective of this work is to review and benchmark the optimization algorithms regarding the time required to achieve the best model fit to operational power plant data. The model was a feedwater heater first-principle model developed in [1]. Table 1 compares different optimisation methods used for adjusting the model to operational data. The second-order trust-region method was shown to be the most optimal for the problem of updating parameters of the heater model developed in [1-2].

Acknowledgements

The author gratefully acknowledges the financial support of the research project N504 493239, funded by the Polish Ministry of Science (MNiI).

Nomenclature

x(t) - state vector used in a general representation of a first-principle model
u(t) - control vector used in a general representation of a first-principle model
y(t) - output vector used in a general representation of a first-principle model
w(t), v(t) - the sequences of independent random variables
θ - the unknown parameter vector used in a general representation of a first-principle model

Abbreviations

FP - First-Principle
DD - Data-Driven
GPS - Generalized Pattern Search
ODE - Ordinary Differential Equations
MADS - Mesh Adaptive Direct Search

References

[1] Barszcz T Czop P 2011 A Feedwater Heater Model Intended for Model-Based Diagnostics of Power Plant Installations, *Applied Thermal Engineering* vol 31 (8-9) pp 1357-1367
[2] Barszcz T and Czop P 2012 Presentation of a Virtual Power Plant environment and its application with combined first-principle and data-driven models intended for the diagnostics of a power plant - part I *SIMULATION Transactions of The Society for Modeling and Simulation International, Simulation* vol 88(2) pp139-166
[3] Barszcz T Czop P 2011 Estimation of feedwater heater parameters based on a grey-box approach. *International Journal of Applied Mathematics and Computer Science* vol 21 (4), pp 703-715
[4] MATHWORKS 2007 *Inc., Matlab/Simulink* (Natick US)
[5] Ljung L 1999 *System identification - Theory for the User* (Prentice-Hall, USA)
[6] Torczon VJ 1997 On the Convergence of Pattern Search Algorithms *SIAM Journal on Optimization* vol 7(1), pp 1–25
[7] Lagarias JC Reeds JA Wright MH and Wright PE 1998 Convergence Properties of the Nelder-Mead Simplex Method in Low Dimensions, *SIAM Journal of Optimization* vol 9(1) pp. 112-147
[8] Coleman TF and Li Y 1996 An Interior, Trust Region Approach for Nonlinear Minimization Subject to Bounds *SIAM Journal on Optimization* vol 6(2) pp 418–445
[9] Coleman TF and Li Y 1994 On the Convergence of Reflective Newton Methods for Large-Scale Nonlinear Minimization Subject to Bounds *Mathematical Programming* vol 67(2) pp 189-224

Approximation methods applied in assessment of valve system fatigue failure

G Wszołek[1] P Czop[2] A Skrobol[1]

[1] Institute of Engineering Processes Automation and Integrated Manufacturing Systems, Silesian University of Technology
[2] AGH University of Science and Technology, Department of Robotics and Mechatronics

E-mail: grzegorz.wszolek@gmail.com

Abstract. This paper proposes an analytical tool that supports the design process of a disc spring valve system used in hydraulic dampers. The proposed analytical tool obtains a key design characteristic of a valve, which is the flow rate and the corresponding maximum stress level in the stack of plates. The tool is prepared based on the cases produced by a first-principle model using a finite element approach. The finite element model was calibrated based on experimental results to provide accurate results in the entire range of input parameters.

1. Introduction

This paper proposes an analytical tool that supports the design process of a disc spring valve system used in hydraulic dampers [1]. Such a valve is also representative of a broader family of valve systems used in many other industrial applications. The tool accelerates the valve system design process. The paper considers a valve system that combines a number of circular metal plates, referred to further in the paper as a stack of plates. The proposed analytical tool obtains a key design characteristic of a valve, which is the flow rate and the corresponding maximum stress level in the stack of plates, as a function of pressure load. The stress level enables the ranking of valve settings for the durability criterion. The calculation process is based on the response of a nonlinear data-driven model which approximates the a priori simulated cases to cover the complete ranges of input design parameters, namely the number of plates, their thickness and diameter. The cases are produced by a first-principle model using a finite element approach. The model was calibrated based on experimental results to provide accurate results in the entire range of input parameters. The advantage of the analytical tool is the possibility to immediately provide a pressure-stress-flow characteristic of a valve instead of repeating time-consuming calculations for each new setting of input parameters. The objectives of the paper are as follows: to (i) prepare the Artificial Neural Networks (ANNs) to approximate simulation data, (ii) validate the ANNs and (iii) demonstrate an example of the rapid design of a hydraulic damper valve system with the use of the ANNs.

[1] Konarksiego 18, 44-100 Gliwice, Poland

2. Problem description

A conventional hydraulic damper that contains a specific type of valve system consisting of a number of circular metal plates, referred to further in the paper as a stack of plates (see Figure 1), is taken into consideration. This valve is part of a piston that is kinematically forced to move up and down in a liquid-filled cylinder. During its movement, the piston divides the cylinder space into two chambers – the vacuum chamber and the compression chamber. The flow of the fluid from one cylinder chamber to the other is restricted by the valve – stack of plates. The pressure difference between the two parts of the cylinder creates the resultant force that acts as a pressure on the bottom surface of the lower plate of the valve system. This resultant pressure causes bending of the stack of plates and valve opening. It should be noticed that the height of the opening depends not only on the pressure difference between the vacuum and compression chambers, but on the stiffness of the stack of plates as well. The opening height is then the first major characteristic of the valve system because, by controlling the fluid flow, it indirectly controls the damping force of the hydraulic damper [2]. The second characteristic of the valve is its durability, which denotes the number of 'open-close' cycles that can be performed by the valve before its destruction. It was earlier described by [3-4] that the maximum stress level that exists in the plates when the applied load creates deflection of the stack limits the durability of the system. A compression valve that works only in one direction is presented in Figure 1a. The direction of the piston movement inside the cylinder and the direction of the fluid flow are marked by arrows. Interested readers can find a more detailed description of the valve system design in [2].

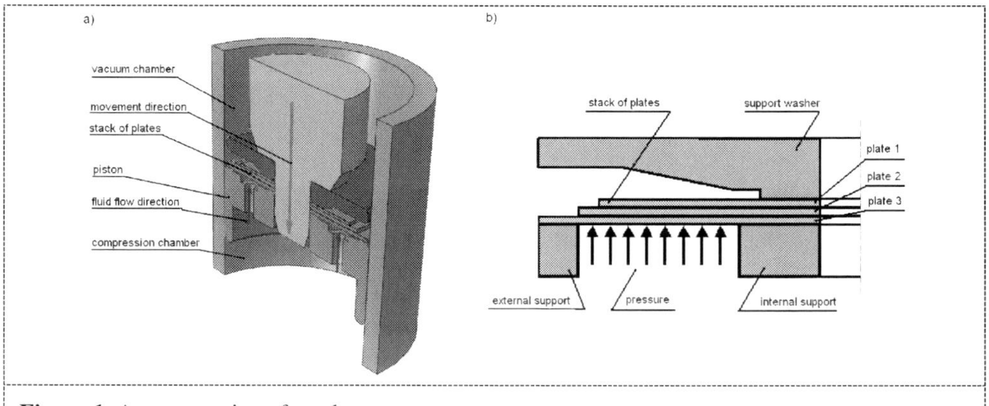

Figure 1. A cross section of a valve system

Designing the valve stack of plates is then an optimization process where the criteria are both damping force and valve system durability characterized by maximum stress level in the stack of plates [4]. To facilitate the valve system design process, a model of the system can be created to simulate the behaviour of the stack of plates under applied loads. This model should be capable of reproducing the essential properties of a valve system during operation in a shock absorber, i.e. stack opening and stress level in the plates.

A special non-linear Finite Element Model, further referred to as the FE Model, has been developed by [4]. This proposal uses Finite Element discretization to model the behaviour of a circular plate stack. The model allows users to improve understanding of the contribution of a stack of plates' corresponding components, including the plastic deformation phenomenon and simulation of complex cases. As an output of the analysis, the height of the stack opening as well as the maximum value of von Mises stress level in each plate can be obtained. The model was prepared in the Abaqus software environment. The FE Model was correlated with experimental data and its ability to solve the problem presented above was confirmed [4]. The mentioned FE Model, however, has several disadvantages, such as (i) the long time required for model preparation and (ii) for analysis, especially when the analysis is performed in several steps, (iii) the high cost of commercial software required to perform the analysis and (iv) special requirements concerning the workstation on which the software can run. All these features become significant when optimization or identification tasks have to be performed, since the fitness function evaluation requires a large number of plate stack configurations to be solved. The hydraulic valve optimization process can be illustrated by the scheme presented in Figure 2.

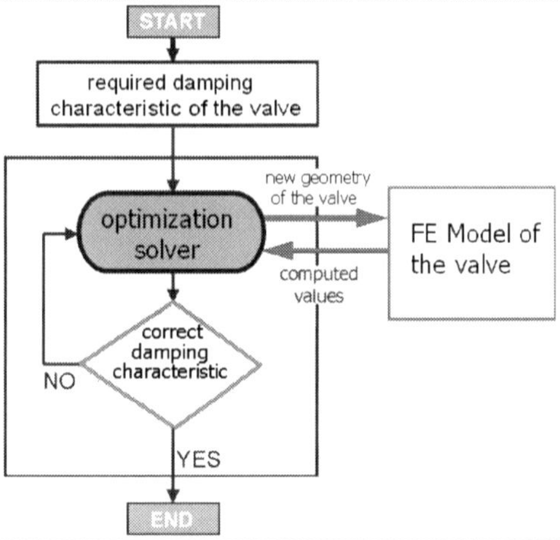

Figure 2. Scheme of the conventional and rapid valve system design process

The main disadvantage of such an approach is that the time needed for simulating the behavior of the valve is very long. A very time consuming optimization process occupies the workstation and commercial software for a long period of time – analysis of 20 different settings of the valve requires more than 30 min. During one optimization run there is a need of analysis of more than 500 different valve designs what extends the optimization process to many hours. Because of what mentioned above, there is a need of speed up of the valve system optimization process and, if possible, decrease the negative impact of disadvantages mentioned above. One way to accelerate the optimization process is to improve the optimization algorithm fitness function evaluation by replacing the FE Model solution of the boundary-value problem by an approximation of the solution as presented in Figure 3.

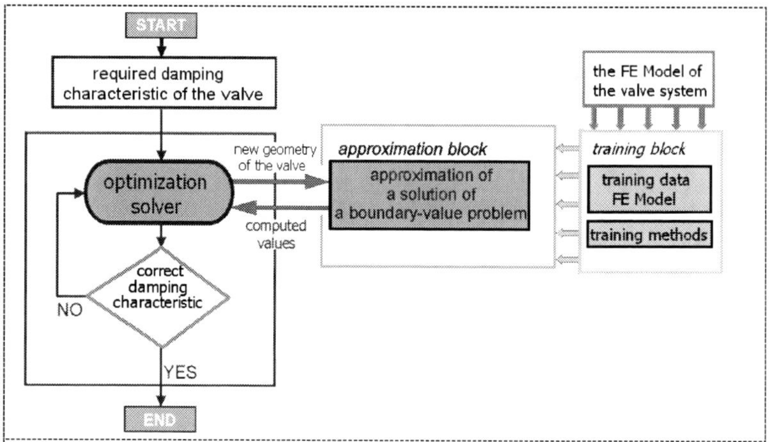

Figure 3. Scheme of a process of rapid design of the valve system

3. Approximation method

Different types of artificial neural networks (ANNs) are well recognized as an efficient and easy-handling tool for several engineering tasks, like identification of defects, optimization or classification tool, etc. The ANNs, especially Feed Forward Neural Networks (FFNNs), are already identified as very good approximators of different boundary-value problems [5].

The Artificial Neural Networks (ANNs) have been chosen as an approximator because of several advantages:

(i) They are a universal tool that is applicable for a wide range of engineering problems (linear, non-linear).

(ii) They are able to generalize; it is possible to train the ANN in such a way that the interpolation and approximation error are at the same level.

(iii) ANNs are easy to operate and can be used by engineers in industry conditions.

(iv) ANNs are resistant to the poor quality of the data set that participates in their creation (e.g. non-uniform data distribution within the range).

(v) It is possible to re-train existing ANNs in the future and update their parameters according to the newest measurements/calculations of the valve system.

The feed forward neural networks with sigmoid activation functions trained using back propagation algorithms are used in this paper for approximation of the key design characteristics of the valve system. The FFNNs were created and trained using the Matlab Neural Network Toolbox, thus interested readers can find more details concerning the architecture and training method in [6]. The Approximation Tool approximates the stack behaviour in order to obtain a key design characteristic of a valve, in other words, the opening height and the corresponding maximum stress level in the stack of plates as a function of pressure load. The stack of plates already described in previous paragraphs is considered. The model shown in Figure 1b consists of a pyramidal stack of three circular plates. The stack is clamped between the support washer and the bottom internal and external support. The thicknesses of all plates as well as the outer diameter of Plate 1 and Plate 2 are the design variables. The outer diameter of the bottom plate (Plate 3) is constant. These design variables, as well as the value of pressure applied to the bottom surface of Plate 3, are the inputs of the Approximation Tool. The outputs are the opening height of the stack of plates and maximum values of von Mises stress in all three plates. Thus the length of the input vector of the approximation model is N=6 and the length of the output vector is L=4. The vector x is given to the input of the approximation model and, as a result of the approximation, the output vector y is obtained. Vectors x and y are then described by Eq. (1).

$$\mathbf{x} = [x_1, \ldots, x_N]; \qquad y = [y_1, \ldots, y_L]; \tag{1}$$

where N is the number of input and L is the number of output variables. The input and output vectors collected together create the pair of vectors. The total number of 4178 pairs is prepared to train the networks. The data obtained with use of a non-linear FE Model, already described in previous paragraphs, is randomly divided into three sets. The size of these sets is presented in Table 1. The range of design variables is presented in Table 2. The input and output vectors collected in the training set, and which are presented to the network during the training process, adjust the network parameters. During the learning process there is a risk of 'over-learning' of the network. This means the interpolation error of the training set will decrease but, at the same time, the generalization abilities of the ANN will become worse. To avoid 'over-learning' the approximation error of the validation set is observed. The third set of data – the testing set – is prepared to give neutral information about the approximation abilities of the ANN. This data has no effect on the training process.

Table 1. Size of sets used for training and testing of FFNNs

	Training set	Validation set	Testing set	TOTAL
No of pairs	2801	692	685	4178

Table 2. Design variables for approximation task

	Thickness Plate 1 [mm]	Thickness Plate 2 [mm]	Thickness Plate 3 [mm]	Out. diam. Plate 1 [mm]	Out. diam. Plate 2 [mm]	Press. [bar]
min/max	0.15/0.35	0.15/0.35	0.15/0.35	10/22	10/22	2.5/50

The multi input – single output (MISO) neural networks are chosen to approximate the given problem. For each design variable of the valve system (opening height, maximum stress level in Plates 1, 2 and 3), a separate approximator is prepared. Although this attempt increases the complexity of the training process, it has several advantages:

- The approximator of each design characteristic is independent from other approximators.
- This allows tuning of the approximator responsible for opening height approximation without influencing other networks.
- This tuning can be performed by additional runs of the training process.

When all outputs are given by one approximator (MIMO), such additional runs, which should improve the response for one output, can lead to 'over-learning' of the rest of the network outputs, and improving the error of the response of one output can negatively affect others. The results of approximation of the opening height, max von Mises stress in Plates 1, 2 and 3 are presented in Table 3. The FFNNs were trained by the Levenberg-Marquardt method [5-6]. This method is a combination of the gradient descent method and Newton's Method. The approximator performance is judged with the help of Mean Square Errors. Three different errors for each approximation problem are measured – training set error (MSEL), validation set error (MSEV) and testing set error (MSET). It should be mentioned that the level of the MSE depends on the range of the values that are compared. Thus for opening height approximation, when the range of output variables is between 0.01 and 1.03mm, the MSE values will be at a very low level. In cases of stress approximation, the MSE value will be at a higher level.

Table 3. The results of approximation of the opening height, max von Mises stress in Plates 1, 2 and 3

	Opening height approximation	Stress in Plate 1 approximation	Stress in Plate 2 approximation	Stress in Plate 3 approximation
MSEL	1.32e-6	1.03e2	2.68e2	4.84e2
MSEV	1.34e-6	1.21e2	3.34e2	5.84e2
MSET	1.43e-6	1.62e2	3.00e2	5.41e2

4. Case study

The goal is to design a stack of plates of a hydraulic damper valve system to obtain a pressure-opening curve within the range presented in Figure 4.

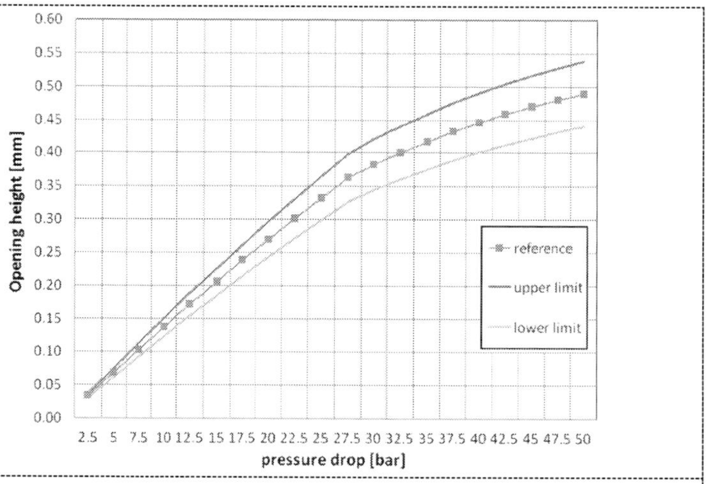

Figure 4. The required opening height – pressure drop characteristic of a damper valve system with 20 points used for fitness function evaluation

The pressure-opening curve was transferred from the pressure-flow characteristic according the formula (2)

$$q = Cd \cdot A \cdot \sqrt{\frac{2p}{\rho}} \tag{2}$$

Where $Cd = 0.6$ is a discharge coefficient, $\rho=960\ kg/m^3$ is fluid density and A is the flow area given by equation (3)

$$A = 2\pi R \cdot H \tag{3}$$

Where R is the radius of the piston valve and H is the opening height. The Matlab Optimization Toolbox™ is used in the optimization process. The Genetic Algorithm (GA) is chosen since this optimization solver can handle optimization problems with nonlinear, linear, and bound constraints, and does not require fitness functions to be differentiable or continuous [5]. The optimization solver changes the design variables, which are the diameters of Plate 1 and 2 and thickness of all three plates within the range presented in Table 1. To evaluate the fitness function, the opening height is calculated by the Approximation Tool based on the design of the stack of plates proposed by the optimization solver. The FFNN described in the previous paragraph is used by the Approximation Tool to approximate the opening height for 20 pressure values from 2.5 bar to 50 bar. The optimization process was performed several times in order to avoid the influence of starting point selection on the optimization results. The time needed to perform one optimization process is up to 1 min. Thus the cost of the optimization task repetition is negligible. Figure 5 shows the pressure-opening curves for different designs found in six optimization processes.

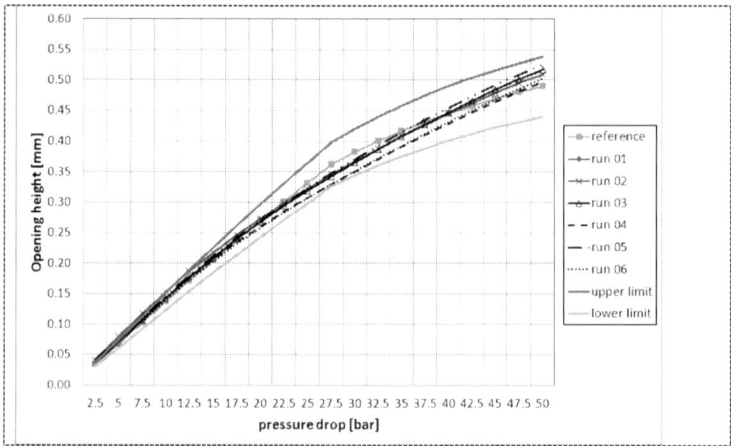

Figure 5. The achived opening height – pressure drop characteristic of a damper valve system with 20 points used for fitness function evaluation

It can be noticed that the results of optimization are within the specified range with one exception – the pressure-flow curve of Run 02 is outside the specification for small flows. For all selected designs, the maximum von Mises stress level was calculated in order to find the best design in terms of fatigue durability. Based on the results of the optimization process and stress calculations, the best option – Run 05 – was chosen. Results achieved with the help of the ANNs were then compared with the results obtained by the FE Model. Figure 6a presents the opening height – the pressure drop characteristic of the valve stack design No. 5 approximated by FFNN and calculated by the FE Model.

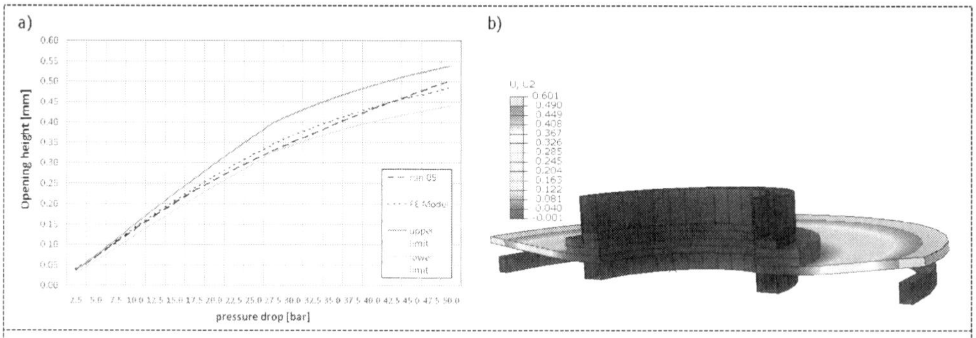

Figure 6. a) The opening height – pressure drop characteristic calculated by the FE Model and approximated by FFNN; b) Deformation of the valve stack calculated by the FE Model

Deformation of the valve stack, provided by the FE Model, is presented in Figure 6b. All six designs found by the optimization algorithm, are presented in Table 4.

Table 4. Predicted maximum von Mises stress and failure rate of each design

	Stress in Plate 1 [MPa]	Stress in Plate 2 [MPa]	Stress in Plate 3 [MPa]	Fatigue (the worst case) [number of cycles]
Run 01	2 810	5 430	2 734	N/A
Run 02	992	1 775	2 717	N/A
Run 03	1 392	1 314	2 099	0.2 mln
Run 04	1 091	3 785	3 287	N/A
Run 05	842	1 534	1 580	1.6 mln
Run 06	958	1 630	3 191	N/A

Fatigue is the progressive and localized structural damage of a valve disc that occurs when a disc is subjected to cyclic pressure loading. High cycle fatigue strength, above 10^3 cycles and typically below 10^8 cycles, is considered in case of valve systems. This type of fatigue is described by stress-based parameters which mainly contribute to crack or disc damage. The maximum stress values should be less than the ultimate tensile stress limit, and may be below the yield stress limit of the material. Fatigue failure appeared in the form of propagating cracks from the middle of the outer ring of the intake valve due to the very high pressures (Figure 7).

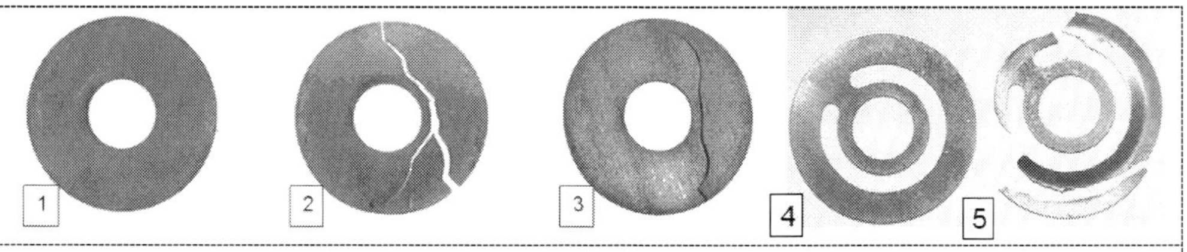

Figure 7. Different failures of hydraulic damper valve system plates

A fatigue prediction model neglects microphysical disc properties and disc contact conditions for simplification. Disc fatigue is a complex transient phenomenon influenced by disc shape (solid or perforated material), geometry (thickness), temperature, surface finish (anisotropy after rollover), presence of oxidizing or insert chemicals, thermal and chemical treatment, residual stresses, contact (fretting), dislocations etc. There are linear and nonlinear fatigue hypotheses which are applicable to a constant load cycles. Fatigue life at variable amplitude is predicted by using material properties from constant amplitude laboratory tests obtained with linear or nonlinear damage accumulation hypothesis. Such tests were performed using a servo-hydraulic equipement to load the disc at specific displacmenet to reporduce the conditions in a hydraulic damper operation. In order to characterize the sensitivity to fatigue, a large number of disc stacks were tested at continuous sinusoidal cycling with a constant amplitude and frequency. The obtained Wöhler diagram (Figure 8) allows to correlate the nuumber of cycles until disc damage and the stress in the plates (Table 4).

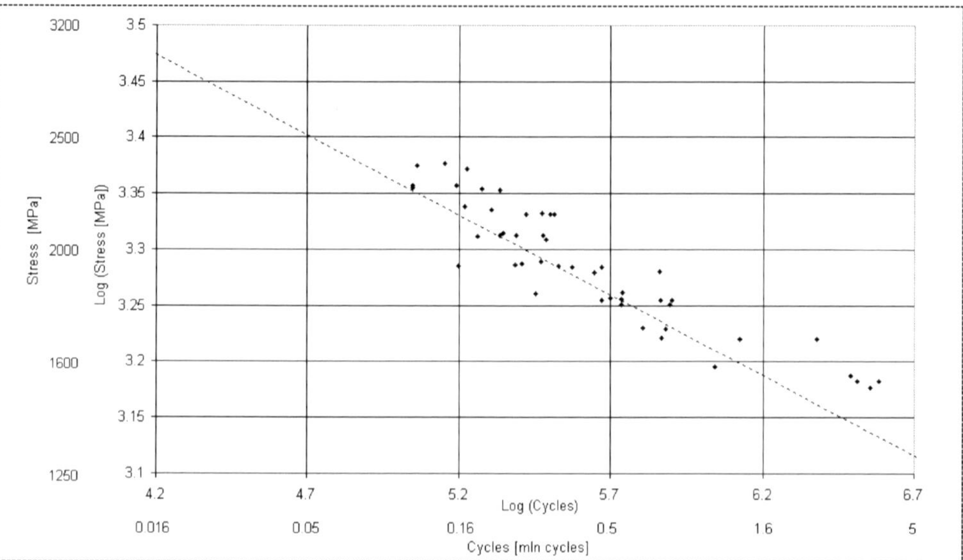

Figure 8. Experimental Wöhler curves

Figure 9 shows that the criterion of disc damage is the first drop in the maximum observed force measurmet.

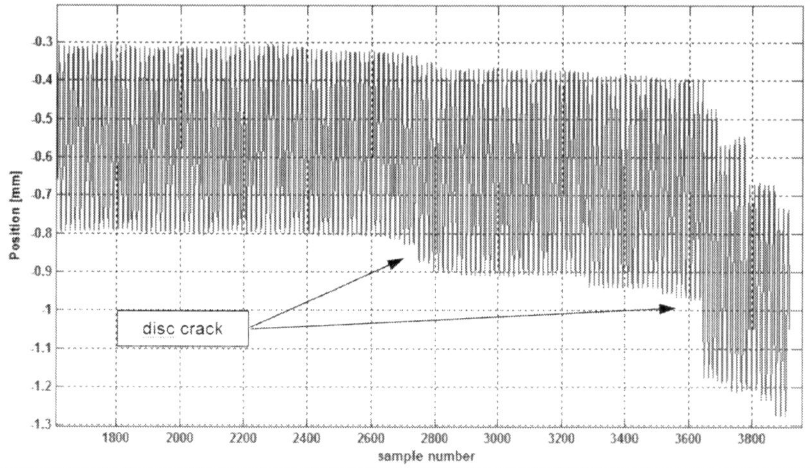

Figure 9. Displacement signal of an exemplary end phase of a test

5. Conclusions

The paper describes the proposal, implementation and example of application of the Feed Forward Neural Networks intended to support the design process of a disc spring valve system used in hydraulic dampers. The FFNNs allow users to obtain a key design characteristic of such a valve, in other words, flow rate and the corresponding maximum stress level in the stack of plates as a function of pressure load, plate thickness and diameter. The paper considers a valve system that consists of a combination of washers configured as a stack of circular plates. The number of plates, their diameter and thickness directly affect the damping force of a shock absorber and the stress level in plates. Due to a significant non-linear relationship between the applied load and opening of the stack, the most popular are computational models of the stack which use Finite Element (FE) spatial discretization and assume axial symmetry of the stack of plates. The Finite Element Method (FEM) provides accurate results, however, it has two disadvantages: (i) the time required for computation and model preparation (pre-processing) strongly depends on the complexity of the model geometry. Additionally, (ii) the FE analysis itself is time consuming – analysis of 20 different settings of the valve requires more than 30 min. of analysis (excluding model preparation and result reading). The model requires a commercial license and an efficient workstation to complete simulation in a reasonable amount of time. The license needs to be maintained even if it is not used for simulation purposes. A proposal of the authors is to improve the design process and availability of simulation results by replacing FEM on-line calculations with the use of a data model approximating simulated cases. This allows computationally expensive analyses to be avoided by construction of an approximation model that will be able to replace high-fidelity code during optimization. This paper proposes the final step in fatigue risk assessment of a valve system. The obtained stress is mapped into the fatigue domain, to predict the number of cycles until disc damage conditions. The validation is performed based on the approximation error of the training and validation set, and generalization abilities measured by MSE of the testing set of data. A valve design process depends on fitness function minimization, which was evaluated using the FFNN. The valve stack designs, selected during the optimization process, were then ranked with respect to durability. A final design that fulfils all requirements was indicated.

Acknowledgement

This work was supported by Polish Ministry of Science (MNiI) [research project N N504 494239]

References

[1] Czop P Skrobol A Sławik D Wszolek G 2012 A nonlinear, data-driven model applied in the design process of disc-spring valve systems used in hydraulic dampers, *SIMULATION Transactions of The Society for Modeling and Simulation International* Published online before print April 19, 2012, doi: 10.1177/0037549712441976

[2] Dixon J C 2007 *The shock absorber handbook* (Wiley England)

[3] Czop P Slawik D Sliwa P Wszolek G 2009 Circular plates theory applied to modeling of intake valves used in shock absorbers. *Journal of Achievements in Materials and Manufacturing Engineering* 33(2) pp 173–180

[4] Czop P Slawik D Sliwa P 2010 Static validation of a model of a disc valve system used in shock absorbers. *International Journal of Vehicle Design* 53(4) pp 317–342

[5] Burczynski T Skrobol 2005 A Coupled evolutionary algorithm and artificial neural network in defects identification. *In: Bathe, K.J. (ed.) Third MIT Conf. On Computational Fluid and Solid Mechanics* pp 122–1226

[6] Math-Works 2011 *Matlab-Simulink documentation* http://www.mathworks.com

The Optical Properties of CdSe Quantum Dots by Using Spray-Atomization Method

C H Rosmani[1,2], M Rusop[1,3], S Abdullah[2,3]
[1]Nano-SciTech Centre (NST), Institute of Science
[2]Faculty of Applied Sciences
[3]Nano-Electronic Centre (NET), Faculty of Electrical Engineering
Universiti Teknologi MARA (UiTM), 40450 Shah Alam Selangor D.E. Malaysia

E-mail: rosmaniey88@gmail.com

Abstract.

Cadmium Selenide (CdSe) quantum dots (QDs) is inorganic material by using spray-atomization method which is the novelty to find out the optical properties for the CdSe QDs. The Selenium (Se) precursor and Cadmium (Cd) precursor were prepared first. Se precursor by using sodium sulfite aqueous was mixed with selenium (Se) powder. For Cd precursor was used cadmium chloride (CdCI) as the Cd precursor. From previous research, CdSe QDs was obtained by using capping agent such as tri-n-octylphosphine oxide (TOPO) and trioctylphosphine (TOP). These capping agent are hazardous to environment and human. By using spray-atomization method it is more safe and economically. The photoluminescence (PL) was used to investigate the optical properties and to investigate the energy band gap from PL result. The field emission scanning electron microscopy (FESEM) was used to know the surface morphology of CdSe QDs. By PL result, the energy band gap was calculate and the comparison was investigate between the size of particle and the energy band gap. This important in this paper is to investigate the optical properties of CdSe QDs by using sprays-atomization method and to relate with the particle size.

Introduction

Cadmium Selenide (CdSe) is inorganic semiconductor material, CdSe was very potential in optoelectronic devices application such as solar cells, Light Emitted Diode (LED) and biosensors [1, 2]. From the previous research, CdSe QDs was clarified by using Chemical Vapour Deposition (CVD) method, spray-pyrolysis, hydrothermal and electrochemical [3-5]. CdSe QDs was used in many field in solar cells application, Light Emitted Diode (LED) and sensors [6, 7] . Quantum dots were well known as the smallest particle which have in nano size. Nano size basically in around 100 nm which is also known as nanoparticle or quantum dots

[1] rosmaniey88@yahoo.com

(QDs), but QDs have unique properties when it is smaller the photoluminescence properties also change from red shift to blue shift [6, 8].

When the colour changed, the wavelength also changed. CdSe QDs have quantized energy which has 1.7 eV direct energy band gap [7-9]. The emission and absorption wavelengths of quantum dots depend on the nanoparticles diameter and called it as quantum confinement [10]. Quantum dots consists of narrow fluorescence and size distribution and has high quantum yield of emission [11].

From previous research , CdSe QDs were prepared by using in many capping agent as the stable colloidal of CdSe QDs to prevent irreversible aggregation, coagulation and fusion of the quantum dots [12-14]. The capping agent from previous basically paraffin liquid, oleic acid [14], vegetables oil [15] and tri-n-octylphosphine oxide (TOPO) [16]. TOPO chemical was hazard, unsafe and not friendly to environment and human [17, 18].

Material

Selenium powder was purchased from Sigma Aldrich with purity 99.5%, Cadmium Chloride (CdCI) was purchased from Sigma Aldrich, Sodium sulfite and ammonium hydroxide was purchased from Sigma Aldrich.

Experimental Set-Up

The glass substrates were washed by using acetone and methanol. In this project, Se precursor must be prepared first .The preparation of Se precursor; 8.0 g of Se powder was mixed in excess 0.2 M of sodium sulfite aqueous solution to completely dissolve Se powder and was heated at 70 ^{0}C in for 8 h with magnetic stirring. For Cd precursor, the 0.1M of cadmium chloride (CdCI) solution was added to 0.1 M of sodium selenosulfate solution. Finally the pH solution was titrated to 12.0 by ammonium hydroxide solution to get stable clear solution. The mixture between Se precursor and Cd precursor were added to become Cadmium Selenide (Case). The solution was using spray–atomization method which is the parameter was variables from 10 min to 50 min. The characterization in this project was used FESEM for surface morphology and photoluminescence spectroscopy (PL) for optical properties. The droplet size was depend on many parameter such as flow rate, nozzle structure and spray distance. But in this paper the main parameter was time of spray at certain time to optimize the optical properties of CdSe QDs.

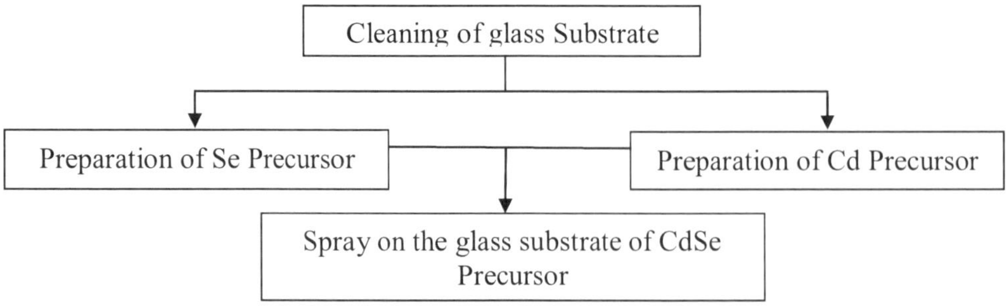

Diagram 1 The Flow Chart of The Overall Experiment

Result and Discussion

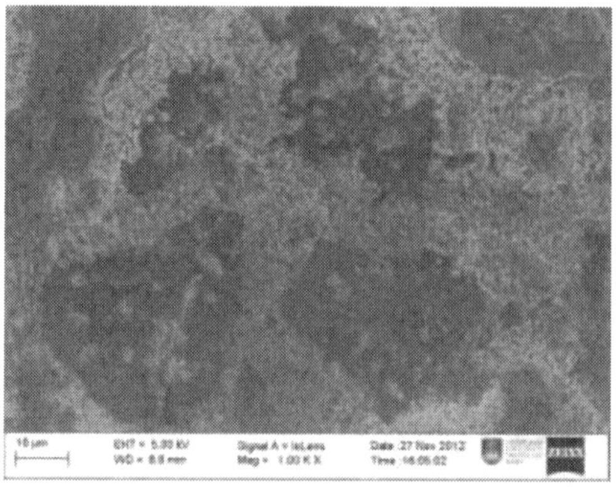

(a)

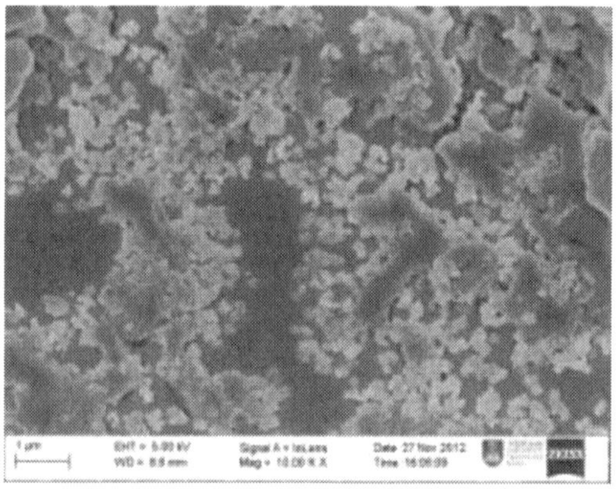

(b)

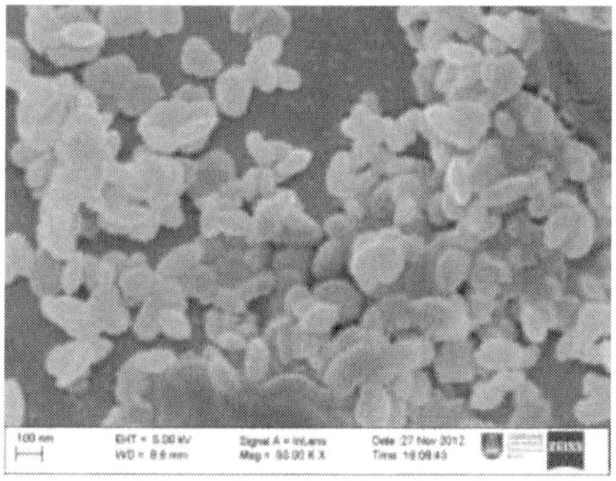

(c)

Figure 1 The FESEM image for CdSe QDs at 25 min of spray process.

To study the surface morphology and structures, the field emission scanning electron microscopy (FESEM) was performed. The Figure 1 is for surface morphology by using FESEM with magnification 1kx, 10kx and 50kx. This magnification from the overall surface with magnification 1kx and was zoom in 10k and 50kx to know the size and shape of structures of CdSe QDs. The size of particle become small when the time of spray-atomization increase, this is because the growth of particle increase when the time increase and the size of particle become small [3]. From the previous study, the particle was growth when the time of spray increases. The CdSe precursor was spray on the glass substrate at 10 s and was left for 20 s for to form crystalline CdSe QDs [3]. When sodium selenosulfite was used as a Se source to produce CdSe QDs via spray-atomization, sulphate was inevitable formed which could contribute to crystalline size of CdSe Qds [3]. Morphology was zoom at different magnification because to know the surface morphology of CdSe QDs at different view.

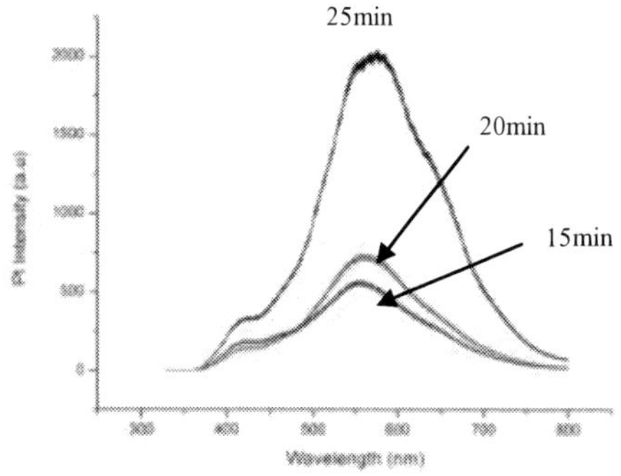

Figure 2 Photoluminescence result from time 15 min, 20 min and 25 min.

To study optical properties the CdSe QDs, CdSe QDs was observing by using photoluminescence spectroscopy (PL), the higher peak was observed at 25 min, second at 20 min and third 15 min. The PL shows the spectra of the QDs at room temperature with laser source 325 nm. The experimental result show that the quantum dots samples have different intensity and wavelength effects of the radius difference. As the reaction time increase the intensity of PL increases and the shift start to change from left to right which is start to red shift. The intensity at 25 min was higher than other because of the particle growth higher than other and the size of particle was increases [3]. From Figure 2, the wavelength was increase from 15 min to 25 min from 556 nm to 575 nm and as the time increase from 15 min to 25 min, the shift was changed from left to right, which was toward to red shift . Red shift has higher wavelength and higher in size. The intensity was optimize at time 15 min to 25 min

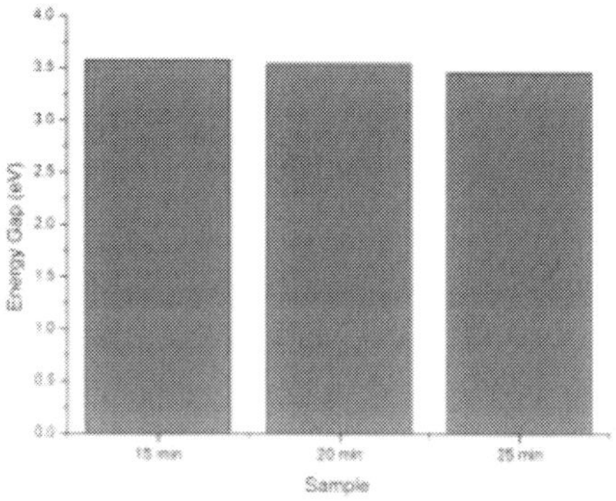

Figure 3 The energy gap for each parameter at different time

Time (min)	Particle Size (nm)	Energy Band Gap (eV)
15	63.21	3.58
20	69.66	3.54
25	70.81	3.46

Table 1 The comparison of particle size and energy band gap for CdSe QDs in different parameter.

From the peak of PL result, and the Figure 3 show the energy band gap was calculated to know the relation between photoluminescence and energy gap. The energy band gap was calculated and plot with the diagram Figure 3 and Table 1. From the result, when time of spray increases the energy gap was decrease. From the theory, when energy gap decrease the nanoparticles size were also increase. It is related with the wavelength from PL result the wavelength was shift from left to right which is the wavelength increases. As the wavelength increase, the energy gap increases because energy band gap inversely proportional with wavelength. The result showed the particle size was increase when time of spray-atomization increases. As the size of nanoparticle was reduced, the role of surface increases [19]. From the result above, when particle size become smaller, the energy gap were increase. From quantum theory, when the size of particle smaller, the energy gap will be increase [20].

Conclusion

CdSe QDs can be obtained by using spray-atomization method which was using the time as the variables parameter. The optical properties of CdSe QDs were defined by photoluminescence (PL) result and surface morphology by using FESEM. From the all result, it can be state that the time of spray of CdSe was gave the effect on the size and optical properties to CdSe QDs. The particle size of CdSe quantum dots was increase as the time of spray was increase, as the particles growth very fast.

Acknowledgement

I would like to express my gratitude to Faculty Applied Science and RMI, UiTM for the financial and facilities support for this work.

References

1. Singh, S., et al., 2011 CdSe nanoparticles grown via radiolytic methods in aqueous solutions *Radiation Physics and Chemistry* **80**(6) 736-741.
2. Chen, S., et al.,2009 Trioctylphosphine as Both Solvent and Stabilizer to Synthesize CdS Nanorods *Nanoscale Res Lett* **4**(10) 1159-1165.
3. Im, S.H., Y.H. Lee, and S.I. Seok,2010 Photoelectrochemical solar cells fabricated from porous CdSe and CdS layers *Electrochimica Acta* **55**(20) 5665-5669.
4. Wang, L., et al., 2010 High quality zinc-blende CdSe nanocrystals synthesized in a hexadecylamine–oleic acid–paraffin liquid mixture *Materials Chemistry and Physics* **120**(1) 54-60.
5. Wang, J. and H. Han, 2010 Hydrothermal synthesis of high-quality type-II CdTe/CdSe quantum dots with near-infrared fluorescence *J Colloid Interface Sci* **351**(1) 83-7.
6. P.Nemec, et al.,2000 Light-controlled growth of CdSe nanocrystalline films prepared by chemical deposition.
7. Araki, Y., et al.,2007 Green light emitting diodes with CdSe quantum dots
8. Zhou, X., et al.,2007 Photoluminescence of CdSe and CdSe/CdO·nH2O core/shell nanoparticles prepared in aqueous solution *Optical Materials* **29**(8) 1048-1054.
9. Takeshi Ota, et al.,2000 Micro-photoluminescence from CdSe quantum dots *Journal of Crystal Growth* **214/215** 778-781.
10. Mi, W., et al., 2012 Temperature dependent synthesis and optical properties of CdSe quantum dots *Ceramics International.*
11. Yordanov, G.G., H. Yoshimura, and C.D. Dushkin, 2008 Fine control of the growth and optical properties of CdSe quantum dots by varying the amount of stearic acid in a liquid paraffin matrix *Colloids and Surfaces A: Physicochemical and Engineering Aspects* **322**(1-3) 177-182.
12. Sharma, S.N., et al.,2012 Ligand-dependent transient absorption studies of hybrid polymer:CdSe quantum dot composites *Solar Energy Materials and Solar Cells* **100** 6-15.
13. Neeleshwar, S., et al., 2005 Size-dependent properties of CdSe quantum dot *Phs. ysical Review B*, **71**(20).
14. Liao, Y.-f. and W.-j. Li,2007 Synthesis of CdSe quantum dots via paraffin liquid and oleic acid. *Journal of Zhejiang University SCIENCE A* **9**(1) 133-136.
15. Sun, M., et al.,2009 Shape evolution of CdSe nanocrystals in vegetable oils: A synergistic effect of selenium precursor *Colloids and Surfaces A: Physicochemical and Engineering Aspects* **350**(1-3) 91-100.
16. Yang, D., Q. Chen, and S. Xu,2007 Synthesis of CdSe/CdS with a simple non-TOP-based route *Journal of Luminescence* **126**(2) 853-858.
17. G.Yordanov, G., C. D.Dushkin, and E. Adachi,2008 Early time ripening during the growth of CdSe nanocrytals in liquid paraffin *Colloids and Surfaces A: Physicochem. Eng. Aspects* **316**: 37-45.
18. Yang, D., Q. Chen, and S. Xu, 2007 Synthesis of CdSe/CdS with a simple non-TOP-based route
19. Mao, H., et al., 2005 Photoluminescence investigation of CdSe quantum dots and the surface state effect *Physica E: Low-dimensional Systems and Nanostructures* **27**(1-2) 124-128.
20. D.Patidar, et al.,2008 ENERGY BAND GAP AND CONDUCTIVITY MEASUREMENT OF CdSe THIN FILMS *Chalcogenide Letters*, 2008. **5**(2) 21-25.

Uncertainty in the analysis of the overall equipment effectiveness on the shop floor

M P Rößler[1] and E Abele

Institute of Production Management, Technology and Machine Tools, Technische Universität Darmstadt, 64287 Darmstadt, Germany

E-mail: roessler@ptw.tu-darmstadt.de

Abstract. In this article an approach will be presented which supports transparency regarding the effectiveness of manufacturing equipment by combining the fuzzy set theory with the method of the overall equipment effectiveness analysis. One of the key principles of lean production and also a fundamental task in production optimization projects is the prior analysis of the current state of a production system by the use of key performance indicators to derive possible future states. The current state of the art in overall equipment effectiveness analysis is usually performed by cumulating different machine states by means of decentralized data collection without the consideration of uncertainty. In manual data collection or semi-automated plant data collection systems the quality of derived data often diverges and leads optimization teams to distorted conclusions about the real optimization potential of manufacturing equipment. The method discussed in this paper is to help practitioners to get more reliable results in the analysis phase and so better results of optimization projects. Under consideration of a case study obtained results are discussed.

1. Introduction

The methods of lean manufacturing derived from the Toyota production system were developed to optimize the efficiency of production systems by constantly removing everything from the value stream which adds no direct value to the customer [1, 2]. So one of the results is to free up capacity and so to create more value from existing resources with fewer additional costs [3]. To achieve this ambition in literature three "inhibitors" that indicate why the performance of a production system can be impaired, have to be overcome [4]. The first inhibitor is variability, secondly waste and thirdly inflexibility, where the driving force behind the development of lean management was and still is the elimination of waste [5, 6, 7]. In view of the above an enhanced method will be presented in this article, which will help practitioners to identify and uncover waste as well as variability on the shop floor concerning manufacturing equipment to ensure high level productivity and low costs.

The appraisal of the overall equipment effectiveness (OEE) measure of a specific manufacturing equipment usually is performed by systematically analyzing a specific time period of the working time of that device, machine or plant and so to derive conclusions how well this equipment is running and to identify losses that impair the effectiveness [8]. This analysis should first be applied to the bottlenecks of a value stream that affect throughput or at costly areas of manufacturing systems [9]. By increasing the effectiveness of a bottleneck, the capacity of the whole value stream increases with

[1] To whom any correspondence should be addressed.

it until a new bottleneck occurs. This identification of e. g. a bottleneck machine can be conducted via value stream analysis which is not part of this article.

2. The overall equipment effectiveness analysis

2.1. Background and calculation

The concept of systematically measuring the overall effectiveness of a plant firstly was described by the Japan Institute of Plant Maintenance (JIPM) as one of five strategies of the Total Productive Maintenance (TPM). As a main goal here the maximization of the plant effectiveness is stated. This is to be done by examining the inputs to the production processes and identifying and eliminating the losses associated with these inputs to maximize the output. To have a measure how effective a plant is running the determination of the equipment effectiveness was stated [10].

The OEE is defined as the percentage of the planned production time, in which defect free products are produced. To identify the reasons for deviations to a value of 100 %, three categories of losses have to be considered [11]:

$$\text{OEE rate} = \text{availability rate} \cdot \text{performance rate} \cdot \text{quality rate} \tag{1}$$

$$= \frac{\text{full speed defect free operating time [h]}}{\text{planned production time [h]}}$$

The following table 1 gives an overview over the categories of losses and main factors of influence addressed by the analysis of the OEE. These losses were adapted to discrete part manufacturers, which feature slightly differences to the original classification of the eight losses in the OEE out of the field of process industry; therefore see Suzuki [10].

Table 1. Loss categories and main factors of influence to the determination of the overall equipment effectiveness.

Loss categories	Main influencing factors
Availability	1. Failure
	2. Changeover
Performance	3. Small stops
	4. Reduced speed
Quality	5. Scrap
	6. Rework

2.2. Data acquisition and analysis

The elevation of the data necessary to perform an OEE analysis can be performed by the use of plant data collection systems, or for less capital-intensive equipment manually by analysing the operator's logs [8].

When firstly implementing and performing an OEE analysis in a specific production environment for instance in the course of an optimization project at a discrete part manufacturer, data necessary usually is collected by the operators at the specific manufacturing equipment over a representative, defined period of time (e. g. one week). In that case operators have to document every relevant activity at their manufacturing equipment such as tool breakdown, changeover, oil refill, scrap, rework, etc. and also the time period associated with that activity in a semi-formal log. Regarding the operator's recording process, different facts and circumstances have to be considered [12]:

- The data recording process itself leads to disruptions in operation cycles,
- The raising of the frequency of short-term events, such as order confirmations, minor stops or answering short questions can overburden the operator,

- The necessary data must be recorded immediately after completion of a task. Therefore the covered tasks must not be too delicate or too complex,
- Recording errors arise from self-written logs rather than from foreign surveys.

The current state of the art in OEE analysis with the use of manually collected data is done as illustrated in figure 1. This following approach results out of the fact described above, that not all short-term events could be captured by the operators. These short-term events are performance losses such as small stops and reduced speed. So these losses usually have to be calculated. The quality loss for scrap also has to be calculated by multiplication of the number of defect parts with the corresponding cycle time.

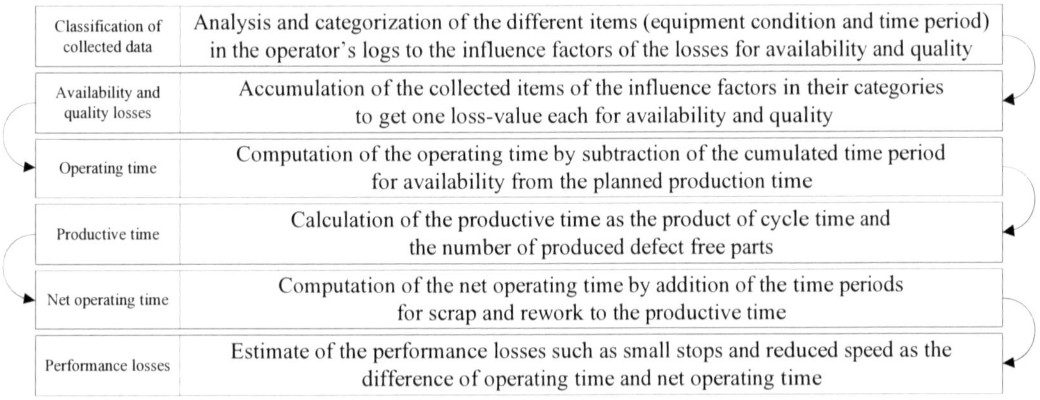

Figure 1. Current state of the art approach in OEE analysis for using manually collected data.

As a result of this approach the so called OEE water fall chart can be derived, see figure 2 [13]. Here the performance fractions for small stops and reduced speed were presumed to be 1:1. The data used as a basis to derive that chart was collected in a case study performed at the Center for Industrial Productivity at the TU Darmstadt, see paragraph 4.2. Therefore a special software tool was developed.

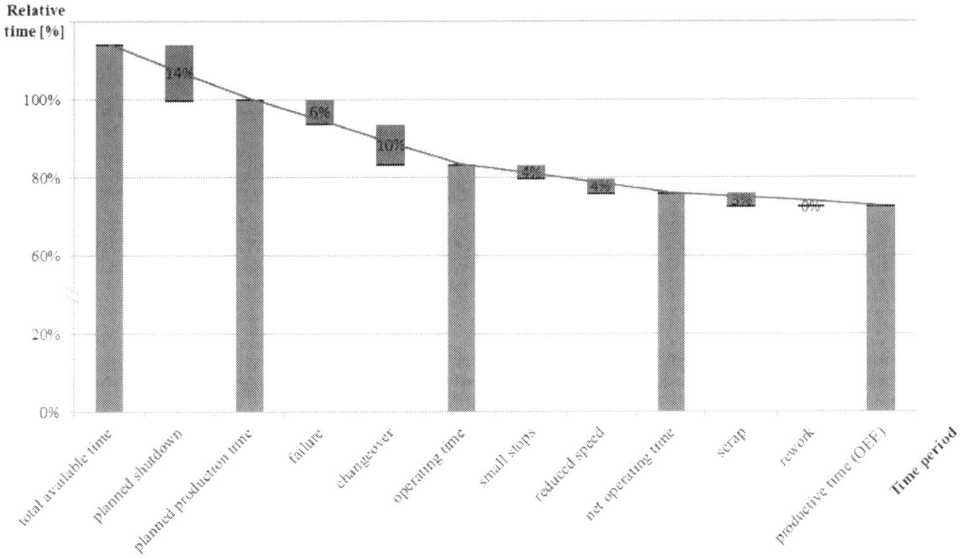

Figure 2. Visualization of an OEE analysis using the water fall chart.

After the visualization of collected data with the water fall chart usually optimization potential is deduced. As a result of this analysis it is common to initiate specific optimization projects in the potential fields of loss, e.g. optimizing the changeover processes with the method of Single Minute Exchange of Die (SMED) [14].

2.3. Limitations of the OEE analysis

The method described above is used when data capturing takes place manually. If a plant data collection system is utilized instead, the OEE analysis can be performed semi-automatically. As a result of different influences the key performance indicator (KPI) of the OEE underlies statistical fluctuations. Therefore several examples can be stated:

- Human perturbation in the recording process: Especially time-referenced vagueness e.g. caused by predefined resolution in form sheets or the data collection process itself,
- Discrepancies in classification of losses: If for example a plant data collection system is employed at a machine tool without additional input from the operator this system is not in position to decide whether an interruption is either performed for changeover or caused by a tool defect. So for deriving the data necessary and in the right quality to perform an OEE analysis mostly human interaction is inevitable,
- Recognition issues, esp. of quality-related problems directly at the process: In the case of not using a plant data collection system a possible reason why fuzziness must be included in the process is that in practice not all quality defects which could lead to scrap are recognized directly at for example a machine. A not obvious quality defect caused by those instable processes often firstly is recognized at quality gates or in the course of customer reclamation after the OEE analysis is done already. In that case the operator doesn't take a specific note in his log and so the KPI gets distorted,
- Missing recognition of sporadic occurring characteristically constitutional changes: Another distorting factor to the robustness of the OEE analysis, no matter what data collection principle is used is that the OEE analysis is based upon a specific time period that is considered during data collection. In this period perhaps characteristically constitutional changes of the analyzed manufacturing equipment do not occur. These effects are not considered in the analysis if the time period is chosen too short,
- Further the proportioning of all losses comprised in an analysis also depends on the actual product mix produced on the emblazed manufacturing equipment.

In the case of a semi-automatically data collection for instance a manufacturing execution system with the capability of interfacing machine controls could be used. Such systems usually are capable of deriving OEE analyses [15].

As carried out in this paragraph, no matter what method of data acquisition is implemented, the data used to derive OEE analyses contains fuzziness and uncertainty in the process of data collection. To include this variability in the process of the OEE analysis, the use of fuzzy logic is suggested here.

3. Fuzzy set theory

As one possible methodology regarding the mathematical handling of uncertainty, fuzzy logic has been developed, based on the theory of fuzzy sets according to Zadeh. Zadeh defines a fuzzy set as a class of objects, having a range of degrees of membership. This set is characterized by a membership function $\mu(x)$ which assigns each object class a membership degree of between zero and one. That means that an object is not only located inside or outside of the set, but that this also may be included "a little" in the set. The membership to this set is gradually staged [16, 17]. Two fuzzy numbers respectively their membership functions are shown in figure 3.

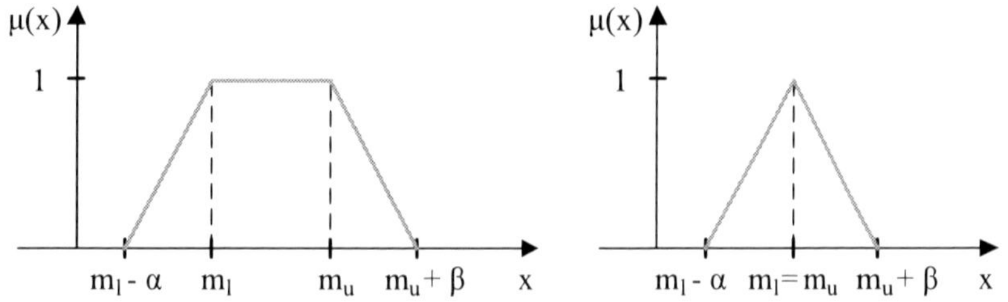

Figure 3. Visualization of a trapezoidal and a triangular fuzzy number type LR.

The intervals described by these membership functions can be specified in different ways. In literature the standard type and the LR type are distinguished, at which in this paper the LR type will be used further. The formal notation for fuzzy numbers is displayed with a tilde over the literal [18, 19, 20]. In this context the membership functions of fuzzy numbers (e. g. $\tilde{M}$, $\tilde{N}$) can be described using a lower value (m_l, n_l), an upper value (m_u, n_u), a lower spread (α, γ) and an upper spread (β, δ).

$$\tilde{M}_{LR}=(m_l;m_u;\alpha;\beta)_{LR} \tag{2}$$

Working with fuzzy numbers requires elementary fuzzy algebra. The operations stated next are used to execute different arithmetic procedures [18, 19].

Addition
$$\tilde{M}\oplus\tilde{N}=(m_l+n_l;m_u+n_u;\alpha+\gamma;\beta+\delta)_{LR} \tag{3}$$

Negation
$$\ominus\tilde{M}=(-m_u;-m_l;\beta;\alpha)_{LR} \tag{4}$$

Subtraction
$$\tilde{M}\ominus\tilde{N}=(m_l-n_u;m_u-n_l;\alpha+\delta;\beta+\gamma)_{LR} \tag{5}$$

Multiplication
$$\tilde{M}\otimes\tilde{N}=(m_l n_l;m_u n_u;m_l\gamma+n_l\alpha-\alpha\gamma;m_u\delta+n_u\beta+\beta\delta)_{LR} \tag{6}$$

Division
$$\tilde{M}\oslash\tilde{N}=\left(\frac{m_l}{n_u};\frac{m_u}{n_l};\frac{m_l\delta+n_u\alpha}{n_u(n_u+\delta)};\frac{m_u\gamma+n_l\beta}{n_l(n_l-\gamma)}\right)_{LR} \tag{7}$$

To facilitate the modeling of fuzzy numbers for the use on a shop floor environment further the triangular form will be preferred. Triangular fuzzy numbers are easy to utilize and give a relatively good representation of the situations under observance. More complex membership functions (than the triangular one) cause greater computational complexity without adding significant benefits [21, 22].

4. Including uncertainty in the OEE analysis

4.1. Methodology of the Fuzzy-OEE analysis

In this paragraph the methodical enhancement of the current state of the art in OEE analysis due to the inclusion of uncertainty will be proposed. This enhancement is suggested to overcome the influences of different circumstances as stated in paragraph 2.3. The method proposed here is based on the state of the art approach of the OEE analysis and enhances it after the process of the classical loss data collection, see figure 4. Principally there are three possibilities how to capture and include additional data about variability to the process analysis:

1. Estimated general factor for mainly manually collected constitution data,
2. Specific factors for single loss categories based on precise and measurable facts,
3. Specific factors for single loss categories with the use of qualified expert estimates.

The inclusion of directly measurable influencing factors to variability mostly is not possible in practice because no additional data is available. At this time it has to be decided what principle should be used to overcome the circumstances in each specific situation, e. g. as described in paragraph 2.3.

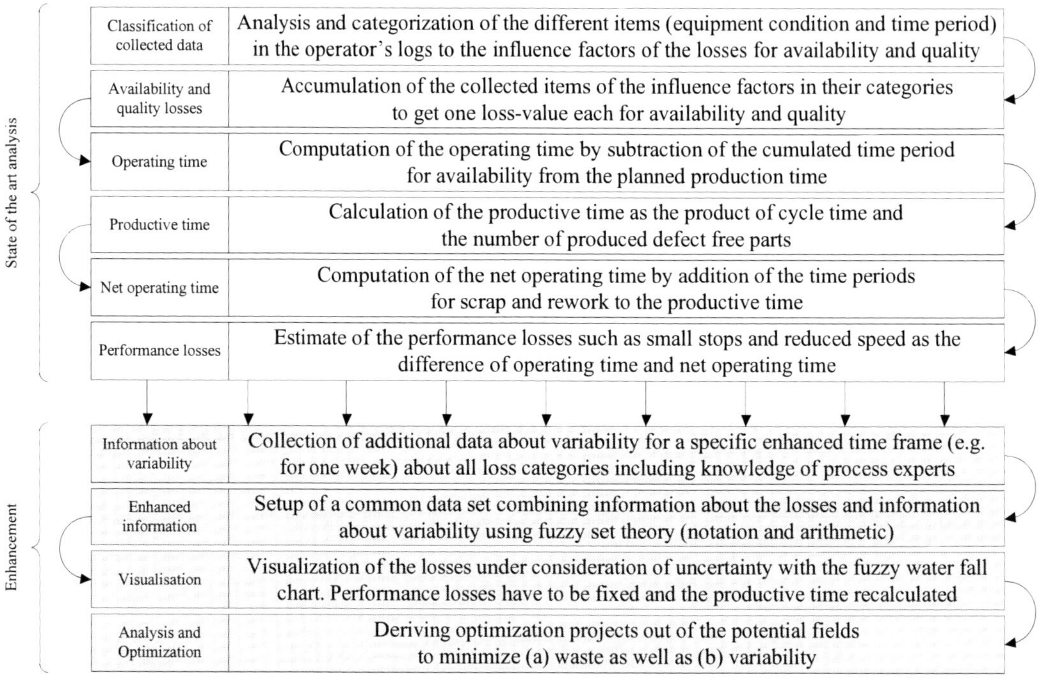

Figure 4. The enhanced method of Fuzzy-OEE analysis (optimized for using manually collected data).

To obtain enhanced information about the losses a combination of crisp loss data out of the state of the art analysis and variability information out of the enhanced analysis has to be performed. This can be realized by converting the crisp loss data into fuzzy numbers and the belated addition of the variability information for each category out of the enhanced analysis using the fuzzy set theory as described in paragraph 3.

For visualizing the enhanced information the use of the fuzzy OEE water fall chart is recommended. Because the derivation of this form of visualization is a complex task which cannot be performed manually, the use of software that is able to handle fuzzy arithmetic is advised.

The last step of the enhanced method is to derive optimization projects out of the potential fields of loss to sustainable remove waste as well as variability. To give an example how this method could be realized in practice and how it benefits the process of the optimization the next paragraph contains an illustrative example.

4.2. Case study at the Center for Industrial Productivity

For the first time the method of the Fuzzy-OEE analysis has been applied at the Process Learning Factory CiP at the Technische Universität Darmstadt. In this realistic production environment with machining and assembly areas methods for optimizing production processes are explored, applied and taught. These methods are taught in an interactive way so a sustainable learning is achieved for students and industry staff [23, 24].

In this production environment the method of Fuzzy-OEE analysis is utilized at a turning machine which usually is used to manufacture plungers for pneumatic cylinders. According to this method the first six steps contain the state of the art in OEE analysis, which has to be performed before information about variability can be included. The data capturing in this very first phase of the enhanced analysis is executed by manually recording of data by the use of an operator's log. Here the times for failure, changeover and scrap are captured. Rework is not possible at this kind of machine. The times for small stops and reduced speed are calculated backwards, see table 2.

Table 2. Input factors and results of the enhanced method of Fuzzy-OEE analysis at a turning machine, which has been observed over an eight hours shift using an operator's log for data recording. During this time a total amount of 123 faultless parts with an average cycle time of 148.5 seconds are manufactured, eight parts are defect.

Time slice	Recorded time [h]	Variability added [h]	Resulting relative time including variability [%]
Total available time	8	±0	(1.14;1.14;0;0)
Planned shutdown	1	±0	(0.143;0.143;0;0)
Planned production time	7	±0	(1;1;0;0)
Failure	0.45	-0.045 / +0.2	(0.0643;0.0643;0.01;0.03)
Changeover	0.725	-0.0725 / +0.2	(0.104;0.104;0.01;0.03)
Operating time	5.825	±0	(0.832;0.832;0.06;0.02)
Small stops	0.255	±0	(0.0364;0.0364;0;0)
Reduced speed	0.255	±0	(0.0364;0.0364;0;0)
Net operating time	5.315	±0	(0.759;0.759;0.06;0.02)
Scrap	0.23	-0.023 / +0.4	(0.0329;0.0329;0;0.06)
Rework	0	±0	(0;0;0;0)
Productive time	5.085	±0	(0.726;0.726;0.11;0.02)

After performing the state of the art OEE analysis in the first phase, according to figure 4, Information about variability has to be collected subsequent. To raise this kind of information, in paragraph 4.1 three possibilities were suggested which are applied here individually for each loss category.

Due to the fact that data collection is performed manually in this case an estimated general factor of 10 % of the recorded time is taken into account as variability for the loss fraction of failure. Here an extra charge of 33 % is added to the variability range because maintenance was performed the day before the analysis. Due to an interview with a maintainer the usual availability rate of the machine sinks rapidly after one week and so failure time could rise up to 45 minutes per shift.

The second observed influence factor is the changeover time, here also a general factor of 10 % is considered to be a realistic fluctuation caused by human perturbation. Additionally to that proportion the maximum possible time for changeover was calculated and considered. This is the cumulated time for changeover if after every lot a change of dies has to be performed.

The time fraction for scrap is also extended by a factor of 10 % for human perturbation, additionally to that an investigation at a quality gate comes into play. Due to the fact that not all quality issues are recognized at the turning machine, some parts with not obvious quality defects are classified as faultless parts by the operator. So they directly increase the OEE. At the quality gate after the machining area in average up to 10 parts per shift are discharged because of quality defects. This fact has to be taken into account as a variability increase at the scrap time in the Fuzzy-OEE analysis.

As a next step the crisp data captured by the operator has to be converted into fuzzy numbers and the fuzzy variability data has to be added using fuzzy arithmetic. To visualize this enhanced data the use of the fuzzy water fall chart is suggested. In this case the OEE ratio lies between 61.6 % and 74.6 % with a maximum membership degree at 72.6 %, see figure 5 and 6.

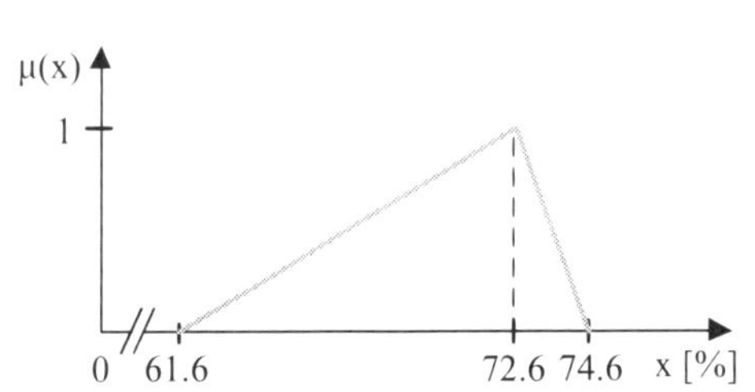

Figure 5. Visualization of the OEE in the selected case study as a fuzzy number type LR.

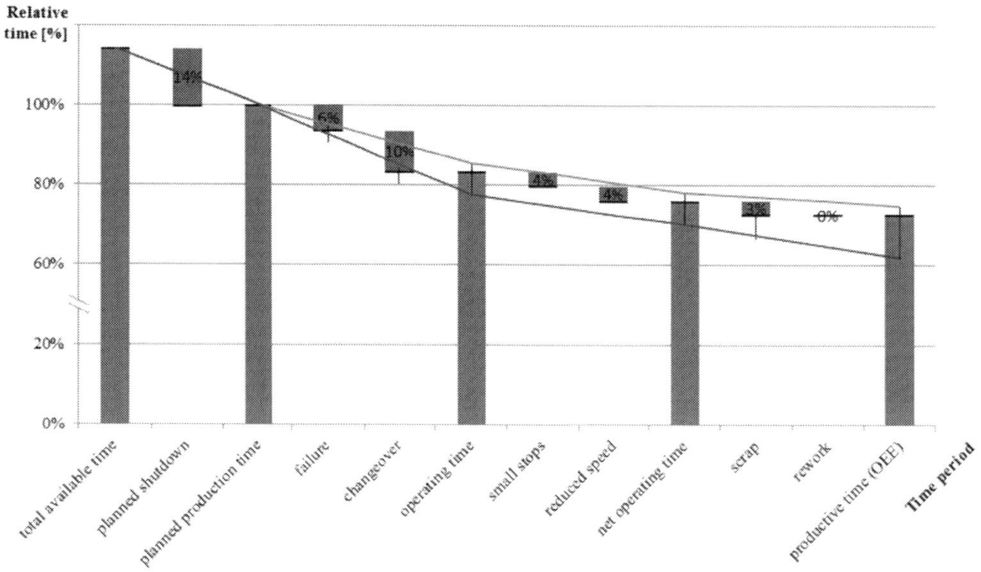

Figure 6. Visualization of a Fuzzy-OEE analysis at a turning machine using the fuzzy water fall chart.

As the last step of the enhanced method optimization projects have to be initiated in the potential fields to minimize waste as well as variability. At the present case study the major waste lies in the field of changeover. Here the method of SMED can be applied to lower the associated time losses. To also minimize the variability included in the process, the major observable potential lies in the area of scrap. Here for instance a quality initiative can be specifically started to minimize these fluctuations. Only under consideration of uncertainty and fuzziness, the potential in this field of action becomes transparent and improvable.

5. Conclusions

In this article the consideration of uncertainty is proposed due to the implementation of the OEE analysis on the shop floor of production companies. This is done to overcome the various deficiencies of the current state of the art. The state of the art in OEE analysis only considers the identification of waste as one of three performance inhibitors. With the use of an enhanced method proposed here more potential can be raised out of given facts considering also a second inhibitor, namely variability. To operationalize this claim in practice, the enhanced method is introduced and discussed by means of a case study. In this case study it is displayed that it is possible to bring more transparency into the analysis and so identify an expanded optimization potential with more reliable results. It is shown, that with the inclusion of the fuzzy set theory in an enhanced method the effectiveness of the OEE analysis can be augmented.

References

[1] Shingo S 1981 Study of the toyota production system: from an industrial engineering viewpoint (Tokyo: Japanese Management Association)
[2] Ohno T 1988 Toyota production system: beyond large-scale production (Tokyo: Diamond, Inc.)
[3] Braggaley B 2006 Using strategic performance measures to accelerate lean performance *Cost Man.* **20** 36-54
[4] Liker J K 2004 The toyota way: 14 management principles from the world's greatest manufacturer (New York: McGraw-Hill)
[5] Womack J P, Jones D T and Roos D 1990 The machine that changed the world: the story of lean production (New York: Free Press)

[6] Sim K L and Rogers J W 2009 Implementing lean production systems: barriers to change *Man. Res. News* **32** 37-49
[7] Bhasin S 2012 Performance of lean in large organisations *J. Man. Syst.* **31** 349-57
[8] Productivity Development Team 2004 Oee for operators: overall equipment effectiveness (New York: Productivity Press)
[9] Hansen R C 2001 Overall equipment maintenance: a powerful production / maintanance tool (New York: Industrial Press, Inc.)
[10] Suzuki T 1994 Tpm in process industries (New York: Japan Institute of Plant Maintanance)
[11] Nicholas J M and Soni A 2006 The portal to lean production: principles and practices for doing more with less (Boca Raton: Taylor & Francis Group)
[12] Bokranz R and Landau K 2006 Produktivitätsmanagement von Arbeitssystemen (Stuttgart: Schäffer-Poeschel Verlag)
[13] Ahmad M and Benson R 1999 Benchmarking in the process industries (Rugby: Institution of Chemical Engineers)
[14] Shingo S 1986 Zero quality control: source inspection and the poka-yoke system (New York: Productivity Press)
[15] Kletti J 2006 MES Manufacturing execution system (Berlin: Springer Verlag)
[16] Zadeh L A 1965 Fuzzy Sets *J. Inf. Contr.* **8** 338-53
[17] Zimmermann H J and Gutsche L 1991 Multi-criteria analyse (Berlin: Springer-Verlag)
[18] Chen S J, Hwang C L and Hwang C Z 1992 Fuzzy multible attribute decision making (New York: Springer-Verlag)
[19] Geldermann J, Sprengler T and Rentz O 2000 Fuzzy outranking for environmental assessment. Case study: Iron and steel making industry *Fuzz. sets syst.* **115** 45-65
[20] Goumas M and Lygerou V 2000 An extension of the promethee method for decision making in fuzzy environment *J. Oper. Res.* **123** 603-13
[21] Driankov D, Hellendoorn H and Reinfrank M 1993 An introduction to fuzzy control (Berlin: Springer-Verlag)
[22] Braglia M, Frosolini M and Zammori F 2009 Uncertainty in value stream mapping analysis *J. Log. Res. Appl.* **12** 435-53
[23] Abele E, Cachay J and Wennemer J 2011 Kompetenzentwicklung und Führung bei Verbesserungsprozessen in der Produktion *Industrie Management* **27** 14-18
[24] Abele E, Anderl R, Brungs F and Mosch C 2011 Materialflusssimulation in der schlanken Produktion *Productivity Management* 48-51

Meta-analysis of the relationship between TQM and Business Performance

Ahmad, M.F[1,2], Zakuan, N[2], Jusoh, A.[2], Tasir, Z.[3] and Takala, J.[4],
1 Department of Production and Operation, Faculty of Technology Management and Business, Universiti Tun Hussein Onn Malaysia (UTHM), Parit Raja, Batu Pahat, Johor, Malaysia
2 Department of Management, Faculty of Human Resource and Management, Universiti Teknologi Malaysia (UTM), Skudai, Johor, Malaysia
3 Department of Foundation of Education, Faculty of Education, Universiti Teknologi Malaysia (UTM), Skudai, Johor, Malaysia
4 Department of Engineering Science and Industrial Management, University of Vaasa, Finland

E-mail: mohdfauzi@uthm.edu.my

Abstract. Meta-analysis has been conducted based on 20 previous works from 4,040 firms at 16 countries from Asia, Europe and America. Throughout this paper a meta-analysis, this paper reviews the relationships between TQM and business performance amongst the regions. Meta-analysis result concludes that the average of rc is 0.47; Asia (rc=0.54), America (rc=0.43) and Europe (rc=0.38). The analysis also shows that Asia developed countries have greatest impact of TQM (rc=0.56). However, the analysis of ANOVA and t-test show that there is no significant difference amongst type of country (developed and developing countries) and regions at p=0.05. In addition, the average result of rc^2 is 0.24; Asia (rc^2=0.33), America (rc^2=0.22) and Europe (rc^2=0.15). Meanwhile, rc^2 in developing countries (rc^2=0.28) are higher than developed countries (rc^2=0.21).

1.0 Introduction

The concept of total quality management (TQM) has been developed as a result of intense global competition [1]. Firms that manage the international trade in global competition have put emphasis on TQM philosophy, procedures, tools and techniques. Juran [2] defines TQM as philosophy aimed at achieving business excellence through the application of tools and technique, as well as the management of soft aspects, such as human motivation in work. Furthermore, Demirbag et.al [3] defines TQM as a management philosophy aims to contribute continuous improvement in the organization with the participation of all employees to achieve customer satisfaction by producing better, cheaper, faster and safer than competitors. The role of TQM in improving business performance is broadly agreed in the literature and empirical study [4],[5],[6]. TQM helps to manage the firm to improve the effectiveness and business performance to achieve world class status for the past two decades [7]. However, the study of mediators and moderators is neglected and is referred to less frequently in literature review. Contingency theory proposes that organizations can create congruence between organizational structure and environmental uncertainty will achieve higher performance result (Ellis et.al., 2002). It means that there is "no a best way" to manage an organization, but the best management base on the circumstances and the environment. Thus, one of the objectives of this study is to empirically analyse and compare the affect the relationships amongst TQM and business performance in difference environment base on the difference type of country and the region.

2.0 Methodology

An extensive literature search was designed to identify and retrieve primary empirical studies relevant to the project's major research question. The databases searched were SpringerLink, Emerald, Taylor & Francis, ScienceDirect, Elsevier, ProQuest and Google Scholar. The descriptor TQM was used when possible; otherwise, it was searched as a keyword. The publication period is between 2003 until 2012. First, researchers find-out the study that used Structural Equitation Modeling (SEM) and regression in their methodology to identify the rc value between TQM and business performance as shown in figure 1. Based on the data, normality test was conducted to determine the type of test. In this study, ANOVA and t-test are used to test the differences between the groups. In the next section, researchers will discuss the result of this study.

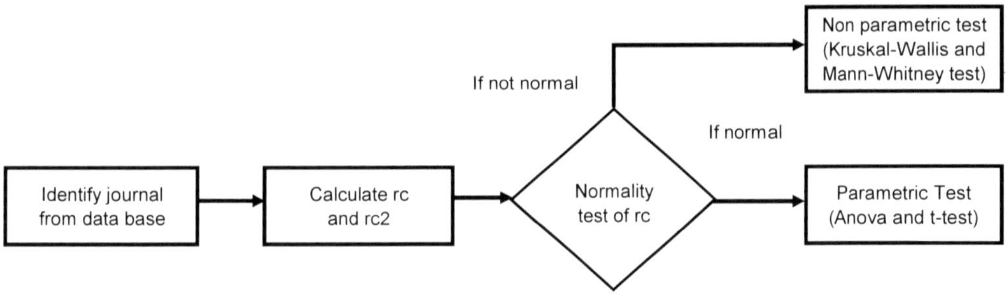

Figure 1: Flow-chart of Methodology

3.0 Result

The next step is to examine how strong the relationship between TQM and performance indicators. A Meta-analysis has been conducted based on 20 previous works from 4,040 firms at 16 countries from Asia, Europe and America. The methods of measuring the relationships in previous study are Structural Equation Modeling (SEM) and linear regression [9],[10]. The regression coefficient (rc) shows the strength of the relationships. Analysis have been made to understand the strength of the relationship, which is categorized into five levels; very weak (0.00-3.0), weak (0.31-0.50), moderate (0.51-0.70), strong (0.71-0.90) and very strong (0.91-1.00). Furthermore, percentage ratio of each category has been calculated to understand the weighted ratio of each category. The analysis results show the regression coefficient (rc) of the relationship is between 0.12 (weak) and 0.77 (strong). The overall mean for rc is 0.46, which is categorized as weak. In particular, the weighted ratio of category are 30% (very weak), 40% (weak), 20% (moderate), 10% (strong) and very strong (0%) as shown in Table 1. This result indicated that 70% of the results show weak relationship. It can be concluded that there are still inconsistent result of the relationship although most of the results are significant. On the other hand, the majority of the results show that the relationships are still weak and moderate. Based on Table 1, the country has been divided into two groups; developing and developed countries. First, normality test are conducted to confirm whether the data is normal. Kolmogorov-Smirnov test result shows that p value is 0.58 ($p \geq 0.05$), it means that rc value is normal data. Further analysis has been made to compare the relationship between developed and developing countries to confirm whether there are any differences based on rc value as shown in Table 1. The result shows that there is no significant difference between the rc among developed and developing countries based on t-test. However, the rc average value of developed countries is 0.49 compared to developing countries, 0.44. It means that the developed countries have higher impact of TQM practices compare to developing countries. Furthermore, the comparative analysis was carried out among three regions; America, Europe and Asia based on rc value in Table 1. One-way ANOVA is used to analyze the differences among the regions. The result shows that there is no significant difference among the regions. However, it is surprising that Asia region has the highest rc among the regions. Further analysis with two way ANOVA have been conducted to confirm whether there are any interaction between the type of country and the region. The result indicates p value is 0.993 ($p>0.05$), which indicates no significant interaction between them. Overall result shows that type of countries and regions have no significant effect to the strength of the relationship between TQM and business performance. However, Asia countries particularly developed countries have shown higher impact of performance than other countries as shown in Figure 2.

The next is to examine how much variance effect in firm performance caused by TQM. Rc squared (rc^2) is a measure of the amount of variability in firm performance that is explained by TQM [11]. Based on Table 1, rc^2 has been calculated of each country for further analysis. The overall average result of rc^2 is 0.24, which can be categorized as very weak. It means that firm performance can be accounted for 24% of variation in TQM. ANOVA and t-test result show that there is no significant difference amongst type of country (developed and developing countries) and regions. It is surprising that Asia developing countries have the highest (rc^2=0.33) and it shows that these countries are in the right track in TQM implementation. However, the European developed countries show the lowest value (rc=0.14). Further analysis with two way ANOVA have been conducted to confirm whether there are any interaction between the type of country and the regions. The result indicates p value is 0.929 ($p>0.05$), which indicates no significant interaction between them.

4.0 Discussion

It was expected that impact of TQM practices among developed countries would be superior. It is proven true , based on the analysis result. This study indicates that developed countries have higher relationship performance compared to developing countries. It means that developed countries have higher impact of TQM implementation. However, the study reveals that there is no significant different of relationships between developed countries and developing countries. Based on this study, developed countries in Asia have showed a higher impact of TQM performance. It explains that Asia countries such as Korea and Japan have not only higher TQM performance [12],[13] but also higher impact of TQM practices on their business performance. It is expected that these countries will continuously dominate the world market through TQM practices and customer satisfaction [14].

5.0 Conclusion

The main objectives of this study are to examine the relationship between TQM and business performance with extensive meta-analysis. Meta-analysis result concludes that the average of rc is 0.47; Asia (rc=0.54), America (rc=0.43) and Europe (rc=0.38). The analysis also shows that Asia developed countries have greatest impact of TQM (rc=0.56). In contrast, European developing countries have the lowest (rc=0.35). In addition, the average result of rc2 is 0.24; Asia (rc^2=0.33), America (rc^2=0.22) and Europe (rc^2=0.15). Meanwhile, rc^2 in developing countries (rc^2=0.28) are higher than developed countries (rc2=0.21). It has proved that relationship between TQM and business performance are based on contingency theory namely type of region and country. European countries should focus on how to improve the impact of TQM on business performance in the future.

Table 1: The relationship between TQM and performance indicator based on regression coefficient

year	authors	country	region	rc	very weak	weak	moderate	strong	Performance	rc2
2005	Bou (2005)	Spain **	Europe	0.27	X				Multi indicators	0.07
2006	Demirbag (2006)	Turkey *	Europe	0.24	X				Multi indicators	0.06
	Feng (2006)	Australia and Singapore **	Asia	0.62			X		Quality	0.38
	Jun (2006)	America**	America	0.20	X				Employee	0.04
2007	Joiner (2007)	Australia**	Asia	0.63			X		Multi indicators	0.40
	Tari (2007)	Spain**	Europe	0.47		X			Quality	0.22
2008	Costa (2008)	Spain**	Europe	0.45		X			Multi indicators	0.20
	Fynes (2008)	Ireland**	Europe	0.28	X				Multi indicators	0.08
	Jung (2008)	America-Mexico**	America	0.43		X			Multi indicators	0.18
	Projogo (2008)	Korea**	Asia	0.43		X			Quality	0.18
2009	Salaheldin (2009)	Qatar*	Asia	0.74				X	Multi indicators	0.55
2010	Chang (2010)	Taiwan*	Asia	0.27	X				Employee	0.07
	Fotopoulos (2010)	Greece**	Europe	0.41		X			Customer	0.17
	Miyagawa (2010)	America**	America	0.67			X		Multi indicators	0.45
	Sadikoglu (2010)	Turkey*	Europe	0.47		X			Multi indicators	0.22
	Zakuan (2010)	Malaysia and Thailand*	Asia	0.77				X	Customer	0.59
2011	Agus (2011)	Malaysia*	Asia	0.68			X		Multi indicators	0.46
	Konecni (2011)	Germany**	Europe	0.41		X			Multi indicators	0.17
	Valmohammadi (2011)	Iran*	Asia	0.28	X				Multi indicators	0.08
2012	Wang (2012)	Taiwan*	Asia	0.45		X			Customer	0.20
	Frequency of each category				6	8	4	2		
	Weighted Ratio				30%	40%	20%	10%		

* Developing countries and **Developed countries

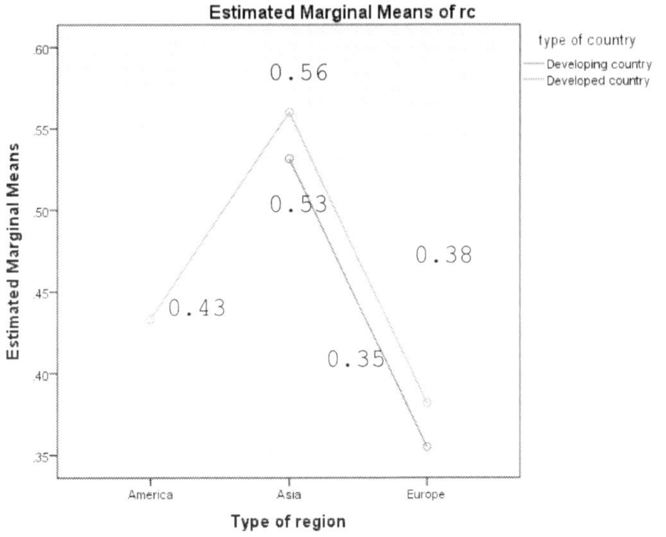

Figure 2: rc comparison between type of country and region

References:

[1] T. Laosirihongthong, "A Comparative Study of Implementation of Manufacturing Strategies in Thai and Indian Automotive Manufacturing Companies," *Industrial Engineering*, vol. 24, no. 2, 2005.

[2] J. M. Juran, *Juran's Quality Handbook*. Singapore: McGRAW-HILL, 1998.

[3] M. Demirbag, E. Tatoglu, M. Tekinkus, and S. Zaim, "An analysis of the relationship between TQM implementation and organizational performance: Evidence from Turkish SMEs," *Journal of Manufacturing Technology Management*, vol. 17, no. 6, pp. 829-847, 2006.

[4] B. Fynes, S. de Búrca, and J. Mangan, "The effect of relationship characteristics on relationship quality and performance," *International Journal of Production Economics*, vol. 111, no. 1, pp. 56-69, Jan. 2008.

[5] Prajogo and S. W. Hong, "The effect of TQM on performance in R&D environments: A perspective from South Korean firms," *Technovation*, vol. 28, no. 12, pp. 855-863, Dec. 2008.

[6] I. Sila, "Examining the effects of contextual factors on TQM and performance through the lens of organizational theories: An empirical study," *Journal of Operations Management*, vol. 25, no. 1, pp. 83-109, Jan. 2007.

[7] V. Arumugam, K.-B. Ooi, and T.-C. Fong, "TQM practices and quality management performance: An investigation of their relationship using data from ISO 9001:2000 firms in Malaysia," *The TQM Journal*, vol. 20, no. 6, pp. 636-650, 2008.

[8] S. Ellis, T. Almor, and O. Shenkar, "Structural contingency revisited: toward a dynamic system model," *Emergence*, vol. 4, no. 4, pp. 51-85, 2002.

[9] N. M. Zakuan, S. M. Yusof, and T. Laosirihongthong, "Total Quality Management & Business Excellence Proposed relationship of TQM and organisational performance using structured equation modelling," *Total Quality Management*, vol. 21, no. March 2012, pp. 37-41, 2010.

[10] A. Agus and Z. Hassan, "Enhancing Production Performance and Customer Performance Through Total Quality Management (TQM): Strategies For Competitive Advantage," *Procedia - Social and Behavioral Sciences*, vol. 24, pp. 1650-1662, Jan. 2011.

[11] A. Field, *Discovering Statistics Using SPSS*, 2nd ed. London: Sage Publication, 2009.

[12] A. G. Garvin, *Managing Quality*. New York: The Free Press, 1988.

[13] M. F. B. Ahmad and S. M. Yusof, "Comparative study of TQM practices between Japanese and non-Japanese electrical and electronics companies in Malaysia: Survey results," *Total Quality Management & Business Excellence*, vol. 21, no. 1, pp. 11-20, Jan. 2010.

[14] M. Ahmad, S. Yusof, and N. Yusof, "Comparative study of quality practices between Japanese and non-Japanese base electrical and electronic companies in Malaysia: A Survey," vol. 47, no. A, pp. 75-89, 2008.

Optimal performance of single-column chromatography and simulated moving bed processes for the separation of optical isomers

Bijan Medi[1], Monzure-Khoda Kazi[1] and Mohammad Amanullah[2*]

1 Nanyang Technological University, School of Chemical and Biomedical Engineering, 62 Nanyang Drive, Singapore 637459, Singapore

2* Corresponding author. Qatar University, College of Engineering, Department of Chemical Engineering, P. O. Box 2713, Doha, Qatar

E-mail: Aman@qu.edu.qa

Abstract. Chromatography has been established as the method of choice for the separation and purification of optically pure drugs which has a market size of about 250 billion USD. Single column chromatography (SCC) is commonly used in the development and testing phase of drug development while multi-column Simulated Moving Bed (SMB) chromatography is more suitable for large scale production due to its continuous nature. In this study, optimal performance of SCC and SMB processes for the separation of optical isomers under linear and overloaded separation conditions has been investigated. The performance indicators, namely productivity and desorbent requirement have been compared under geometric similarity for the separation of a mixture of guaifenesin, and Tröger's base enantiomers. SCC process has been analyzed under equilibrium assumption i.e., assuming infinite column efficiency, and zero dispersion, and its optimal performance parameters are compared with the optimal prediction of an SMB process by triangle theory. Simulation results obtained using actual experimental data indicate that SCC may compete with SMB in terms of productivity depending on the molecules to be separated. Besides, insights into the process performances in terms of degree of freedom and relationship between the optimal operating point and solubility limit of the optical isomers have been ascertained. This investigation enables appropriate selection of single or multi-column chromatographic processes based on column packing properties and isotherm parameters.

1. Introduction

Continuous chromatographic separation is the method of choice for large-scale production due to its higher productivity and robustness [1; 2]. Simulated moving bed (SMB), introduced in 1960 [3], has become the bench mark in this regard. Backed by the celebrated triangle theory [4], it is possible to identify the region of complete separation and find a relation between decision variables and performance indicators in a relatively straightforward manner. On the other hand, single-column chromatography (SCC) although does not appear to be competitive against SMB, it is still attractive due to higher flexibility, lower capital investment and quick analysis [5]. In this paper, under equilibrium theory assumptions, we present mathematical analysis of SMB and SCC under linear and nonlinear ranges of operation. For Langmuir isotherm, new analytical results are utilized for obtaining the triangular region of the SMB process [4] and elution profiles of the SCC process [6]. Thanks to equilibrium theory, it is possible to highlight optimal operating points and theoretical degrees of freedom for both processes without elaborate numerical methods and multi-objective optimization algorithms.

2. Problem statement

The schematic diagrams of the SMB and SCC processes are given in Figs. 1(a) and 1(b), respectively. It is assumed that:

- The column cross-sectional area, overall void fraction, and maximum allowable pressure drop are the same for both units.
- Complete separation and complete recovery are sought in both SCC and SMB units.
- Only physical constraints arising from maximum allowable pressure drop are considered.

Complete recovery necessitates that we collect only two fractions as products in SCC, and complete separation suggests that the purity values must be 100% in both extract and raffinate streams. The degrees of freedom are equal to the number of independent decision variables. We will derive the performance indicators and other important parameters in terms of the independent variables.

Under the assumption of complete separation and recovery, performance indicators productivity and desorbent requirement can be defined as:

$$Pr \equiv \frac{\text{Mass of feed processed per unit time}}{\text{Mass of stationary phase}} \quad Dr \equiv \frac{\text{Overall desorbent flow rate}}{\text{Mass of feed processed per unit time}} \tag{1}$$

The pressure drop in the column is calculated using Darcy's law [5]:

$$\Delta P = \frac{\phi u L \mu}{d_p^2} = \alpha u L \tag{2}$$

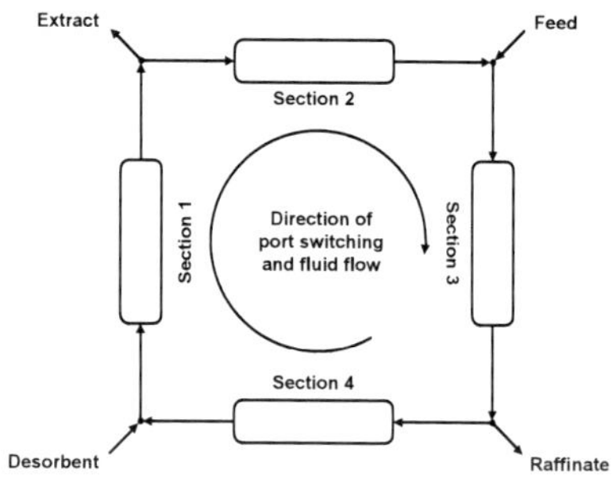

(a)

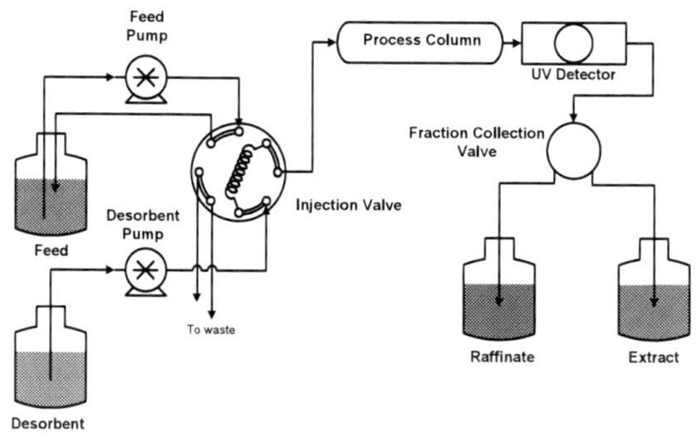

(b)

Figure 1. 1(a). An SMB process with four sections and four identical columns (one column per section); 1(b). A single-column chromatographic process for a binary separation.

3. Simulated moving bed

3.1. Linear isotherm

For SMB, the productivity and the dimensionless flow rates are defined as [2]:

$$Pr = \frac{c_T^F Q^F}{(1-\varepsilon)\rho_s V_T}; \quad m_j = \frac{Q^j t^* - \varepsilon V}{(1-\varepsilon)V} \tag{3}$$

Q^F is proportional to $(m_2 - m_3)$. For a linear isotherm maximum productivity and minimum desorbent consumption occur when $m_1 = H_2, m_2 = H_1, m_3 = H_2, m_4 = H_1$. Therefore,

$$Pr = \frac{c_T^F(m_3 - m_2)}{4\rho_s t^*} = \frac{c_T^F(H_2 - H_1)}{4\rho_s t^*} \tag{4}$$

Sum of all the flow rates in all sections of the SMB unit at the optimal operating point is obtained as:

$$\Sigma Q^j = \frac{4\varepsilon V}{t^*}\left(1 + \frac{(1-\varepsilon)}{\varepsilon}\frac{(H_2 + H_1)}{2}\right) \tag{5}$$

From Eq. 5, we can derive t^*. Maximum value of Q^j is limited by the maximum allowable pressure drop that the stationary phase can withstand.

$$\Delta P_{max} = \Sigma \Delta P_j = \alpha \frac{\Sigma Q^j}{A}\frac{L}{4} \tag{6}$$

Cycle time can be obtained in terms of t^* as:

$$t_{cy} = 4t^* \frac{\alpha L^2}{\Delta P_{max}}\left(\varepsilon + (1-\varepsilon)\frac{(H_2 + H_1)}{2}\right) \tag{7}$$

Substituition of t^* to Eq 3 gives Pr as:

$$Pr = \frac{c_T^F}{\rho_s}\frac{\Delta P_{max}}{\alpha L^2}\frac{(H_2 - H_1)}{\left(\varepsilon + (1-\varepsilon)\frac{(H_2+H_1)}{2}\right)} \tag{8}$$

Desorbent requirement at the optimal operating point for a linear isotherm is found as:

$$Dr = \frac{Q^F + Q^D}{c_T^F Q^F} = \frac{1}{c_T^F}\left(1 + \frac{m_1 - m_4}{m_3 - m_2}\right) = \frac{1}{c_T^F}\left(1 + \frac{H_2 - H_1}{H_2 - H_1}\right) = \frac{2}{c_T^F} \tag{9}$$

3.2. Langmuir isotherm

The values of dimensionless flow rates at the optimal operating point for Langmuir isotherm are given as [6]:

$$m_1 = H_2; \quad m_2 = \frac{\omega_2^F H_1}{H_2}; \quad m_3 = \frac{\omega_2^F\left(H_1\left(H_1 - \omega_1^F\right) + \omega_1^F\left(H_2 - H_1\right)\right)}{H_1\left(H_2 - \omega_1^F\right)} \tag{10}$$

$$m_4 = \frac{1}{2}\left\{m_3 + H_1 + K_1 c_1^F(m_3 - m_2) - \sqrt{\left[m_3 + H_1 + K_1 c_1^F(m_3 - m_2)\right]^2 - 4m_3 H_1}\right\} \tag{11}$$

Substituition of these values and taking same steps as in the case of linear isotherm provides expressions for Q^F, t_{cy}, Pr and Dr as:

$$Q^F = A\frac{\Delta P_{max}}{\alpha L}\frac{(1-\varepsilon)\frac{\omega_1^F\omega_2^F}{H_1 H_2}\frac{(H_2-H_1)^2}{H_2-\omega_1^F}}{\left(\varepsilon + (1-\varepsilon)\frac{\sum m_j}{4}\right)}; \quad t_{cy} = \frac{\alpha L^2}{\Delta P_{max}}\left(\varepsilon + (1-\varepsilon)\frac{\Sigma m_j}{4}\right) \tag{12}$$

$$Pr = \frac{c_T^F}{\rho_s}\frac{\Delta P_{max}}{\alpha L^2}\frac{\frac{\omega_1^F\omega_2^F}{H_1 H_2}\frac{(H_2-H_1)^2}{H_2-\omega_1^F}}{\left(\varepsilon + (1-\varepsilon)\frac{\sum m_j}{4}\right)}; \quad Dr = \frac{1 + \frac{H_1 H_2}{\omega_1^F\omega_2^F}\frac{(H_2-\omega_1^F)(H_2-m_4)}{(H_2-H_1)^2}}{c_T^F} \tag{13}$$

4. Single-column chromatography

4.1. Linear isotherm

We adopt similar methodology as SMB to obtain equivalent expressions for SCC. As shown in Fig. 2(a), maximum productivity at complete separation and recovery is achieved at touching-band condition, where there is no overlapping or gap between eluted peaks. In such conditions, we have $t_{cy} = 2t_p = \frac{2V_{inj}}{Q^D}$. Assuming complete separation and complete recovery, for single-column chromatography and using Darcy's law for Q^D, the productivity is:

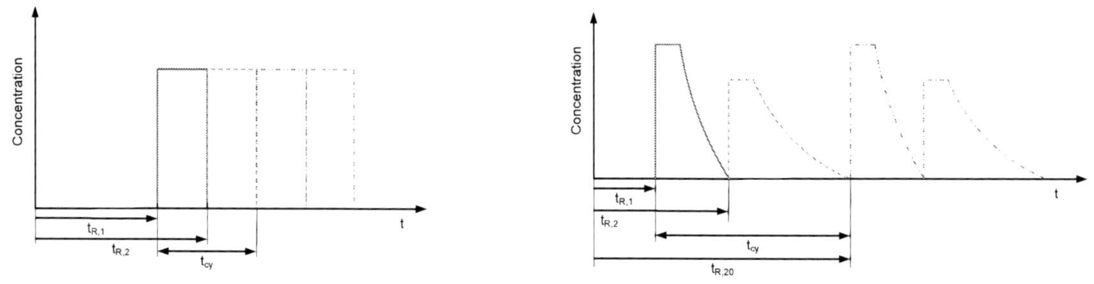

(a) (b)

Figure 2. Touching-band eluted peaks of the SCC process: (a) linear isotherm; (b) Langmuir isotherm

$$Pr = \frac{c_T^F V_{inj}}{(1-\varepsilon)\rho_s V_T t_{cy}} = \frac{c_T^F Q^D}{2(1-\varepsilon)\rho_s V_T} = \frac{c_T^F}{\rho_s}\frac{\Delta P_{max}}{\alpha L^2}\frac{1}{2(1-\varepsilon)} \tag{14}$$

Under the linear condition, the retention time of component i is defined as

$$t_{R,i} = \frac{\varepsilon V_T}{Q^D}\left(1 + \frac{(1-\varepsilon)}{\varepsilon}H_i\right) \tag{15}$$

Under the condiiton of complete separation and complete recovery, the value of cycle time must be equal to twice the difference between retention time values of components 2 and 1. Substituiting Q^D as a function of maximum allowable pressure drop:

$$t_{cy} = 2(t_{R,2} - t_{R,1}) = \frac{2(1-\varepsilon)V_T}{Q^D}(H_2 - H_1) = \frac{\alpha L^2}{\Delta P_{max}}\left[2(1-\varepsilon)(H_2 - H_1)\right] \tag{16}$$

Given that $t_p = V_{inj}/Q^D$, injection volume can also be represented by: $\frac{V_{inj}}{V_T} = (1-\varepsilon)(H_2 - H_1)$. To compare with SMB, an average feed flow rate can be defined for single-column chromatography as $Q^F = V_{inj}/t_{cy}$. Substituting t_{cy} and V_{inj}, we obtain:

$$Q^F = A\frac{\Delta P_{max}}{\alpha L}\frac{1}{2} \tag{17}$$

Similarly, like the case of SMB, desorbent requirement is obtained as:

$$Dr = \frac{Q^D t_{cy} + V_{inj}}{V_{inj} c_T^F} = \frac{Q^D \frac{2V_{inj}}{Q^D} + V_{inj}}{V_{inj} c_T^F} = \frac{3}{c_T^F} \tag{18}$$

4.2. Langmuir isotherm

Similar approach can be taken to obtain explicit values for productivity and desorbent requirement of SCC under Langmuir isotherm. Maximum productivity at complete separation and recovery must be sought at touching-band condition In such a condition, we have incipient baseline separation between peaks, which imposes an additional constraint on the problem. Here we need explicit relations for the retention time values of two components and a criterion on incipient baseline separation. A typical chromatogram is shown in Fig. 2(b). The relations for productivity, desorbent requirement, feed flow rate, and cycle time for SCC and SMB under linear and Langmuir isotherms are summarized in Tables 1 and 2, respectively. Note that a detail derivation of Pr, Dr, Q^F and t_{cy} for SCC with Langmuir isotherm is beyond the scope of this paper.

Table 1. Summary of important equations obtained for SCC and SMB subject to linear isotherm.

Parameter	SCC	SMB
Pr	$\frac{c_T^F}{\rho_s}\frac{\Delta P_{max}}{\alpha L^2}\frac{1}{2(1-\varepsilon)}$	$\frac{c_T^F}{\rho_s}\frac{\Delta P_{max}}{\alpha L^2}\frac{(H_2-H_1)}{\left(\varepsilon+(1-\varepsilon)\frac{(H_2+H_1)}{2}\right)}$
Dr	$\frac{3}{c_T^F}$	$\frac{2}{c_T^F}$
Q^F	$A\frac{\Delta P_{max}}{\alpha L}\frac{1}{2}$	$A\frac{\Delta P_{max}}{\alpha L}\frac{(1-\varepsilon)(H_2-H_1)}{\left(\varepsilon+(1-\varepsilon)\frac{(H_2+H_1)}{2}\right)}$
t_{cy}	$\frac{\alpha L^2}{\Delta P_{max}}[2(1-\varepsilon)(H_2-H_1)]$	$\frac{\alpha L^2}{\Delta P_{max}}\left(\varepsilon+(1-\varepsilon)\frac{(H_2+H_1)}{2}\right)$

Table 2. Summary of important equations obtained for SCC and SMB subject to Langmuir isotherm.

Parameter	SCC	SMB
Pr	$\frac{c_T^F}{\rho_s}\frac{\Delta P_{max}}{\alpha L^2}\frac{1}{(1-\varepsilon)\left[1+\frac{H_1H_2}{\omega_1^F\omega_2^F}\left(\frac{H_2-\omega_1^F}{H_2-H_1}\right)^2\right]}$	$\frac{c_T^F}{\rho_s}\frac{\Delta P_{max}}{\alpha L^2}\frac{\frac{\omega_1^F\omega_2^F}{H_1H_2}\frac{(H_2-H_1)^2}{H_2-\omega_1^F}}{\left(\varepsilon+(1-\varepsilon)\frac{\sum m_j}{4}\right)}$
Dr	$\frac{2+\frac{H_1H_2}{\omega_1^F\omega_2^F}\left(\frac{H_2-\omega_1^F}{H_2-H_1}\right)^2}{c_T^F}$	$\frac{1+\frac{H_1H_2}{\omega_1^F\omega_2^F}\frac{(H_2-\omega_1^F)(H_2-m_4)}{(H_2-H_1)^2}}{c_T^F}$
Q^F	$A\frac{\Delta P_{max}}{\alpha L}\frac{1}{1+\frac{H_1H_2}{\omega_1^F\omega_2^F}\left(\frac{H_2-\omega_1^F}{H_2-H_1}\right)^2}$	$A\frac{\Delta P_{max}}{\alpha L}\frac{(1-\varepsilon)\frac{\omega_1^F\omega_2^F}{H_1H_2}\frac{(H_2-H_1)^2}{H_2-\omega_1^F}}{\left(\varepsilon+(1-\varepsilon)\frac{\sum m_j}{4}\right)}$
t_{cy}	$\frac{\alpha L^2}{\Delta P_{max}}(1-\varepsilon)\left((H_2-\omega_1^F)+\frac{\omega_1^F\omega_2^F}{H_1H_2}\frac{(H_2-H_1)^2}{H_2-\omega_1^F}\right)$	$\frac{\alpha L^2}{\Delta P_{max}}\left(\varepsilon+(1-\varepsilon)\frac{\Sigma m_j}{4}\right)$

5. Results and discussion

Equations summarized in Tables 1 and 2, show that at optimal points, there are one-to-one similarities between SCC and SMB process in terms of corresponding parameters, namely productivity, desorbent requirement, feed flow rate, and cycle time. The similarities are not limited to linear isotherm and are clearly extended to the case of Langmuir isotherm. Furthermore, it is worthy of attention that the equations obtained for Langmuir isotherm can be simplified to the case of linear isotherm either at very low feed concentrations or where Langmuir equilibrium constants are set to zero ($K_i = 0$). Equivalently, we can set $\omega_1^F = H_1$ and $\omega_2^F = H_2$.

It is also important to note that meaningful groups of parameters are identified in the results and they have their counterparts for SCC and SMB as given in Tables 1 and 2. Looking at the productivity for example, the first group carries the effect of feed concentration. The second group carries the effect of pressure drop and geometry, and the third one carries the effect of isotherm parameters.

We have investigated the effects of change in feed concentration and Henry constants on the process peformance by simulation case studies. We have considered two case studies: case study A: separation of a mixture of guaifenesin enantiomers and case study B: separation of Tröger's base enantiomers. Isotherm parameters and physical parameters are given in Table 3.

Table 3. Isotherm parameters and physical properties used in the case studies. Note that for the linear isotherms, K_i values are simply set to zero.

Parameter	Case A (guaifenesin) [7]	Case B (Tröger's base) [8]	Unit
H_1	1.41	2.18	-
H_2	3.49	6.45	-
K_1	0.0135	0.065	L/g
K_2	0.0550	0.39	L/g
D	1	1	cm
L	10	10	cm
ε	0.704	0.704	-
α	7.59×10^8	7.59×10^8	Pa.s/m^2
ρ_s	2027.03	2027.03	g/L

5.1. Effects of feed concentration

Analytical expressions reported in Tables 1 and 2 suggest that for fixed geometry and constant physical properties, the only degree of freedom is the feed concentration for both SCC and SMB units at optimal points. Feed concentration has a positive effect on productivity for both SCC and SMB under linear or Langmuir isotherm. For linear isotherm, the effect is just linear. For Langmuir isotherm however, productivity is a complex, but still increasing function of feed concentration as can be seen in Figs. 3(a) and 4(a), respectively. Nevertheless, the rate of change in productivity for Langmuir isotherm is less than that of the linear isotherm.

It is worthy of attention that for case A, the productivity of SCC is higher than that of SMB for the entire range of concentration, but for case B, this order is reversed. This finding suggests that theoretically, SCC can be better than SMB in terms of productivity for the separation of certain enantiomers.

Eqs. 9 and 18, clearly show that single-column chromatography demands more desorbent than SMB under linear range of operation. For Langmuir isotherm, the same argument holds for the case studies considered here as can be seen in Figs. 3(b) and 4(b), although drawing a general conclusion for Langmuir isotherm is not straightforward. Moreover, desorbent requirement appears as a sole function of feed concentration for linear isotherm, but it is affected by isotherm parameters for Langmuir isotherm. In contrast with the common practice of considering cycle time as an independent decision variable, for a fixed set of physical properties and geometry, it is a constant for linear isotherm and a sole function of feed concentration for Langmuir isotherm.

5.2. Effects of pressure drop and overall length

The effect of maximum allowable pressure drop is the same for all cases; productivity linearly increases with increasing allowable pressure drop. On the other hand, productivity is always inversely related to the square of the overall length of the unit. Therefore, ideally, shorter columns are favored to increase productivity. Besides, the fact that productivity appears as a function of the length of unit necessitates that a fair comparison between SCC and SMB must be made at a fixed overall length. Conversely, column cross-sectional area has no effect on productivity and may be used as a scaling parameters. On the other hand, desorbent requirement is not affected by any of these parameters.

Feed flow rate has similar functionality when it is compared to productivity. The major difference is that the cross-sectional area of unit directly affects the feed flow rate. Therefore, feed flow rate not

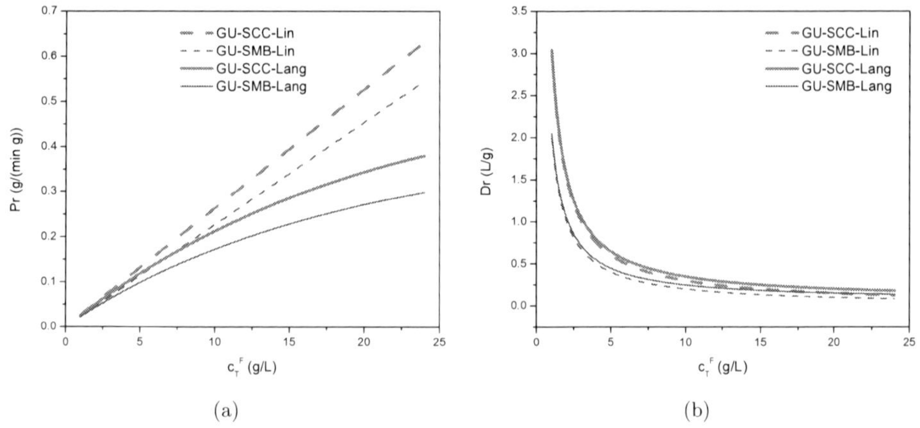

Figure 3. Performance indicators vs. feed concentration for guaifenesin (case A): (a) productivity; (b) desorbent requirement.

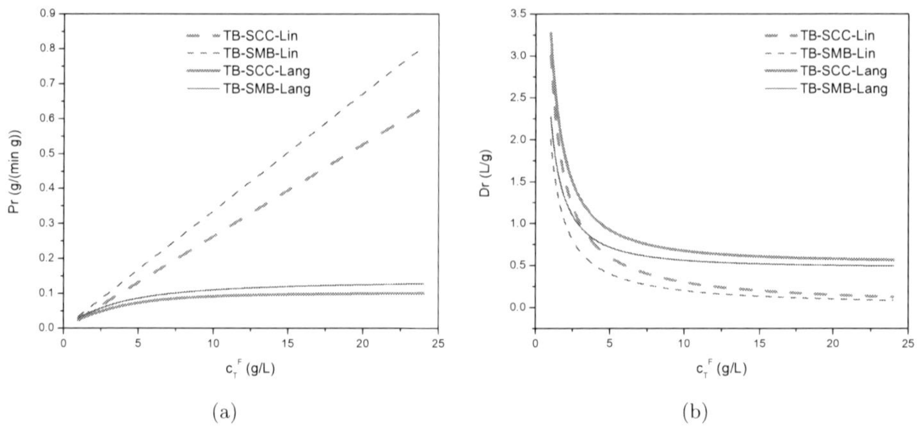

Figure 4. Performance indicators vs. feed concentration for Tröger's base (case B): (a) productivity; (b) desorbent requirement.

only is related to the geometry of process through length, but also a function of cross-sectional area, which means feed flow rate for the two processes must be compared at a fixed A/L ratio. Besides, these results affirm that shorter columns are ideally favored for more feed throughput.

It must be emphasized that optimal cycle time varies with the square of the length of the unit. This means, in practical applications, minimum possible switching time might become the active constraint when the length of the unit is used as a decision variable to increase productivity. The lower limit may be dictated by the response time of switching valves and communication devices.

5.3. Effects of Henry constants

Tables 1 and 2 indicates that, Henry constants have a complex effect on the productivity of SMB, but relatively simpler effect on the productivity of SCC. In fact, they have no effect on the productivity of SCC in linear range of operation. This may result in a higher productivity for SCC compared to SMB if for example, highly retentive or low-selectivity materials are processed.

Figs. 5 and 6 show that, productivity increases as the difference between two Henry constants increases. It is worthy of attention that for a small region, where the difference between Henry constants is small, it is possible that theoretically the productivity of SCC will be greater than that of SMB.

In linear range of operation, desorbent requirement is solely a function of feed concentration. For Langmuir isotherm, it is also a function of isotherm parameters, but not any other parameter. This contrasts the common belief that desorbent requirement must be a function of desorbent flow rate. Moreover, the results reveal that in contrast to productivity, desorbent requirement is completely independent of the process scale.

For desorbent requirement, the difference between SCC and SMB is small for small Henry constants, but it approaches a certain limit when H_2 increases as can be seen in Fig. 7. Moreover, SMB always has a smaller desorbent requirement compared to SCC.

For SCC, cycle time is related to difference between Henry constants, but for SMB, it is a function of the sum of Henry constants. For a Langmuir isotherm, a similar but more complex functionality is observed. This finding shows that two different phenomena decide on the optimal cycle time values for SCC and SMB.

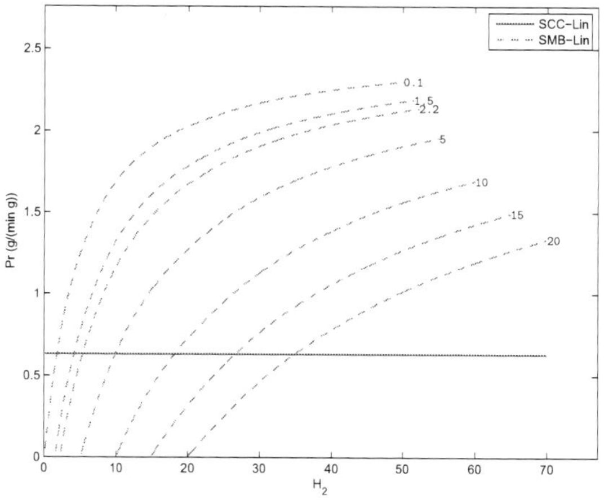

Figure 5. Productivity vs. Henry constants for linear isotherm. Numbers next to the lines are H_1 values. c_T^F is 24 g/L. (Results are the same for the two case studies).

On the other hand, for SCC, similar to cycle time, injection volume is no longer an independent decision variable For linear isotherm, it is just a function of the length of unit, physical properties, and Henry constants. For Langmuir isotherm however, the effect of feed concentration is added and therefore, injection volume also becomes a function of feed concentration.

6. Concluding remarks

In this paper, analytical investigation of the separation behavior in single-column chromatography and simulated moving bed at their optimal operating points for the case of complete separation and recovery has been presented. It is shown that performance indicators namely, productivity and desorbent requirement can be expressed in similar terms for these processes under equilibrium theory assumptions.

Feed concentration has been identified as the only degree of freedom when geometry and physical properties are kept constant. Hence, all other parameters, such as cycle time, feed flow rate, and injection volume can be expressed as a function of feed concentration for a fixed problem. In addition, feed concentration has positives effect on both performance indicators (i.e., it increases productivity and decreases desorbent requirement) in the case studies investigated in this work and therefore, the optimal operating point must occur at the maximum feed concentration determined by solubility.

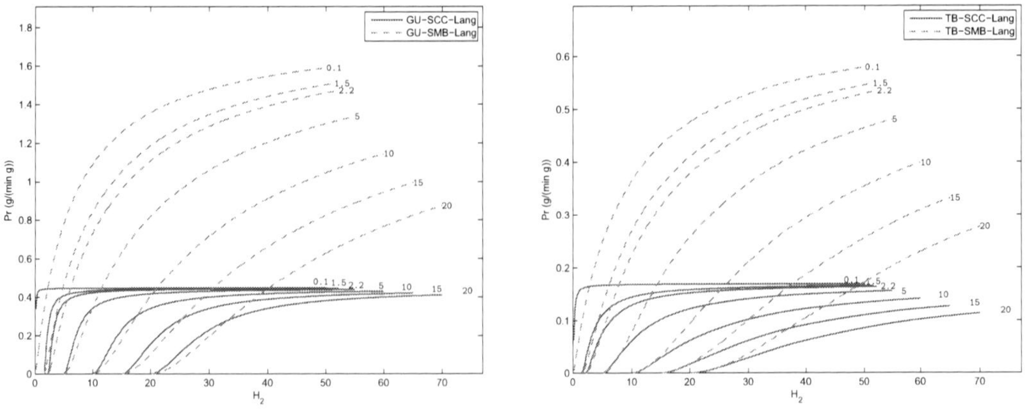

(a) (b)

Figure 6. Productivity vs. Henry constants for Langmuir isotherm: (a) guaifenesin (case A); (b) Tröger's base (case B). Numbers next to the lines are H_1 values. c_T^F is 24 g/L.

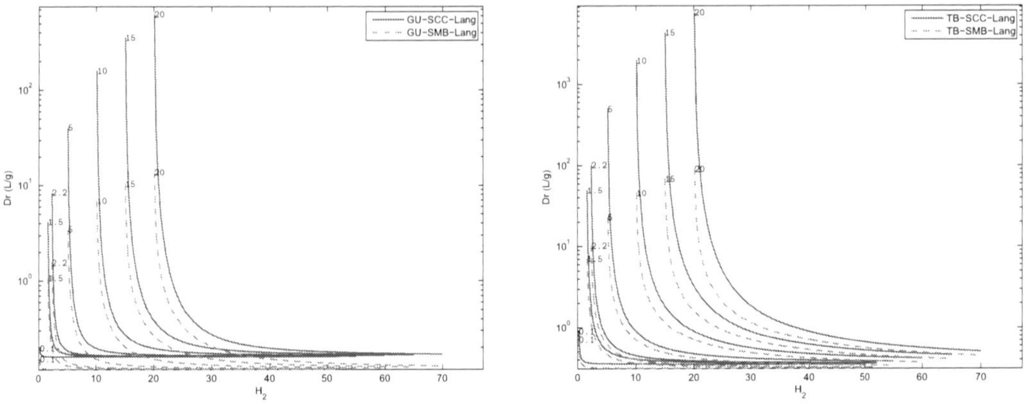

(a) (b)

Figure 7. Desorbent requirement vs. Henry constants for Langmuir isotherm: (a) guaifenesin (case A); (b) Tröger's base (case B). Numbers next to the lines are H_1 values. c_T^F is 24 g/L.

It is also noted that in some ranges of Henry constants, single-column chromatography can outperform simulated moving bed in terms of productivity, but it is always inferior in terms of desorbent requirement.

Results showed that productivity is proportional to the inverse of the square of column length. Therefore, shorter columns are favored for increasing the productivity. Use of shorter columns implies smaller switching time which is constrained by switching valves' hardware limitation.

Nomenclature

A	column cross sectional area $[m^2]$
c_T^F	total feed concentration [g/L]
Dr	desorbent requirement [L/g]
d_P	particles diameter [μm]
H_i	Henry constant of component i [-]

K_i	equilibrium constant of component i in Langmuir isotherm [L/g]
L	overall length [cm]
m_j	dimensionless flow-rate ratio [-]
Pr	productivity [g/(min g)]
ΔP	pressure drop [bar]
Q	volumetric flow rate [mL/min]
t^*	switching time [s]
t_{cy}	cycle time [s]
t_p	pulse width $= V_{inj}/Q_D$ [s]
$t_{R,20}$	time where the tail of more retained compound ends [s]
$t_{R,i}$	retention time of component i [s]
u	superficial velocity [cm/s]
V	volume of one column (SMB) [m^3]
V_{inj}	injection volume [μL]
V_T	overall volume [m^3]
v	interstitial velocity [cm/s]
z	axial coordinate [m]

Greek letters

α	lumped physical parameter defined as $\alpha = \phi\mu/d_p^2$ [Pa.s/m^2]
ε	overall void fraction of column [-]
ϕ	resistance parameter [-]
ω	characteristic parameter [-]
μ	viscosity [Pa.s]
ρ	density [g/L]

Subscripts and superscripts

1	less retained compound
2	more retained compound
ax	axial
D	desorbent
F	feed
i	component index
in	inlet
j	section
p	pulse width
s	solid/stationary phase

References

[1] Mazzotti M, Storti G and Morbidelli M 1997 *Journal of Chromatography A* **769** 3

[2] Rajendran A, Paredes G and Mazzotti M 2009 *Journal of Chromatography A* **1216** 709

[3] D Broughton B and C Gerhold G 1961 Continuous sorption process employing fixed bed of sorbent and moving inlets and outlets

[4] Mazzotti M 2006 *Journal of Chromatography A* **1126**

[5] Cox G B 2005 *Preparative Enantioselective Chromatography* 1st ed (Noida: Blackwell)

[6] Rajendran A and Mazzotti M 2011 *Industrial & Engineering Chemistry Research* **50** 352

[7] Francotte E, Richert P, Mazzotti M and Morbidelli M 1998 *Journal of Chromatography A* **796** 239

[8] Amanullah M, Abel S and Mazzotti M 2005 *Adsorption* **11**

Quality Service Analysis and Improvement of Pharmacy Unit of XYZ Hospital Using Value Stream Analysis Methodology

Jonny and Januar Nasution

School of Industrial Engineering, Faculty of Engineering, Bina Nusantara University, Jl. KH Syahdan No. 9, Palmerah, Jakarta, 11410, Indonesia

E-mail: jonny@binus.ac.id

Abstract. Value stream mapping is a tool which is needed to let the business leader of XYZ Hospital to see what is actually happening in its business process that have caused longer lead time for self-produced medicines in its pharmacy unit. This problem has triggered many complaints filed by patients. After deploying this tool, the team has come up with the fact that in processing the medicine, pharmacy unit does not have any storage and capsule packing tool and this condition has caused many wasting times in its process. Therefore, the team has proposed to the business leader to procure the required tools in order to shorten its process. This research has resulted in shortened lead time from 45 minutes to 30 minutes as required by the government through Indonesian health ministry with increased %VA (valued added activity) or Process Cycle Efficiency (PCE) from 66% to 68% (considered lean because it is upper than required 30%). This result has proved that the process effectiveness has been increase by the improvement.

1. Introduction

Based on Indonesian Hospital Quality Service Standard, the quality service of pharmacy unit can not be separated from the holistic quality service of a hospital and therefore, the hospital should focus on patient safety through providing high quality medicines to its patients as required by the practitioners in the hospital.

In a hospital, pharmacy services could include services to outpatient and inpatient with their unique characteristic. For outpatients, the pharmacy services starts from consultation (during in outpatient unit) to medicine handover (in pharamcy unit), whereas for inpatients, the pharmacy services starts from instruction given by responsible practioners through inpatient visit to medicine handover to the patient s responsible nurse.

The kind of services given to patient are determined by what kind of the patient served by the hospital that may include 1) general or 2) company s patient or insurance s patient. Each of these services may vary from one to another based on the responsibility bearer.

On the other hand, the type of precribed medicines also determines the kind of pharmacy services that may also include 1) patent or 2) self-produced medicine. For the second type, the pharmacy should produce self-produced medicine by mixing several patent medicines as required by practioners whereas for the first type, the pharmacy only needs to pack prescribed medicine as required by the practioners.

The Indonesian Health Ministry has determined that lead time for pharmacy service related to self-produced medicine should be less than 30 minutes to ensure the patient s quick recovery. Therefore, XYZ Hospital should also obey this standard in order to stay acknowledged by the government. Up to now, the pharmacy service related to self-produced medicine in XYZ Hospital was still above 30 minutes. This condition has triggered many complaints from patients and urged this research in order to evaluate dan improve the process by deploying value stream analysis methodology. In order to get focus, this research is designed to focus only on self-produced medicine using sample around 30 general outpatients with same diagnostic health problem. These samples are drawn from busiest scheduled internist outpatient unit with gastro diagnosed disease due to its longest lead time triggered by number of self-produced medicines up to 30 to 40 capsules per patient [6].

Based on Value Stream Analysis methodology, this research is designed 1) to depict the current state value stream mapping (VSVSM) of the self-produced medicine process in pharmacy unit, 2) to identify the root causes of longer lead time and their solutions, 3) to develop the future state value stream mapping (FSVSM) of the process, 3) to implement the solution to ensure the condition is improved as stated in its future state value stream map.

2. Literature Review

In 1996, Womack and Jones has introduced lean lean thinking as alternative solution to quality improvement for many companys including hospital. Actually, lean thinking has prior introduced by Taichi Ohno through Toyota Way. This concept is designated to make the process become more flexible in order to adapt the fluctuation of demand, decrease waste, optimize process, improve process control and optimally ulitize human resources. To implement this concept, there are several methodology including value stream mapping (VSM) [15].

Value stream mapping (VSM) is used to map value and non value added activities conducted to produce product or service to its customers. This VSM tool is one of improvement tools that help the organization to visualize the process including both material and information flow to improve product and service by finding its waste and source [11]. To apply this methodology, we can use the following methodology:

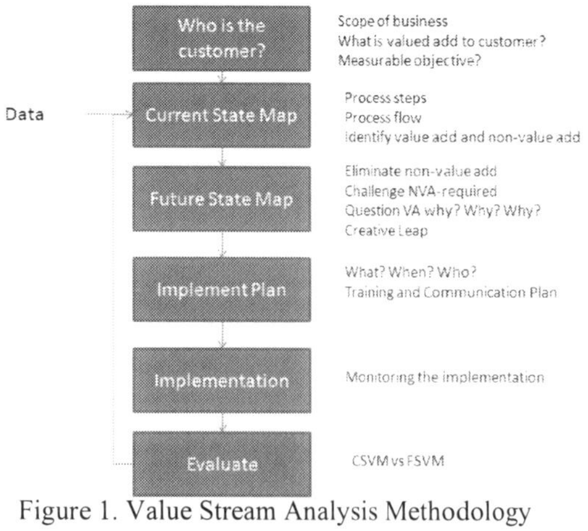

Figure 1. Value Stream Analysis Methodology

From the above methodology, there are 6 steps that we should follow in order to improve the process. First of all, we determine the scope of business process, what is value add to customer and measurable objectives. For this research, the scope of business process is self-produced medicine to general patient (Internist Outpatient). The measurable objectives used to ensure whether the improvement has taken place or not are lead time, value add time (using cycle time [2]), non value add time, and percentage value add time or known as Process Cycle Efficiency (PCE) [12]. First indicator named as lead time is measured by adding time used to perform both value and non added value activities by pharmacist and its staffs. Secondly, value added time is measured by adding all cycle time performed by each value added activity and same to non value added time but from all non value added activities. Lastly, percentage of lead time or called as Process Cycle Efficiency (PCE) is measured by deploying the following formula:

$$PCE = \frac{VA}{Leadtime} x100\% \quad (1)$$

In second step (current state map), The authors visualize process flow and identify value add and non-value add activities using related data. In this research, the value add activities are outpatient admission, preparation, quality assurance I by pharmacist, production, quality assurance II by pharmacist, and handover. Whereas, non-value add activities are those activities including waiting time for preparation, waiting time for quality assurance I by pharmacist, waiting time for production, waiting time for quality assurance II by pharmacist, and waiting time for handover.

In third step, from current state, the authors develop future state map by considering improvement ideas to eliminate non-value added, challenge to NVA required, question to VA and creative leap. In this research, the author try to decrease NVA due to the impossibility to eliminate NVA.

In fourth step, the authors develop implement plan and implement it by conducting training and communication plan. Furthermore, the authors also monitor its implementation and evaluate to ensure the improvement has taken place.

3. Result

The following diagram is current state value stream mapping for high level process:

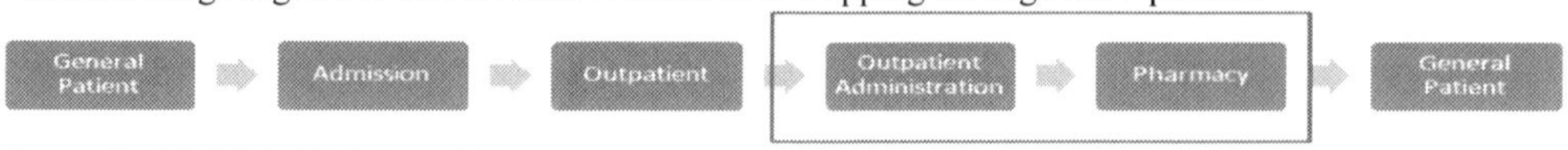

Figure 2. CSVSM: High Level Process

The above figure has shown that general patients are required to register themselves through admission before having consultation with the practioners. After that, they should complete the payment before they go to pharmacy to get their medicines. The self-produced medicine (mixing patent medicine required by practioners done by pharmacist)process that is targeted to improve starts from outpatient administration (after consultation) to pharmacy (handover) with detailed as follows:

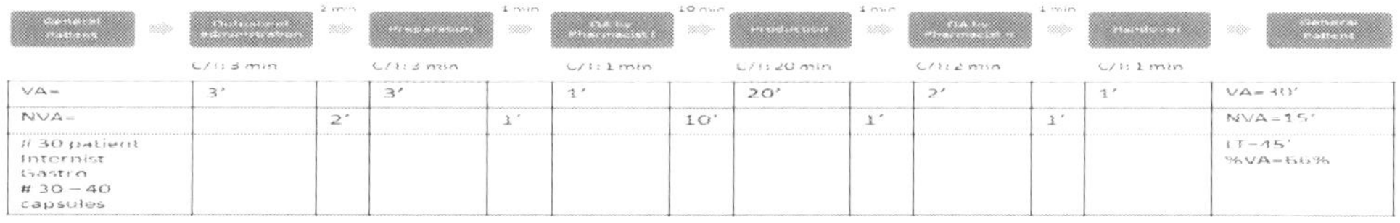

Figure 3. CSVSM: Low Level Process

The above data and information flow are based on observation conducted in XYZ hospital by using research design as depicted in the following table.

Table1. Research Design

Indicator	Current State	Future State
Number of sampel	30 patients from Interniste Outpatient with Gastro diagnosed disease	30 patients from Interniste Outpatient with Gastro diagnosed disease
Number and types of prescribed medicines	30-40 capsules wtih 2-3 type of prescribed medicines	30-40 capsules wtih 2-3 type of prescribed medicines
Observation day	30 day with in Internist Outpatient during busiest schedule in XYZ hospital	30 day with in Internist Outpatient during busiest schedule in XYZ hospital

The observation result from the above research design is also provided from the following table:

Table 2. Observation: Current State Result

OBSERVATIONAL FORM
CURRENT STATE. #CAPSULE. 30 - 40, INTERNIST
PHARMACY INSTALLATION - BINA HUSADA HOSPITAL

No	Date	#Types	# Capsule	Outpatient administration	Waiting	Preparation	Waiting	QA by Pharmacist I	Waiting	Production	Waiting	QA by Pharmacist II	Waiting	Handover	VA	NVA	Total	%
1	May 2, 2012	2	30	3,09	2,55	3,37	0,86	0,96	9,58	20,00	1,32	1,60	0,85	1,04	29,02	15,15	44,17	66%
2	May 2, 2012	3	30	3,47	2,49	3,96	0,86	0,82	12,18	19,04	0,99	1,55	1,31	0,95	28,83	17,83	46,66	62%
3	May 2, 2012	3	30	2,52	2,32	2,15	1,02	1,34	10,95	24,00	1,03	2,82	1,38	0,99	32,84	16,69	49,52	66%
4	May 2, 2012	2	30	2,85	2,66	3,62	0,82	0,87	12,72	17,00	0,79	2,34	0,71	1,15	26,69	17,70	44,39	60%
5	May 9, 2012	2	40	3,42	1,79	3,88	0,81	1,26	11,59	21,12	1,39	2,73	0,88	0,87	32,41	16,46	48,88	66%
6	May 9, 2012	2	30	2,87	2,08	2,36	1,21	1,05	8,25	20,13	1,39	2,34	0,97	0,91	28,74	13,90	42,64	67%
7	May 9, 2012	2	40	2,53	1,69	3,92	1,20	0,79	7,21	22,14	0,77	2,82	0,70	1,27	32,21	11,57	43,77	74%
8	May 9, 2012	3	40	2,79	2,52	3,62	1,32	0,79	9,95	21,47	1,06	2,07	0,80	0,98	30,75	15,65	46,40	66%
9	May 11, 2012	2	30	3,11	2,41	2,45	1,33	1,00	10,28	18,58	0,94	1,93	0,96	1,14	27,07	15,92	43,00	63%
10	May 11, 2012	3	30	3,32	1,75	3,92	1,05	0,83	9,96	19,06	1,37	2,89	1,20	1,12	30,02	15,33	45,35	66%
11	May 11, 2012	2	40	3,76	2,69	3,46	0,86	0,93	12,87	19,18	0,87	2,19	0,70	0,87	29,51	17,99	47,50	62%
12	May 11, 2012	2	30	2,96	1,50	2,93	0,85	1,31	7,28	18,42	1,33	2,77	0,77	1,23	28,40	11,74	40,13	71%
13	May 16, 2012	2	30	3,23	1,67	2,54	1,04	0,91	12,47	17,44	0,85	2,38	1,06	0,82	26,49	17,09	43,59	61%
14	May 16, 2012	2	30	2,52	2,67	3,30	1,18	1,37	12,98	17,24	0,72	1,92	0,87	0,94	26,36	18,43	44,79	59%
15	May 16, 2012	3	40	2,12	1,68	2,36	1,18	1,36	9,21	16,17	0,97	2,85	1,32	1,05	24,86	14,35	39,21	63%
16	May 16, 2012	2	30	2,51	2,75	3,87	0,99	1,14	7,53	15,29	0,87	2,75	1,10	1,26	25,56	13,24	38,80	66%
17	May 18, 2012	2	40	2,93	2,40	3,12	1,08	1,00	11,70	23,47	1,37	2,07	0,93	1,19	32,59	17,48	50,07	65%
18	May 18, 2012	2	40	3,18	1,61	3,70	1,33	0,97	10,85	24,63	0,87	1,77	0,70	0,92	34,25	15,37	49,62	69%
19	May 18, 2012	2	30	3,11	2,11	3,99	1,38	1,09	12,87	20,00	0,73	2,10	0,75	0,80	30,30	17,84	48,14	63%
20	May 18, 2012	2	40	2,09	2,10	3,98	1,19	0,70	6,99	24,40	1,14	1,57	1,23	1,43	32,75	12,65	45,40	72%
21	May 23, 2012	2	30	3,45	1,94	3,64	0,98	0,99	12,59	25,00	1,15	1,95	0,98	1,31	35,04	17,64	52,68	67%
22	May 23, 2012	2	30	3,37	2,00	3,72	0,93	1,31	8,79	16,58	1,07	2,34	0,74	0,79	27,32	13,54	40,86	67%
23	May 23, 2012	2	30	2,52	1,79	3,01	0,87	1,23	7,17	19,82	0,82	1,83	0,91	1,32	28,42	11,56	39,98	71%
24	May 23, 2012	2	40	3,58	2,28	3,30	1,06	0,90	7,64	22,71	1,31	2,09	0,89	1,06	32,57	13,19	45,76	71%
25	May 25, 2012	2	40	2,20	1,93	2,31	0,89	1,15	9,03	22,49	1,36	2,49	0,91	0,87	30,64	14,11	44,75	68%
26	May 25, 2012	2	40	3,73	2,79	3,63	0,77	1,37	9,76	21,02	1,31	2,52	1,31	1,41	32,27	15,95	48,22	67%
27	May 25, 2012	3	30	3,92	2,45	2,07	1,18	0,76	7,91	22,76	0,72	1,79	0,97	1,29	31,30	13,22	44,53	70%
28	May 25, 2012	3	30	2,92	2,01	2,92	1,09	1,13	6,87	20,65	1,38	1,95	0,79	1,30	29,58	12,14	41,72	71%
29	May 30, 2012	3	40	3,38	2,18	2,08	1,22	0,80	8,79	21,00	1,31	2,63	1,06	0,81	29,89	14,55	44,45	67%
30	May 30, 2012	2	30	2,86	2,58	3,31	1,25	1,09	12,46	20,00	0,99	2,68	1,07	1,28	29,95	18,36	48,31	62%
				3,01	2,18	3,22	1,06	1,04	9,95	20,36	1,07	2,26	0,96	1,06	29,89	15,22	45,11	66%
				3,00	2,00	3,00	1,00	1,00	10,00	20,00	1,00	2,00	1,00	1,00	30,00	15,00	45,00	66%

Figure 3depicts that the research focus is ranged from outpatient administration and pharmacy. In details, the processing including several more steps such as outpatient administration, preparation, QA by pharmacist I, Production, QA by pharmacist II and Handover. Also from the diagram we can see that the lead time required 45 minutes for the patient to get the medicine which is far more than the standard (30 minutes). This was caused by nonvalue added time about 15 minutes. Furthermore, this resulted in %VA (PCE) only 66% by deploying PCE formula as follows:

$$PCE = \frac{30}{45} x100\% = 66\% \quad (2)$$

Then, the authors listed the value add time and found that production and preparation time should be improved in order to shorth the non value add time of waiting time for those two processes. In order to depict what happen in pharmacy unit, we develop spaghetti diagram as follow:

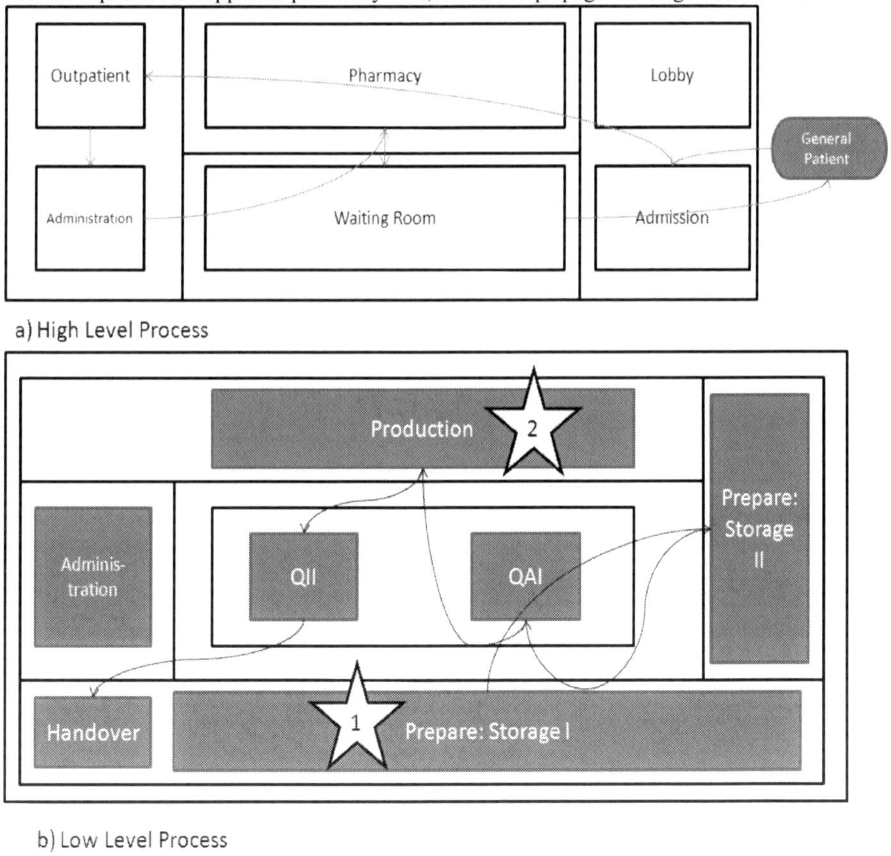

Figure 4. Spaghetti diagram: a) high level process and b) low level process

From the above figure, there are 2 problems identified which have caused longer lead time. First is preparation and second is production. Therefore, we proceeded with root cause analysis as shown in the following diagram:

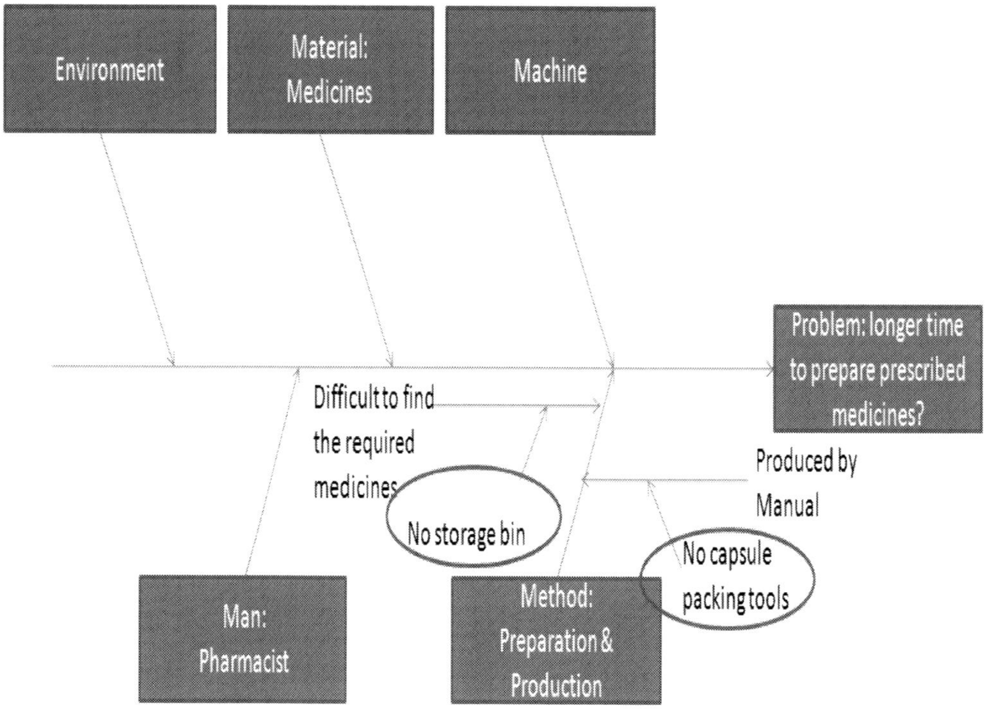

Figure 5. Root Cause Analysis

From figure 5, we can see that no storage bin and no capsule packing tools have caused longer lead time. Therefore, we plan and implement the improvement initiative as shown in table 3 as follows:

Tabel 3. Implementation Plan

No	Solution	Root Cause	Due Date	PIC	Location
1	Storage Bin	No Storage Bin	Oct – Nov 2012	Head of Pharmacy Unit	Pharmacy Unit
2	Procure capsule packing tools (w/ training)	No capsule packing tools	Oct – Nov 2012	Head of Pharmacy Unit	Pharmacy Unit

From table 3, the team has made initiative to procure about 50 storage bin and 2 packing tools in order to help the staffs to improve the process. In order to implement this initiative, the team also conducted training and communication forums to ensure that all staffs can use the tools properly. After conducting the required training, all staffs are seemed ready to implement the tools for shortening the lead team of their process.

The results are shown as follow:

1) Initiative 1: Storage bin

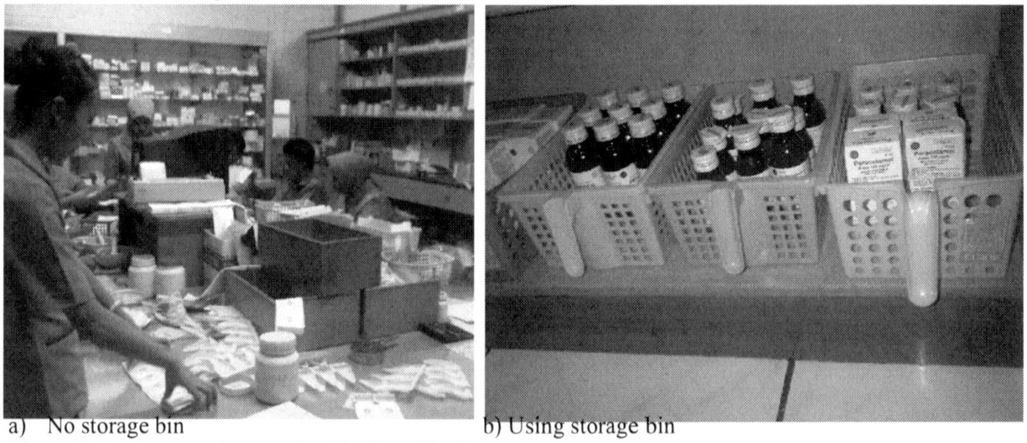

a) No storage bin b) Using storage bin

Figure 6. Implementation result: a) before, b) after

Before implementation, medicines were stored without storage bin (as shown in fig. 6.a). This has caused confusion when the staffs are about to pick up the patient medicine from the shelf. From patient safety perspective, this arrangement may also trigger high risk to false medication. Therefore, the team has decided to procure storage bin. There were about 50 storage bins used to place the medicine based on its medicine name and its maintenance is properly maintained by rolling schedules of the staff. After improvement was taken placed, medicines were stored by storage bins (as shown in fig. 6b).

2) Initiative 2: capsule packing tool

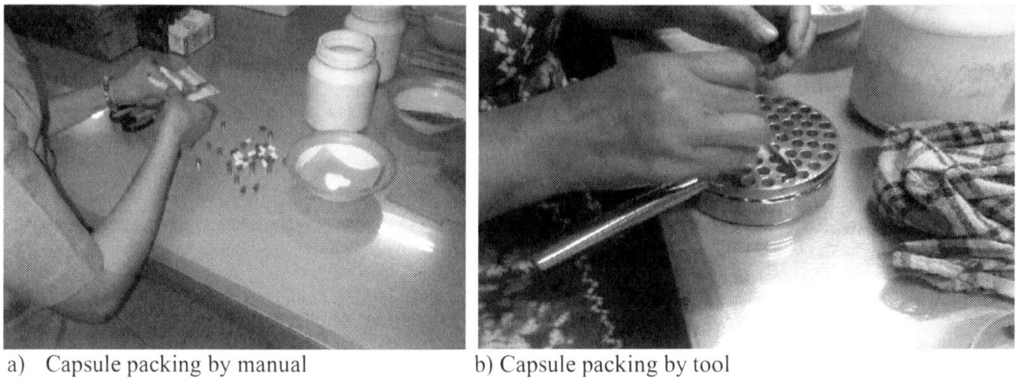

a) Capsule packing by manual b) Capsule packing by tool

Figure 7. Implementation result: a) before, b) after

Before implementation, capsules were packed manually by hand (fig. 7a). This has triggered long time to complete the patient s requirement. During observation, by preparing medicine in this way has caused customer to wait about almost 1 hour. This condition, of course, has triggered high number of complaints filed to XYZ hospital. Therefore, the team also decided to improve this process by procuring 2 capsule packing tools as shown in fig. 7b. After improvement, capsules were packed by using the tools and it has proven to cut the process about 50% and also decrease the patient s complaint files.

After implementation, the authors evaluated the result which is confirmed as its future state value stream mapping as follow:

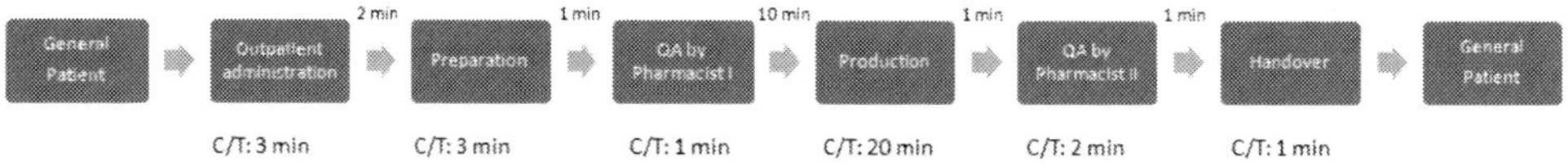

VA=	3′		2′		1′		10′		2′		1′	VA=19′
NVA=		1′		1′		5′		1′		1′		NVA=9′
# 30 patient Internist Gastro # 30 – 40 capsules												LT=28′ %VA=68%

Figure 8: Future State Value Stream Mapping.

The above figure is depicted by observation data during the implementation is taken place which can be described by following table:

Table 4. Observation: Future State Result

OBSERVATIONAL FORM
FUTURE STATE: #CAPSULE: 30 - 40, INTERNIST
PHARMACY INSTALLATION - BINA HUSADA HOSPITAL

No	Date	#Types	# Capsule	Outpatient administration	Waiting	Preparation	Waiting	QA by Pharmacist I	Waiting	Production	Waiting	QA by Pharmacist II	Waiting	Handover	VA	NVA	Total	%
1	July 4, 2012	2	30	2,84	0,90	1,75	0,61	1,06	3,93	11,23	0,95	1,61	1,31	1,30	18,49	7,90	26,39	70%
2	July 4, 2012	2	40	1,89	1,36	1,42	0,86	1,16	3,23	10,29	1,35	2,05	1,38	1,48	16,81	8,18	24,99	67%
3	July 4, 2012	2	30	2,96	0,95	1,78	0,86	0,78	6,11	10,23	0,99	2,36	0,88	1,41	18,12	9,78	27,89	65%
4	July 4, 2012	2	40	2,45	1,05	2,10	1,38	0,78	4,64	11,12	1,37	2,37	1,11	1,06	18,83	9,54	28,37	66%
5	July 6, 2012	3	40	2,11	1,06	2,25	1,07	1,18	4,25	11,24	1,27	2,22	1,37	1,25	19,00	9,02	28,02	68%
6	July 6, 2012	3	30	1,97	0,74	2,57	0,94	0,95	4,43	11,25	1,26	2,65	1,28	0,94	19,39	8,65	28,04	69%
7	July 6, 2012	2	40	1,23	0,80	2,36	0,96	1,47	6,29	11,73	1,20	1,75	0,97	1,41	18,56	10,22	28,78	64%
8	July 6, 2012	2	40	2,57	1,10	1,94	1,35	0,80	4,37	10,70	1,35	2,79	0,99	1,43	18,80	9,17	27,97	67%
9	July 11, 2012	2	30	2,44	0,72	2,51	1,26	1,45	3,22	10,50	1,20	2,12	1,32	1,12	19,02	7,71	26,73	71%
10	July 11, 2012	2	40	1,96	1,39	2,67	0,77	1,43	6,86	10,40	1,15	1,48	1,00	1,49	17,94	11,18	29,12	62%
11	July 11, 2012	2	40	1,77	0,89	2,33	1,14	1,29	4,61	8,33	1,32	2,61	1,33	0,82	16,33	9,29	25,62	64%
12	July 11, 2012	2	30	2,18	0,72	2,55	0,72	1,15	6,79	10,23	0,80	1,99	1,01	0,95	18,10	10,03	28,13	64%
13	July 13, 2012	3	30	2,42	0,82	1,84	0,95	1,26	4,19	12,63	0,96	1,91	1,39	0,75	20,05	8,32	28,38	71%
14	July 13, 2012	3	40	3,03	1,18	2,74	0,93	0,87	3,56	11,52	0,94	1,68	1,34	1,10	19,83	7,95	27,79	71%
15	July 13, 2012	2	40	1,65	1,07	2,48	1,28	1,33	3,67	9,82	1,22	1,48	0,84	1,00	16,76	8,08	24,84	67%
16	July 13, 2012	2	30	2,03	1,34	2,85	1,32	1,32	5,50	9,76	1,18	1,86	1,08	1,38	17,82	10,41	28,23	63%
17	July 18, 2012	3	40	2,53	1,15	1,70	1,07	1,21	4,34	10,01	0,88	1,55	0,96	1,44	17,00	8,40	25,40	67%
18	July 18, 2012	2	30	1,84	0,74	2,93	0,72	1,31	5,19	10,26	1,22	1,43	0,95	1,30	17,77	8,82	26,59	67%
19	July 18, 2012	2	30	2,75	0,82	2,26	1,39	0,87	3,96	9,04	0,87	2,78	1,16	1,37	17,71	8,20	25,91	68%
20	July 18, 2012	2	30	1,85	1,24	2,41	1,39	0,92	4,88	9,45	1,09	2,54	0,73	0,86	17,16	9,34	26,50	65%
21	July 20, 2012	3	40	2,35	1,18	2,98	0,88	1,27	3,45	10,31	1,31	2,36	0,90	1,22	19,28	7,72	27,00	71%
22	July 20, 2012	2	40	2,07	0,75	2,50	1,30	0,75	6,08	10,78	1,35	1,53	1,07	0,81	17,63	10,55	28,19	63%
23	July 20, 2012	2	40	2,15	1,20	1,70	0,77	1,21	4,79	11,93	0,76	1,49	0,91	1,21	18,47	8,44	26,91	69%
24	July 20, 2012	2	40	2,85	0,96	2,80	1,38	1,38	5,09	11,23	1,30	1,45	1,09	1,07	19,71	9,82	29,53	67%
25	July 25, 2012	3	40	2,13	0,80	1,88	1,18	1,49	6,33	10,23	0,83	1,94	0,89	1,49	17,67	10,03	27,70	64%
26	July 25, 2012	2	30	2,70	1,07	2,80	1,21	0,86	5,98	11,67	0,71	1,75	1,23	1,08	19,78	10,21	29,99	66%
27	July 25, 2012	2	30	2,07	1,15	1,79	0,90	1,47	3,53	12,16	0,98	1,95	1,26	1,32	19,44	7,82	27,26	71%
28	July 25, 2012	3	30	1,83	0,96	2,40	0,73	0,96	5,23	12,44	1,35	1,87	0,73	1,32	19,50	8,99	28,49	68%
29	July 27, 2012	2	30	2,84	0,90	2,26	0,71	1,48	6,10	11,76	1,16	1,69	0,77	1,30	20,03	9,64	29,67	68%
30	July 27, 2012	2	40	2,41	0,95	2,20	0,93	0,80	4,89	12,34	0,76	3,00	1,04	1,25	20,74	8,57	29,30	71%
				2,26	1,00	2,29	1,04	1,14	4,85	10,82	1,10	2,01	1,08	1,20	18,53	9,07	27,59	67%
				2,00	1,00	2,00	1,00	1,00	5,00	11,00	1,00	2,00	1,00	1,00	19,00	9,00	28,00	

After implementation, we can see that the lead time has reduced to 30 minutes for the patient to get the medicine below required standard (30 minutes) consisting undenieable nonvalue added time about 9 minutes and value added time 19 minutes. Furthermore, this resulted in %VA or PCE improved to 68% by deploying PCE formula as follow

$$PCE = \frac{19}{28} x100\% = 68\%$$

4. Conclusion

Based on the result, this research has resulted 1) current state value stream mapping (VSVSM) with information that lead time 45 minutes, value added 30 minutes, non value added 15 minutes and %VA or PCE 66% , 2) future state value stream mapping with information that lead time 28 minutes, value added 19 minutes, non value added 9 minutes with %VA or PCE 68%, 3) improvement by procuring storage bin and packing capsule tool to overcome root causes as no storage bin and no capsule packing tool. The implication of this research is that the staff should maintain the usage of storage bin and capsule packing tool in order to maintain the improved %VA or PCE by providing Standard Operating Procedure.

Acknowledgment
The authors would like to thank Bina Nusantara University for financial support of research grant No. 113/ DRIC/ IV/ 2012 and XYZ Hospital for its technical support during research in XYZ Hospital.

References

[1] Arthur, Jay 2011 *Lean Six Sigma for Hospitals* (New York, McGraw-Hill Companies)
[2] Barnes, Ralph M 1980 *Motion and Time Study Design and Measurement Work* (USA, John Wiley and Sons)
[3] Chase, Aquilano, Jacobs 2001 *Operations Managemen for Competitive Advantage* (New York, McGraw-Hill Companies)
[4] Graban, Mark 2006 *Lean Hospitals: Improving Quality, Patient Safety, and Employee Satisfaction* (New York, CRC Press)
[5] Heizer, Render 2001 *Operations Management* (USA, Prentice Hall)
[6] Koelling, Etiel, Mahapatra, Messner, Grov 2013 *Value Stream Mapping the Emergency Department* (Available online via <http://www.iienet.org/uploadedfiles/SHS/Resource_Library/Details/180.pdf> [accessed January 27, 2013])
[7] Liker, Jeffrey K. 2004 *The Toyota Way*. (New York, McGraw-Hill Companies)
[8] Nahmias, Steven 2005 *Production and Operation Analysis* (New York, McGraw-Hill Companies)
[9] Ravindran, Philips, Solberg 1987 *Operations Research: Principles and Practice* (Singapore: John Wiley & Sons)
[10] Rother M. & Shook 2003 *Learning to See: Value Stream Mapping to Create Value and Eliminate Muda* (Brookline: The Lean Enterprise Institute)
[11] Vincent Gaspersz 2007 *Organizational Excellence* (Jakarta Gramedia)
[12] Wachter, Robert M 2008 *Understanding Patient Safety* (New York, McGraw-Hill Companies)
[13] Winston, Wayne L 1991 *Operations Research: Applications and Algorithms* (USA: PWS-KENT Publishing Company)
[14] Womack, J.P. and Jones, D.T. 1996 *Lean Thinking: Banish Waste and Create Wealth in your Corporation* (New York: Simon & Schuster)

Electrical properties of carbon nanotubes synthesis by double furnace thermal-CVD technique at different temperatures on porous silicon template

F.S. Husairi[1,2,a], S.A.M Zobir[1,2], M. Rusop[1,3], S. Abdullah[1,2]

[1]*NANO-SciTech Centre (NST), Institute of Science;*
[2]*Faculty of Applied Sciences;*
[3]*NANO-ElecTronic Centre (NET), Faculty of Electrical Engineering;*
Universiti Teknologi MARA (UiTM),
40450 Shah Alam, Selangor, Malaysia
[a]*mhusairifadzilah@yahoo.com*

Abstract. Multiwalled Carbon Nanotubes (MWCNs) were synthesized using double-furnace thermal chemical vapor deposition technique at 700 - 900 ^{0}C on porous silicon nanostructure (PSiNs). Palm oil used as a carbon natural source, ferrocene as a catalyst and nitrogen gas as a carrier gas. The precursor were vaporized at 475 ^{0}C carried by nitrogen gas which flow at constant rate 150 sccm/min. Carbon nanotubes characterized by using Raman Spectroscopy and field emission scanning electron microscopy (FESEM) to check its structure and crystallites before tested with I-V probe. Au contact used as a metal contact deposited on CNTs layer. Carbon nanotubes (CNTs) with uniform diameter were found grown on porous silicon for each temperature used. Based on micro-Raman spectroscopy result, the peak of carbon nanotubes (around 1 300 to 1 600 nm) was detected. The *I-V* characteristic of CNTs deposited had different profile when deposited at different temperature.

1. Introduction

A carbon nanotubes is a carbon in tube-shape material and having a diameter in nanometer scale. This carbon consist a hollow cylindrical-shape molecules with graphitized walls [1]. Carbon nanotubes (CNTs) attracted a lot of attention in wide researches' application today especially in electronics application such as sensor [2], field emission [3], hydrogen storage [4] etc. CNTs also are a good material in term of thermal stability and mechanical properties. Electrical characteristics studied of important material to measure the capability of material in electronic field application. It can be explained in term of resistance, conductance and capacitance measurement. Generally, CNTs divided in two types: single-walled (SWCNTs) and multi-walled (MWCNTs). The SWCNTs are consists a simple nanostructure with single rolled layer (concentric tubes) of graphite, while the MWCNTs are consist multiple rolled layers of graphite (various SWCNTs). The SWCNTs can be either semiconducting or metallic materials because its depend on the chiral vector (*n*, *m*; where *n* and *m* are integer) [5, 6]. Different with SWCNTs, the MWCNTs could be form by combination of metallic and semiconducting nanotubes element [7].

In this work, porous silicon nanostructures were used as a substrate to grow carbon nanotubes. Porous silicon is Si substrates after partial electrochemical etching in hydrofluoric acid (HF) based material. Porous silicon nanostructures (PSiNs) ideal substrate for grow self-oriented nanotubes on large surfaces and CNTs produced was better than on plain Si [8]. Carbon nanotubes (CNTs) generally can be synthesized by using a few techniques, likes arc discharge [9], laser ablation [10], chemical vapor deposition [3] and others. The selection method is important in order to produce CNTs with predictable and reproducible properties. In addition, the controllable CNTs diameter and chirality would allow the exploitation in electrical, thermal and mechanical properties in various applications. In this work, thermal-CVD method was choose because CNTs can be synthesized at lower temperatures (500-900 ^{0}C) [3] and it is a simply method. CNTs synthesized at different deposition temperatures on porous silicon to study the effect of structures and morphology of CNTs to their electrical properties because each parameter will affect nanomaterial properties. This work found that the structural of CNTs affected by temperature and I-V curve show the relationship between structural and electrical properties of CNTs.

2. Experiment detail

2.1 Electrochemical Etching Process

Firstly, porous silicon nanostructures (PSiNs) was prepared by using electrochemical etching method using p-type Si wafer [100] with 0.4-2 Ω cm resistivity and 330 ± 40 μm thickness. A square dimension was cleaved and placed in cell using a piece of aluminum foil as a back contact. The etching process was set up at t = 20 minute and J = 20 mA/cm^2. Ethanoic hydrofluoric acid (HF) 48% and absolute ethanol (C_2H_5OH) at ratio 1:1 with illumination by halogen lamp was used as a based solution in etching process. After porous form on silicon, it was rinsed by DI water and dried using nitrogen gas. The surface topography of PSiNs was investigated by using FESEM.

2.2 Chemical Vapor Deposition (CVD) Method

The carbon nanotubes synthesized by using double-furnace CVD method [11] set up like in figure 1. 5% wt of ferrocene which used as catalyst mixed in palm oil and placed in alumina boat in a double furnace (Furnace 1). PSiN, which used as a template, place on alumina boat in the quartz tube at the centre of Furnace 2. Rubber stopper used at the end of quartz tube to avoid the carbon gases react with oxygen and produce carbon dioxide. Nitrogen gases was used as a gas carrier flow at constant flow rate 150 sccm/min from end of quartz at Furnace 1 to Furnace 2. The Furnace1 was heated to 475 ^{0}C and 700-900 ^{0}C at Furnace 2. The deposition time set at 1 hour and followed with annealing process for 30 min. The characteristics of CNTs on porous silicon analyzed by using FESEM (JOEL), micro Raman spectroscopy (Horiba Jobin Yvon, LabRam HR800) and I-V probe.

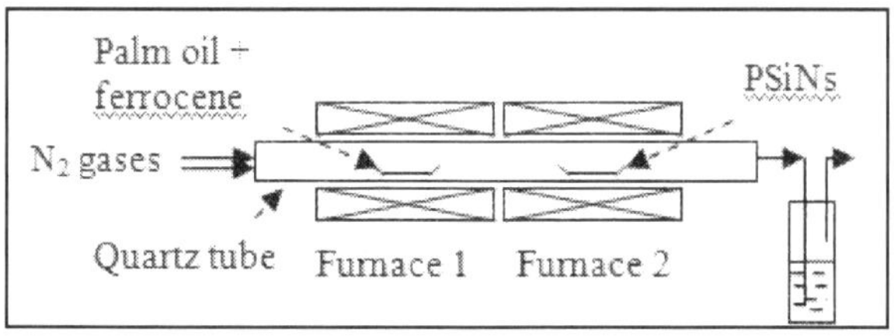

Figure 1. Double furnace thermal-chemical vapor deposition (TCVD) set-up to produced CNTs.

3. Result and Discussion

Field emission scanning electron microscope (FESEM-JOEL) was used to investigate the surface structure of porous silicon nanostructures (PSiNs) and CNTs. Figure 2(a) show the FESEM image of PSiNs surface structures prepared by electrochemical etching obtain at J_A = 20 mA/cm^2 and t = 20 minutes view by at magnification 50k. FESEM results show micrometer-size of pore form on the surface of p-type silicon with the columnar pores is not regular. These pores know as nano pores because, the size of pore produced were between 15 ~ 30 nm and its represent by a dark area (small figure) in the figure 2(a).

We were successful deposited multiwall carbon nanotubes (CNTs) on porous silicon by using double furnace thermal-CVD at different deposition temperatures. Figure 2 show CNTs on porous silicon grow viewed using magnification 1k and 100k. We can see that the density of CNTs grow on Si substrate increase respectively with temperatures. At lower temperature (figure 2(b)), less CNTs produced because the thermal energy at furnace 2 not suitable for carbon source and catalyst activity to form a CNTs.

Figure 2. The FESEM image of (a) PSiNs surface produced by electrochemical etching process and multi-wall CNTs grow on porous silicon substrate at temperatures various temperatures; (b) 700, (c) 750, (d) 800, (e) 850 and (f) 900 ^{0}C.

When temperatures increase, dense CNTs synthesized on PSiN substrate. Smooth nanotubes produced at temperature 750 - 900 ^{O}C. Temperature at 750 ^{O}C is good parameter to synthesized good quality of CNTs because small diameter and smooth tubes produced (figure 2(c)). It can be supported by Raman result in figure 3. Besides that, uniform nanotubes in bundle success synthesized at this temperature. Based on FESEM image, we also can observed that the diameter of tube change with temperatures. The CNTs have large diameter in the range 75-90 nm at temperatures 850 ^{O}C (in figure 2(e)). CNTs produced at 900 ^{O}C had a small diameter, but more amorphous carbon produced. The multi-walled carbon nanotubes have diameters of 15-30 nm indicated in figure 4(f). The type of CNTs grown on porous detected by using Raman spectroscopy (figure 5).

Raman spectroscopy used to measure Raman scattering spectra of nanotubes which are important for CNTs characteristics [12]. Figure 4 shows the Micro-Raman spectra for the CNTs grown on porous silicon substrate at temperature 700 – 900 ^{O}C. The spectrum is dominated by two peaks (D and G) located at 1530 cm^{-1} (D line) and 1570 - 1650 cm^{-1} (G line). In the high frequency spectra, a G-band representing crystalline graphitic carbon and D-band indicates lattice distortions in the curve graphene sheets, tube ends, etc [13]. The carbon G and D peaks for the samples at that position substantiate that are multi-walled (MWCNT) type were synthesized [14]. No lower frequency RBM peaks which lies from 100 to 400 cm^{-1} can be attributed single-walled CNTs (SWCNT) detected on this sample.

From figure 4 it was observed that the intensity of G-peak is greater than D-peak with sharp peak indicated that good graphitization in nanotubes successes synthesis and the graphitization increase by temperature observed. The G-peaks intensity highest at deposition temperature 750°C observed, that means good graphitization of CNTs produced at this temperature. At 700 ^{O}C, 750 ^{O}C and 850 ^{O}C, the D- and G-peak observed at 1350 cm^{-1} and 1580 cm^{-1}. At temperature 800 ^{O}C (Fig. 3(c)), D- and G-peak shifted to higher wave number (1416 cm^{-1} and 1642 cm^{-1}). This shifted happen maybe cause by defects in the tube [15].

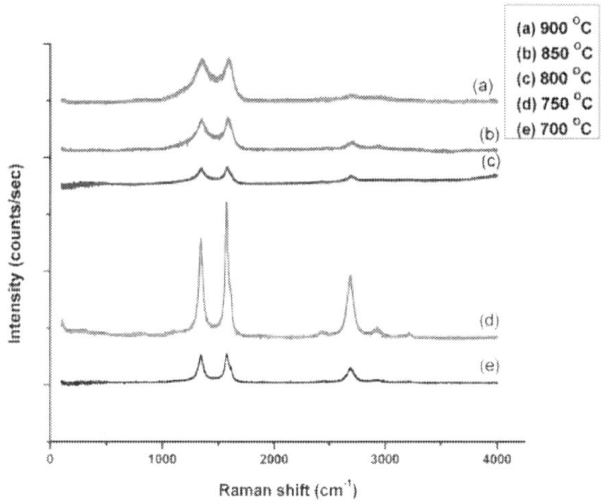

Figure 3. Raman spectra of MWCNTs on PSiNs substrate deposited at different temperatures.

In CNTs sample, the relative intensity of the D band with respect to the G band can be used to measure the concentration of defect [16]. When I_D/I_G ratio low, the CNTs with graphite structures were produced more compare to defect structures. Table 1 below show the I_D/I_G ratio of CNTs deposited on Si substrate. At 750 ^{O}C, the high graphite structures produced compare to other samples because the relative intensity I_D to I_G is lowest.

Table 1. I_D/I_G ratio of CNTs deposited at different temperatures.

Deposition Temperature (°C)	I_D/I_G
700	0.9705
750	0.7079
800	0.9516
850	0.9352
900	0.9812

I-V characteristics of CNTs samples come by using 2-point probe *I-V* testing. Au contact used as a metal contact sputter on CNTs layer about 60 nm to reduce the contact resistant between CNTs layer and *I-V* probe. Figure 4 show the *I-V* characteristics samples of CNTs samples after 3V potential applied on its. All samples show Ohmic profile with linear line produced when graph current versus voltage plotted. This can be concluded that Au performs a good contact with CNTs layer because the barrier is so low. The current flow is high even though low potential applied (0-5V) during the testing for all sample especially sample prepared at 750 °C. This can reduce the applied potential in device application.

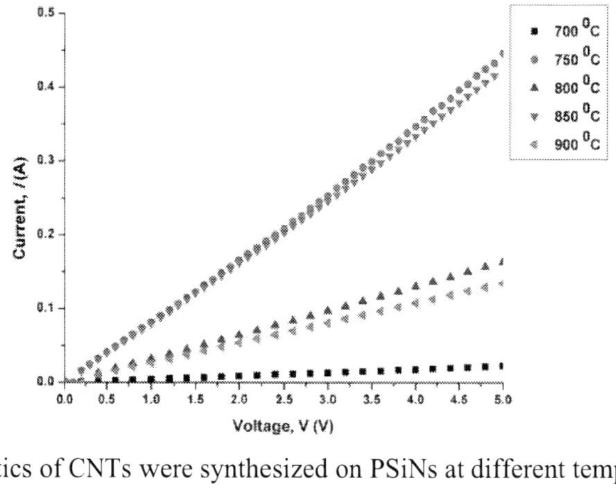

Figure 5. *I-V* characteristics of CNTs were synthesized on PSiNs at different temperatures.

The resistance dramatically drops by temperature showed in Table 2. Commonly, the conductivity of nanomaterial depends on crystallinity and doping or impurities in nanometerial. For CNTs, another factor can influence that property is diameter of tube. From table 2 below, we can see that the conducting of CNTs increase when temperature depositing increase. This is due to the diameter and graphene structures of CNTs were synthesized. CNTs synthesized at temperature 850 °C possess large tubes diameter, so the number of wall (N_{walls}) in tube is large because based on literature N_{walls} increases at larger CNT diameter [17]. The high conductivity of align CNTs produced effected by graphene structures existence more compare with defect. These help the carrier moving easily through graphitic interfacial layer [18] because lower barrier resistance.

Table 2. The resistance of CNTs deposited at different temperatures.

Deposition temperature (°C)	Resistance (Ω)
700	250.00

750	11.11
800	30.30
850	12.05
900	37.04

4. Conclusion

We have successful synthesized of multi-walled carbon nanotubes by thermal chemical vapor deposition on porous silicon nanostructures substrate made by electrochemical etching method. CNTs produced on porous silicon substrate proved that it can be used with semiconductor devices to fabricate integrated circuits (IC). *I-V* characteristics of CNTs affected by structures and quality produced. CNTs synthesized on porous silicon at temperature 750^0C possess more graphite structures and fewer defects. It was directly influence the electrical property. Conductivity of CNTs increases when the diameter of nanotubes increases.

5. References

[1] R. Angelucci, R. Rizzoli, V. Vinciguerra, M. Fortuna Bevilacqua, S. Guerri, F. Corticelli, and M. Passini, 2007, *Physica E: Low-dimensional Systems and Nanostructures,* vol. 37, 11-15.
[2] J. A. Robinson, E. S. Snow, and F. K. Perkins, 2007, *Sensors and Actuators A: Physical,* vol. 135, 309-314.
[3] J. Li, W. Lei, X. Zhang, B. Wang, and L. Ba, 2004, *Solid-State Electronics,* vol. 48, 2147–2151.
[4] Y. Ye, C. C. Ahn, C. Witham, B. Fultz, J. Liu, A. G. Rinzler, D. Colbert, K. A. Smith, and R. E. Smalley, 1999, *Applied Physics Letters,* vol. 74, 2307-2309.
[5] E. T. Thostenson, Z. Ren, and T.-W. Chou, 2001, *Composites Science and Technology,* vol. 61, 1899-1912.
[6] J. Kong and S. U. D. o. Chemistry, *Carbon nanotubes: synthesis, integration and properties*: Stanford University, 2002.
[7] O. K. Varghese, P. D. Kichambre, D. Gong, K. G. Ong, E. C. Dickey, and C. A. Grimes, 2001, *Sensors and Actuators B: Chemical,* vol. 81, 32-41.
[8] T. W. Ebbesen and P. M. Ajayan, 1992, *Nature,* vol. 358, 220-222.
[9] Z. Shi, Y. Lian, F. H. Liao, X. Zhou, Z. Gu, Y. Zhang, S. Iijima, H. Li, K. T. Yue, and S.-L. Zhang, 2000, *Journal of Physics and Chemistry of Solids,* vol. 61, 1031-1036.
[10] Y. Zhang and S. Iijima, 1999, *Applied Physics Letters,* vol. 75, 3087-3089.
[11] S. A. Bakar, S. Muhamad, P. S. M. Saad, S. A. M. Zobir, R. M. Nor, Y. M. Siran, S. A. M. Rejab, A. J. Asis, S. Tahiruddin, S. Abdullah, and M. R. Mahmood, 2011, *Defect and Diffusion Forum,* vol. 312-315 906-911.
[12] P. M. Fauchet, 1998, *Journal of Luminescence,* vol. 80, 53-64.
[13] J. Dian, A. Macek, D. Nižňanský, I. Němec, V. Vrkoslav, T. Chvojka, and I. Jelínek, 2004, *Applied Surface Science,* vol. 238, 169-174.
[14] A. B. Suriani, A. A. Azira, S. F. Nik, R. Md Nor, and M. Rusop, 2009, *Materials Letters,* vol. 63, 2704-2706.
[15] R. A. Afre, T. Soga, T. Jimbo, M. Kumar, Y. Ando, M. Sharon, P. R. Somani, and M. Umeno, 2006, *Microporous and Mesoporous Materials,* vol. 96, 184-190.
[16] R. M. Malek Abbaslou, J. Soltan, and A. K. Dalai, 2010, *Applied Catalysis A: General,* vol. 372, 147-152.

[17] N. Chiodarelli, O. Richard, H. Bender, M. Heyns, S. De Gendt, G. Groeseneken, and P. M. Vereecken, 2012, *Carbon,* vol. 50, 1748-1752.

[18] Y. Chai, A. Hazeghi, K. Takei, H.-Y. Chen, P. C. H. Chan, A. Javey, and H.-S. P. Wong, 2012, *IEEE Transactions On Electron Devices* vol. 59, 12-19.

Acknowledgement

The author would like to thank the Universiti Teknologi MARA (UiTM) and the Malaysian Government for their support and funding.

Acceptance of Internet Banking Systems among Young Managers

Mohd Shoki Md Ariff[1], Yeow S.M.[1], Norhayati Zakuan[1*], Ahamad Zaidi Bahari[1]

1 Department of Management, Faculty of Human Resource and Management, Universiti Teknologi Malaysia (UTM), Skudai, Johor, Malaysia

E-mail: norhayatimz@utm.my

Abstract. The aim of this paper is to determine acceptance of internet banking system among potential young users, specifically future young managers. The relationships and the effects of computer self-efficacy (CSE) and extended technology acceptance model (TAM) on the behavioural intention (BI) to use internet banking system were examined. Measurement of CSE, TAM and BI were adapted from previous studies. However construct for TAM has been extended by adding a new variable which is perceived credibility (PC). A survey through questionnaire was conducted to determine the acceptance level of CSE, TAM and BI. Data were obtained from 275 Technology Management students, who are pursuing their undergraduate studies in a Malaysia's public university. The confirmatory factor analysis performed has identified four variables as determinant factors of internet banking acceptance. The first variable is computer self-efficacy (CSE), and another three variables from TAM constructs which are perceived usefulness (PU), perceived ease of use (PE) and perceived credibility (PC). The finding of this study indicated that CSE has a positive effect on PU and PE of the Internet banking systems. Respondents' CSE was positively affecting their PC of the systems, indicating that the higher the ability of one in computer skills, the higher the security and privacy issues of PC will be concerned. The multiple regression analysis indicated that only two construct of TAM; PU and PC were significantly associated with BI. It was found that the future managers' CSE indirectly affects their BI to use the internet banking systems through PU and PC of TAM. TAM was found to have direct effects on respondents' BI to use the systems. Both CSE and the PU and PC of TAM were good predictors in understanding individual responses to information technology. The role of PE of the original TAM to predict the attitude of users towards the use of information technology systems was surprisingly insignificant.

Keywords: Computer Self-Efficacy (CSE), Behavioural Intention (BI), Technology Acceptance Model (TAM) and Internet Banking

1.0 Introduction

Researches on examining determinants of computer technology acceptance are becoming a central field of study in electronic mediated environment. In Malaysia, many researches have been conducted in internet banking services that addressed specific understanding on internet banking acceptance among Malaysians, including intention of potential young users of undergraduate students to use the services [1]. It was reported that 89% of internet users in Malaysia are youths and young adults [2] and most of them are students of tertiary education who are regarded as future users of internet banking. In examining users' acceptance of information systems, the technology acceptance model or TAM [3] appears to be the most widely accepted model among information systems researchers [4] because of its parsimony [5, 6] and it has better empirical support [4, 7]. In TAM, two determinants of perceived usefulness (PU) and perceived ease of use (PE) are widely used to determine acceptance of information systems and they are valid in predicting the individual's acceptance of various corporate IT [8]. Review on the acceptance of internet banking systems researches involving TAM and BI suggested that users' perceived credibility (PC) of the systems, should be considered to better predict BI of potential users to use the systems [1, 4 and 9]. Further, the external variables in TAM model, such as CSE, are proposed [10] and CSE has been extensively used by researchers to better explain how individual reacts to information systems [1, 4, 7, 11]. Therefore, this study develops a technology acceptance model for internet banking system based on the modified version of TAM to examine the effects of CSE and PU, PE and PC on the BI to adopt the systems. The model developed extends the understanding on how potential young users response to information technology systems from Malaysia's undergraduate students perspective.

2.0 Literature Review

TAM was introduced by Davis [3], which stressed on adoption degree of a person in using a technology and aspects that affect his or her acceptance or intention to use the technology. TAM suggests that users' adoption of information technology systems is determined by their intention to use the systems, which in turn is determined by their beliefs towards the systems. Users' beliefs about information technology systems are influenced by their perceived usefulness (PU) and perceived ease of use (PE) of the systems. PU is the extent to which a person believes that using a particular system will enhance his or her job performance [3, 4] and it has been found to have positive effects on BI to use internet banking [4, 7]. PE indicates the degree to which a person believes that using a particular system would be free of effort [3, 4] and it has been found to have positive effects on BI [4, 5, 7]. Ajzen and Fishbein [12] suggested that a person's adoption or actual behavioural could be determined by considering his or her prior intention along with his beliefs for a given behaviour. Thus, BI serves as an indicator of a person's readiness to perform certain behaviour. It measures a person's relative strength of intention to perform behaviour [3], such as his intention to use information technology systems. Current researches involving TAM in examining acceptance of internet banking systems suggest that additional variables in TAM are required to derive a better understanding on determinant factors influencing the decision of bank users to use the systems. Thus, PC is added into the classical TAM [1, 4, and 9]. PC is more related to one's judgment on the privacy and security issues of the internet banking systems. Security refers to the protection of information or systems from unsanctioned intrusions or outflows and privacy refers to the protection of various types of data that are collected during users' interactions with the internet banking system. PC has been found to have a significant positive effect on behavioural intention [4].

One key benefit of employing TAM to understand system usage behavior is that it provides a framework to examine the effects of external variables, such as individual differences, on the system usage [4, 13]. Individual differences, such as CSE, is considered as one of the external variable in TAM model [1, 4, 10] and it plays a vital role in shaping an individual's feeling and behaviour [11]. CSE is the judgment of one's ability to use a computer [11] and this ability would lead to more favourable behavioural intention through its influence on PU and PE [1, 4, 10].

3.0 Validity and Reliability

A questionnaire of 29 items was developed based on previous researched to measure CSE, and PU, PE and PC of TAM and BI. In Section A of the questionnaire, seven items proposed by Compaeu and Higgins [11[were adapted to measure CSE. These items are widely used and able to judge an individual ability to use a computer as well as the extent to which this ability influences his or her BI to accept an information system [1, 4, 11]. Six items for each PU and PE of the original TAM were adapted from Davis [3] as they are able to determine individuals' judgement on the extent to which use of an information system will enhance his or her job performance and free of effort [3]. PC was included in the TAM model with five items based on the work of Pikkarainen *et al.* [9], to examine judgement of an individual on issues of privacy and security in internet banking system [4, 9]. BI is included in the final model of TAM [10] and five questions (Section C) were proposed [3, 10] and adapted in this study. The five items are sufficient and possess high internal consistency to determine individuals' BI [3, 10].

275 students (94.8% response rate as the pre-determined sample size was 290 for 5% margin error) who are pursuing their degree in Technology Management in a Malaysia's public university participated in this research. The students were chosen using stratified random sampling procedure based on their programs and year of study. In this study, the KMO measure of sampling adequacy for CSE and TAM (PU, PE and PC) and BI were above 0.70, supported by Bartlett's test of Sphericity of 0.00, allowing the research to proceed with factor analysis. Based on the Total Variance Explained of Confirmatory Factor Analysis (CFA), four dimensions with eigenvalues greater than 1.0 – CSE, and PU, PE and PC of TAM - have been extracted with the cumulative percentage of 72.699%. All the seven items of CSE, six items of PU and PE respectively and five items of PC were retained based on the results on component matrix with factor loading ≥ 0.5. For BI, only one dimension in the initial solution with eigenvalues greater than 1.0 has been extracted (cumulative percentage of 69.164%) and all the five items were accepted based on the result on component matrix with factor loading ≥ 0.5.

Reliability coefficient assessing the consistency of the entire scale, in which *Cronbach's Alpha* is the most often used; and it is generally agreed that Cronbach's Alpha should exceed 0.70 [14]. In this study, the alpha value for CSE was 0.910 and the values for PU, PE, PC and BI were 0.927, 0.929, 0.919 and 0.888 respectively. Thus, construct of CSE, TAM and BI used in this study was reliable and valid for further analysis.

4.0 Result and Analysis

The multiple regression analysis, as presented in Table 1, indicated that CSE had a positive effect on PU (Regression 1 - β 0.493, t 9.357, Sig. 0.000) and PE (Regression 2 - β 0. 397, t 7.144, Sig. 0.000) and PC (Regression 3 - β 0. 160, t 2.685, Sig. 0.008) of the Internet banking systems. Thus, H1a and H1b were supported. However, H1c was rejected.

Table 1: Results on the effect of Computer Self-Efficacy on TAM (Perceived Usefulness, Perceived Ease of Use and Perceived Credibility) of the behavioural intention to use the internet banking system

Hypothesis		Regression 1	Regression 2	Regression 3
	(Constant)	**CSE** 2.106	**CSE** 2.431	**CSE** 2.808
H1a: Computer self-efficacy will have a positive effect on perceived usefulness of the behavioural intention to use Internet banking.	**PU (*β*)** t-value p-value VIF	0.493 9.357 0.000** 1.000		
H1b: Computer self-efficacy will have a positive effect on perceived ease of use of the behavioural intention to use Internet banking.	**PE (*β*)** t-value p-value VIF		0.397 7.144 0.000** 1.000	
H1c: Computer self-efficacy will have a negative effect on perceived credibility of the behavioural intention to use Internet banking	**PC (*β*)** t-value p-value VIF			0.160 2.685 0.008** 1.000
Note: * $p<0.05$; ** $p<0.01$	F statistic R^2	87.546 0.243	51.044 0.158	7.208 0.026

The result of multiple regression analysis in Table 2 indicated that only two constructs of TAM – PU and PC – were significantly associated with BI (Regression 4). PU of β 0.272, t 3.696, Sig. 0.000 and PC of β 0.411, t 7.150, Sig. 0.000 denoting that users' perceived usefulness and perceived credibility of internet banking systems will positively influence their intention to use the systems. Thus, H2a and H2c were supported. H2b was rejected because PE was insignificant to BI.

Table 2: Results on the effect of PU, PE, and PC of TAM on Behavioural Intention

Hypothesis		Regression 4
	(Constant)	**BI** 1.285
H2a: Perceived usefulness will positively affect the behavioural intention to use internet banking.	**PU** (*β*) t-value p-value VIF	0.272 3.696 0.000** 2.090
H2b: Perceived ease of use will positively affect the behavioural intention to use internet banking.	**PE** (*β*) t-value p-value VIF	-0.037 -0.486 0.628 2.281
H2c: Perceived credibility will positively affect the behavioural intention to use internet banking.	**PC** (*β*) t-value p-value VIF	0.411 7.150 0.000** 1.277
Note: * p<0.05; ** p<0.01	F statistic R^2	38.580 0.299

The total effects of CSE and TAM on BI was tabulated in Table 3. It was found that the respondents' CSE indirectly affects their BI to use the internet banking systems through PU and PC of TAM. The total indirect effect is accounted 0.200. TAM was found to have direct effects on respondents' BI to use the systems. The total direct effect is 0.683. Thus, it can be concluded that PU and PC of TAM were found to have direct influences on users' BI to use the internet banking systems.

Table 3: Results of the total effects of Computer Self-Efficacy and TAM on Behavioural Intention

	Total effects of CSE on BI	Total effects of TAM on BI
The young managers of internet banking	Indirect effects =(0.493*0.272) + (0.160*0.411) **=0.200**	Direct effects =0.272 + 0.411 **=0.683**

5.0 Discussions

This study examined the formation of CSE, PU, PE and PC of TAM and BI to predict users' acceptance of information technology systems, specifically internet banking system. The finding indicated that CSE and PU, PE and PC of TAM are determinants of users' acceptance of internet banking systems, suggesting that the previous researches findings on this matter is well supported [1, 4]. The positive effects of external variable of CSE on PU and PE [1, 4, 7] of TAM are consistent with previous research findings. This indicated that users' PU and PE of internet banking system will influence their acceptance of and intention to use the system. The positive impact of security and privacy of PC on users' intention to use internet banking service suggest that young users are very concern on these issues when they are considering using the service. However, the positive effect of CSE on PC is inconsistent with the work of Wang, et.al. [4]. It was found that the potential young users' CSE indirectly affects their BI to use the internet banking systems through PU and PC of TAM [4]. This suggests that users' CSE, through PU and PC, is important external variable when predicting their intention to use internet banking system. The positive effects of PU of TAM on BI show consistency with past researches [4, 7]. The positive and stronger effect of PC than PU [4] highlights that issues of privacy and security of extended TAM are important construct of information systems acceptance. These issues are continuously and consistently concerned by users and they will strongly influence intention of users to use internet banking services. However, the role of PE is insignificant, which is very inconsistent with the previous research findings [1, 3, 4]. It is recommended that future researches should consider other variables of individual differences of external variables, which may contribute to the insignificant effect of users' PE of information technology systems and their intention to adopt the systems.

6.0 Conclusions

The result of this research revealed that CSE, and PU, PE and PC of TAM are determinants of users' BI to use internet banking systems. The inclusion of CSE and PC with PU and PE in the TAM model contributed to the better understanding on the intention of potential users to adopt internet banking systems. The indirect relationships between CSE and BI through PU and PC of TAM indicated that potential users' CSE is important in influencing their intention to use the systems. The insignificant effect of PE on BI (PE: Sig. 0.628, p-value > 0.05) should be further validated as most researches revealed that users' perceived ease of use of information systems did affect their BI to use the systems.

References

[1] Amin, H. (2007). Internet banking adoption among young intellectuals. *Journal of Internet Banking and Commerce, 12* (3), 1-13.

[2] Haque, A. and Khatibi, A., (2005). E-shopping: Current Practices and Future Opportunities towards Malaysian Customer Perspective. Journal of Social Sciences, Vol. 1, Issue 1, pp. 41-46.

[3] Davis, F. D. (1989) Perceived usefulness, Perceived Ease of use and user acceptance of information technology, *MIS Quarterly,* Vol. 16 (2), 319-340.

[4] Wang, Y. S., Wang, Y.M., Lin, H. H. and Tang T.I. (2003) Determinants of user acceptance of Internet banking: An empirical study, *International Journal of Service Industry Management,* Vol. 14 (5), 501-519.

[5] Agarwal, R. and Prasad, J. (1999), "Are individual differences germane to the acceptance of new information technologies? Decision Sciences, Vol. 30 No.2, pp. 361-91.

[6] Davis, F., Bagozzi, R.P. and Warshaw, P.R. (1989), "User acceptance of computer technology: a comparison of two theoretical models", *Management Science,* Vol. 35 No. 8, pp. 982-1003.

[7] Guriting, P., and Ndubisi N.O. (2006). Borneo online banking: Evaluating customer perceptions and behavioural intention. *Management Research News*, 29 (1/2), 6-15

[8] Doll, W.J., Hendrickson, A. and Deng, X. (1998), "Using Davis's perceived usefulness and ease-of-use instruments for decision making: a confirmatory and multi-group invariance analysis", Decision Science, Vol. 29 No. 4, pp. 839-69.

[9] Pikkarainen, T., Pikkarainen K., Karjaluoto, H. and Pahnila S. (2004) Consumer acceptance of online banking: an extension of technology acceptance model. *Internet Research,* 14(3), 224 235.

[10] Venkatesh V, Davis FD (1996). A model of the antecedents of perceived ease of use: development and test. *Decis. Sci.* 27(3), 451- 481.

[11] Compeau, D. R., and Higgins, C. A. (1995). Computer self-efficacy: Development of a measure and initial test. *MIS Quarterly, 19*, 189-211.

[12] Ajzen, I. and Fishbein, M. (1980), Understanding Attitudes and Predicting Social Behavior, Prentice-Hall, Englewood Cliffs, NJ.

[13] Hong, W., Thong, J.Y.L., Wong, W.M. and Tam, K.Y. (2001), "Determinants of user acceptance of digital libraries: an empirical examination of individual differences and system characteristics", Journal of MIS, Vol. 18 No. 3, pp. 97-124.

[14] Sekaran, U. (2000). *Research Methods for Business*, John Wiley & Sons, United States of America, 28-35.

Syntheses of crosslinked latex nanoparticles using differential microemulsion polymerization

N F Hassmoro[1,2,*], **M Rusop**[2,3] **and S Abdullah**[1,2]

[1] Faculty of Applied Sciences;
[2] NANO-SciTech Centre, Institute of Science;
[3] NANO-ElecTric Centre, Faculty of Electrical Engineering;
Universiti teknologi MARA (UiTM), 40450 Shah Alam, Selangor, Malaysia

*E-mail: nurfarahinhassmoro@gmail.com

Abstract. The differential microemulsion polymerization was used to synthesize latex nanoparticles. In this paper, 1, 3-butylene glycol dimethacrylate (1, 3-BGDMA) was used as a crosslinker respectively 1-5 weight% of monomer total. Butyl acrylate (BA), butyl methacrylate (BMA), and methacrylic acid (MAA) was used as the monomer. The thin film of latex nanoparticles were prepared by using spin coating method and have been dried at 100°C for 5 minutes. The amount of the crosslinker added in the polymerization was optimized and we found that the particle sizes fall in the range of 30-60 nm. The structural morphology of the uncrosslinked latex represented the most homogeneous image compared to the crosslinked latex. The effect of the amount of crosslinker on the particle sizes investigated by the Zeta-sizer Nano series while Atomic Force microscopy (AFM) was used to study the structural properties of latex nanoparticles.

1. Introduction

Latex with size less than 100 nm have received an increasing attention due to the wide range of application such as coating, paint, adhesive, and textile [1,2]. The common method to prepare the latex nanoparticles are emulsion polymerization [3,4] and microemulsion polymerization. Microemulsion polymerization has been actively studied since 1980's. It can produce stable dispersion of polymer latexes with particle sizes in nanosize range [5-7]. Latex nanoparticles have an extremely large specific surface area which can provide high smoothness, strong adhesion and lasting latex dispersion stability. Crosslinking technology has been introduced in the polymerization method since it can improve the stabilities of latex and extend their applications. Recently, a number of crosslinkers and crosslinking technologies have been reported elsewhere [8,9]

In the present work, 1,3-butylene glycol dimethacrylate (1,3-BGDMA) was used as a crossliker during the polymerization of the latex nanoparticles. 1,3-BDGMA are chosen since it has a good stain resistance and solvency. The objective of this study is to investigate the effect of the addition of crosslinker on particle sizes and surface morphology of latex nanoparticles via differential microemulsion polymerization through Zeta-sizer Nano series and Atomic Force microscopy (AFM).

[*] nurfarahinhassmoro@gmail.com

2. Methodology

Figure 1 shows the flow chart of the research methodology.

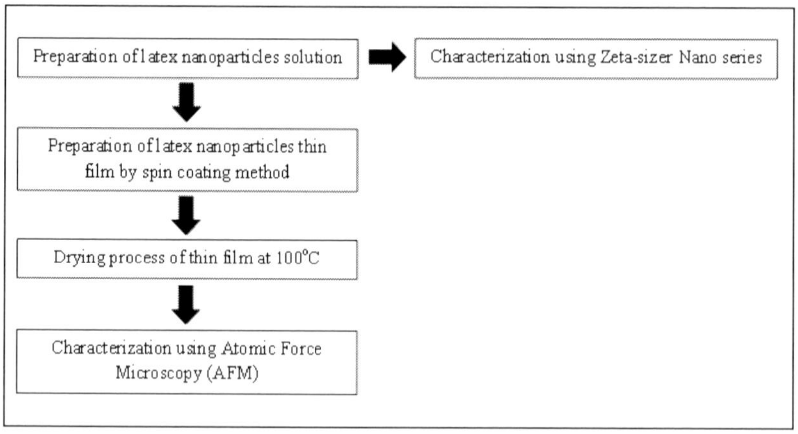

Figure 1. Flow chart of research methodology

2.1. Materials

The crosslinking agent 1,3-butylene glycol dimethacrylate (1,3-BGDMA) was purchased from Sigma Aldrich. Butyl methacrylate (BMA), butyl acrylate (BA), methacrylate acid (MAA), N,N,N,N',N'-tetramethylethylenediamine (TMEDA), 2-amino-2methyl-1propanol (AMP), and 1-pentanol were purchased from Merck without further purification. Ammonium persulfate (APS), sodium dodecyl sulphate (SDS) and ammonia solution 25 % brought from R&M chemical were used as received. Deionized water was used throughout the experimental work.

2.2. Polymerization

Detailed polymerization recipes and the characterization of the samples are summarized in Table 1. As the usual method of polymerization, all the reactions were carried out in a 500 cm^3 three necked round bottom flask with a reflux condenser, a thermometer, a magnetic stirrer, and the nitrogen inlet to dissolve the oxygen [2]. Microemulsion composed of 0.3 g of monomer content, 1.0 g SDS, 0.1 g of 1-pentanol, and 78.0 g of de-ionized water was heated to 40 °C and stirred about 300 revolutions per minute (rpm) with a gentle flow of bubble nitrogen. Ammonia solution was used to adjust the pH in the range of 9 – 10. TMEDA and APS that was dissolved in water were added into the microemulsion. The rest of monomer was added drop wise into the microemulsion and stirring continue to 2 – 3 hours to complete the polymerization process. The microemulsion was cooled down at room temperature and the pH was adjusted using AMP solution to 9.2 – 9.5.

Table 1. Summarized of polymerization recipes and the characterization of the samples

Sample	BMA (g)	1,3-BGDMA (g)	Particle Size (nm)	Surface roughness (nm)
IC0	7.7	-	37.07	3.78
IC1	7.6	0.1	39.29	6.99
IC2	7.5	0.2	47.50	3.52
IC3	7.4	0.3	47.37	11.24
IC4	7.3	0.4	57.13	6.41
IC5	7.2	0.5	37.66	5.81

2.3 Thin Film
Glass substrate (2 cm x 2 cm) was cleaned to get a good adhesion of coating. The substrates were cleaned using acetone and methanol. The latex nanoparticles solution was coated onto the glass substrate using spin coating method. The spin coating was done in 3000 rpm at 30 seconds. Each sample was dried at 100°C for 5 minutes to vaporize the solvent and removed the contamination.

3. Results and discussion

3.1. Particle sizer
Particles size measurement was carried out using Zeta-sizer Nano series at 25°C and the reading was taken triplicate at 173° scattering angle. Figure 2 shows the average particle sizes of latex nanoparticles with various amount of crosslinker. The results revealed that particle sizes changed when the amount of 1,3-BGDMA was increased. Without crosslinking the average particle sizes is 37.07 nm. As the crosslinking agent was introduced into the polymerization the particle sizes increased to 39.29 nm at 0.1 g of 1,3-BGDMA. The average particle sizes for 0.2 g, 0.3 g, and 0.4 g, are 47.50 nm, 47.37 nm, and 57.13 nm respectively. The particle sizes for 0.5 g of 1,3-BDGMA was reduced to 37.66 nm and this might be due to the increasing number of primary particles. As the crosslinker increased, the rate of nucleation increased and thus increased the number of primary particles, which lead to decreased in the diameter of the latex nanoparticles [10].

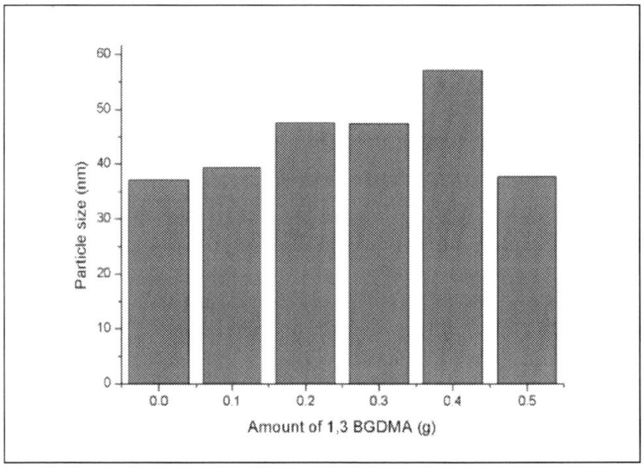

Figure 2. Average particle size of latex nanoparticles at different amount of 1,3-BGDMA

3.2. Atomic Force Microscopy (AFM)
AFM XE-100 Park System was used to capture the image morphology and topography of the surface latex nanoparticles and to obtain detailed information about the structure of the latex film. Figure 3(a) shown the three dimensional image of uncrosslinked latex with surface roughness 3.78 nm. The uncrosslinked latex represented the most homogeneous structure compare to the others image. The image totally change when the crosslinking agent were introduced into the polymerization as shown in Figure 3(b-e) and the surface roughness are 6.99, 3.52, 11.24, 6.41, and 5.81 nm respectively. Figure 3(d) gives a higher surface roughness and from the image the peak and the valley are bigger compare to the others. This observation suggests that the addition of crosslinking agent was fast enough to significantly inhibit the homogenization of the film [11, 12].

Figure 3. AFM image of (a) IC0, (b) IC1, (c) IC2, (d) IC3, (e) IC4, and (f) IC5

4. Conclusion

As a conclusion, crosslinked latex nanoparticles were successfully prepared via differential microemulsion polymerization with particle sizes in the range of 30-60 nm. The 1, 3-butylene glycol dimethacrylate (1,3-BDGMA) was used as a crosslinker and the amount was varied from 1-5 weight% of monomer total. The average particle sizes increased as the crosslinker were added from 0-4 weight% of monomer total. The particle sizes reduced to 37.66 nm at the 0.5 g of 1,3-BDGMA and this might due the increasing number of primary particles. The structural morphology of the uncrosslinked represented the most homogeneous image. The addition of crosslinking agent was significantly retarding the homogenization of the film. Therefore, we have found that the addition of crosslinking agent has sensitively effect the particle sizes and the surface morphology of latex nanoparticles.

Acknowledgement
Authors are gratefully acknowledged for the financial support to the MyBrain15 Sponsorship Programme, Malaysian Government and also to the Research Management Institute, Universiti Teknologi MARA (UiTM), Shah Alam, Selangor, Malaysia.

References

[1] Liubov Shtykova, Mariam Masuda, Carina Eriksson, Kristina Sjodin, Elisabeth Marling, Fredrik Schlyter and Magnus Nyden 2008 *Prog. Org. Coa.t* **63** 160-166
[2] Harald Wutzel and Wolfgang M. Samhaber 2007 *Monatsh. Chem.* **138** 357-361
[3] Yuanchun Mu, Teng Qiu and Xiaoyu Li 2009 *Mater. Lett.* **63** 1614-1617
[4] Xuefeng Hu, Jieyu Zhang and Wantai Yang 2009 *Polymer* **50** 141-147
[5] Weihua Ming, Frank N. Jones and Shoukuan Fu 1998 *Polym. Bul.l* **40** 749-756
[6] Xiao-Jun Xu and Leong Ming Gan 2005 *Curr. Opin. Colloid In.* **10** 239-244
[7] Chaiwat Norakankorn, Qinmin Pan, Garry L. Rempel and Suda Kiatkamjonwong 2009 *Eur. Polym. J.* **45** 2977-2986
[8] Shaobing Wu, Jon D. Jorgensen, Allen D. Skaja, Jonathan P.Williams and Mark D. Soucek 1999 *Prog. Org. Coat.* **36** 21-33
[9] Guangfeng Liu and Peng Liu 2010 *Colloid Surface A* **354** 377-381
[10] Xin Chen, Zhimin Chen, Guang Lu, Weifeng Bu, and Bai Yang 2003 *J.Colloid Inter. Sci.* **264** 266-270
[11] R. Joshi, E. Lefevre, C. Patel, T. Provder, R. Crombez, W. Shen and F. Jones 2008 *J. Therm. Anal. Calorim.* **93** 19-26
[12] Ravi G. Joshi, Theodore Provder, Paul Ziemer, Wenjing Mao, Weidian Shen, and Frank N. Jones 2009 *J. Coat. Technol. Res.* **6** 47-65

IOP Conf. Series: Materials Science and Engineering **46** (2013) 012026 doi:10.1088/1757-899X/46/1/012026

Synthesis and Temperature Dependence of I-V Characteristic of Spin-Coated Nanostructured ZnO on P-Type Silicon

K. A. Eswar[1, 2, 4], F.S. Husairi[2, 3], S. A. Mohamad[1, 2], A. Azlinda[2, 3], M. Rusop[3, 4] and S. Abdullah[2, 3]
[1]Universiti Teknologi Mara (UiTM) Cawangan Sabah, 88997 Kota Kinabalu, Sabah Malaysia
[2]Faculty of Applied Sciences, Universiti Teknologi MARA (UiTM), 40450 Shah Alam, Selangor, Malaysia
[3]NANO-SciTech Centre (NST), Institute of Science, Universiti Teknologi MARA (UiTM), 40450 Shah Alam, Selangor, Malaysia
[4]NANO-ElecTronic Centre (NET), Faculty of Electrical Engineering, Universiti Teknologi MARA (UiTM), 40450 Shah Alam, Selangor, Malaysia

E-mail: kevinalvin86@gmail.com

Abstract. P-type silicon was used as a substrate of deposition of nanostructured ZnO by sol-gel spin coating technique, while zinc acetate, diethanolamine, and isopropyl were used as starting material, stabilizer and solvent respectively. Surface morphology was studied by Field Emission Scanning Electron Microscopy (FESEM). It was found that nanostructured ZnO was distributed uniformly over the P-type Silicon substrate. This was supported by atomic force microscopy (AFM) image which shows the all the surfaces are covered by nanoparticle of ZnO. X-ray diffraction (XRD) was employed to analyse the structural and crystalline of nanostructured ZnO. Three peak (100), (002), and (101) correspond to nanostructed ZnO are appear. Temperature dependence I-V characteristic was investigated in range of 25 K to 300 K by using Close Cycle Cryostat and using Helium as cooler agent. It is found that the resistance were decrease toward the higher temperature. The result shows that resistivity gradually decrease due to increases of temperature.

1. Introduction

ZnO semiconductor is very interesting due to its electrical and optical properties. Its direct wide band gap energy of ~3.37 eV large binding energy of ~60 meV makes it very interesting. These properties make it potentially in optoelectronic device such as ultraviolet photoconductive sensors [1-3], gas sensors [4, 5], UV laser [6], and thin film transistor [7]. Besides, it is also important in bio-medical application [8]. There are several method have been used to deposits ZnO on various substrate includes pulsed laser deposition (PLD) [9], DC reactive magnetron [10], activated reactive evaporation [11], RF magnetron sputtering [12] and sol-gel method[6, 7, 13-15]. Among them, sol-gel deposition method is very interesting due to low-cost, simple and highly controlled way [16]. Structural of znO in different thickness by sol-gel deposition method were studied by Shariffudin *et al* [17]. They found that 42.7 nm of thickness gave the lowest resistivity. Electrical properties of ZnO on glass prepared by dip-coating in room temperature have been studied and the result shows that the withdraw speed influence the resistivity of ZnO thin film and the result shows high withdraw speed produce more resistivity [13]. Besides, structural and electrical properties ZnO thin film on silicon and quartz also studied by Zhao *et al* [9] based on substrate temperature and oxygen pressure. They found that resistivity increases substrate temperature and oxygen pressure. In this study, the nanostructured ZnO were deposited on silicon substrate by sol-gel spin coating method. Atomic force microscopy

(AFM), field emission scanning electron microscopy (FESEM) and X-Rah diffraction (XRD) were used to study the structural and surface morphologies of nanostructured ZnO. Current versus voltage were investigated using close cycle cryostat.

2. Experimental

Zinc acetate as a starting material, diethanolamine (DEA) as stabilizer, and isopropyl as a solvent were used in this work, while P-type silicon (Si) with orientation (100) was used as a substrate. In other to prepared P-type silicon substrate, cleaning process was started by using acetone and methanol to remove any residual on the substrate surface. Then, dilute HF, mixed of HF40% and di-water with ratio 1:10 was used for etching the surface of Si. After that, the etched Si washed by di-water before dried by nitrogen gas. All cleaning process was done in ultrasonic bath.

ZnO Sol-gel of 0.5 M concentration was prepared by dissolved zinc acetate dihydrate into isopropyl. Then, diethanolamine was added slowly to produce clear solution. Furthermore, the solution was heated for an hour at 60°C followed by aged for 24 hour at room temperature. Nanostructured ZnO was deposited on P-type Si substrate by spin coating method. Rotations of spin coater were kept at 300 rpm within 60 s while 10 drops of precursor were dropped on substrate to produce thin film. Then, nanostructured ZnO thin film were dried at 150°C within 10 minutes. The process of deposition by spin coating and drying were repeated to enhance the thickness of thin film before annealed at 500°C for an hour.

X-ray diffraction (XRD) characterization with Cu Kα radiation was employed to analyze the crystalline and diameter size of nanostructured ZnO. Field Emission Scanning Electron Microscopy (FESEM), JOEL JSM-J600F was used in other to investigate the surface morphology of nanostructured ZnO. I-V temperature dependence was studied by close-cycle cryostat and Helium as a cooling agent. Temperature ranges of 25 K to 300 K were used in this work. Gold metal contact was deposited on nanostructured ZnO by sputter coater before connected to the circuit.

3. Result and Discussion

3.1. Structural and morphological properties

Figure 1 shows the FESEM image of nanostructured ZnO deposited on silicon by spin coating. Based on figure 1, it is found that ZnO thin films are composed by nanoparticle of ZnO. Nanoparticles of ZnO were distributed uniformly over the silicon surface. Average sizes of nanoparticles of ZnO are found about ~ 61.92 nm obtained by direct measurement from FESEM Images. It is believed that the ZnO nanoparticles are product of reaction between zinc acetate and isopropyl which is stabilized by diethanolamine [14]. Minrui *et al* [14] were studied the reaction of zinc acetate in isopropyl and stabilized by diethanolamine. In sol-gel synthesis, hydrolysis and condensation are reactions that involved in sample preparation [16]. They proposed that $[CH_3COOZn]^+$ ion, $[CH_3COO]^-$ ion, H^+ ion, and $(OH)^-$ ion are produced from hydrolysis of zinc acetate as shown in (1). Then, when DEA added to the solution, the reaction is described in (2). We are suggest that ZnO is resulted from decomposition of $Zn(C_2H_4O)_2NH$ after heat treatment.

$$Zn(CH_3COO)_2.2(H_2O) \longrightarrow [CH_3COOZn]^+ + [CH_3COO]^- + 2H^+ + 2(OH)^- \quad (1)$$

$$[CH_3COOZn]^+ + HN(C_2H_4OH)_2 \longrightarrow CH_3COOH + Zn(C_2H_4O)_2NH + H^+ \quad (2)$$

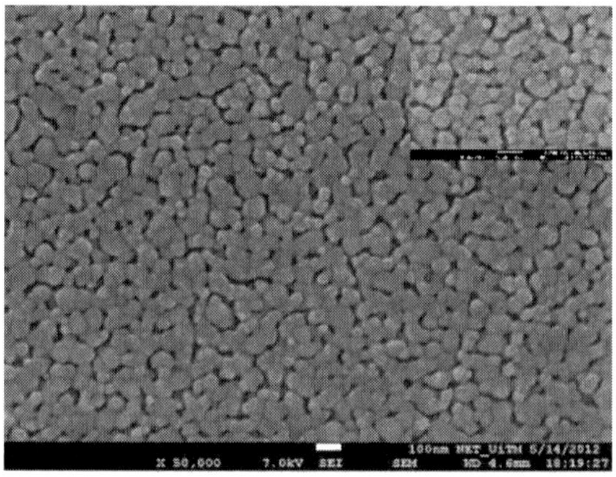

Figure 1: FESEM image of nanostructured ZnO on P-type Silicon. (Inserted picture was magnified up to 150 kX).

Two-dimensional and three-dimensional AFM image of nanostructured ZnO thin film are shown in figure 2. In other to analyze the surface morphology and grain size, XEI imaging software was used that covered 2μmx2μm area. The result of roughness averages of 4.977 nm shows nanostructured ZnO thin film has a smooth and dense surface morphology [14]. The particle size of ~62.4 nm was found by direct measurement using the XEI imaging software. So, the particle sizes measured are almost same to particle size obtained from FESEM image.

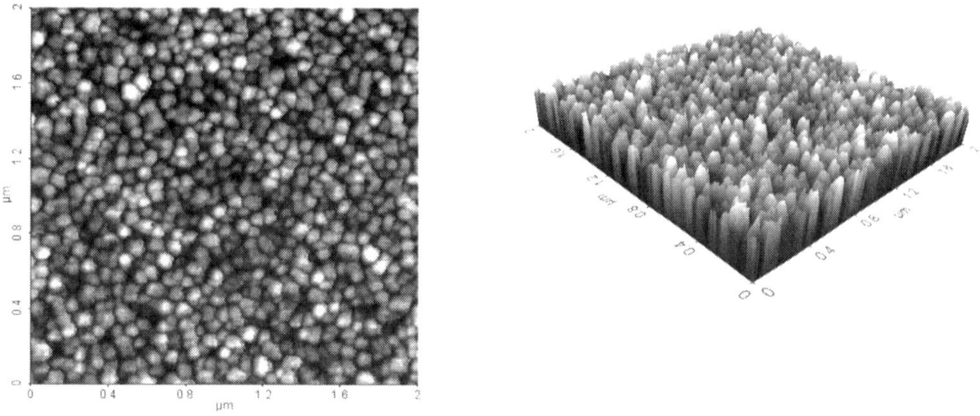

Figure 2: AFM images in two-dimensional and three-dimensional of nanostructured ZnO on silicon.

X-ray diffraction spectra of nanostructured ZnO was shown in figure 3. By referring to JCPDS No. 36-1451, three peaks of 32.044°, 34.662°, and 36.447° are corresponding to (100), (002), and (101) plane [17-19]. The low intensity of these three peaks might be the thicknesses of nanostructured ZnO films are very thin and it shows that crystalline randomly oriented [17] due to the peak are almost same in term of intensity. Besides, silicon peaks appears in high intensity prove that the silicon substrate more detected compared to nanostroctured ZnO.

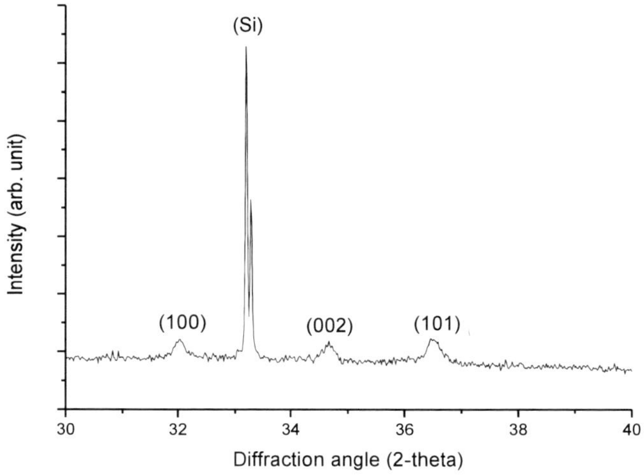

Figure 3: XRD spectra of nanostructured ZnO on P-type silicon.

3.2. Temperature dependence of I-V characteristic

In other to study the temperature dependence of I-V characteristic of nanostructured ZnO, gold metal contact, copper wire and silver paste was used to complete the circuit. Then, forward bias was applied to the circuit while close cycle cryostat systems were used as temperature controller. Figure 4 shows the I-V curve in different temperatures ranging of 25 K to 300 K. Generally, current induced were increase due to increases of temperature. Tangents represent reciprocal of resistance or conductances of each peak were listed in table 1. Based on table 1, resistivity were obtained by using equation (3) and (4), and then displayed in figure 5. Symbol V, I, and R are represents voltage, current and resistance respectively. The result shows the magnitude of current is increasing significantly at a given voltage as increases of temperature.

$$\mathbf{V = IR} \tag{3}$$

$$\frac{\mathbf{I}}{\mathbf{V}} = \frac{\mathbf{1}}{\mathbf{R}} \tag{4}$$

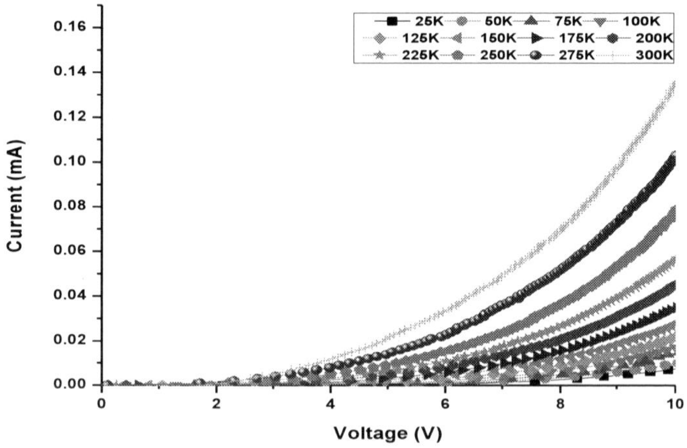

Figure 4: I-V curve in different temperatures.

By increasing the temperature, penetration barrier will be decreased due to increases of thermal energy [20]. So, the carrier concentration becomes higher and increasing the conductance. It also attributed by the change of current transport process from generation recombination at low and intermediate temperature to thermionic emission at higher temperature as suggested by Mtangi *et al* [21]. So, obviously temperature influences the I-V characteristic of nanostructured ZnO.

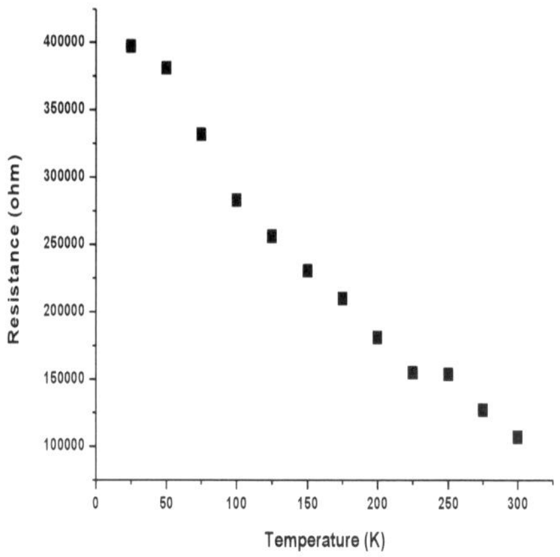

Figure 5: Resistance of nanostructured ZnO in different temperatures

Table 1: Conductance (G) obtained from tangent of I-V curve

Temperature (K)	Conductance ($m\Omega^{-1}$ or mS)
25	2.52E-03
50	2.63E-03
75	3.01E-03
100	3.53E-03
125	3.91E-03
150	4.34E-03
175	4.76E-03
200	5.52E-03
225	6.45E-03
250	6.50E-03
275	7.87E-03
300	9.35E-03

4. Conclusions

Nanostructured Zno was successfully synthesized on P-type silicon by very simple and low-cost method. Based on FESEM and AFM images, it is found nanostructured ZnO thin film were composed by nanoparticles. The decomposition of zinc acetate in solvent and condensation during heat treatment were produced an averages of ~ 62 nm nanoparticle. Temperature I-V characteristic shows that conductance increase when the temperature increases as found in Table 1.

5. References

[1] Mamat M H, Ishak N I, Khusaimi Z, Zahidi M M, Abdullah M H, Muhamad S, Sin N D M, and Mahmood M R 2012 *Japanese Journal of Applied Physics,* **51**, 06FF03
[2] Mamat M H, Khusaimi Z, Zahidi M M, and Mahmood M R 2011 *Japanese Journal of Applied Physics,* **50**, 06GF05
[3] Mamat M H, Sahdan M Z, Khusaimi Z, Ahmed A Z, Abdullah S, and Rusop M 2010 *Optical Materials,* **32**, 696-699
[4] Yi S H, Choi S K, Jang J M, Kim J A, and Jung W G 2007 *J Colloid Interface Sci,* **313**, 705-10
[5] Hieu N V, Le D T T, Khoang N D, Quy N V, Hoa N D, Tam P D, Le A T, and Trung T 2011 *International Journal of Nanotechnology,* **8**, 174-187
[6] Liu A, Zhang J, and Wang Q 2010 *Chemical Engineering Communications,* **198**, 494-503
[7] Kim K, Song Y-W, Leem J, Lee S Y, and Kim S 2009 *Journal of the Korean Physical Society,* **55**, 140-143
[8] Foo K L, Kashif M, Hashim U, Ali S M U, and Willander M 2012 *2012 International Conference on Biomedical Engineering (ICoBE)*, (Penang)
[9] Zhao J-L, Li X-M, Bian J-M, Yu W-D, and Gao X-D 2005 *Journal of Crystal Growth,* **276**, 507-512
[10] Ma Q-B, Ye Z-Z, He H-P, Wang J-R, Zhu L-P, and Zhao B-H 2007 *Vacuum,* **82**, 9-14
[11] Yuvaraj D and Rao K N 2008 *Vacuum,* **82**, 1274-1279
[12] Choi C-H and Kim S-H 2005 *Journal of Crystal Growth,* **283**, 170-179
[13] Malek M F, Zakaria N, Sahdan M Z, Mamat M H, Khusaimi Z, and Rusop M 2010 *2010 International Conference on Electronic Devices, Systems and Applications (ICEDSA2010)*, (Kuala Lumpur)
[14] Wang M, Wang J, Chen W, Cui Y, and Wang L 2006 *Materials Chemistry and Physics,* **97**, 219-225
[15] Mamat M H, Khusaimi Z, Musa M Z, and Rusop M 2010 *6th International Conference on Electrical and Computer Engineering ICECE*, (Dhaka, Bangladesh)
[16] Znaidi L 2010 *Materials Science and Engineering: B,* **174**, 18-30
[17] Shariffudin S S, Salina M, Herman S H, and Rusop M 2012 *Transactions on Electrical and Electronic Materials,* **13**, 102-105
[18] Yang X D, Su S C, Xu Y, and Mei T 2011 *Advanced Materials Research,* **311-313**, 1271-1276
[19] Min S K, Kwang G Y, Jae-Young L, Soaram K, Giwoong N, Dong-Yul L, Jin S K, and Jong S K 2011 *Journal of the Korean Physical Society,* **59**, 2354
[20] Zakirov A S, Yuldashev S U, Cho H D, Lee J C, and Kang T W 2011 *Journal of the Korean Physical Society,* **59**, 482
[21] Mtangi W, Auret F D, Nyamhere C, Janse van Rensburg P J, Chawanda A, Diale M, Nel J M, and Meyer W E 2009 *Physica B: Condensed Matter,* **404**, 4402-4405

Experimental investigation of sound absorption properties of perforated date palm fibers panel

A K Elwaleed[1], N Nikabdullah [1,2], M J M Nor[2], M F M Tahir[2], R Zulkifli[2]

[1]Institute of Space Science, UKM, Malaysia
[2]Department of Mechanical and Materials Engineering, UKM, Malaysia

E-mail: elwaleed@ukm.my

Abstract. This paper presents the sound absorption properties of a natural waste of date palm fiber perforated panel. A single layer of the date palm fibers was tested in this study for its sound absorption properties. The experimental measurements were carried out using impedance tube at the acoustic lab, Faculty of Engineering, Universiti Kebangsaan Malaysia. The experiment was conducted for the panel without air gap, with air gap and with perforated plate facing. Three air gap thicknesses of 10 mm, 20 mm and 30 mm were used between the date palm fiber sample and the rigid backing of the impedance tube. The results showed that when facing the palm date fiber sample with perforated plate the sound absorption coefficient improved at the higher and lower frequency ranges. This increase in sound absorption coincided with reduction in medium frequency absorption. However, this could be improved by using different densities or perforated plate with the date palm fiber panel.

1. Introduction

Vegetable fiber is one of the varieties of natural fibers obtained from stems, leaves, roots, fruits and seeds of plants. However, from commercial and technological points of view, cotton, kenaf, sisal, flax, palm, coir, arecanut and banana fibers acquire utmost significance, since reinforced plastics, strings, cords, cables, ropes, mats, brushes, hats, baskets and fancy articles such as bags are manufactured with those fibers [1].

The date palm (Phoenix dactylifera) is one of the most cultivated palms around the world. It geographically covers the deserts from the Atlantic coastline of Mauritania to India and from the Mediterranean Sea to about 15° in Africa. The main date-producing countries of the world are Iraq, Saudi Arabia, Egypt, Iran, Algeria, Pakistan and the Sudan. The date palm in Sudan is common in the Northern Sudan along the Nile [2]. Date palms have a fibrous structure, with four types of fibers: leaf fibers in the peduncle, baste fibers in the stem, wood fibers in the trunk and surface fibers around the trunk [3]. Riahi et al. [4] investigated the application of date-palm fibers filters as porous medium for the tertiary domestic wastewater treatment. Al-Sulaiman [5] evaluated the performance of the date palm fiber as wetted pads in evaporative cooling.

Most practical sound absorbing products used in the building construction industry consist of glass or mineral-fiber materials. However, the growing concern about the potential health risks popularly seen as being associated with glass or mineral fiber materials provides an opportunity to develop for sound absorption panels made of natural fibers. Many researches have been conducted in developing particle composite boards using agricultural wastes [6,7]. Yang et al. reported that the sound absorption coefficient of rice straw-wood particle composite boards are higher than other wood-based materials in the 500-8000 Hz frequency range, which is caused by the low specific gravity of composite boards, which are more porous than other wood-based materials [6]. From the view of environmental protection, natural bamboo fibers were used for sound absorbing purposes. Impedance tube measurement of the bamboo fiber samples, revealed similar properties to that of glass wool. Bamboo material formed into a fiber board yields a superior sound absorption property when compared to plywood material of similar density [8]. Ersoy and Kucuk [9] experimentally investigated the sound absorption feature of tea leaf fiber as an industrial waste material. The good acoustic absorption aspect of that fiber with respect to other absorbers was noticed. Coconut is one of the most important harvests in Malaysia. Coir fiber from coconut husk is one of the hardest natural fibers having high amount of lignin. The sound absorption attribute of coir fiber was investigated previously in Automotive Research Group laboratories, Universiti Kebangsaan Malaysia. Those studies covered experimental observations in reverberation room [10] and using impedance tube [11].

The acoustic properties of panels can be improved by using perforated plate design. Davern [12] investigated the effect of the airspace layers, perforated plate, and porosity on the acoustic properties of materials. He found that the porosity of the perforated plate and the density of the porous material would significantly affected the acoustic impedance and sound absorption coefficient of the panel, in which case, the frequency band near the resonance frequency achieved high acoustic absorption. Lee and Chen [13] found that the acoustic absorption of multilayer materials is better with a perforated plate backed with airspaces. Fouladi et al. [14] studied the enhancement of coir fiber acoustical absorption using perforated plates and air gap layers.

The research on the acoustical properties of the date palm fibers started at the University of Khartoum [15]. The sound absorption coefficient of the date palm fiber was simulated using Delany and Bazley model based on the flow resistivity of the fibers. Elwaleed et al. [16] studied experimentally the variation of sound absorption coefficient against frequency for date palm fiber sample at normal incidence as measured by the impedance tube (Figure 1). They studied the effect of effect of varying the air gap distance, between the date palm fiber sample and the back wall, on sound absorption coefficient. It can be seen that introducing air gap between the sample and the rigid backing of the impedance tube can be useful for increasing the absorption coefficient at some lower frequencies range.

The aim of this research is to study the effect of using perforated plate on the sound absorption of date palm fiber.

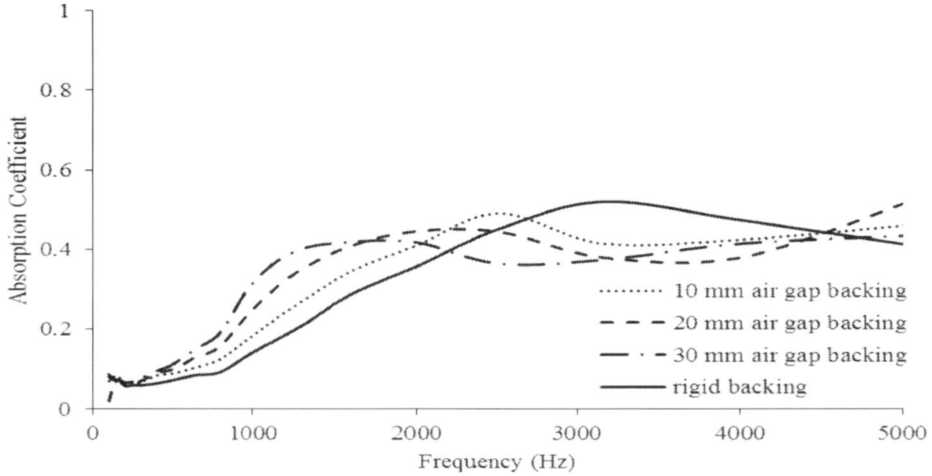

Figure 1. Absorption coefficient for the date palm fiber with and without air-gap [16]

2. Materials and Methods

2.1. Date Palm Fiber (DPF).

The fiber is collected from amplexicaul, the sheathing leaf base, which surrounds the stem. It has a netted structure, which is covered by soft or ground tissues (Figure 2). The amplexicaul is carefully collected from the tree. It is dried in shade at room temperature for a period of two days in order to remove any excess moisture content. The pulp (parenchyma) which is present on the individual fiber is removed by combing. Finally, the fiber is scraped to remove the pulp completely. Numbers of 30 samples were selected to obtain an average value for density and fiber diameter. The average diameter and the average density of the fibers are 0.408 mm and 919 kg/m^3, respectively.

Two plastic molds were fabricated to prepare samples of two different sizes equivalent to the diameters of the impedance tubes to cover the low and high frequency range of measurements. Figure 3 shows the plastic molds and samples of date palm fibers. The thickness and the density of the prepared samples are 30 mm and 77 kg/m^3, respectively.

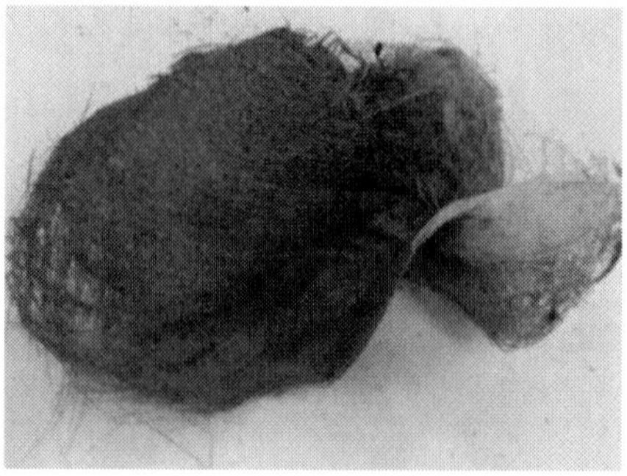

Figure 2. Date palm amplexicaul

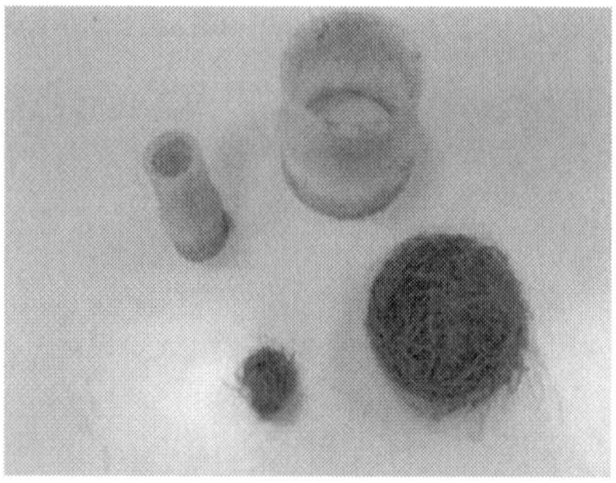

Figure 3. Plastic molds and date palm fiber samples

2.2. Experimental Measurement in Impedance Tube

The experiment was conducted using two impedance tubes of 28 mm and 100 mm diameters, noise generator, two channel data acquisition system 01 dB, two ¼ in microphones type GRAS-40BP in each tube, software package SCS8100. The measurements were made based on ISO 10534-2 standard [17]. The microphones' sensitivity was calibrated using calibrator type GRAS-42AB at 114 dB level and 1 kHz. Photo of the system is shown in Figure 4. Noise generator transmitted a random noise into the tubes. Interior sound pressure spectrum was measured by the two microphones and transfer functions between them were calculated. The acoustical absorption coefficient was calculated from these transfer functions and distances between the microphones and date palm fiber sample. The frequency span of experiment was 100–5000 Hz with 3 Hz resolution. Before running the experiment the two impedance tube microphones were calibrated relatively to each other using the standard switching technique. This was based on mounting the sample in the sample holder and conducting the measurement to make sure that the sound field inside the tube is well defined.

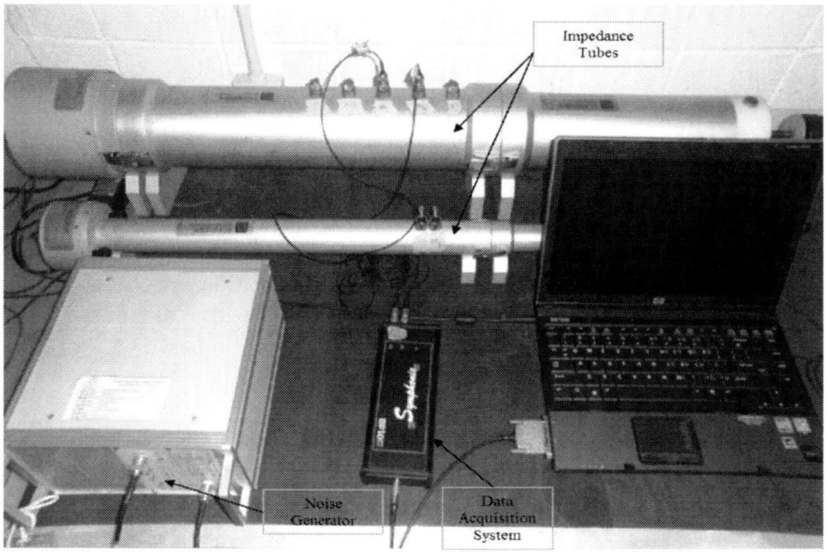

Figure 4. Sound absorption experimental set-up

Perforated plate (PP) was used to enhance the sound absorption of the fiber. The PP used in this study was made of Aluminum (Figure 5). The experiment was conducted for the panel without air gap, with air gap and with PP facing. Three air gap thicknesses of 10 mm, 20 mm and 30 mm were used between the date palm fiber sample and the rigid backing of the impedance tube. Figure 6 shows a sketch of the set-up used for the measurement of the sound absorption of the sample and the PP inside the impedance tube. The flowchart of Figure 7 shows the method of running the experiment.

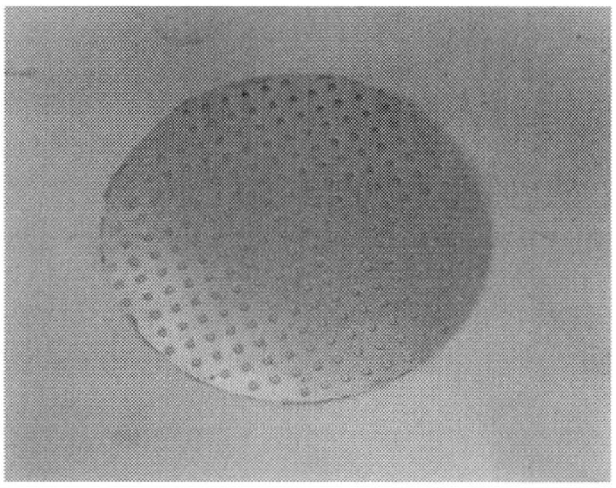

Figure 5. Aluminum perforated plate (PP)

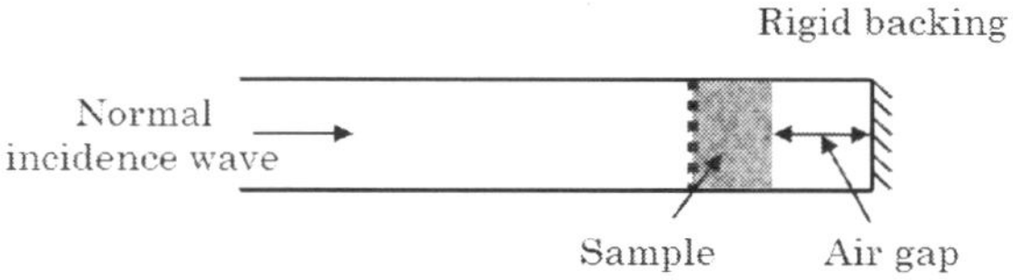

Figure 6. Sketch of the palm date fibers sample and the PP inside the impedance tube

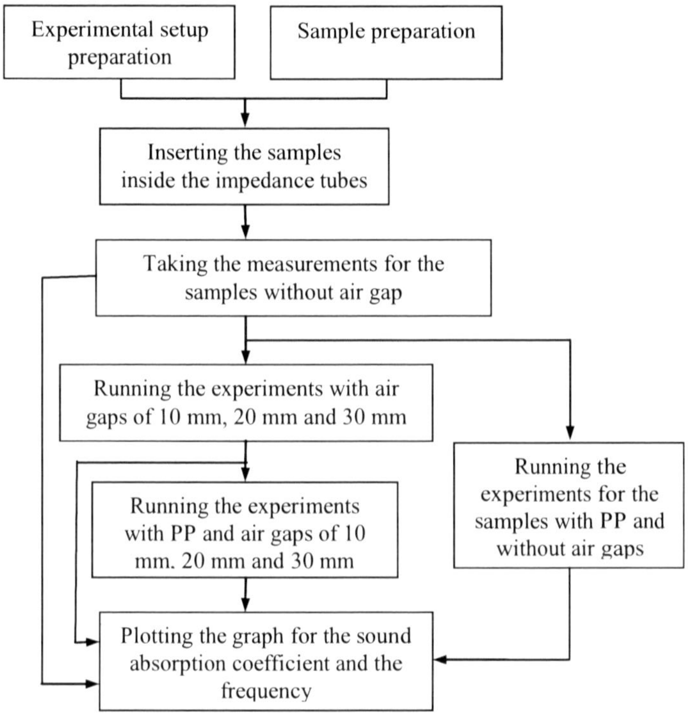

Figure 7. Flow chart showing the method of running the experiment

3. Results and Discussion

The effect of using perforated plate facing on the sound absorption of the palm date fiber is shown in Figure 8. It can be observed that the sound absorption coefficient increased between 1000 Hz and 3000 Hz by shifting the peak toward the lower frequency range. However, the sound absorption coefficent decreased above 3000 Hz and the peak decreased by 4%.

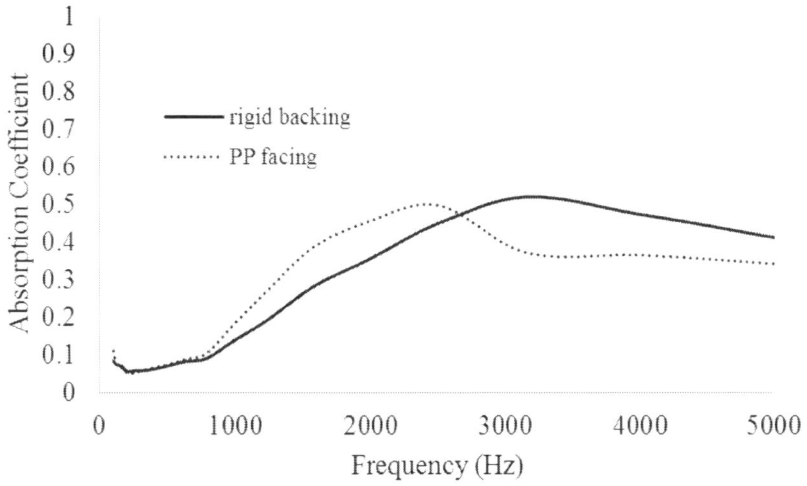

Figure 8. Absorption coefficient for the date palm fiber with and without air-gap

The results from Figure 9 show the sound absorption for the palm date fiber with riggid backing, PP facing, 10 mm air gap backing and a combination of PP facing and 10 mm air gap backing. The results show that the best performance for improving the sound absrorption at the low frequency range can be obtained using the palm date fiber combied with the PP facing and the 10 mm air gap backing. However, this is concided with a decrease in the sound absorption at the medium frequency range.

Further increasing the air gap thickness between the date palm fiber and the impedance tube riggid backing results in more improvement in the sound absorption of the date palm fiber in the lower and higher frequency ranges (Figure 10). It can be seen that icreasing the air gap thickness can be useful for increasing the absorption coefficient at some lower frequencies range. The general indication is that air-gap shifts the resonance absorption towards the low frequency range. Below 1500 Hz frequency there is improvement in sound absorption coefficient for all air gap thicknesses. It can be observed that the increase in the air gap thickness moved the peaks toward lower frequencies and improved the low frequencies absorption and above the 4000 Hz (Figure 10). However, that increase coincided with reduction in medium frequency absorption. Table 1 shows the increase of the sound absorption coefficient at 1257 Hz for all air gap thiknesses. At this frequency the increase of the air gap by 10 mm, 20 mm and 30 mm increase the sound absorption coefficient by 31%, 61% and 64%, respectively. As reported by Fatima and Mohanty [18] this increase is due to loss of acoustical wave energy of transmitted wave in the presence of sample-air passage and of reflected wave from rigid backing, through air-sample passage in the propagation of acoustical wave.

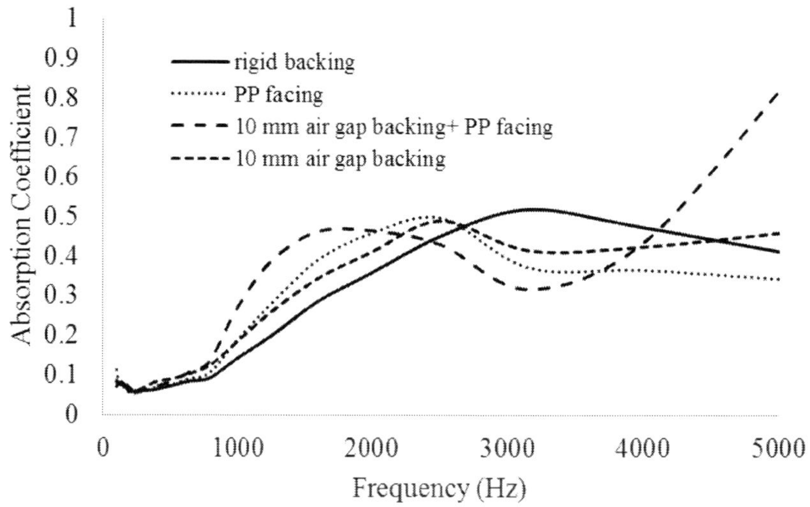

Figure 9. Sound absorption for the palm date fiber with riggid backing, PP facing, 10 mm air gap backing and a combination of PP facing and 10 mm air gap backing

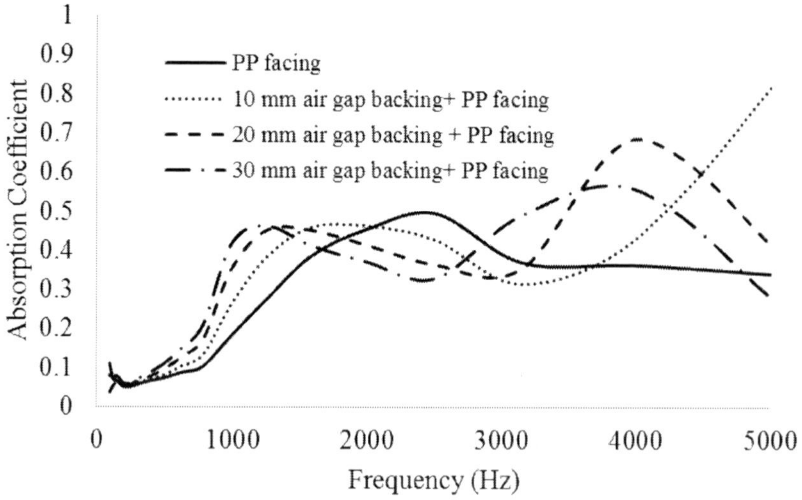

Figure 10. Absorption coefficient for the palm date fiber with PP facing for different air gap backing thicknesses

Table 1. Effect of increasing the air gap on the sound absorption coefficient for palm date fiber with PP facing

Air Gap (mm)	Absorption Coefficient at 1257 Hz	Increase of Absorption Coefficient at 1257 Hz (%)
0	0.28	-
10	0.39	31
20	0.45	61
30	0.46	64

4. Conclusions

Improvement of sound absorption of date palm fiber using perforated plate was investigated in this study. A constant thickness sample was considered in this study. An enhancement in the sound absorption in the lower frequency range was achieved by facing the sample with a perforated plate and backing the sample with air gap of different thicknesses of 10 mm, 20 mm and 30 mm. The increase in the air gap thickness moved the peaks toward lower frequencies and improved the low frequencies absorption and above the 4000 Hz. However, that increase coincided with reduction of absorption in medium frequency range. The performance of the date palm fiber can be improved by increasing the density of the sample and using plate of different perforation.

5. Acknowledgements

The authors would like to thank the Universiti Kebangsaan Malaysia for sponsoring this research under project GGPM-2012-073.

6. References

[1] Murali Mohan Rao K and Mohana Rao K 2007 *Compos. Struct.* **77** 288-295.
[2] Khristova P, Kordsachia O and Khider T 2005 *Bioresource Technol.* **96** 79-85.
[3] Kriker A, Debicki G, Bali A,. Khenfer M M and Chabannet M 2005 *Cement Concrete Comp.* **27** 554–564.
[4] Riahi K, Mammoub A and Thayer B 2009 *J. Hazard Mater.* **161** 608–613.
[5] Al-Sulaiman F 2002 *Energ. Convers. Manage.* **43** 2267-2273.
[6] Yang H-S, Kim D-J and Kim H-J 2003 *Bioresource Technol.* **86** 117-121.
[7] Khedari J, Nankongnab N, Hirunlabh J and Teekasap S 2004 *Build. Environ.* **39** 59-65.
[8] Koizumi T, Tsujiuchi N and Adachi A 2002 *The development of sound absorbing materials using natural bamboo fibres*, In: Brebbia CA, De Wilde WP, editors. High performance structures and composites 4. High performance structures and materials, Witpress 157–66.
[9] Ersoy S and Kucuk H 2009 *Appl. Acoust.* **70** 215–20.
[10] Zulkifli R, Nor M J M, Tahir M F M, Ismail A R and Nuawi M Z 2008 *J. Appl. Sci.* **8** 3709–14.
[11] Fouladi M H, Ayub Md and Nor M J M 2011 *Appl. Acoust.* **72** 35–42.
[12] Davern W A 1977 *Appl. Acoust.* **10** 85-112.
[13] Lee F -C and Chen W –H 2001 *J. Sound Vib.* **248** 621–634.
[14] Fouladi M H, Nor M J M, Ayub Md and Leman Z A 2010 *Appl Acoust 71* 241–249.
[15] Ali M M and Mohammed M O 2010 *Palm Date Fibres as Sound Absorber*, Bachelor Thesis, Department of Mechanical Engineering, Faculty of Engineering, University of Khartoum, Sudan.
[16] Elwaleed A K, Mohamed N A N, Nor M J M, Tahir M F M and Zulkifli R 2012 *The 4th International Conference on Noise, Vibration and Comfort (NVC 2012),* 26-28 November 2012, Kuala Lumpur, Malaysia.
[17] ISO 10534-2, Determination of sound absorption coefficient and impedance in impedance tubes – part 2: transfer function method; 1998
[18] Fatima S and Mohanty A R 2011 *Appl. Acoust.* **72** 108–114.

Thermal Arc Spray Overview

Muhamad Hafiz Abd Malek[1] , Nor Hayati Saad[2], Sunhaji Kiyai Abas[3] and Noriyati Mohd Shah[4]

[1,2,3]Faculty of Mechanical Engineering, Universiti Teknologi MARA, 40450 Shah Alam, Selangor, Malaysia
[4]Norimax Sdn Bhd, No 2, Jln TPP 5/17, Taman Perindustrian Puchong, Seksyen 15, 47160 Puchong, Selangor, Malaysia

Email:muhamadhafiz8607@yahoo.com

Abstract: Usage of protective coating for corrosion protection was on highly demand during the past decade; and thermal spray coating played a major part during that time. In recent years, the thermal arc spray coating becomes a popular coating. Many big players in oil and gas such as PETRONAS, EXXON MOBIL and SHELL in Malaysia tend to use the coating on steel structure as a corrosion protection. Further developments in coating processes, the devices, and raw materials have led to expansion of functional coatings and applications scope from conventional coating to specialized industries. It is widely used because of its ability to withstand high process temperature, offer advantages in efficiency, lower cost and acts as a corrosion protection. Previous research also indicated that the thermal arc spray offers better coating properties compared to other methods of spray. This paper reviews some critical area of thermal spray coating by discussing the process/parameter of thermal arc spray technology and quality control of coating. Coating performance against corrosion, wear and special characteristic of coating are also described. The field application of arc spray technology are demonstrated and reviewed.

Keyword:
Thermal spray, aluminium, zinc, coating, wire arc, corrosion

1. Introduction

Corrosion will occur when four things present; anode, cathode, return current path and electrolyte (either water or wet soil). Corrosion will occur when electron flow through electrolyte at different current potential. Potential difference is caused by galvanic action or electrolyte experiences differential aeration or impurities in metal leading to pitting corrosion [1]. There are many consequences of corrosion and failed handling it will result loss of profit (replacement of corroded component), safety issue and damage company reputation. Problem of corrosion on steel structure can be solved by using cathodic protection or proper coating.

Coating technology nowadays has become one of the advanced knowledge and technique in modern country. By manipulating coating material and method, many surface problems could be figured out, and would lead to the saving cost in term of production and maintenance [2]. Since 1900s, thermal spray coating was used, when oxy-acetylene torches were altered for powder and wires application [3]. In 1984 Conoco Hutton TLP used 200 microns of thermal arc spray aluminium with sealer on risers for offshore platform in North Sea and after eight years, the coating on riser was still in good condition [4] .That was the first time application of thermal arc spray used in major offshore

[1] Corressponding Author, Tel : +60147213716

structure. After that, the technology was developed in industry with massive development on material, equipment and application.

Thermal spray process can be classified into two groups: combustion type and electrical type (Figure 1)

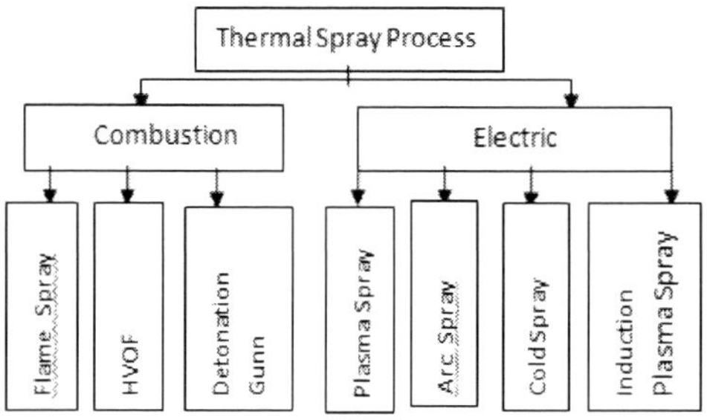

Figure 1: Thermal Spray Method [5]

Recently, thermal spray (wire arc) is considered by many industries as the most selected method for corrosion protection due to its economic benefits in term of production and maintenance. A cost comparison was discussed [5] to determine the highest production cost which is the flame spraying process as illustrated in Table 1.

Table 1: Comparison Cost between Flame Spray and Arc Spray Thermal Coating [5]

Characteristic	Flame spraying	Arc spraying
Liquefaction of zinc, kg/h	18	35
Energy costs for liquefying zinc, $/h	≈28	≈25
Effective use, min/h	40	40
Effective use, kg/h	12	23
Energy costs, $/m² ($/kg equals $/m² at 100 μm zinc thickness)	2.33	0.11
Spraying speed, m²/h for 100 μm zinc	12	23
Wage costs, $/m² at 60 $/h	4.17	2.17
Total costs, $/m²	6.50	2.28

2. Process and Variable Parameters of Thermal Arc Spray Technologies

To optimize coating technology usage, the system should consider substrate, coating properties, process, and its application. Those elements are important for development of an economical coated-surface engineering [2].

Thermal arc spray technology involves four main parts namely 1) Spray gun - used to receive the wire and shoot the melted material onto substrate, 2) Air Compressor–to accelerate melted material, 3) Blasting pot – for garnet space when blasting preparation is performed, 4) Thermal Arc Spray Machine – to supply current to wire so that arc can be created [6].

Using thermal arc spray process, two consumable coating materials, in wire form, are fed into spray gun. By creating an arc through receiving the current, it enables the wire to be melted and the melted material will be accelerated by compressed air around 4-6 bar [6]. The acceleration will be towards substrate in high velocity, so that molten spray will be solidified and turned as coating on the substrate [7] (as shown in Figure 2). Because of principle of thermal arc spray function, coating material must be electrically conductive so that arc can be created and the material can be melted. The selection of the coating material depends on customer requirement but generally they require zinc, zinc alloy, aluminium or aluminium alloy. The applied arc spray parameters are listed in Table 2.

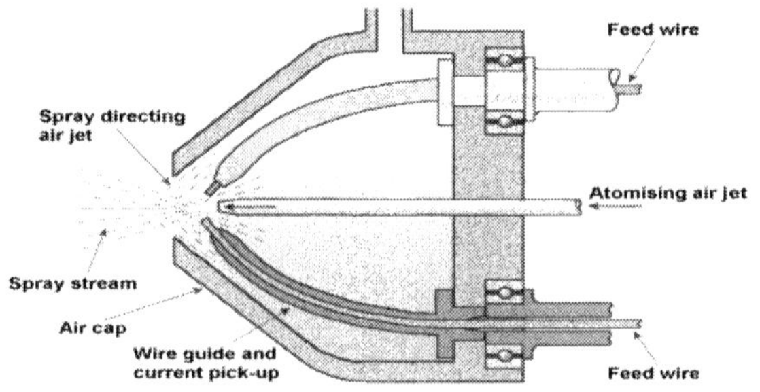

Figure 2: Arc Spraying Process [6]

Table 2: Characteristic or Arc Spray Process Parameter [1, 8, 9, 10]

Thermal Source	Electric Arc
Particle temperature	Up to 5000 ^{0}C
Particle velocity	100 m/s –300 m/s
Feed Stock Form	ɷ 1.6mm-3.2mm
Max. Spray Rate	16 kg/hr
Spray distance	15cm-25cm
Power input	6-80 kW
Air flow rate	Up to 60 m^3/h

Before thermal arc spray coating is performed, substrate surface must be cleaned properly. The surface will be blasted by using compressed-air abrasive blasting; since this method is effective for cleaning of mill-scaled and surface preparation for good adhesion of thermal spray coating [11]. Continues impact on steel surface using abrasive particle in high velocity will ensure removal of mill-scaled; efficiency of the process can be up to 100% [11]. In addition of cleanliness efficiency, substrate surface roughness is an important consideration for proper coating bonding. For thermal arc spray, coarse angular surface profile is needed around 110 microns [12] to provide greater adhesion bonding condition and hinder rust spot formation [11]. Before blasting is applied, the surface of substrate must be checked visually and cleaned from any contamination and excessive rust. The abrasive itself shall be free from oil, grease, moisture or any impurities so that abrasive blasting can be performed in an optimum condition [13]. Normally abrasive which used is Garnet 20/40 (non metallic material). It has proved that it can provide desired surface finish and surface roughness [7]. Non metallic abrasive is more preferable compared to the metallic material; metallic abrasive often result in energy being given off as sparks which may cause surface defect. After blasting process, the following conditions must be confirmed before coating can be applied on the substrate surface: 1) Surface finish must be at minimum SA 2.5 according to ISO 8501-1 as it can be measured by visually inspected; 2) Surface roughness must be between 75 -110 microns , this method requires a Replica Tape method. The replica tape is attached on substrate surface which was blasted and is rubbed with a circular ended tool until circular dark colour appears. Then, the replica tape is removed and measured using a dial gauge[12]; 3) Salt contamination of surface shall not exceed 20 mg/m^2 using Elcometer 138 based on ISO 8502-6 ; [14] 4) Dust level shall not exceed more than rating 1 using cellophane tape according to ISO 8502-3 [15]. Abrasive blasting shall not be performed if humidity of air is greater than 85% or steel temperature is below 3^{0}C above dew point of surrounding air [7]. Failure to comply with these conditions will affect the quality of surface coating which may lead to surface cracking. It may require coating team to re-work and re-blast the substrate which will increase the production cost. Figure 3 shows process flow of substrate surface preparation.

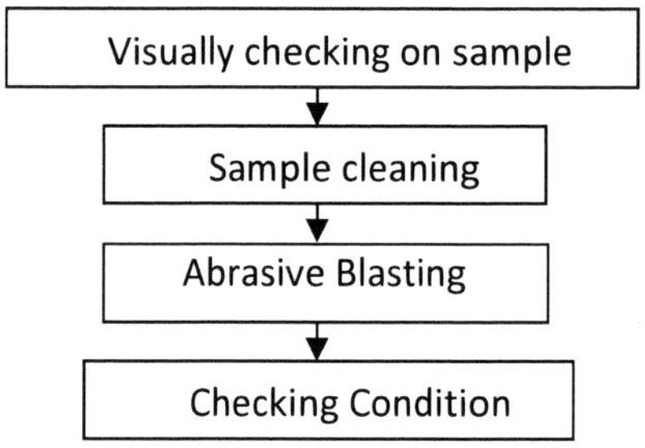

- Surface Finish ≥ 2.5 SA
- Surface roughness; 75-110 microns
- Salt contamination ≤ 20 mg/m^2
- Dust level at Rating 1

Figure 3: Process flow of surface preparation

The coating process will be performed if relative humidity of air is less than 85% [7]. Humidity can be measured using hygrometer. Relationship between air temperature, dew point and relative humidity can be observed in Figure 4.

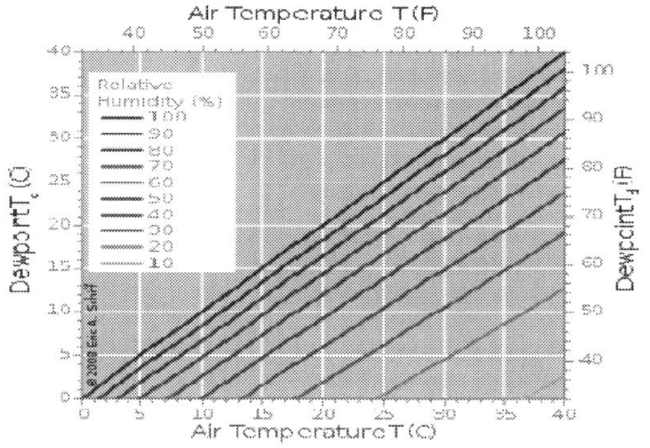

Figure 4: Relationship between air temperature, dewpoint and relative humidity [16]

From Figure 4, the Dewpoint temperature, T_{dp} and Relative Humidity, RH can be calculated using Equation 1 and 2. This conversion can be used as long as relative humidity is above 50% [16].

$$T_{dp} = T - \left[\frac{100-RH}{5}\right] \qquad (1)$$

$$RH = 100 - 5[T-T_{dp}] \qquad (2)$$

After blasting is done, at least one layer of coating must be performed on substrate within four hours. If holding period exceeds 4 hours, the surface needs to be cleaned and re-blasted [7, 17].

Skilled sprayer is required to perform the coating work. Normally spraying shall be performed in block pattern of 2ft squares. A sprayer nozzle stand-off distance shall be at maximum of 25 cm since excessive stand-off distance could produce more porous coating due to cooling and deceleration of particle projected on substrate surface. Too close stand-off distance may lead to poor coverage on coating thickness due to overheating and producing internal stress inside the coating [5, 18, 19]. So, it is important to find optimum stand-off distance since it will influence adherence of coating bonding [19]. The spraying shall overlap with the previous pass to achieve required thickness i.e. 200 microns. The angle of spray nozzle must be perpendicular to the substrate surface for better quality coating. Overall coating needs to be completed within 8 hours after blast cleaning process. Before subsequent layer of coating is conducted, the coated layer should be inspected visually to check for any contamination [7].

3. Qualification Test and Quality Control

The coating thickness needs to be measured after spraying process. Since melted material of thermal spray turns to solid state once it hits the surface metal, so it is easier to measure coating thickness by using a magnetic or electronic thickness gauge [20]. Bend test is required to check any coating failure. It means if major cracking with lifting appears on substrate surface, the coating is considered failed and correction action must be taken [21]. The bend test requires at least 3 samples (150mm x 50mm x 1.8mm). The sample needs to be bent at 90 degrees so that any defect on the surface can be observed immediately as illustrated in Figure 5 [21]. The other test is called pull-off test, and the test is important to determine adhesion of metal coating onto the substrate with a recommended minimum value is 7.0 Mpa [20]. Dolly is used to pull the coating on plate and epoxy glue acts as adhesion medium between the dolly and thermal spray coating [21]. An example of a test panel with a size of 1 m^2 was selected for a particular test [21]. Both tests (bend test and pull-off test) must be performed once per $100m^2$.

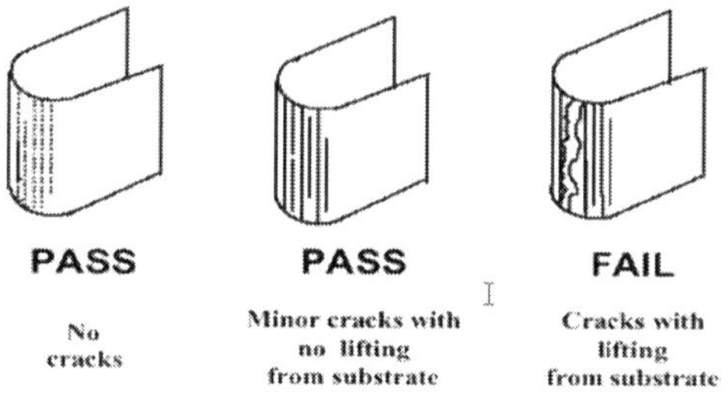

Figure 5: Bend Test Determination [22]

A sealer is normally applied on the coating as additional protection after thermal spray is performed and to seal the coating from any porosity [23]. It must be carried out within 24 hours after thermal spraying, but it is better to be done immediately after thermal spraying to reduce any opportunity of moisture or pollutants creation resulting penetration reduction of sealer [24]. Usage of sealer is essential since it can perforate due to low viscosity, so extend the life of substrate and leading to reduction of total area of the exposed metal. Normally sealer is epoxy (up to 120 ^{0}C) or silicone (up to 450 ^{0}C) [21]. Theoretically, thickness of applied sealer is 38-40 microns. However it may be inspected at 30X power magnification to ensure porosity is filled by sealer [25]. The sealer also provides smooth surface and attractive look. Sealer could be pigmented in colour code and make it easy for coating inspector to differentiate between coating and sealer [4].

As precaution during coating, coating process must be carried in a closed chamber to avoid reaction between sprayed material and component of ambient air [8]. Sprayed particle will oxidize if reaction occurs with the components of ambient air which leads to reduction of corrosion resistance of the coating [8, 26].

4. Material Usage and Characteristic of Thermal Arc Spray

Factors which need to be considered before applying coating are; operating temperature, structural application, environment condition and surface preparation aspects [1].

In thermal spray technology, zinc, aluminium and its alloy are commonly used as coating material. Both materials give excellent corrosion protection in natural environments. Generally, aluminium and its alloy are used as coating material for arc spray while zinc used in flame spray method [4]. The condition can be vice versa depending on client requirement.

Zinc coating provides better corrosion control in most environments as a protective coating. The performance is due to its ability to form dense coating. Zinc density is 7.14 $g.cm^{-3}$; two times higher compared to aluminium (2.70 $g.cm^{-3}$). Zinc also has a low rate of corrosion, which is around 10 to 100 times below than ferrous metal. Because of high sacrificial action of zinc, it becomes a good corrosion prevention for steel [5]. Furthermore, the sacrificial action of zinc produces insoluble corrosion product which will block any empty void in coating (porosity) and prevent the corrosive media to directly contact with substrate [27]. So it will become more resistant to mechanical damage. This action, however will lead to rapid depletion of coating [3]. Naturally, zinc performs in the range of 6 – 12 pH, which is more suitable in alkaline [1].

In industrial environment such as in freshwater, marine environment, under thermal insulation or hot surface area, thermal spray aluminium is the most chose due to its superior performance. American Welding Society has concluded that aluminium as an outstanding material based on the report, *Corrosion Test of Flame Sprayed Coated Steel-19 Year [*28]. Aluminium has been found as the most effective for steel protection in offshore field [29]. Aluminium performance pH is in the range of 4-11 [1]; it can withstand either in acidic or alkaline condition. Aluminium coating advantages come from its ability to form passive film formation, thus giving a lower corrosion rate [3]. However its sacrificial protection is greatly low due to this reason. Aluminium also offers less weight and cheap price compared to zinc [30]. One of the key factor aluminium is more preferable rather than zinc because of its high temperature resistance, up to 550 ^{0}C. It has lower tendency to build up pits and blister due to formation of aluminium oxide when it interacts with air, thus preventing further corrosion [4]. Coating thickness from 100 microns till 250 microns are normally used for corrosion protection [4]. The process temperature is from 200°C – 500 °C. If the process temperature is more than 550 ^{0}C (i.e. 550 ^{0}C – 900 ^{0}C), a heat treatment process is required before the aluminium can be sprayed onto the substrate surface. The heat treatment process is necessary to partially diffuse aluminium into steel surfacet [31]. Minimum thickness of 150 micron is needed to ensure enough inter-diffusion of aluminium with the substrate during heat treatment.

Table 3 shows coating properties comparison between flames spray (FS) and arc spray (AS). It justifies that the arc spray application produces better coating quality for oil and gas industry.

Table 3: Coating Properties [32]

Characteristics & Properties	FS	AS
Porosity (%)	10.97	8.85
Oxide (%)	14.64	15.01
Hardness (HV_{100g})	244	295
Wear rate ($mm^3/m)x10^4$	2.51	0.83

From Table 3, it can be explained that by using the arc spray method, the produced coating have the highest hardness due to low porosity (8.85%) and also high percentage of oxide content (15.01%).

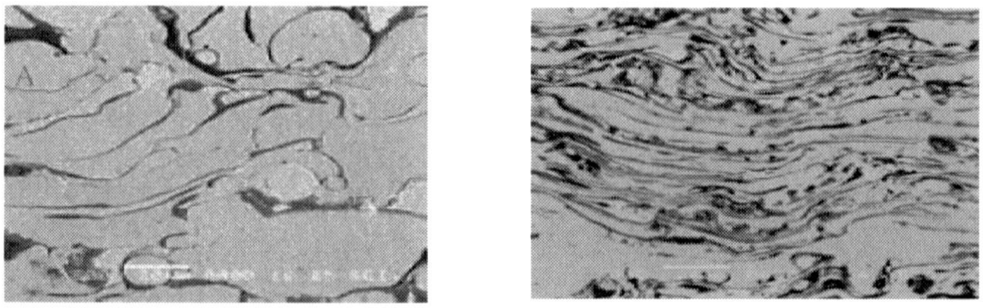

Figure 6: Image of (a) Flame spray and (b) arc Spray of Ni-20Cr using SEM [32]

Generally, temperature and velocity will affect porosity level. Increasing particle velocity will result dense microstructure [33]. The arc spray method normally produces less porosity coating compared to flame spray technique as shown in Figure 6. The reason is that the particle velocity of flame spray is slightly lower around 180 m/s compared to particle velocity of arc spray 300 m/s [10]. A study [33] has suggested that the chemical composition of coating influences substrate hardness since hard phase is formed during solidification state. The phase is due to the presence of Boron element in the coating material.

5. Coating Performance

Currently, thermal spray coatings are frequently applied on offshore structure such as for valves, risers, and platform above sea level. Thermal arc spray aluminium had been proven of providing corrosion protection over than 30 years by using a thickness of only 200 microns [4]. Aluminium coating and zinc coating both can offer complete protection of corrosion but the usage of coating differs in term of thickness applied. America Welding Society has reported that zinc coating thickness used is at minimum of 300 microns [30]. In offshore platform, mostly coating is used in atmospheric zone and tidal zone [3]. Combination of coating with sealer will enhance lifetime and performance of thermal spray aluminium coating due to reduction of exposed area when porosity is filled by the sealer. Due to ease in predicting coating lifetime, it makes maintenance programme easy. It provides an advantage in terms of cycle cost even though the initial investment cost is high; therefore, it leads to cost saving. As an example in 1978, US Navy launched an evaluation method of testing the corrosion performance of thermal spray using aluminium at 515 ^{0}C for steam valves. As reported after 8 years, 3000 steam valves had shown no sign of corrosion and the result benefited in term of cost saving by USD 800,000 [1]. This gives advantages in reduction of corrosion allowance. Currently steel structure in splash zone is designed with corrosion allowance from 3 mm until 12 mm [4]. In theory, using thermal arc spray with aluminium offers zero corrosion allowance, and this could save another production cost. Furthermore, thermal arc spray coating can dry immediately after coating is applied, and the steel structure is ready in a short time [29]. The simplicity of process makes the process and quality control easy to handle.

Table 4: Corrosion rate comparison between coated steel (TSA) and uncoated surface [5]

Temperature (^{o}C)	Corrosion Rate (mm/year)	
	Thermal Spray Coating	Bare Steel
60	0.0511	0.1872
70	0.3025	0.1718
80	0.0454	0.1099

Table 4 and Figure 7 show comparisons of corrosion rate performance between bare steel (without coating) and structure with thermal spray coating. From the result, thermal arc spray aluminium offers a great corrosion protection. Based on America Petroleum Institute (API) 581, the estimated corrosion rate for carbon and alloy steel in marine environment is 0.13 mm/year at 60 ^{0}C, so the difference is about -0.0789 mm/year giving a signal that the thermal spray aluminium coating is a better method to reduce corrosion on steel structure [5, 22].

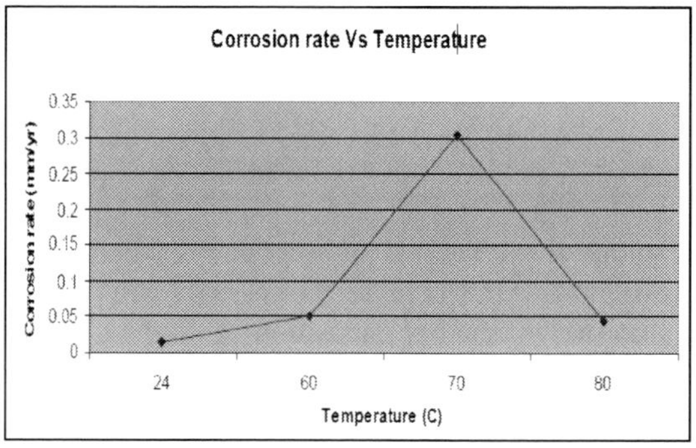

Figure 7: Graph of corrosion rate versus temperature [5]

6. Application of Thermal Spray Coating

Thermal spray coating has been used mostly as a corrosion protection. Other than offshore field, some power plants also applied the coating for their machinery such as boiler tube, gear box, shaft and crankshaft [8]. Examples of products are listed in Table 5.

Table 5: Thermal spray coating Application [1]

Application	Material	Thermal Spray Process
Pipe section for salt water pump	Aluminium	Arc Spray
Bridge	Zinc, Zinc-Aluminium, Aluminium	Flame or Arc Spray
Chemical & Water Storage Tanks	Aluminium, Zinc	Flame or Arc Spray
Piping in Water Plant	Aluminium	Arc Spray
Cooling Water Pump	Zinc, Zinc-Aluminium	Arc Spray

Application of thermal spray coating is not only on steel structure, it also can be used on concrete to protect steel rebar (Figure 8). Theoretically, corrosion damage of steel rebar in reinforced concrete structure can be prevented due to high pH (more than 12) electrolyte contained in the pores of concrete [34]. The condition can ensure passivation of the steel surface. However due to unfavourable environment, repair work and presence of carbonation/chloride contamination, [35] corrosion mechanism can occur and in worst case, corrosion on reinforced steel can take up five times the volume of steel (in case concrete is spalling) [34, 36]. In 1997, the thermal arc spray zinc was applied on steel reinforced concrete structure located at Arabian Gulf since the structure experience severe problem. After one and half year, there is no sign of rebar corrosion [34].

In this coating type, bonding between coating and concrete can be maximized by ensuring concrete is dry, concrete shall be pre-heated at least at 50 ^{0}C and coating usage should be in multiple pass [1].

installed so that patching of the structure can be carried out. After that fitting of electrical contact for zinc coating is performed and arc spraying of the zinc anode is applied **[36].**

Figure 8: Arc spray application on concrete at San Luis Pass Bridge, Texas **[35].**

7. Conclusion

In summary, thermal arc spray coating should be considered to be chose as a coating for an effective corrosion protection medium. Zinc and Aluminium are both suitable in atmospheric condition. However aluminium is the most selected due to its property and advantages offered. Further research need to be conducted to analyze the entire potential of arc spray technology as a corrosion protection so that massive improvement in coating technology can be explored. Extensive research in tidal zone or immersion zone may be carried out for new development in thermal arc spray technology. New material usage may be tested in future research.

8. Acknowledgements

This research is supported by Research Management Institute, Universiti Teknologi MARA, Malaysia under RIF grant no: 600-RMI/DANA 5/3/RIF(98/2012). The authors wish to thank Mr Ong Chong Hup for giving permission to publish this paper

9. References

[1] K.S.Soo and Hasnal Bakiyah Mohd Nor November 2000 *Thermal Spray Coating For Corrosion Protection - Breakthrough in Corrosion Protection and its Control* Kuantan Pahang p 1-9
[2] E.Lugscheider & T.Weber December 1990 IEEE Transaction on Plasma Science *Plasma spraying- An Innovative Coating Technique: Process Variants and Applications* **18** 968
[3] Yantao Li, Jianguo Liu, Jizhou Duan, and Baorong Hou December 2006 Material Performance *Thermally Sprayed Aluminium and Zinc Coatings for Tidal Zone Cathodic Protection of Offshore Platform Pile Leg* p16
[4] Halimah Pit & I.J Rippon access2012 *Potential Application of TSA in SSB/SSPC* p2-5
[5] Mohd Fazril Irfan Mohd Fuad 2010 *Thermal Spray Coating for Corrosion Under Insulation Prevention* Universiti Teknologi Petronas vol 1-4 pp1-45
[6] Ong Chong Hup 2010 *Norimax:Corrosion Control Materials and Consultancy Services in Thermal Spray Coating* Norimax Sdn Bhd Puchong,Selangor vol 2 p 8
[7] Norimax Sdn Bhd 2008 *Thermal Spray Coating Procedure Specification-Arc Spray* Norimax ISO 9001: 2008 p 1-7
[8] H.D.Steffens, Z.Babiak, M.Wewel December 1990 IEEE Transaction on Plasma Science *Recent Developments in Arc Spraying* **18** 974-975
9] Elliott R. Sampson March 2000 Material Performance *Thermal spray Application for Power Plant Components* pg 62
[10] J R Davis 2001 Surface Engineering for Corrosion & Wear Resistance *Process Comparison* **7** 183-193
[11] Dr R Hudson 2000 UK *Surface Preparation for Coating* National Physical Laboratory vol.2 p2-3
[12] ISO 8503-5 2003 *Preparation of steel substrates before application of paints and related products-Replica tape method for the determination of the surface profile* p 1-7

[13] ISO 8502-3 1993 *Preparation of steel substrates before application of paints and related products-Assessment of dust on steel surfaces prepared for painting* p 1-3
[14] ISO 8502-6 1995 *Preparation of steel substrates before application of paints and related products-Extraction of soluble contaminants for analysis* p 1-5
[15] ISO 8502-3 2007 *Preparation of steel substrates before application of paints and related products-Visual Assessment of surface cleanliness*" p 7-13
[16] MG Lawrence 2005 *The relationship between relative humidity & the dew temperature in moisture air: A simple conversion and application* Bull. Am. Meteorol. Soc pp225–233
[17] Steve Luterek acces2012 *Corrosion Protection of Duplex Stainless Steel at Elevated Temperature in an Offshore Environment by Thermal Spray Application of Aluminium* HFW Industries Philadelphia,Newyork pg 18
[18] M S A RAHIM, S Nor Hayati and H.Luay Bakir 2009 International J. Precision Technology *Plasma Spray ceramic coating and measurement of developed coating behaviour* **1** pg 170
[19] Abdul Rahim Mahamad Sahab, Nor Hayati Saad, Salmiah Kasolang and Juri Saedon 2012 Procedia Engineering 41 *Impact of Plasma Spray Variables Parameters on Mechanical and Wear Behaviour of Plasma Spray* $Al_2O_3\%wtTiO_2$ *Coating in Abrasion and Erosion Application* 1689-1695
[20] Carol S Baxter 1997 Material Performance *Thermal Spray Inspection Method and Training Requirement* pp 31-32
[21] Shell Global Solution International 2008 *Thermal Spray Coatings of Aluminium and 85/15 Zinc Aluminium Alloy* 5 13
[22] J A Ellor, W T Young, and J Repp 2004 *Thermal Spray Metal Coatings to Protect Steel Pilings: Final Report & Guide* Ocean City Research Group Washington Transportation Research Board Vol 2 pp9-17
[23] Tony Cunningham & Ron Avery 2000 Material Performance *Sealer Coatings for thermal sprayed aluminium in the offshore industry* pg 46,
[24] The Association of Metal Sprayers access 2012 *Sealing and painting of sprayed Aluminium and Zinc sprayed coatings* p 2
[25] Petronas Technical Standard 2004 *Technical Specification Thermal Spraying of Coatings of Zinc, Aluminium and Their Alloy* p 13
[26] A P Newbery & P S Grant 2006 Journal of Materials Processing Technology 178 *Oxidation during electric arc spray forming of stee*l pp 259-269
[27] Eklavya Calla & S C.Modi 2000 Process Corrosion its Mitigation and Preventive Maintenance *Longlife Corrosion Protection of Steel by Zinc-Aluminium Coating Formed by Thermal Spray* p 1-8
[28] American Welding Society access 2012 *Corrosion Test of Flame Sprayed Coated Steel-19 Years Report* Part 1 p 10
[29] Association of Metal Sprayers Birmingham 1993 *Code of Practice for Protective Coating of Iron and Steel Structures Against Corrosion* Section 1 p 1-2
[30] Association of Metal Sprayers Birmingham 1993 *Code of Practice for Protective Coating of Iron and Steel Structures Against Corrosion* Section 3 p 1-2
[31] The Association of Metal Sprayers access 2012 *Thermal spray coatings to combat and corrosion at elevated temperatures* p 3
[32] Nuchjira Dejang, Pittaya Kuntasudjai and Sukanda Jiansirisomboon 2006 *Effect of Thermal Spray Process on Ni-20Cr* Chiang Mai University pg 1-3
[33] V R S Sa Brito 2012 Journal Material and Design *Corrosion Resistance and Characterization of Metallic Coating Deposited by Thermal Spray On Carbon Steel pp 285-286*
[34] Michael Knepper 1999 Proceedings Eurocorr99 *Lifetime Extension of Thermally Sprayed Zinc Anodes Corrosion Protection of Reinforced Concrete Structures By Using Organic Coating* pg 1-10
[35] Steven F. Daily 2003 Concrete Repair Buletin *Galvanic Cathodic Protection of Reinforced and Prestressed Concrete Using a Thermally Sprayed Aluminium Coating* pg 12-13
[36] Michael Knepper & Jochen Spricstersbach 1999 Proceeding UTSC 1999 *Thermally Sprayed Zinc Coating for the corrosion protection of concrete structure* , p 1-2

3D Machine Vision and Additive Manufacturing: Concurrent Product and Process Development

Ismet P. Ilyas
Design Engineering department
Politeknik Manufaktur Negeri Bandung
Jl. Kanayakan 21, Dago, Bandung 40135, INDONESIA

e-mail: ismetpi@polman-bandung.ac.id

Abstract. The manufacturing environment rapidly changes in turbulence fashion. Digital manufacturing (DM) plays a significant role and one of the key strategies in setting up vision and strategic planning toward the knowledge based manufacturing. An approach of combining 3D machine vision (3D-MV) and an Additive Manufacturing (AM) may finally be finding its niche in manufacturing. This paper briefly overviews the integration of the 3D machine vision and AM in concurrent product and process development, the challenges and opportunities, the implementation of the 3D-MV and AM at POLMAN Bandung in accelerating product design and process development, and discusses a direct deployment of this approach on a real case from our industrial partners that have placed this as one of the very important and strategic approach in research as well as product/prototype development. The strategic aspects and needs of this combination approach in research, design and development are main concerns of the presentation.

1. Introduction

Currently, the global trends and requirements toward product quality (aesthetics and ergonomics) become more and more complex and precise. Furthermore, global competition provides a significant impact and new challenge to the industries to continuously seek integrated manufacturing solutions in order to improve process performance as well as to accelerate product development cycle. Small and/or large product changes certainly will require changes on each component of the products. In this situation, a time requirement to conduct the changes more and more faster as well as time requirement to manufacturing processes without ignoring determined product quality [1].

To overcome with the above challenges as well as to maintain and accelerate product development cycle, two new members in advanced manufacturing technologies: 3D Machine Vision (3D-MV) and additive manufacturing (AM), have able to show their capability and offer its reliability as a solution since they are commercialized. The 3D-MV is a technology that in general enables to model a 3D surface of a product through a technique called 3D-scanning which can then be utilized to manufacture the product rapidly [2]. While AM technology, which is also known as direct digital manufacturing (DDM), has a capability to manufacture a geometrically-complex 3D part directly from 3D CAD solid model without tooling required [3,4].

Referring to the advancement of these technologies, a combination deployment of the 3D-MV and AM at Politeknik Manufaktur (POLMAN) Bandung – The Bandung Polytechnic for Manufacturing - has been accepted as an important part in supporting an advanced design and manufacturing process, especially for a geometrically complex product which difficult or impossible be manufactured through a conventional route. Engineering and design process at POLMAN Bandung which is also supported with advanced CAD/CAM/CAE technologies enable to provide solution in design concept optimization of the product prior to manufacture process and its production planning. However, in many cases, manufacturing a complex 3D physical prototype with a real 1:1 scale are still highly required and cannot be ignored because a visualization as well as an evaluation of a 3D physical part still an effective media compare to 2D or 3D scaled model.

2. Product Development

Recently, there are two common methods/approaches in engineering and product development: conventional and non-conventional [2]. In conventional method, product development starts with generating a 3D CAD model (wireframe/surface/solid) of the product utilizing CAD software. Through a concept modeling to realize a designed product, a format of 3D CAD model can be converted to CAE model for numeric simulation modeling and CAM model in order to generate cutting tool-path (NC-code). The conventional approach unfortunately cannot be empowered for re-engineering process, simulated and optimized the existing product design without any support data/information from CAD format. Therefore, it is required a non-conventional method that can be utilized to capture and record geometrical data of the product as well as to generate a numeric concept model that can be deployed in CAE and CAM system. This is known as a reverse engineering (RE). With the deployment of the 3D-MV in product design and development, RE is expected to grow further in the year to come.

3. Definition and Basic Process

3.1. 3D Machine Vision (3D-MV)

As a concept being familiar currently, RE can be overview as an analysis process toward [5]:

1. Identification of product design information and its relation,
2. Design representation format as well as a higher level accuracy of abstraction, and
3. Realization of an assigned product design

In general, a main goal of RE is to generate a product concept model from its existing physical model. Therefore, a feature combination technique in the 3D-MV (3D scanning with a software package) has a capability to reconstruct the model is highly required [6]. In 3D-MV application, 3D scanning is deployed to capture a coordinate of 3D points (point-cloud) on the product surface. Product geometrical representation in point-cloud format is an initial step for surface parametric development. The following Figure 2 shows a deployment of RE in product design and development.

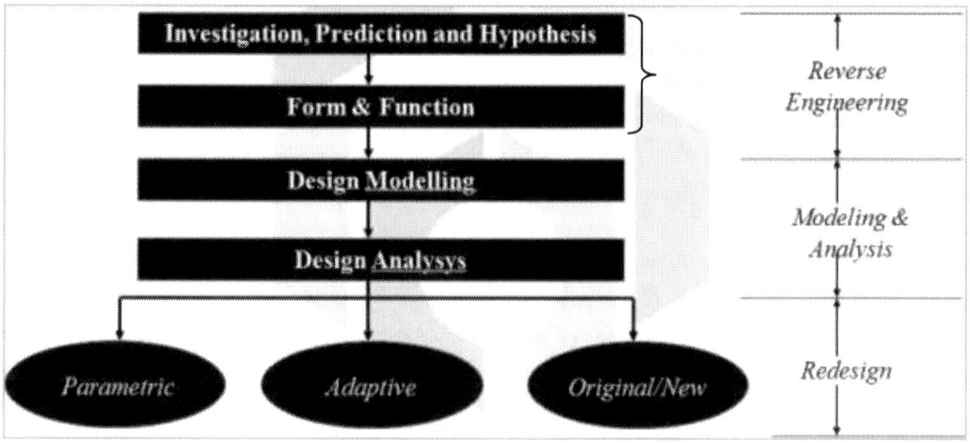

Figure 1. RE Deployment [1]

In product design and development at POLMAN Bandung, the deployment of RE to the 3D-MV implementation is derived into the following seven (7) steps procedure (Figure 2).

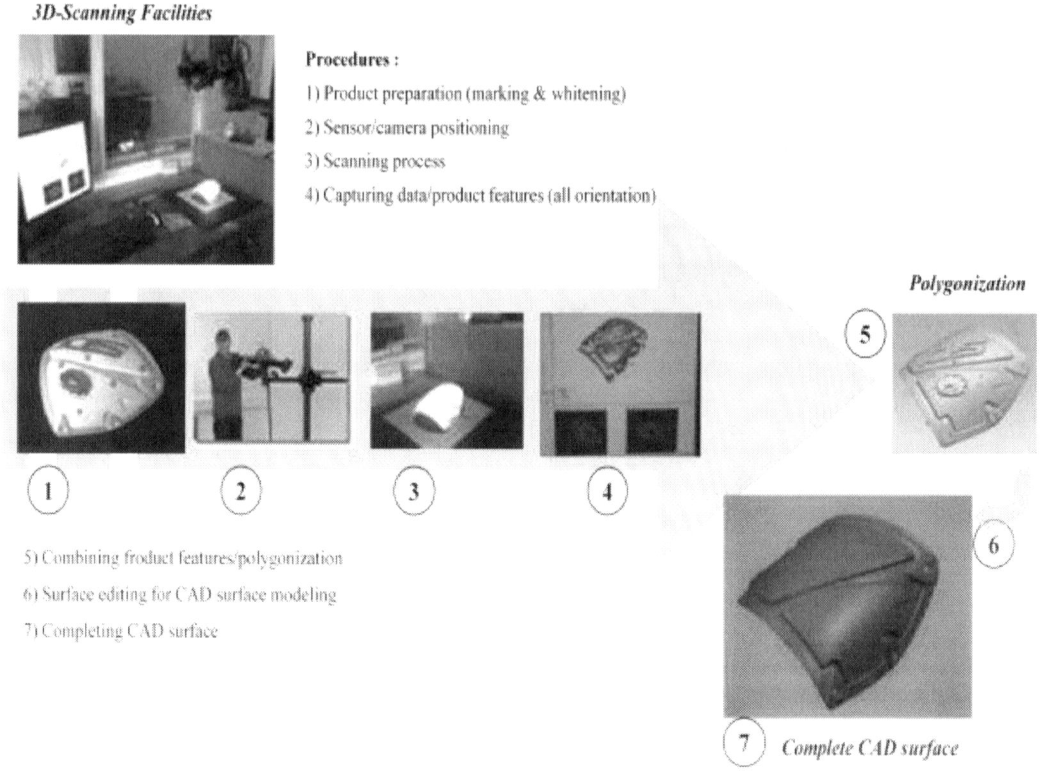

Figure 2. A Basic Procedure of RE

As illustrated, the whole 7 step procedures actually are simplified and grouped into 3 main phases: Scanning Process, Processing and Modelling, and Model Verification.

Phase 1: Scanning Process
As it mentioned before, a main aim in RE is to develop a physical concept model of a product. With a current advanced technology in digitizing and data capturing, scanning is a process to capture a 3D surface data from undefined product geometry. In this scanning process, a 3D-MVis utilize to capture and record data/information of product surfaces in a numeric format in order to generate a matrix of 3D coordinate points (step 1 – 4).

A 3D-MV at POLMAN Bandung - ATOS I Optical 3D scanning system (Figure 3) is an independent facility which is deployed for an accurate 3D scanning process utilizing optical scanning technology. The hearth of this 3D-MV is that its software package enables to manipulate capturing data and convert them to a format that can be utilized to generate a 3D CAD surface model (IGES, ASCII, STL) or even NC code [5].

Figure 3. 3D-MV at POLMAN Bandung: ATOS I Optical 3D Scanning System

Phase 2: Processing and Modeling
Processing and modelling captured scanning data is an important phase. Mostly, captured scanning data always require correction and modification (ie. reducing a number of points, fixing scanning results, merging different scanning results – polygonization) in order to provide a replica and a variety geometrical data of the scanned product (step 5). Thus, modified data is then imported to a reconstruction software package for converting into CAD surface concept model, which is furthermore utilized for required post-processing such as CAD/CAM/CAE and prototyping (step 6).

Phase 3: Model Verification
With the utilization of 3D CAD surface model for post-processing, a next important phase is to verify the whole processes through product quality inspection. In this case, a generated concept of the 3D CAD surface model from the 3D-MV scanning is evaluated by comparing it with a master CAD model of the product so that any deviation occur on product geometry caused by post-processing can be evaluated (step 7).

3.2. Additive Manufacturing (AM)
Additive Manufacturing (AM) has a unique capability to manufacture physically a 3D part/product with a relatively complex geometry directly from a 3D CAD solid model without tooling requirement [3,4]. Currently, the AM have become an integrated part in computer-based design process and its deployment in industry spreads widely. With an advanced development of the technology and

material, the current capability of the AM fulfils the demands of different applications. The main general classifications of the AM application are to develop [8,9]:

1. prototype: design concept visualization, design verification & optimization, design review
2. tooling: rapid tooling (RT)
3. end/functional product: rapid manufacturing (RM)

A variety development of the AM technology is very wide and each of which deploys specific method/technique and material to manufacture a required finished part. However, three basic processes shown in Figure 4 below are the basic steps for all available AM technology. The first step, utilizing CAD software, is to generate a 3D CAD solid model. The second is then to convert generated 3D CAD model to a format model that contains a stack of 2D cross-section layer on Z-axis. Thus , the third (final) step is to develop a physic 3D model of the part/product based on a defined cross-section layer [8].

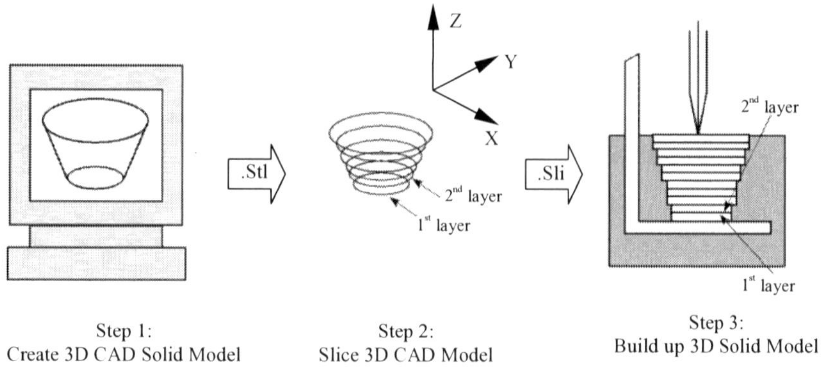

Figure 4. DDM Basic Processes [8]

Fused Deposition Modeling (FDM) from Stratasys is the AM technology deployed at POLMAN Bandung (Figure 5a). The technology in FDM develops a 3D solid part/product by extruding a thin layer of thermo plastic (ie. ABS, PC) paste through thermal extrusion nozzle (Figure 5b). To build up each layer, nozzle extrudes plastic paste which is relates to defined cross-section and then extrude next layer on top of the previous layer.

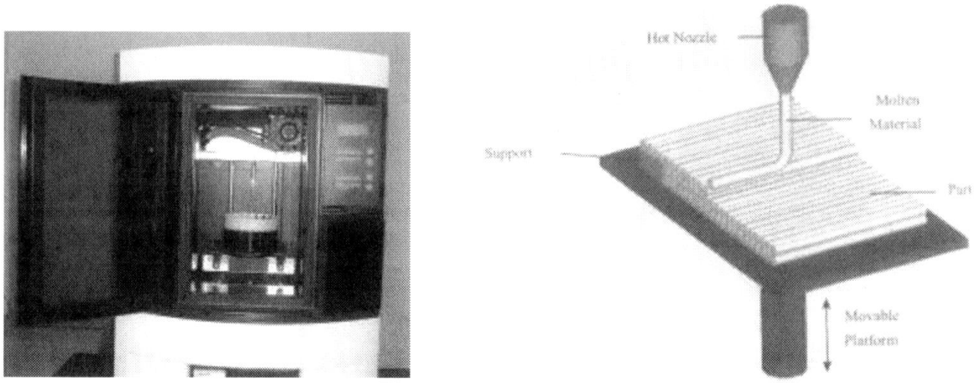

a. FDM facility at POLMAN Bandung

b. Basic FDM Process [8]

Figure 5. Fused Deposition Modeling (FDM)

4. POLMAN Bandung Experiences

4.1. 3D Machine Vision (3D-MV) Implementation

As it mentioned previously, POLMAN Bandung equips its RE process with a 3D-MV which has a capability to transform a required real-scale model. Since 2006, the availability of the 3D-MV as well as other advanced computer-based modelling technologies has improved a quality and efficiency of product development process from design up to manufacturing process at POLMAN Bandung. With optimizing its capability, the implementation of 3D-MV at POLMAN Bandung can go beyond inspection of surface defects. The following are four main application areas of the 3D-MV implementation at POLMAN Bandung:

1. Product duplication: no drawing and manufacturing documentation available.
2. Product Re-engineering: design analysis and modification for performance and quality improvement
3. Rapid prototyping and tooling (RP&T) development
4. Quality inspection.

To get more understanding, Figure 6 to Figure 9 show a sample of real cases which represent the implementation of the 3D-MV in related above application areas. Basically, every case in each application area has various requirements, from fixing design information up to design modification.

1. Duplication & Modification

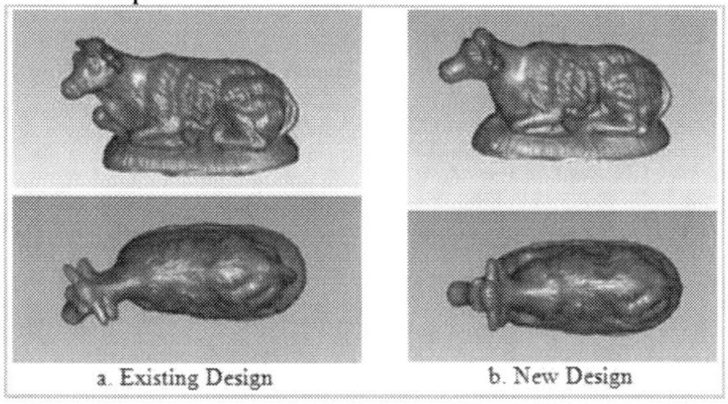

Remark:

Modify the existing design to simplify the manufacturing the mold

Figure 6. Ornament

2. Rapid Manufacturing & Tooling (NC-Code)

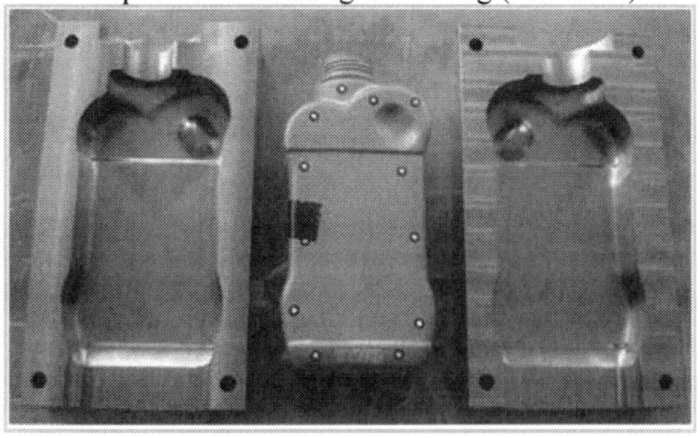

Remark:

Rapid generation of NC-Code (tool path) format directly from 3D surface modeling of the available product.

Figure 7. Oil Container

3. Part Quality Inspection

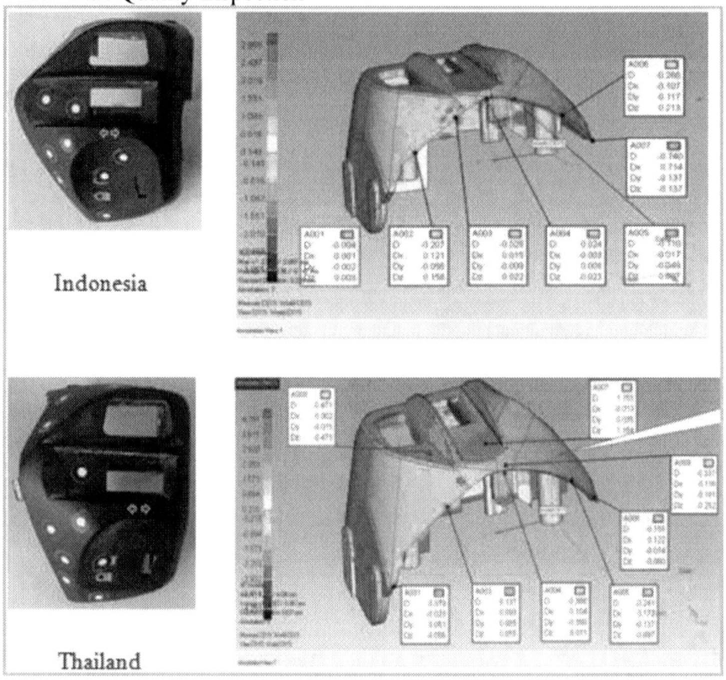

Remark:

Inspect & evaluate two results of the products from two different manufacturer (measured & compared against master model)

Figure 8. Product Comparison

4. Part & Tool Quality Inspection

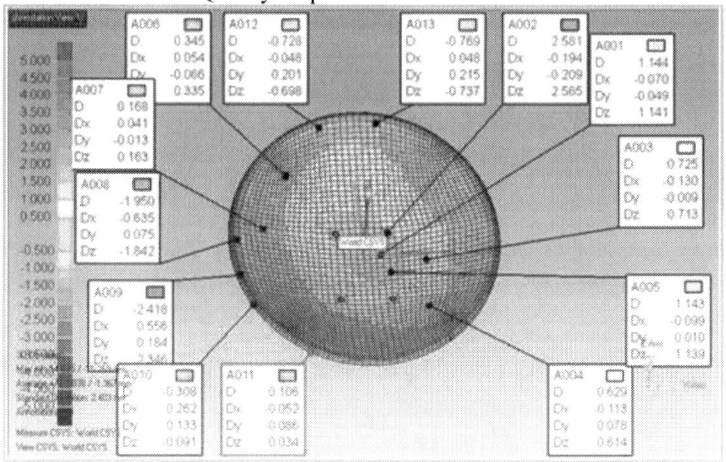

Remark:

Inspect & evaluate the products as a result of the designed tool

Figure 9. Geometry Deviation

4.2. Additive Manufacturing (AM) Implementation

To be advance in product design and development, POLMAN Bandung recently deploys AM technology - Fused Deposition Modelling (FDM) - to rapid development of prototyping for design verification. **Error! Reference source not found.** and **Error! Reference source not found.** present some sample cases of main implementation of AM technology at POLMAN Bandung.

IOP Conf. Series: Materials Science and Engineering **46** (2013) 012029 doi:10.1088/1757-899X/46/1/012029

1. Prototyping 1

Remark:

Prototyping a new design of the container regarding to the exceeding spare of the lit.

Figure 10. Refreshment Container

2. Prototyping 2

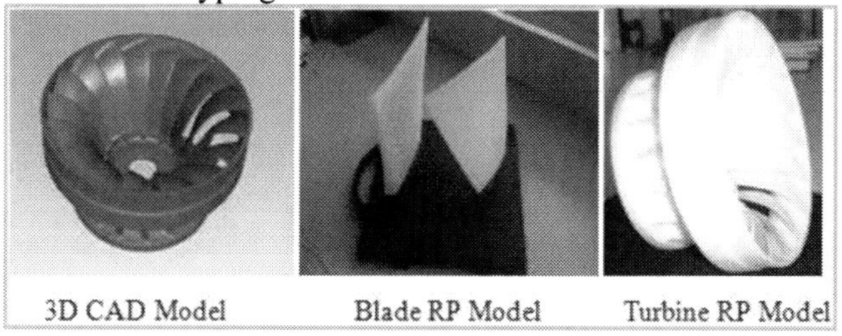

Remark:

Prototyping a new design of the Francis Turbine (redesign the blade)

Figure 11. Francis Turbine

4.3. 3D-MV and AM Expansion of System

Although the 3D-MV and AM technology at POLMAN Bandung currently utilized as a means of ensuring accuracy and product quality of the product independently, initial steps have been taken to enable deviation/defect accumulated by the systems to be used for improvement of product accuracy and quality. A consideration to expand the utilization of the 3D-MV and AM in product design and development is to integrated these technologies in predicting optimum quality and measuring the defects/deviation before they actually occur. The following Figure shows a general scheme of combining 3D-MV and AM technology in designing and developing a quality part/product.

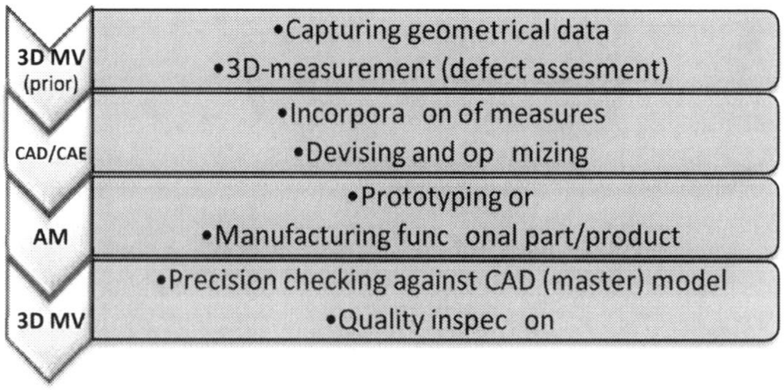

Figure 12. Combination Use of 3D-MV and AM

Of course the implementations of both 3D-MV and AM have no value if they cannot provide benefits and profits. With a proper methodology and technology structure, the implementation of the 3D-MV and AM at POLMAN Bandung has provided tangible and in-tangible benefits in term of both process and result/outcome, as follow:

1. Process: ability of providing proper technical data for design as well as manufacturing process of discontinuous products; overcome a limitation of CAD in product modification and improve its manufacturing process; improve process time through design documentation improvement.
2. Result/outcome: can be utilized to redesign a product in order to improve quality and performance.

Furthermore, the benefits of the 3D-MV and AM technology in product design and development highly depend on a starting point of the knowledge to implement the technology. If the 3D-MV and AM are deployed where technology structure of design and manufacture has not yet well established, the transition to deploy the technology will be relatively difficult and the relatively huge efforts are required. On the other hand, if the technology structures are available and well established, the deployment of the 3D-MV and AM technology in design and manufacturing is relatively easy.

Despite of the benefits, there are challenges in deploying the 3D-MV and AM technology. These technologies are not simple, fast, and an economic to be implemented. There are important issues need to be considered, such as a matriculation or introduction to the technology so that they can be integrated into the existing system, and cost justification for its implementation [10]. As the implementation of the 3D-MV and AM technology is becoming more and more intense at POLMAN Bandung, there are a trend of positive impact toward a certain technology to continuously improve the process of design and manufacturing at POLMAN Bandung.

5. Conclusions

After over viewing the basic aspects of the 3D-MV and AM technology as well as the implementation procedures at POLMAN Bandung, it is required to know and understand that the 3D-MV and AM technology can provide benefits as well as profits in design and manufacturing process of a product. More significantly, further knowledge and understanding of the 3D-MV and AM technology are highly required to improve a system integration of design process and manufacturing. In other words, implementing structured basic principles of the 3D-MV and AM technology properly becomes a key-point as well as an important step in achieving main objective for development process and product quality improvement.

References

[1] Sokovic, M, et al, 2003, RE (Reverse Engineering) as Necessary Phase by Rapid Product Development, *12th Conf. Of Achievements in Mechanical and Materials Engineering*, 7–10 December 2003, Zakopane Poland, 825–830..

[2] Bradley, C, et al, 2005, Advances in the Field of Reverse Engineering, Computer Aided Design and Application, Vol. 2, No. 2, 697-706

[3] Kruth, J.P, Leu, M.C dan Nakagawa, T, 1998, Progress in Additive Manufacturing and Rapid Prototyping. Annual CIRP Manufacturing Technology, 47 (2), 525-540

[4] Dalgarno, K.W dan Wright, C.S, 2003, Approaches to Processing Metals and Ceramics through the Laser Scanning of Powder Beds, SME Technical Papers

[5] Sokovic, M, 2004, 3D-Scanning as Main Phase by Reverse Engineering, Proceedings of the Workshops on Monitoring of Cutting Equipment and Processes using Acoustic Signals, MONACO, Department of Manufacturing Engineering, Budapest, 51–57.

[6] Herbertson, T, 2003, Reverse engineering, in Fourth International Conference on Industrial Tools ICIT 2003, April 8th–12th, Bled, Slovenia, 419–422.

[7] Varady, T., et al, 1997, Reverse Engineering of Geometric Models – An Introduction, Computer Aided Design, Vol.29, No.4, April, 255–268.
[8] Pham, D.T dan Dimov, S.S, 2001, Rapid of Rapid Prototyping and Tooling, London, UK, Springer-Verlag Manufacturing: the Technologies and Applications
[9] Folkestad J.E dan Johnson, R.L, 2002, Integrated Rapid Prototyping and Rapid Tooling, Integrated Manufacturing Systems, 13 (2), 97
[10] Ferreira, J.C., et al., 2001, Integrated Product and Tooling Development via Reverse Engineering Methodologies and Rapid Prototyping Techniques, in 3 Congresso Brasilerio de Gestao de Desenvolvimento de Produto, 25–27 Setembro de, Florianopolis, Brasil

Production and Supply Optimization in PT. Goodyear Indonesia Tbk.

Khristian Edi Nugroho Soebandrija[1] and Stephanus Reynaldo[2]
Faculty Member[1] and Student[2]
Industrial Engineering Department, Faculty of Engineering,
Kampus Syahdan, Jl. K.H. Syahdan No. 9, Palmerah, Jakarta Barat 11480.
Phone : 021-534-5830, Fax : 021-530-0244

E-mail : Knugroho@binus.edu and KhristianEdi@yahoo.com

Abstract. *The Production and Supply Optimization in PT. Goodyear Indonesia Tbk and its study in this paper provide the purposes to create a model of the Master Production Schedule (MPS). The mentioned MPS complies with company policy, production capacity and supply received. From view point of Industrial Engineering, there are needs for forecasting method effectiveness toward the company production capacity. Furthermore, from view point of Information Systems, there are needs to provide the MPS as compare to prior systems that was done manually resulting in the frequent error at the time of making MPS deductible. Several Methods are used and considered, such as: The method of time series, Mean Square Error (MSE), Mean Absolute Percentage Error (MAPE), the calculation of production capacity, MPS, and designing information systems. By having the combination of views from Industrial Engineering and Information Systems perspective the improvement can be conducted from both theoretical development and its managerial implications.*

Keywords: *Master Production Schedule (MPS), Industrial Engineering, Information Systems, Mean Square Error (MSE), Mean Absolute Percentage Error (MAPE).*

1. Introduction

PT.Goodyear Indonesia.Tbk is a manufacturing company that manufactures tires, and one of the largest tire distributors in Indonesia that are not merely dealing with the domestic market, but export their products overseas. The research will focus on the supply chain in company that has the task to manage the supply, production planning, and product distribution.

Several prior studies have been conducted and represented by major journals that are authored by Segerstedt (2006), Sallehuddin (2007), Gustavsson (2008) and Xie (2004). Those journals are further elaborated by discussion in other related journals, in term of Forecasting, Time Series Method, MRP, MPS and Information Systems.

Furthermore, several related books and articles are discussed as well to support the discussion in both theoretical aspects and managerial implication in its case in PT. Goodyear Indonesia, Tbk. PT. Goodyear Indonesia, Tbk, uses its own methods for its forecasting method, and this company does not specify a particular name for this method. However, it is unknown whether the company forecasting methods have been effective or not. Therefore this company conducted its own forecasting calculations to find it out by using the method of time series. This paper is further elaborated the discussion surrounding this issue of effectiveness through the mentioned time series method and information systems that are used.

First, In term of Forecasting and Time Series Method, the related journals have conducted prior researches that support the discussion in this paper, in particular Segerstedt (2006), Sallehuddin (2007), Gustavsson (2008) and Xie (2004). In term of Forecasting and Time Series Method, the related books have elaborated both aspects. Precisely, Baroto (2002, p.30) in quantitative forecasting method using time as a basis for forecasting. This method is done by creating a pattern of demand, which then analyzed for the search pattern that will make it easier to determine what forecasting methods should be used based on the pattern. How to determine the most convenient form of the pattern is to use the eye or *visual.* (Yulianto, 2006, p.22). After selecting and performing calculations obtained by the method of forecasting accuracy of the calculations or forecasting error will perform. In this study, the accuracy of the calculation used is the Mean Square Error (MSE) and Mean Absolute Percentage Error (MAPE). The latter method, is used to select the best forecasting method based on the level of accuracy. Posterior to choose the methods, then the next step comprises the comparison with the MSE and MAPE method companies to see which method is better.

Second, subsequently, in term of Master Production Schedule (MPS) and Material Requirement Planning (MRP), the related journals have conducted prior researches that support the discussion in this paper. Precisely, the journals that are authored by (Zaidi, 2012, p. 1457) discussed the MPS as relate to MRP, with the surrounding discussion on Inventory Management and Forecasting as two topics that are indispensable in this paper. The research of (Mula, 2006, p. 275) has elaborated this paper from view of MRP that is expanded into Supply Chain, as this paper pertaining PT. Goodyear Indonesia Tbk is facing. The first Mula's view on MRP is supported by prior researches of Hegseth (1984), New and Mapes (1984), Murthy and Ma(1991), Donselaar (1992), Donselaar et al. (2000), and Hatchuel et al. (1997). The second Mula's view on Supply Chain is supported by prior researches of Bertrand and Rutten (1999), Caridi and Cigolini (200), Das and Abdel-Malek (2003).

Third, furthermore, in term of Information System, the related journals have conducted prior researches that support the discussion in this paper. Several factors within the scope of supply chain are deemed important to be considered, to ensure the fulfilment of demand from customers. Those factors are: a. The company is unable to meet the demand that *outsource* or ask factory outside Indonesia to send it. b. Companies outside Indonesia factory produce more than the demand so they distributed it to another subsidiary. c. Companies do not have the machinery to produce the requested product. So that the necessary and correct quality information to assist in production planning. (Gustavsson&Wanstrom, 2008, p.334).

In term of Information System, the related books have elaborated it. Precisely, Marchewka (2010) on Information Technology Project Management, Satzinger (2005) on Object-Oriented Analysis and Design, and Whitten (2004) on Systems Analysis and Design Methods, and not to mention other books that are elaborated in this paper.

2. Methods

2.1. Data Collection

This session provide information on how to gather the data through data collection. The Data are collected randomly, as depicted in the following processes. Prior to start the research, observation and interviews will conducted to the company. This is done in order to get a clear view of business processes and policies that are owned by the company.

Furthermore, the data collection is conducted, in term of the product, the actual production, product inventory, supply, labor day, the engine, the actual production, forecasting the company's methods, and demand. Posterior to data capture of product data, the overall demand of the two categories of products, the consumer and commercial bias are then categorized for domestic and *export* markets. Of each product category is also taken two products, namely 165/80 R13 83T TL DURAPLUS, 175/65R14 82H DURAPLUS, HM G2020 10.00-20 16 TT, and HM XTRA GRIP 7.00-15 TT 10. Data on overall demand and product categories are divided between domestic and export markets, will take the data from the years 2009-2011 the data for forecasting and demand. As for the products of the category of consumer and commercial bias the data request will be taken in 2009-2012. For the actual production will take overall demand data in 2011.

2.2. Times Series Method
Time series method is a forecast method of quantitative using time as the base of the forecast. In general, demand for the future will be influenced by time. To make a forecast, a data demand from the past is needed. This data will be analysis-using time as the parameter and the base. There are four pattern, trend, seasonal, cylical and random (Baroto, 2002, p.30). The identification of the data pattern can be used with visual.(Yulianto,2006, p.22). Furthermore the discussion of Time Series Method are referring to three aspects of 1. Moving Average, 2. Weight Moving Average, and 3. Winter. Respectively, the description and formulas are discussed by 1. Nasution(2003, p.35) and Gasperz (2008, p. 87), 2. Gaspersz (2001, p. 92), 3. Nasution (2003, p.41). Eventually those aspects are represented by the following equations, as the third one is segregated in Trend, Seasonal, or Forecast.

$$\text{Moving Average n-Period} = \frac{\sum(\text{Demand in previous n-period})}{n} \quad (1)$$

$$\text{WMA}(n) = \frac{\sum(\text{Weight for n Period})\,(\text{Actual Demand in n Period})}{\sum(\text{Weight})} \quad (2)$$

$$F_t = \alpha \frac{Dt}{I_{t-m}} + (1-\alpha)(F_{t-1} + T_{t-1}),\ T_t = \beta\,(F_t + F_{t-1}) + (1-\beta)\,T_{t-1},\ It = \gamma \frac{Dt}{F_t} + (1-\beta)\,I_{t-m},\ Ft+1 = (Ft-Tt)\,I_{t+1-m} \quad (3)$$

Where, Ft=Value smoothing, = Smoothing constant ($0< \alpha <1$), Dt= Actual Demand,

I_t=Estimation of seasonal, T_t=Estimation of trend, β=The constant trend smoothing ($0<\beta <1$),

F_{t+1}= Forecast, γ=The constant seasonal smoothing ($0< \gamma <1$)

2.3. The Size of Forecasting Accuracy
Based on Nasution (2003, p.35), the size of the forecasting results which is a measure of forecasting error is a measure of the degree of difference between the results of forecasting and actual demand happening.

2.3.1. Mean Square Error (MSE)
MSE is calculated by summing the squares of all errors of forecasting at each period and dividing by the number of the forecast period. Mathematically, the MSE is defined as follows:

$$MSE = \sum \frac{(At-Ft)^2}{n} \quad (4)$$

2.3.2. Mean Absolute Percentage Error (MAPE)
MAPE is a measure of the relative error. MAPE is usually more significant than MAD because MAPE stated percentage forecast error of the actual demand over a given period that will provide the error percentage is too high or low. Systematically, MAPE is expressed as follows:

$$MAPE = \left(100 / n \right) \sum \left| \frac{At - Ft}{At} \right| \quad (5)$$

2.4. Master Production Schedule (MPS)
This paper elaborated the reason why MPS Model in Table 1 of Segerstedt is chosen and the discussion on comparison with similar model. Precisely, First, for the reason, (Segerstedt, 2006, p.3595) indicated that MPS can also evaluate alternative schedules in capacity requirements, providing input to the MRP system and helps the production manager to generate priorities for production scheduling.
Specifically, (Segerstedt, 2006, p. 3594 and 3595) refers to similar situation in PT. Goodyear Indonesia Tbk, pertaining the mechanism of manually introducing a customer order in the MPS through the form of spreadsheet software. On the other hand, this company also have the its own computer system for materials and production control, or the ERP, in this case SAP/R3. Thus, it is one of the reason that tMPS Model of Segersttedt is chosen, in addition to the feature of Availability to Promise. The mentioned reason by Segerstedt, is supported by Nasution (2003, p. 95), in which production plan stating the size of the aggregate and manufacturing output of a company. Implementation of production planning requires an aggregate product planning disaggregation into the planning for each individual product. MPS is the final statement about the "how" a lot of end items to be produced and "when" should be produced. Second, for the similar model, this paper highlight the MPS as indicated by (Xie, 2004, p. 205 and 206) that has similar model, but emphasize more on the total cost, schedule instability (SI) and service level (SL) of a production inventory system. The MPS drives the material requirements planning (MRP) system and provides the important link between forecasting, order entry and production planning activities on the one hand, and the detailed planning and scheduling of components and raw materials on the other. The similar model, has been elaborated as well by (Mula, 2006, p. 278), with the further consideration of customer satisfaction that is ambiguous in term its maximization. Therefore, Mula's research is supported with prior research by (Lehtimaki, 1987) that studies MPS in a MRP environment that maximizes the fuzzy customer satisfaction level from the perspective of a multi-objective decision problem. The objective of maximizing customer satisfaction is ambiguous and can be modelled using fuzzy set theory.

Table 1 MPS Example

Item No	:		*Description*		:		
Lead Time	:		*Safety Stock*		:		
On Hand	:		*Demand Time Fences*		:		
Lot Size	:		*Planning Time Fences*		:		
	Past Due	*Time Period (Month)*					
		1	2	3	4	5	6
Forecast							
Customer Order							
Project Available Balance							
Available to Promise							
Master Schedule							

3. Data Processing

In the data processing techniques to the industry will start from the manufacture of patterns for each product category and the total demand by the method of time series. This is done to look at the data pattern of each product category and the total demand. Data pattern determination is conducted by visual means. Based on the four patterns in the methods of time series forecasting methods will be taken based on the patterns that emerge.

Two steps in calculation methods are: First, MSE and MAPE is intended to determine the best method and also to see whether the different methods are needed for each product category. Second, the iteration of MSE and MAPE, but based on the company forecasting method data to be compared with first forecasting methods. Both results are then decided on which one is better.
Based on the pattern of demand data to the PT. Goodyear to its products can be seen to have a repeating pattern every year. In addition, the same pattern can be seen on the demand for domestic and *export* markets, as well as to the product type consumer and commercial bias. Then based on the analysis and identification of data patterns visually, it can be concluded that the PT. Goodyear has a seasonal pattern data. There are three types of prediction that can be used, namely, Moving Average, Moving Average Weight, and Winter as discussed in session 2.2
Both the Mean Square Error (MSE) and Mean Absolute Percentage Error (MAPE) conduct a comparison between each product and period used to find the smallest MSE and MAPE. In Table 2 indicated that *winter* has the smallest MSE and MAPE than the moving average and the weight moving average. Therefore, the chosen method is a *winter* method of forecasting by a period of 2 due to having the smallest value of MSE and MAPE than other periods.

Then, Table 3 indicated that the ratio for the smallest MSE and MAPE. *Winter* methods of forecasting have the smallest MSE and MAPE in comparison to the company. So it can be concluded that the method of forecasting winter N = 2 is the best forecasting method for the company. Then based on the method of winter N = 2 will do the calculations for product forecasting.
Furthermore, in Table 4 Based on calculations shown in the table 4 about the efficiency and utilization for each month of the company. Can be seen that the efficiency and capacity utilization per month for the company was good. Moreover, it can be seen that the average of the efficiency and utilization was 90.3% and 89.8% which is a figure which is fairly good. This is because the resulting product is a tire company that not too many differences between each type of tire product.

Table 2 The results of MSE and MAPE Moving Average, Moving Average and the *Winter Weight*

Produk	Periode (Bulan)	*Moving Average*		*Weight Moving Average*		*Winter*	
		MSE	MAPE	MSE	MAPE	MSE	MAPE
Permintaan Total	2	222,443	5,422	7032236,306	1112,846	2,042	0,515
	6	405,881	8,409	1025251,911	425,318	1533,382	15,991
	12	261,071	6,495	290,677	6,890	609,066	9,340
Domestik	2	11,179	8,532	151679,191	1113,542	1,247	1,821
	6	10,553	9,136	21907,494	425,001	10,779	8,207
	12	7,184	7,363	7,811	7,692	19,057	11,371
Consumer Domestik	2	2,900	8,343	42380,170	1117,029	0,071	1,182
	6	2,854	9,365	6206,916	428,382	2,155	7,216
	12	2,309	8,430	2,462	8,708	3,309	8,759
Comercial Bias Domestik	2	3,708	10,597	33054,077	1117,295	0,720	2,369
	6	3,122	10,633	4682,983	423,640	3,937	11,347
	12	2,084	8,238	2,240	8,591	6,877	13,847
Export	2	188,487	5,763	5120834,805	1114,034	3,694	0,791
	6	322,126	8,362	747539,754	425,810	222,049	5,867
	12	218,387	6,775	241,911	7,055	440,428	8,544
Consumer Export	2	114,603	7,308	2202216,087	1116,270	2,199	0,331
	6	171,949	9,439	321513,226	426,792	150,649	7,792
	12	105,747	7,537	118,276	7,794	226,163	9,917
Comercial Bias Export	2	10,295	4,875	485192,686	1102,040	1,698	1,336
	6	9,252	4,750	69023,002	414,409	13,334	5,587
	12	8,422	4,586	8,519	4,605	26,369	7,466

Table 3 Comparison of MSE and MAPE method of Winter and Company

Produk	**Peramalan Perusahaan**		***Winter* 2 bulan**	
	MSE	**MAPE**	**MSE**	**MAPE**
Permintaan Total	786,089	9,597	2,042	0,515
Domestik	22,556	11,195	1,247	1,821
Consumer Domestik	3,910	11,422	0,071	1,182
Comercial Bias Domestik	3,360	10,517	0,720	2,369
Export	549,726	10,701	3,694	0,791
Consumer Export	81,485	6,602	2,199	0,331
Comercial Bias Export	7,058	4,031	1,698	1,336

Table 4 Calculation results of efficiency and utilization

Bulan	**Produksi Aktual**	***Effeciency* (%)**	***Utilization* (%)**
Jan	204,456	91,2	90,7
Feb	212,041	98,1	97,6
Mar	218,841	94,3	93,8
Apr	219,547	97,9	97,4
May	214,921	95,9	95,4
Jun	187,391	83,6	83,2
Jul	212,327	88,4	88,0
Aug	196,481	98,2	97,7
Sep	197,687	85,1	84,7
Oct	191,285	82,4	82,0
Nov	181,071	80,8	80,4
Dec	196,569	87,7	87,2
Total Rata-Rata		90,3	89,8

Table 5 Examples of MPS Total Demand

Item No	: -		*Description*	: Permintaan Total
Lead Time	: -		*Safety Stock*	: 101,287
On Hand	: 10,609		*Demand Time Fences*	: 18
Multiple lot	: 100		*Planning Time Fences*	: 18

	Inventory	*Time Period (Month)*											
		1	2	3	4	5	6	7	8	9	10	11	12
Forecast		203,077	215,813	206,973	211,480	212,616	209,703	210,879	197,325	184,241	177,232	197,294	209,629
Customer Order		203,0084	215,298	206,6258	210,895	212,078	209,43381	210,335	197,437445	185,334	178,54	198,1929898	209,5760936
Supply		3,454	5,704	4,115	4,580	4,087	3,903	3,454	6,208	4,022	3,981	3,247	3,671
Customer Order II		199,554	209,594	202,511	206,315	207,991	205,531	206,881	191,229	181,312	174,559	194,946	205,905
Project Available Balance	10,609	101,287	101,287	101,287	101,287	101,287	101,287	101,287	101,287	101,287	101,287	101,287	101,287
Available to Promise	10,609	101,287	101,287	101,287	101,287	101,287	101,287	101,287	101,287	101,287	101,287	101,287	101,287
Master Schedule		290,233	209,594	202,511	206,315	207,991	205,531	206,881	191,229	181,312	174,559	194,946	205,905
Master Schedule II		290,300	209,600	202,600	206,400	208,000	205,600	207,900	191,300	181,400	174,600	195,000	206,000
Capacity		225,316	217,269	233,363	225,316	225,316	225,316	241,410	201,175	233,363	233,363	225,316	225,316
Difference		-64,984	7,669	30,763	18,916	17,316	19,716	33,510	9,875	51,963	58,763	30,316	19,316

Model MPS was made under the existing standards MPS, but have differences with the supply, customer order II, capacity and difference. Customer order II obtained from the reduction of customer orders with supply and difference obtained from the reduction of capacity by the master schedule. If there is a lack of production in scheduling due to differences in demand and production capacity it will be left. Therefore, it would need the approval of upper management in determining the action to be taken. Nevertheless, based on interviews known that the lack of production can be anticipated with the supply of plant PT. Goodyear outside Indonesia or other policies owned by the company.

Table 6 Examples of MPS per product

Item No	:	-				*Description*		:	C-165 80 R13 DURAPLUS 83T TL				
Lead Time	:	-				*Safety Stock*		:	1,282				
On Hand	:	0,850				*Demand Time Fences*		:	18				
Multiple lot	:	100				*Planning Time Fences*		:	18				
	Inventory	*Time Period (Month)*											
		1	2	3	4	5	6	7	8	9	10	11	12
Forecast		2,308	2,172	2,016	2,288	2,242	3,069	2,936	3,089	3,090	2,766	2,600	2,282
Customer Order		2,323	2,229	2,088	2,312	2,266	3,042	2,822	3,009	3,031	2,710	2,601	2,337
Supply		0,100	0,174	0,087	0,107	0,101	0,094	0,088	0,251	0,160	0,167	0,175	0,183
Customer Order II		2,223	2,055	2,001	2,205	2,165	2,948	2,734	2,758	2,871	2,542	2,426	2,154
Project Available Balance	0,850	1,282	1,282	1,282	1,282	1,282	1,282	1,282	1,282	1,282	1,282	1,282	1,282
Available to Promise	0,850	1,282	1,282	1,282	1,282	1,282	1,282	1,282	1,282	1,282	1,282	1,282	1,282
Master Schedule		2,655	2,055	2,001	2,205	2,165	2,948	2,734	2,758	2,871	2,542	2,426	2,154
Master Schedule II		2,700	2,100	2,100	2,300	2,200	3,000	2,800	2,800	2,900	2,600	2,500	2,200

Table 7 Examples of MPS Results Per Product

Master Schedule	*Time Period (Month)*											
	1	2	3	4	5	6	7	8	9	10	11	12
C-165 80 R13 DURAPLUS 83T TL	2,700	2,100	2,100	2,300	2,200	3,000	2,800	2,800	2,900	2,600	2,500	2,200
C-175 65R14 82H DURAPLUS	2,200	2,000	1,200	1,900	1,500	1,600	1,300	1,900	1,800	2,000	2,600	3,300
CB-10.00-20 HM G2020 16 TT	6,500	5,800	4,200	4,600	4,100	4,000	3,600	6,100	4,100	4,000	3,800	3,900
CB-7.00-15 HM XTRA GRIP 10 TT	1,700	1,200	1,300	1,300	1,400	1,400	1,400	1,300	1,400	1,300	1,300	1,400
Total	13,100	11,100	8,800	10,100	9,200	10,000	9,100	12,100	10,200	9,900	10,200	10,800
Capacity	225,316	217,269	233,363	225,316	225,316	225,316	241,410	201,175	233,363	233,363	225,316	225,316
Difference	212,216	206,169	224,563	215,216	216,116	215,316	232,310	189,075	223,163	223,463	215,116	214,516

Basically, the MPS for a product nearly equal MPS total. However, has the distinction of where the MPS per product comparison will be done by adding up the capacity of *master* production *schedule* of each product based on the category of machine. If there is insufficient production capacity there will be reconsidered based on the company's policy and the final decision on top-level management.

4. Discussion and Information Systems

Information Systems is an organized combination of people, software, hardware, communication networks and data resources that collect, transform, and disseminate information within an organization (McLeod, 2001, p.4). The Information Systems in this paper performed the analysis and design of the new information system will be made. Figure 1 indicates the Deployment Diagram in PT. Goodyear Indonesia Tbk, through the programming language of VB.net with MYSQL database, with the purpose of establishing user friendly features with the characteristics of error reduction and ability to adapt to the system. Other than user features, each department involved in PT. Goodyear Indonesia Tbk, is connected through Local Area Network (LAN).

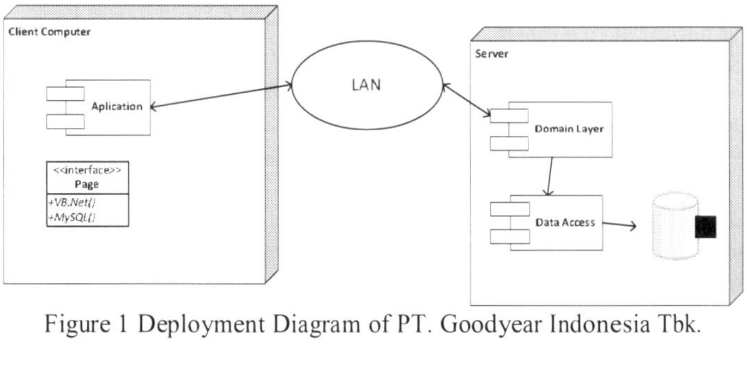

Figure 1 Deployment Diagram of PT. Goodyear Indonesia Tbk.

Furthermore, Figure 2, indicated the Information Systems in term of Class Diagram of PT. Goodyear Indonesia, Tbk, pertaining Product Master (including its SAP Code), is surrounded by Inventory, Supply and Demand, in addition to the TBM, Production Capacity and Plan, and also Product Business Unit (PBU) and its detail of PBU The necessity of the analysis on the company's business processes and then on the supply chain are the focus in which the information system is designed. Which will produce a rich picture, activity diagrams, use case, use case description table event, object, class diagrams and sequence diagrams. Thus the analysis phase has to be completed and continued to the design phase of designing the user interface, sequence diagram for each use case, class diagram design , architecture, and database design, that are indispensable for PT. Goodyear Indonesia.

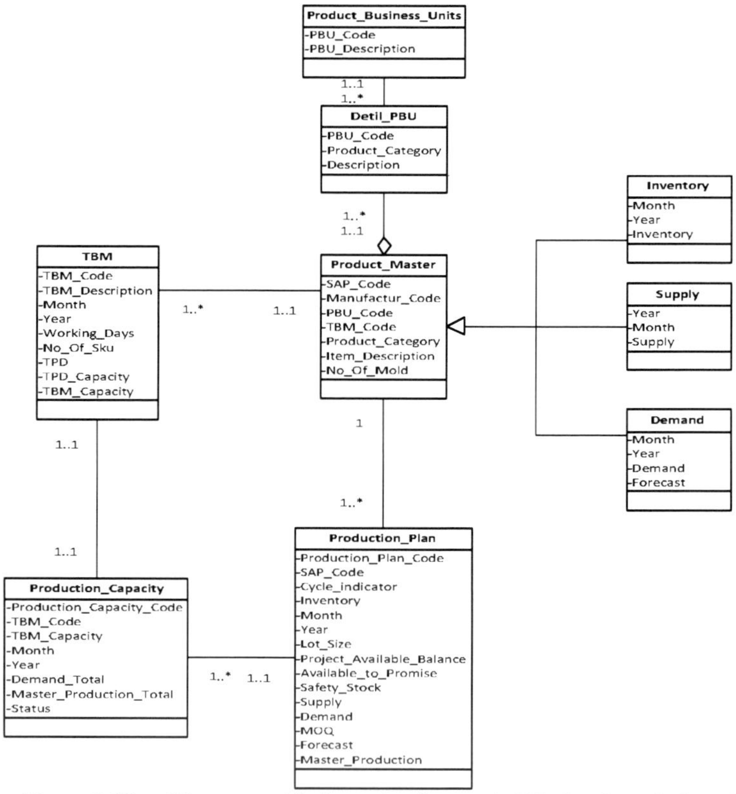

Figure 2 Class Diagram and Information System in PT. Goodyear Indonesia, Tbk.

Based on observations in mind that the supply chain are the focus of research known that the supply chain requires a system that can help improve work processes and reduce the level of error that may occur. This saves processing time through the supply chain in making production scheduling and minimize errors is occuring. Where to support this system, programming language will be used VB.net to get the user interface is user friendly and to be using MySQL database.

Unified Model Language is a standard model of the form and notation developed for developing object-oriented. In general, the depiction of the diagram, the notation used for the show, describing the model is defined by using UML. Using UML analysts and users are able to understand a variety of diagrams that are used for system development project. (Satzinger et al., 2005, p.48)

According to Marchewka (2010, p.364) in a final stage of implementation or the implementation of the project is the result of a system that has been created. Where to implement the new system needed an implementation plan. In the implementation plan there are three approaches that can be done. Direct cutover, an implementation plan that is done by replacing the old with the new system directly. Process of the supply chain itself, beginning at the time of the supply chain receives the data order / demand from the sales. Later sections will examine the supply chain of products that are still available at the warehouse. After ascertaining the amount of products available in the warehouse, then the supply chain publish an order confirmation is sent to the finance section. Parts supply chain will also examine the capacity of the manufacturing production, in order to know the limits of production that can be done. Furthermore the supply chain will directly create production scheduling based on the company's policy, and wait for release from the sales order. After the release order is received by the supply chain will immediately issue orders to the manufacturing production. Parts manufacturing to produce the finished product will send a confirmation that product has been completed. Parts supply chain will then be issued a delivery order for a product that has been made to the warehouse which will then be sent to the customer.

5. Conclusion

The Production and Supply Optimization in PT. Goodyear Indonesia Tbk and its study in this paper provide the purposes to create a model of the Master Production Schedule (MPS). The mentioned MPS complies with company policy, production capacity and supply received. Precisely, in several aspects the conclusion comprises: First, Forecasting winter methods with N = 2 due to get the smallest error rate is better than other methods. And after a comparison with the company forecasting methods, methods of winter N = 2 still has the smallest error. Second, The company does not require different forecasting method because based on the indigo error for each category of products, methods of winter N = 2memiliki lowest error value compare other methods that do the same withwinter N = 2. Third, Based on the calculation of efficiency and utilization where the results obtained 90.3% and 89.8% made the production capacity at its optimal. Fourth, MPS models suggested by the form shown in tables 5 and 6 to demand total and per product. Ultimately as the conclusion for future research, in this paper there are prior research that are indispensable in as the stepping stone for further elaboration of the research in the area of Forecasting, Time Series Method, MRP, MPS and Information Systems, in particular by prior researches of Segerstedt (2006), Sallehuddin (2007), Gustavsson (2008) and Xie (2004).

6. Recommendation

Subsequent to the conclusion and its elaborated aspects in the above paragraph; the recommendation is deployed, not only from the Industrial Engineering Perspective but also from the Information Systems Perspective. Furthermore, the elaboration are the following: From view point of Industrial Engineering, there are needs for forecasting method effectiveness toward the company production capacity. Furthermore, from view point of Information Systems, there are needs to provide the MPS as compare to prior systems that was done manually resulting in the frequent e error at the time of making MPS deductible. Several Methods are used and considered, such as: The method of time series, Mean Square Error (MSE), Mean Absolute Percentage Error (MAPE), the calculation of production capacity, MPS, and designing information systems. By having the combination of views from Industrial Engineering and Information Systems perspective the improvement can be conducted from both theoretical development and its managerial implications. First, Company should update the methods used in the company to become more optimal considering all theoretical and managerial implication. Second, Manufacturing companies should replace the code with the code SAP, as already standardized in the ERP, in this case SAP/R3 for PT. Goodyear Indonesia.

Third, Companies should make a backup data regularly to avoid the undesirable possibility. Fourth, for further research is suggested taking into account data that is not counted in this study, not to mention similar Multinational Companies operating in several countries worldwide, and other aspect of limitation, constraints of this paper that is merely within PT. Goodyear Indonesia Tbk.

7. References

[1] Baroto, T. (2002). *Perencanaan dan Pengendalian Produksi*. Jakarta: Ghalia Indonesia.
[2] Bertrand, J.W.M., Rutten, W.G.M.M., (1999). Evaluation of three production planning procedures for the use of recipe flexibility. *European Journal of Operational Research* 115, 179–194
[3] Caridi, M., Cigolini, R., (2000). Buffering against lumpy demand in MRP environments: a theoretical approach and a case study. Manufacturing Information Systems, *Proceedings of the fourth SMESME International Conference.*
[4] Das, S.K., Abdel-Malek, L., (2003). Modeling the flexibility of order quantities and lead-times in supply chains. *International Journal of Production Economics* 85, 171–181.
[5] Donselaar, K.V., (1992). The use of MRP and LRP in a stochastic environment. *Production Planning and Control* 3, 239–246.
[6] Donselaar, K.V., Nieuwenhof, J.V., Visschers, J., (2000). The impact of material coordination concepts on planning stability in supply chains. *International Journal of Production Economics*. 68, 169–176.
[7] Gaspersz, V. (2001). *Production Planning and Inventory Control Berdasarkan Pendekatan Sistem Terintegrasi MRP II dan JIT Menuju Manufakturing 21*.Jakarta: PT. Gramedia Pustaka Utama.
[8] Gustavsson, M., Wanstrom, C. (2008).*International Journal of Quality & reliability Management,* 26(4), 325-340.
[9] Hatchuel, A., Saidi-Kabeche, D., Sardas, J.C., (1997). Towards a new planning and scheduling approach for multistage production systems. *International Journal of Production Research* 35, 867–886.
[10]Hegseth, M.A. (1984). The challenge of operation yield. *Production and Inventory Management*. 25, 4–10.
[11]Marchewka, J.T. (2010).*Information Technology Project Management.* 3rd Edition.Asia: John Wiley & Sons.
[12]Mathiassen, L., et.al.(2000).*Object Oriented Analysis & Design.*Denmark: Marko Publishing Aps.
[13]McLeod, R. (2001). *Sistem Informasi Manajemen*. Jilid 1.Edisi Ketujuh. Jalarta: Prenhallindo.
[14]Mula, J., Poler, R., Sabater, J.P.G, Lario, F.C. (2006). Models for Production Planning under Uncertainty: A Review. 103. 271-285
[15]Murthy, D.N.P., Ma, L., (1991). MRP with uncertainty: A review and some extensions. *International Journal of Production Economics*. 25, 51–64.
[16] Nasution, A.H. (2003). *Perencanaan dan PengendalianProduksi*.Jakarta: GunaWidya.
[17]New, C., Mapes, J., (1984). MRP with high uncertain yield losses. *Journal of Operations Management* 4, 315–330.
[18] O`Brien, J. A.(2003). *Introduction to Information Systems: Essential for The E-Business Enterprise, 11th Edition.* New York: McGraw-Hill/Irwin.
[19] Pan, L, Kleiner,B.H.(1995). *International Journal of Productivity and Performance Management.* 44(3),4-7
[20]Render, B. , & Heizer, J. (2006). *Operations Management*. 8th Edition. Singapore: Pearson Education, Inc.
[21] Sallehuddin, R. (2007). Forecasting Time Series Data Using Hybrid Grey Relational Artificial Neural Network and Auto Regressive Integrated Moving Average Model. *Neural Network World.* 17(6), 573-600
[22] Satzinger,J. W., et al. (2005). *Object-Oriented Analysis & Design with the Unified Process.*Cambridge, MA: Course Technology.
[23]Segerstedt, A. (2006). Master Production Scheduling and a Comparison of Material Requirement Planning and cover-time planning. *International Journal of Production Research,*44(18-19),3585-3606.
[24] Whitten, J. L., et al. (2004). *Systems Analysis and Design Methods*. New York:McGraw Hill.
[25]Xie,J., Lee, T.S, Zhao, X. (2004). Impact of Forecasting Error on the Performance of Capacitated Multi-Item Production Systems
[26] Yulianto, A.D, Subagyo, (2006).*Jurnal Mesin dan Industri*, 3(2), 21-30

Effect of strain rates on tensile and work hardening properties for Al-Zn magnesium alloys

A L Noradila[1, 3], Z Sajuri[1], J Syarif [1], Y Miyashita[2] and Y Mutoh[2]

[1] Department of Mechanical and Materials Engineering, Faculty of Engineering and Built Environment, Universiti Kebangsaan Malaysia, 43600 UKM Bangi, Selangor, Malaysia
[2] Department of Mechanical Engineering, Nagaoka University of Technology, 1603-1, Kamitomioka, Nagaoka, Niigata 940-2188, Japan
[3] Department of Mechanics, Faculty of Mechanical and Manufacturing Engineering, Universiti Tun Hussein Onn Malaysia, 86400 Parit Raja, Johor

E-mail: noradila.abdlatif@gmail.com

Abstract. The effect of strain rate on the mechanical behavior of Al-Zn magnesium alloys was examined in room temperature from low to high strain rates by using a Universal Testing Machine. Quasi static tensile test was performed in four different strain rates to obtained their effect on tensile properties and work hardening rate using a round shape tensile sample. Two types of Al-Zn magnesium alloys were used in this research study i.e. AZ31 and AZ61 magnesium alloys. The yield stress and tensile strength of AZ31 were found to be strain rate dependent but not for AZ61. The elongation of AZ31 were approximately about 15% for all strain rate levels but for AZ61 the elongations were slightly decreased with increasing strain rates. For all strain rate levels, the work hardening rate of AZ61 were found higher compared to that of AZ31. The change in fracture mode as observed from the fracture surface implies that the fracture mechanisms in AZ31 change as the strain rate increases.

1. Introduction

High specific strength and lightweight properties promote wider applications of magnesium alloys in automotive and aerospace industries. The lightweight property of magnesium alloys entices manufacturer to replace components of denser materials such as steel, cast irons, copper base alloys and even aluminium alloys. In addition, magnesium alloys also possess an excellent mechanical, physical and chemical properties as well as good corrosion resistance [1, 2]. However, in real applications, the chances of the magnesium alloys' made components to collide with other objects (i.e. accident or crash) in different impact velocities could happen in automotive and aerospace applications. The impact could cause damage and leads to catastrophic failure of the materials. Thus, it is important in particular to identify the effect of increasing loading rate on mechanical properties of magnesium alloys to ensure the reliability and durability.

Several studies on the effect of tensile loading rates of mechanical behavior of magnesium alloys are reported. Ahmad and Wei have investigated the tensile properties of die-cast magnesium alloy AZ91D at increased strain rates [1]. Then, the mechanical behavior of an AZ31 magnesium sheet has been investigated by Ulacia et al. at high and low strain rates with elevated temperatures [3]. They found that the strength of the alloys increased with increasing strain rates [1, 3].

In general, aluminium content in magnesium alloys is known to influence the strength of the material with the presence of precipitates in the microstructure. It allows a better work hardening rate of magnesium alloys to indicate the high strength of material with occurrence of many dislocation pile ups against the grain boundaries for delayed fracture [2, 4]. In this study, the effect of strain rates on tensile properties and work hardening behavior of AZ31 and AZ61 magnesium alloys were investigated and compared.

2. Experimental procedure

Two types of extruded magnesium alloys were selected for the quasi static tensile test with increased strain rates. The materials were AZ31 and AZ61 magnesium alloys which have different chemical compositions. The aluminium contents in AZ31 and AZ61 are approximately 3% and 6%, respectively. In both alloys, the content of zinc is approximately 1%. The tensile test samples were cut from extruded bar of 16 mm in diameter to a dumb-bell shape. The gage length and diameter of the sample were 10 and 3 mm, respectively. All samples were subjected to grind polish by using 400 to 1500 grits emery papers to obtain the smooth surface before tensile tests. The quasi static tensile test was conducted with strain rates of 1×10^{-4}, 1×10^{-3}, 1×10^{-2} and 1×10^{-1} sec^{-1}. The strain was measured from the beginning of the test until the specimen break by using high resolution extensometer. The Universal Testing Machine with a capacity of 100 kN was used. All tests were performed in room temperature and lab environment. After the tensile tests, the fracture surface of samples were observed and examined by using a scanning electron microscope (SEM) to identify the mode of fracture.

The data (force, area of cross section, elongation and engineering strain) from the quasi static tensile test at room temperature were used to calculate the true stress and true strain. The true stress and true strain were calculated using formula as in equation (1) and (2).

$$\sigma_T = \frac{F}{A_T} \qquad (1)$$

Here, F is force and A_T is true area of cross section and

$$\varepsilon_T = \ln\left(1 + \frac{\mathrm{d}L}{L_O}\right) \times 100 \qquad (2)$$

where dL is elongation and L_O is original sample length

Based on the stress and strain obtained, the work hardening rates for both materials were identified by using equation (3) as follows [5]:

$$\frac{\mathrm{d}\sigma}{\mathrm{d}\varepsilon} = \frac{\sigma_2 - \sigma_1}{\varepsilon_2 - \varepsilon_1} \qquad (3)$$

$\mathrm{d}\sigma$: change of true stress
$\mathrm{d}\varepsilon$: change of true strain of material

3. Results and discussion

3.1. Tensile test

The tensile properties and behavior of Al-Zn magnesium alloys were listed in table 1 and the stress-strain curves are shown in figure 1, respectively. In overall, AZ61 magnesium alloy exhibited higher tensile strength compared to AZ31. The responses to the changes in strain rate by both materials were found different. Referring to the stress-strain curves in figure 1, the AZ31 shows increment in yield stress and ultimate tensile strength with increasing strain rates. This trend was similar to the results obtained by Ahmad and Wei for die-cast magnesium alloy AZ91D [1]. Study for material of AZ31 + 0.5 Ca was done by Trojanova et al. and for AZ61 magnesium alloys by Yoshida et al. They found that there occurred a rapid rise of flow stress at high strain rates for magnesium alloys [6, 7].

Table 1. Tensile properties of AZ31 and AZ61.

	Strain rate (s^{-1})	σ_y (MPa)	σ_{UTS} (MPa)	Elongation (%)
AZ31 magnesium alloy	1×10^{-4}	173	226	15.7
	1×10^{-3}	191	240	16.2
	1×10^{-2}	217	249	14.4
	1×10^{-1}	204	251	15.4
AZ61 magnesium alloy	1×10^{-4}	205	316	16.2
	1×10^{-3}	219	327	16.6
	1×10^{-2}	228	317	12.0
	1×10^{-1}	221	318	14.4

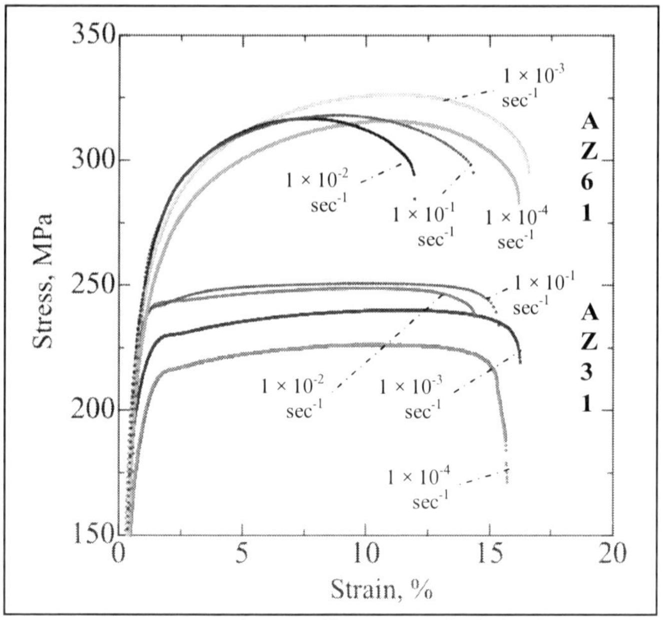

Figure 1. Strain rate effect on tensile behavior.

For AZ31, the elongations of were approximately 15% which showed that there is no influence of strain rates on elongation. However, even though the yield and tensile strengths for all samples were different with increasing strain rate, the ultimate tensile strengths of individual AZ31 sample under certain loading velocity condition was found marginally higher than yield strength where only limited stress increment is allowed after plastic deformation. This indicated that AZ31 magnesium alloy exhibited low work hardening properties. This phenomenon is also mentioned by Srivatsan et.al who studied the tensile behavior of rapidly solidified of magnesium alloy [8]. Further, the yield and maximum tensile stresses increased with increasing strain rate up to 1×10^{-2} sec^{-1}. It is assumed that this was due to the strain rate dependency of critical resolved shear stress (CRSS) of non-basal slip systems for magnesium alloy [3, 9].

In contrast, the strength of AZ61 magnesium alloy was not significantly affected by the increasing strain rates but the elongations of samples were slightly decreased at an increased strain rate. AZ61 exhibited higher work hardening rate as compared to that of AZ31 which indicated by significant difference of ultimate tensile strength to yield stress of AZ61 as shown in figure 1.

3.2. Work hardening

In this study, work hardening rate was used to show the ability of Al-Zn magnesium alloys for work hardens after the occurrence of plastic deformation. Work hardening occurs due to the dislocation movements and dislocation generation against the crystal structure of alloys. Figure 2 shows the trends of true stress and work hardening rates against true strain for both Al-Zn magnesium alloys.

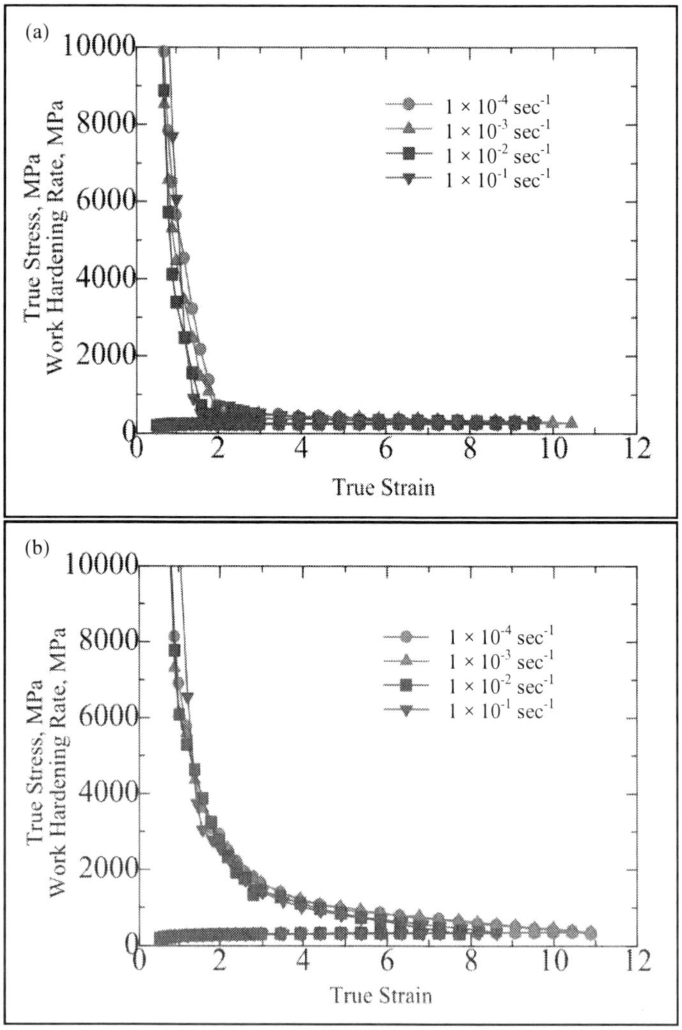

Figure 2. Graph of work hardening rate against increased strain rate of (a) AZ31 (b) AZ61.

In figure 2(a) shows the work hardening rate versus the true strain of AZ31 magnesium alloy. The figure shows that, for all the strain rate conditions, the work hardening occurred only at the early stage of deformation. The hardening process was then diminished just after the yielding which indicates the poor strain hardening behavior of the material. In contrast, work hardening rates for all AZ61 samples in different strain rate conditions were higher compared to that of AZ31 as shows in figure 2(b). The work hardening process in AZ61 continues after yielding point until to the level of ultimate tensile strength. The higher work hardening rate for AZ61 was believed to be attributed by more dislocation pile-up and blocked at the $Mg_{17}Al_{12}$ precipitates and grain boundaries as compared to that for AZ31 in which precipitates were very limited due to high purity of the alloy [2, 10, 11 and 12]. However, among the AZ61 samples, the work hardening rate at 1×10^{-1} sec^{-1} was found slightly higher compare to that of other samples with lower strain rate conditions.

3.3. Fracture Surface of Al-Zn Magnesium alloys

SEM fractographs of samples tested at low and high strain rates are shown in figure 3(a) and 3(b). The micrographs revealed that the sample subjected to low strain rate exhibited brittle intergranular fracture with clear surface steps observed on fracture plane as shown in figure 3(a). These fracture steps and planar surfaces are believed to be the h.c.p. basal plane of crystal structure [13]. The surface morphology is changed in the samples subjected to high strain rate of 1×10^{-1} sec^{-1} as observed in figure 3(b). Here, the material fractured in relatively ductile mode which implies that the fracture mechanisms in AZ31 change as the strain rate increases.

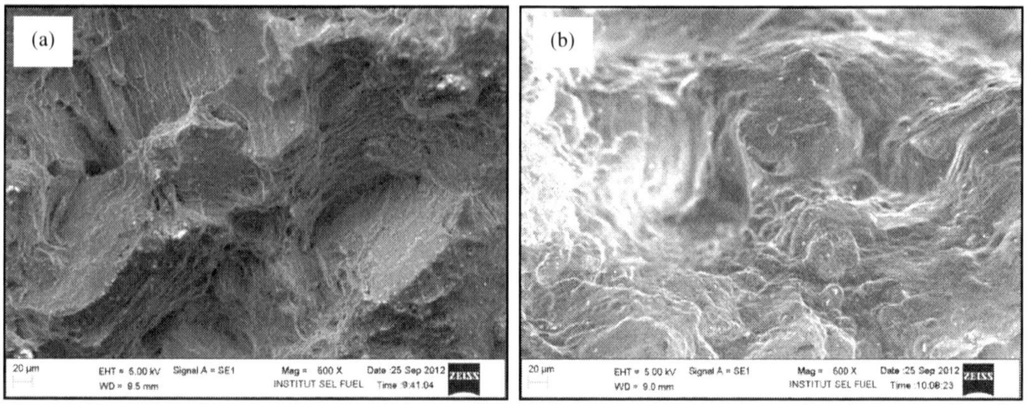

Figure 3. Fracture surface of the AZ31 magnesium alloy at a strain rate of (a) 1×10^{-4} sec^{-1} (b) 1×10^{-1} sec^{-1}.

4. Conclusion

Quasi static tensile test of AZ31 and AZ61 magnesium alloys were investigated. Based on the results obtained, the findings are concluded as listed below:

a. Tensile strength and work hardening rate of AZ61 were higher compared to that of AZ31.
b. Tensile strength of AZ31 was strain rate dependant but not for AZ61.
c. The elongations of AZ31 were approximately 15% for all strain rates whereas, the elongations of AZ61 were slightly decreased as the increased of strain rate.
d. The differences in fracture surface of AZ31 implies the change in fracture mechanism from low to high strain rates.

5. Acknowledgement

The authors acknowledge the supports from the Universiti Kebangsaan Malaysia, Ministry of Higher Education Malaysia (ERGS/1/2011/TK/UKM/02/9), Nagaoka University of Technology and Universiti Tun Hussein Onn Malaysia.

IOP Conf. Series: Materials Science and Engineering **46** (2013) 012031 doi:10.1088/1757-899X/46/1/012031

References

[1] Ahmad I R and Wei S D 2010 *Appl. Mechanics and Materials,* **24-25,** 325-330.
[2] Kainer K U 2003. *Magnesium-alloys and technologies* (WILEY-VCH GmbH & Co. KGaA.)
[3] Ulacia I, Dudamell N V, Galvez F, Yi S, Perez-Prado M T and Hurtado I 2010 *Acta Materialia,* **58,** 2988-98.
[4] Marya M, Hector L G, Verma R and Tong W 2006 *Materials Science and Engineering A,* **418,** 341-356.
[5] Armstrong R W and Walley S M 2008 *Int. Materials Rev.,* **53,** 105-127.
[6] Trojanova Z, Lukac P, Podrabsky T and Pesicka J 2012 *Materials Engineering,* **19,** 12-17.
[7] Yoshida Y, Arai K, Itoh S, Kamado S and Kojima Y 2005 *Science and Technology of Advanced Materials,* **6,** 185-194.
[8] Srivatsan T S, Vasudevan S and Petraroli M 2008 *J. of Alloys and Compounds* **461,** 154-159.
[9] Barnett M R 2003 *Metallurgical and Materials Transactions A,* **34,** 1799-06.
[10] Kim H L and Chang Y W 2011 *Metals and Materials Int.,* **17,** 563-568.
[11] Kapoor R and Nemat-Naseer S 1998 *Mechanics of Materials,* **27,** 1-12.
[12] Tanski T 2012 *J. of Achievements in Materials and Manufacturing Engineering,* **54,** 260-274.
[13] Zhang H X, Wang W X, Wei Y H, Li J Y and Wang J L 2011 *Trans. Nonferrous Met. Soc. China,* **12,** 1225-33.

Enhancing a Multi-body Mechanism with Learning-Aided Cues in an Augmented Reality Environment

Manjit Singh Sidhu

Dept. of Graphics and Multimedia, College of IT, Universiti Tenaga Nasional

manjit@uniten.edu.my

Abstract. Augmented Reality (AR) is a potential area of research for education, covering issues such as tracking and calibration, and realistic rendering of virtual objects. The ability to augment real world with virtual information has opened the possibility of using AR technology in areas such as education and training as well. In the domain of Computer Aided Learning (CAL), researchers have long been looking into enhancing the effectiveness of the teaching and learning process by providing cues that could assist learners to better comprehend the materials presented. Although a number of works were done looking into the effectiveness of learning-aided cues, but none has really addressed this issue for AR-based learning solutions. This paper discusses the design and model of an AR based software that uses visual cues to enhance the learning process and the outcome perception results of the cues.

1. Introduction

Computer simulations are providing unique insights into the way the world works today [1]. Students can now experiment real problem solving task in a virtual world of complex, dynamic systems in a way that was impossible before. Therefore numerous technologies have emerged and are being used in the educational sector to produce a better knowledge society. As such many technologies have emerged and are being used in the educational sector to enhance the learning process. Some of these technologies include the use of software, interactive white boards, multimedia and virtual reality. Most studies found in the literatures also reported the advantages and disadvantages of these technologies [2 - 5]. On the other hand the learning and understanding process of the information presented and learnt from the previously mentioned technologies has not improved much [6]. A more recent technology i.e. augmented reality is becoming popular and being tested by many researchers to see if it could help learners visualize and understand the learning process better. Augmented Reality (AR) is commonly known as the ability to overlay computer graphics onto the real world. Unlike immersive Virtual Reality (VR), AR interfaces allow users to see the real world at the same time as virtual imagery attached to real locations and objects. AR concepts also use tangible user interface (TUI). Unlike physical interface whereby the user may interact with the software via a mouse or keyboard connection, TUI uses wireless interaction. Further details on how this technology works are given in [7]. The motivation of the research aims to look at alternative solutions to visualize engineering problems using AR and visual cues to enhance the learning process. The system developed in this research was used and tested by first year engineering students at UNITEN. User interviews and questionnaires were used to collect users understanding of the subject matter and perceptions of the learning aided cues employed. All of which showed positive response.

2. Visual Cues

According to Bonnie Meyer in her book published in 1975, cueing means the addition of a non-content aspect of prose, which gives emphasis to certain aspects of the semantic content or points out aspects of the structure of the content [8]. Hence, researchers have been identifying and looking at various important cues to aid learners in computer mediated learning environments. The use of multimedia elements in instructional design have also contributed to the importance of using suitable cues to aid in the delivery of multimedia enhanced contents [9] and [10]. Appropriate cues in multimedia enhanced learning environment will reduce students cognitive load [11] and [12]. The advancements in the area of computer graphics and AR have resulted in the use of complex 2D and 3D synthetic computer generated objects in AR-based computer aided learning (CAL) applications. A number of researches have been done incorporating AR technology in various domains of educational and training scenarios [13] and [14]. This project in particular, is looking into, not only using AR technology to assist in the delivery of mechanical engineering education, but also enhancing the learning process with suitable learning-aided cues.

3. Cues in Teaching and Learning Environment

Researchers and developers have begun to realize the importance of cues in aiding learners to better comprehend contents presented in computer aided learning environments. Cues are meant to guide learners cognitive processing rather than providing new information [8]. Visual cueing is the addition of non-content information (e.g., arrows, circles, and coloring) to visual representations. In multimedia based learning, learners are engaged in three important cognitive processes: Selecting, Organizing, and Integrating [11]. The authors believe that the presence of cues is even more significant in learning environments where spatial intelligence and self-paced learning occurs, such as those in an AR-based learning environment. There are various types of cues which have been explored and researched in. Fig. 1 provides (a rather crude and non-exhaustive) summary of these cues.

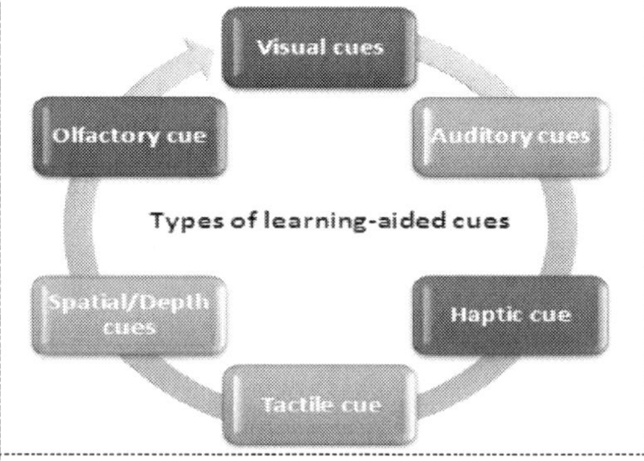

Figure 1. Types of learning aided cues.

4. Augmented Reality in Engineering Education

Augmented Reality uses computer-driven technologies to deliver and augment sensory information to users, thus supplementing and/or replacing real-world sensations [15]. Milgram and Kisihino have described the relationship and taxonomy of AR technology along the Virtuality Continuum (VC) in a Mixed Reality (MR) environment [10]. Fig. 2, shows the VC in which AR can be defined as a class of display where the surrounding real environment is enhanced, overlaid or augmented with computer generated information. An Augmented Reality (AR) system can be defined as having these three properties [16]:

- *Combines real and virtual objects in a real environment,*
- *Runs interactively, and in real time; and*
- *Registers (aligns) real and virtual objects with each other.*

According to Johnson, Levine, Smith and Stone, Augmented reality has strong potential to provide both powerful contextual, in situated learning experiences and serendipitous exploration and discovery of the connected nature of information in the real world [17]. As mentioned in the previous sections, AR has been used in delivering teaching and learning process. Some of the prominent and successful implementations of AR in the various teaching and learning domain, such as geometry, chemistry and biology are listed in [13] and [14]. A major project where AR platform had been implemented in school environment is the ARiSE project, where several European countries took part in this project [18].

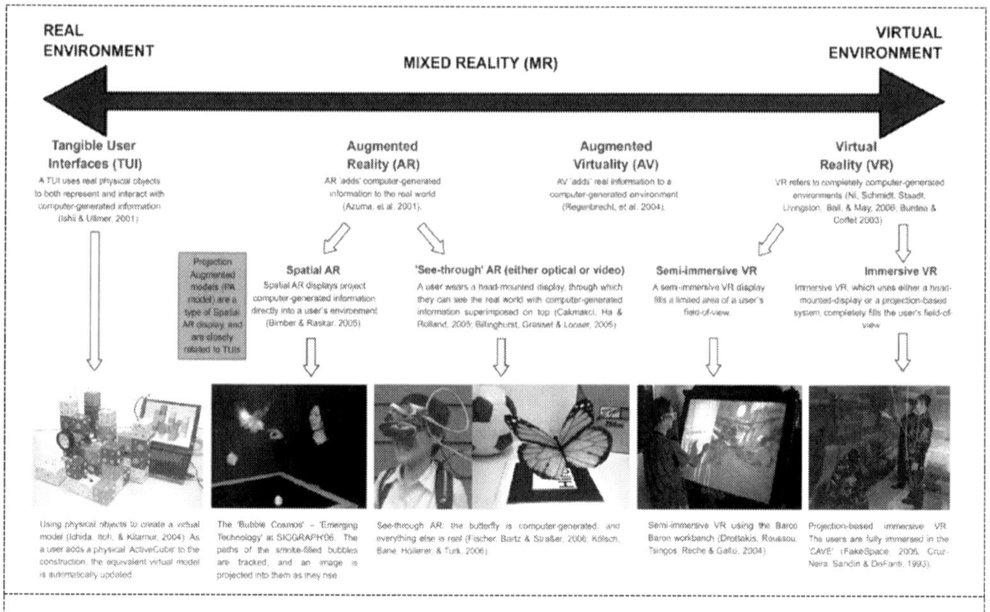

Figure 2. Reality-virtuality continuum [Milgram & Kishino, 1994].

One area where many have used AR as medium of delivering educational content is in the teaching of engineering syllabus. A learning support system based on Web3D and AR technologies have been developed to deliver contents related to engineering subjects such as platonic solids and machines [19]. An AR system for automotive engineering education has been developed with the objective of teaching disassemble/assemble procedures of an automatic transmission [20]. Another research looked into the possibility of using AR in the electrical and mechanical engineering education [21]. In the research paper it was also argued that by combining AR technology with education software such as Matlab and Simulink will improve the pedagogy methods. A prototype Mixed Reality (MR) system where distributed CAVE-based workspaces have been developed to foster collaborative engineering between remote sites [22]. AR-DEHAES, a learning toolkit for engineering graphics at tertiary level has been developed [23]. This toolkit consists of a software application, an explanatory video, a notebook and an augmented book. Another recent project which is similar to the authors proposed project is an AR based simulation of multibody systems developed by [24]. They use a fiducial/marker based AR to augment real world with a virtual model of slider-crank mechanism using an AR video see-through system setup.

5. Theoretical Background

The developed AR system simulates the workings of a multibody system for an engineering mechanics problem. This simulated multibody system is a dynamic behavior of interconnected rigid bodies, each of which may undergo translational and rotational displacements, see Fig. 3. This type of a multibody system is known as the slider and crank mechanism. The slider-crank mechanism is an arrangement of mechanical parts designed to convert straight-line motion to rotary motion or vise-versa [25].

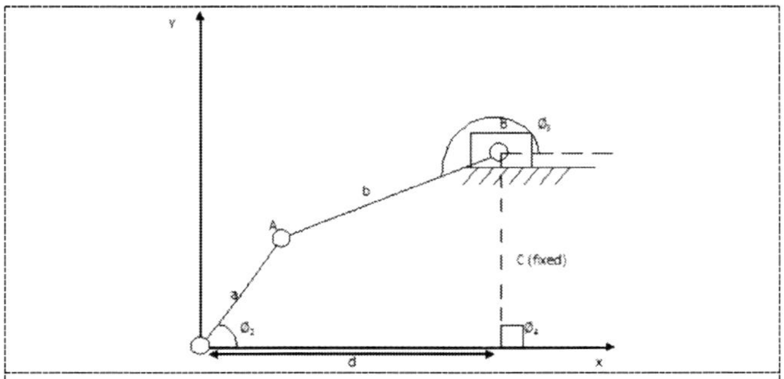

Figure 3. The schematic of the multibody system, the slider and crank mechanism used in the proposed AR-based learning system.

The simplified solution schemes for this slider-crank mechanism (used for the proposed AR system) are given in (1) and (2):

$$\emptyset_3 = \sin^{-1}\left(\frac{a \sin \emptyset_2 - c}{b}\right) \quad (1)$$

$$d = a \cos \emptyset_2 - db \cos \emptyset_3 \quad (2)$$

The AR system developed is based on the monitor based video displays or Window-on-the-world (WoW) display setup, see Fig. 4. This set up uses merged video streams of virtual objects and real world s views, and displays these merged sequences of images on conventional computer monitors.

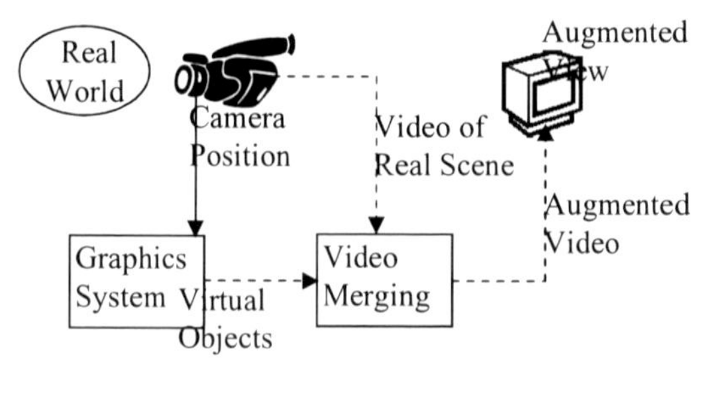

Figure 4. Monitor based AR system.

6. Visual Cues in Augmented Reality System

In this study the researches implemented selected visual cues such as color, text or arrows, audio explaining concepts and animated mechanisms to see if this method could help students in understanding the mechanisms of an engineering problem presented in an AR environment. Since visual cues have not been used and tested in an augmented reality engineering application for learning, in this paper a 3D augmented reality (AR) application for teaching engineering was developed and tested. Seven patterns of visual/auditory sensory cues were embedded into the AR application to see if it could help students visualize an engineering problem i.e. slider crank mechanism. Sensory cues may include visual cues, auditory cues, tactile cues, haptic cues, olfactory cues, and others as mentioned earlier in Fig. 1. Sensory cues play an important role in theories of perception, especially theories of appearance (how things look). The six visual cues (markers) (Fig. 5) employed in this research for visualizing the slider crank problems are 1. Change color, 2. Display slider crank, 3. Display 2D diagram, 4. Display axes, 5. Display trail/path, and 6. Display all mechanisms. Only one auditory cue was used. The markers are used to allow the camera to calculate its respective orientation and distance, necessary for real time renderings. The system is composed of three components: tangible user interface, display component and rendering component. The tangible interface mimics a textbook problem based interface for user interaction. The coordinate information of a particular cue is generated as soon as the camera recognizes the maker being shown and translated into the AR environment. The system setup is shown in Fig. 6.

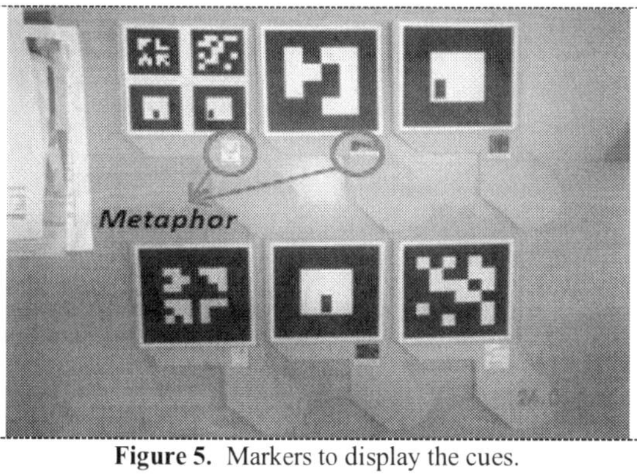

Figure 5. Markers to display the cues.

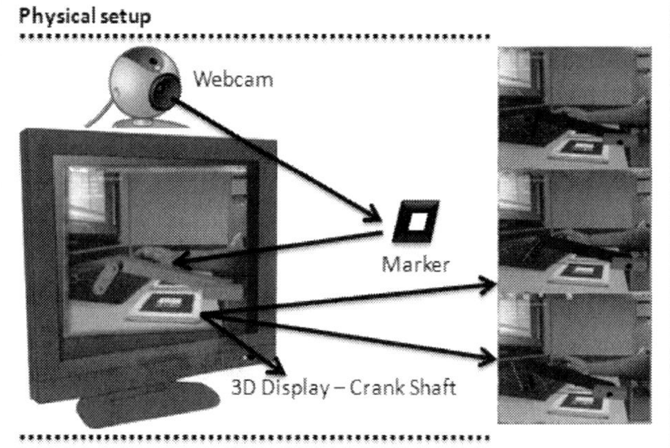

Figure 6. System setup.

7. Visualizing the Engineering Concepts using Cues

This section briefly explains the learning and visualization process of using the visual cues used in this study. As mentioned earlier in section VI, six visual cues and one auditory cue were designed and used to see if students could recall and link the theory learnt earlier in the normal classroom. Each marker is used to perform a different task and display a cue. One of the cues used to display all the engineering mechanisms in the slider crank problem is shown in Fig. 7. The system also narrates the user about the objectives and gives a brief explanation about the problem. Since not all users were familiar by interacting with the system using tangible user interface (TUI), we also designed the keyboard user interface that functions (alphabets used to perform a particular task) the same as the TUI. Examples of using visual cues are shown through Figures 8 9. Figure 8 depicts the cue to show the axes, Figure 9 depicts the cue to show/change the color of the slider crank. The student may also see multiple cues (two or more) by using two markers such as the one shown in Figure 10 (showing the trail and the 2D diagram of the slider crank mechanism).

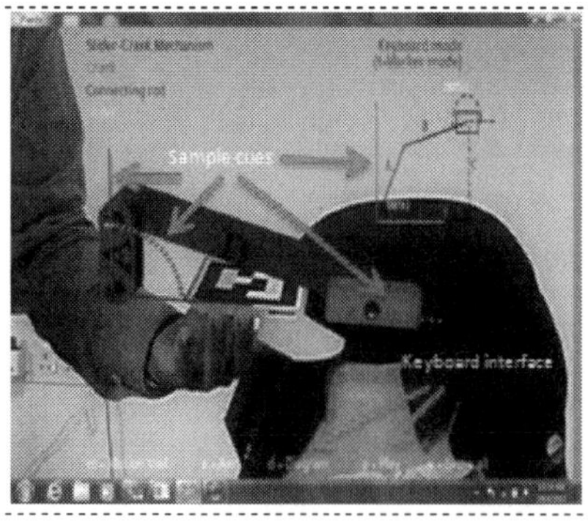

Figure 7. All cues in the system.

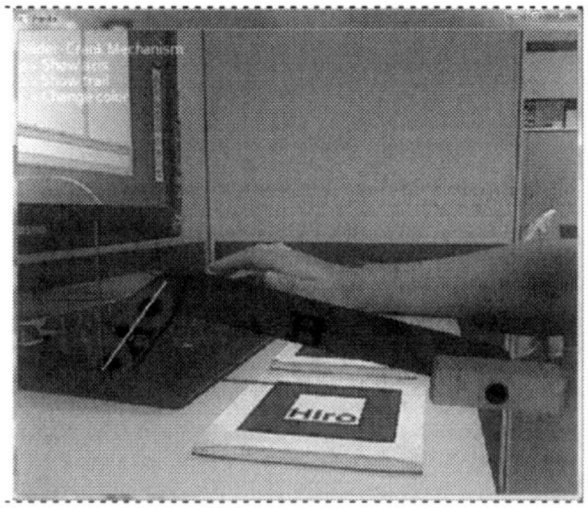

Figure 8. Single cue for showing the axes.

Figure 9. Cue showing the change of color.

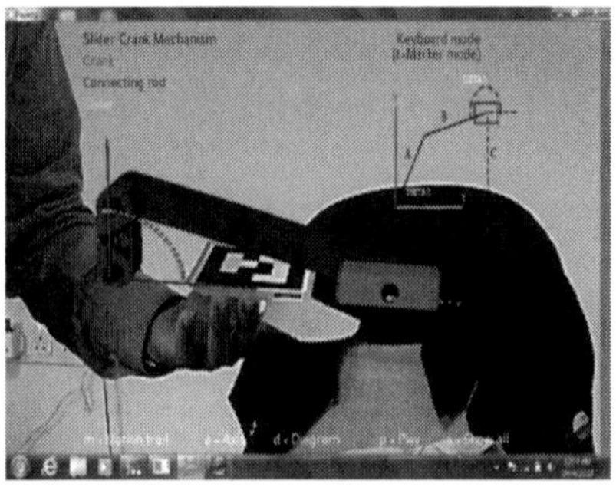

Figure 10. The motion trail and the 2D diagram.

8. Results and Discussion

8.1 Education environment and evaluation

The AR system was tested with first year mechanical engineering students at UNITEN. A total of 37 students participated in the evaluation. Two systems were setup in the computer aided learning multimedia software (CALMUS) research laboratory. Each student spent approximately fifteen minutes in using the system. Interviews and questionnaires were used to measure the usefulness of the visual cues, how well does the system assist them in understanding the engineering concepts and effectiveness of the method in their learning.

8.2 Interviews and questionnaires

To assess the system design, interviews and questionnaires, the most used evaluation techniques, were used to generate more in-depth data on users visualization process and give designers insight into improving the system. An evaluation session was arranged after a lecture where each student was asked to elaborate their thoughts on the hands-on experience in order to identify usability concerns. They were also allowed to present their ideas and share any thoughts for improving the system. A survey was done to obtain user perceptions on the cues based on *Very useful* , *Useful* and *Not useful*

8.3 Results of interviews and questionnaires

The results of the study showed that the weaker students (previously having problems in understanding the subject matter) who used the AR system in this study and had significantly better understanding on both comprehension and visualization of the problem. However some students had problems interacting with the tangible user interface as they were not familiar with the system. In general the students appeared to be satisfied by using the system and felt it helped them in their learning. However they also commented that the system was restricted by technology limitation when the AR system was difficult to be interacted with the markers as they had to hold them for some time. On the other hand they appreciated the use of 3D graphics and animation and the use of alphabets to interact with the system/display a particular visual cue. The students also revealed in the interviews that the AR system was more natural to interact with and looked very much like the real world object. The results in Figure 12, shows a positive impact on all the three aspects of using cues in this study. As a whole most students found all the cues to be *Very useful* and a small number of students found some of the cues were not *useful* . This could be a result of the students being unfamiliar of using tangible user interfaces and they had to spend more time adapting to this sort of user interface rather than concentrating on its contents using cueing method.

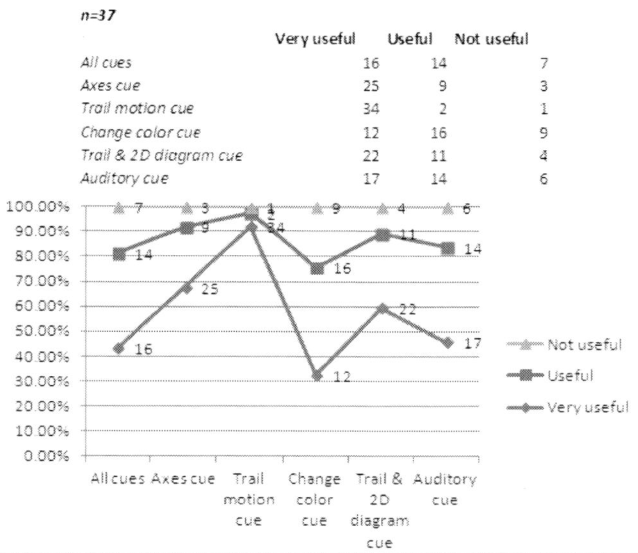

Figure 11. Outcome perception of the cues.

9. Conclusion

Newer technologies are changing the way students learn today. In this study, a new system using augmented reality was developed and tested. Visual cues were employed to see if it could further enhance the learning of engineering problems. The learning aided cues helped students in reducing the cognitive workload when a single cue was used to display a particular engineering concept in the problem presented to them. The study found that the students benefited from using the system as they found it to be very useful and an interesting new method of learning. The technology limitations in this study are the future aims of the study.

Acknowledgement

We would like to acknowledge the considerable contributions of Assoc. Prof. Dr. Faris Tarlochan from the Mechanical Engineering Dept, UNITEN for explaining the concepts of the slider crank mechanisms, Universiti Tenaga Nasional for the financial support for the publication of this paper. Thanks are also due to Dr. Kirandeep Kaur for proof reading this paper.

References

[1] S. Wolfram, "A New kind of science. Champaign, IL: Wolfram Media," pp. 573 and 1072-1073 2002.

[2] V. Oleg, "Fundamentals of Kinematics and Dynamics of Machines and Mechanisms". Oleg Vinogradov CRC Press 2000, Print ISBN:978-0-8493-0257-2, eBook ISBN: 978-1-4200-4233-7, DOI: 10.1201/9781420042337.ch1.

[3] K. John, M. Judy, M. Vance, and Larochelle. "Spatial Mechanism Design in Virtual Reality with Networking. Journal of Mechanical Design". Vol. 124. 2002.

[4] S. Sidhu. Manjit., "Visualizing Curvilinear Cylindrical Engineering Members in 3-D Environment with Motion Path" - NCM2009: 5th International Joint Conference on INC, IMS and IDC (24 - 28th 09), ISBN: 978-0-7695-3769-6, Organizer IEEE, pp: 1358 - 1363. Seoul, Korea 2009.

[5] S. Sidhu. Manjit., "A 3-D Graphical Model for a Four Bar Linkage Mechanism Problem" First IRAST International Conference on Data Engineering and Internet Technology (DEIT), Organized IEEE, SGU & IRAST, ISBN: 978-1-1244-8581-9 pp. 156 – 159. Bali, Indonesia IEEE Explore & SCOPUS-Cited Publication. 2011.

[6] G. Huber., Dietmajer. P., "Four Bar: An educational synthesis and design tool for four bar mechanisms". Proceedings of the 10th World Congress on the Theory Machines and Mechanisms, Finland, pp. 20–24, 2010.

[7] F. Liarokopis., "Multimedia Augmented Reality Interface for E-Learning (MARIE)." World Transactions on Engineering and Technology Education. 1(2). 2002.

[8] de Koning, B. B., Tabbers, H., Rikers, R. M. J. P., & Paas, F., "Attention Cueing as a Means to Enhance Learning from an Animation", Applied Cognitive Psychology, 21(6), 731-746, 2010.

[9] de Koning, B. B., Tabbers, H. K., Rikers, R. M. J. P., & Paas, F., "Attention guidance in learning from a complex animation: Seeing is understanding?". Learning and Instruction, 20(2), 111-122, 2010.

[10] Lijia Lin, Robert K. Atkinson, "Using animations and visual cueing to support learning of scientific concepts and processes". Computer & Education, Volume 56, Issue 3, 650-658.

[11] Mayer, R. E., & Moreno, R., "Nine Ways to Reduce Cognitive Load in Multimedia Learning", Educational Psychologist, 38(1), 43-52, 2003.

[12] Brianna M. Scott, Neil H. Schwartz, "Navigational spatial displays: The role of metacognition as cognitive load", Learning and Instruction, Volume 17, Issue 1, 89-105, February 2007.

[13] S. Zagoranski, S. Divjak, "Use of Augmented Reality in Education", EUROCON 2003 Ljubljana, Slovenia

[14] Augmented Reality in Education, http://wik.ed.uiuc.edu/index.php/Augmented_Reality_in_Education accessed on 23-02-2012

[15] C. J. Jerome., "Effects of spatial and non-spatial multimodal cues on orienting of visual-spatial attention in an augmented environment", Technology-Based Training Research Unit, U.S. Army Research Institute for the Behavioral and Social Sciences, Orlando, FL., 2007.

[16] R. Azuma, Y. Baillot, R. Behringer, S. Feiner, S. Julier, B. MacIntyre, "Recent Advances in Augmented Reality," *IEEE Computer Graphics and Applications* Vol. 21, No. 6, pp. 34-47, Nov/Dec 2001.

[17] L. Johnson., A. Levine, R. Smith, & S. Stone, (2010). The 2010 Horizon Report. Austin, Texas: The New Media Consortium.

[18] A. Balog, C. Pribeanu, & D. Iordache, Augmented Reality in Schools: Preliminary Evaluation Results from a Summer School. Proceedings of WASET International Conference on Technology and Education - ICTE 2007, WASET Vol. 24, Oct. 2007. ISSN 1307-6884. pp.114-117.

[19] F. Liarokapis., N. Mourkoussis., "Web3D and Augmented Reality to support Engineering Education", World Transactions on Engineering and Technology Education, 3(1):11-14, (2004).

[20] I. Farkhatdinov, & J. Ryu, (2003) "Development of educational System for Automotive Engineering based on Augmented Reality". Available from: http://robot.kut.ac.kr/papers/DeveEduVirtual.pdf accessed on 23 February 2011

[21] "Technical Practice on Virtual reality and Augmented reality for Electrical and Mechanical engineering education", http://www.wseas.us/e-library/conferences/2008/venice/edu/edu04.pdf accessed on 23 February 2010.

[22] D. Müller, "Mixed reality learning spaces for collaborative experimentation: A challenge for engineering education and training", International Journal of Online Engineering (IJOE), 2007. 3(4).

[23] J. Martín-Gutiérrez, J. L. Saorín, M. Contero, M. Alcaniz, "AR_Dehaes: An Educational Toolkit Based on Augmented Reality Technology for Learning Engineering Graphics," Advanced Learning Technologies (ICALT), 2010 IEEE 10th International Conference on , vol., no., pp.133-137, 5-7 July 2010

[24] P. Valentini, E. Pezzuti, "Interactive Multibody Simulation in Augmented Reality", ECCOMAS Thematic Conference on Multibody Dynamics, June 29-July 2, 2009.

[25] Slider-crank, http://www.britannica.com/EBchecked/topic/548729/slider-crank-mechanism

Biography

Assoc. Prof. Dr. Manjit Singh Sidhu is currently the Head of Graphics and Multimedia, College of Information Technology (COIT), University Tenaga Nasional. He received his BSc. (Hons) degree in Computer Science from the University of Wolverhampton, UK in 1997 and Masters in Information Technology from University Putra Malaysia in 2000. He completed his PhD. in Computer Science from Universiti Malaya in 2007. He is a Chartered IT Professional Fellow, UK and a member of the British Computer Society. He is a senior member of Institute of Electrical and Electronics Engineers (IEEE), Computer and Communications Society, Malaysian National Computer Confederation (MNCC), and Associate Fellow of the Malaysian Scientific Association (MSA). His research interests include patterns of interactions in multimedia and virtual reality applications, 2-D & 3-D visualization, computer simulations and animation. Anuar Muslimen was a research assistant and now a Tutor attached to the college of foundation studies UNITEN.

The Combined Multi-objective Optimization Design for a Light Guide Rod

Yu-Sen Yang [1], Chun-Yao Shih [2], Hong-Yao Chien [3], Rong-Fong Fung [1,*]
[1] Department of Mechanical & Automation Engineering
National Kaohsiung First University of Science and Technology,
1 University Road, Yenchau, Kaohsiung County 824, TAIWAN
[2] Graduate Institute of Electro-optical Engineering,
National Kaohsiung First University of Science and Technology,
1 University Road, Yenchau, Kaohsiung County 824, TAIWAN
[3] Industrial Master Program on Advanced Headlamp Industry
National Kaohsiung First University of Science and Technology,
1 University Road, Yenchau, Kaohsiung County 824, TAIWAN

Abstract. The light guide rod (LGR) has been popularly used for the vehicles, and the automobile lamp industries need mass production to match this trend. This paper aims to develop a systemic way to find the best parameters' combination for the LGR, and the parameters are usually restricted to some levels and random values. In this paper, the LGR example with two optical performances of illuminance flux and uniformity is to be optimized by use of the real-coded genetic algorithm (RGA) and grey relational analysis (GRA). The illuminance flux and uniformity of the best parameters' combination are obtained and compared with the initial set. Comparisons with Taguchi-Grey can improve 5% of gain and comparisons with Pareto genetic algorithm (PaGA) can improve 1.7% of gain .The combined multi-objective optimization can saving 7% time and it is found that the new proposed method has positive gains in performances.

Keywords: Grey relational analysis (GRA), light guide rod, multi-objective optimization, real-coded genetic algorithm (RGA).

1. Introduction

The light guide rod (LGR) used in daily life is more and more than before, because it has the merit of uniform light and energy saving, and can be designed in any light shape freely. Moreover, it is easily designed in a restrained space and has good performance if one uses the light emitting diode (LED) to be the light source. On the other hand, the LED strip might not be uniform and the cold-cathode fluorescent lamp consumes more power, therefore more and more advanced, fashion vehicles use the LGR to take place of lamp eyebrows or other lightings.

However, material using and parameter choosing of the LGR for the automotive lamp factory are inexperienced so that they are usually not able to find a good setting for the manufacture and design. The LGR's scattering in this study is made by Mie scattering material, which uses small particles in medium. The beam's incident angle is greater than the critical angle of total internal reflection and

[*]Corresponding author: Professor Rong-Fong Fung; Tel.: +886-7-6011000 ext. 2238; Fax: 886-7-6011066

then the light will go out of the matter. The LED placing angle, viewing angle, distance between LEDs, transmittance percent of LGR's outer layer, and density of scattering particles will affect the optical performances. Therefore, a systematical process is needed to deal with every type of parameters' combination and to find out the best one.

In this paper, we propose a new modified of the real-coded genetic algorithm (RGA) and grey relational analysis (GRA). A method to deal with the LGR design problem and the main difference from the combined Taguchi-Grey method [1-2] is to permit some factors with random values within limited bounds. From our method, it is found that the effect of individual factor's response is faster than the RGA with standard GRA. Therefore, our proposed method provides a valuable, credible, non-restricted and faster process to find out multi-objective optimization parameters, even for no reference or experience of the same type products before.

2. Experimental details

These LGR products must meet the characteristics of good performances and product standards. Figure 1 shows the flow chart of this paper, and the experimental details are described as follows.

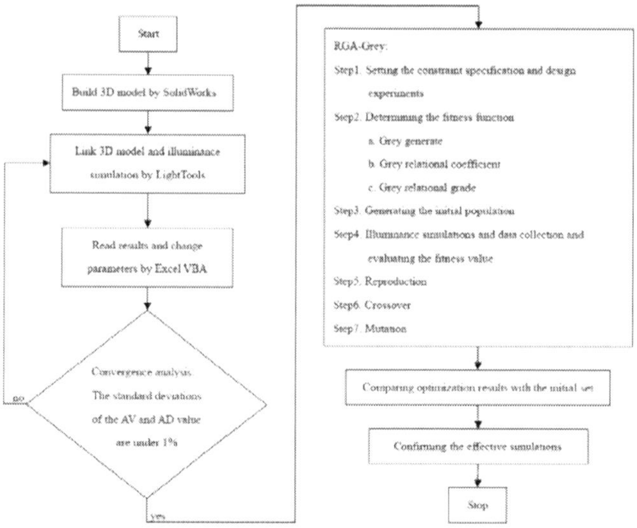

Figure 1. The flow chart of optimization illuminance simulation.

2.1. The 3D Model and standardization

We build up or import the three-dimensional (3D) geometrical model of the LGR by SolidWorks software and then link with LightTools software to perform illuminance simulations, where the position relationships among light source, receiver and the LGR are shown in Figure 2.

Because the LGR is a large-sized and complex shape, it will take long time to perform illuminance simulations. In this paper, the optimized results are expected to be applied to other kinds of LGR. With the design standardization of the LGR, it provides a useful basis model and can be applied directly. Therefore, we do not need to perform all the process over again for another new LGR. In this way, it can accelerate the development, and reduce the cost and timing of the LGR research.

To simplify and represent the LGR model adequately, we take a 2-LED part as a standard one from the origin model, which is designed for 64 LEDs. The width of the standard model is 23.5 mm and height is 9 mm. In this paper, the optimization design is performed first in the standard model, and then applies the results to deal with the origin model and validates this standard model effective. The position relationships among light source, receiver and the standard LGR are shown in Figure 3.

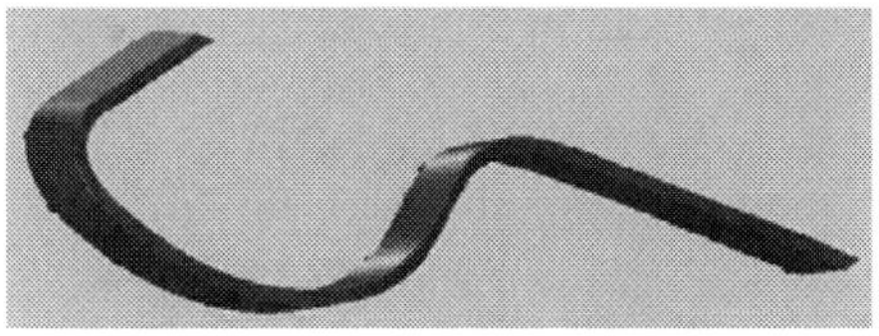

Figure 2. The Light guide rod.

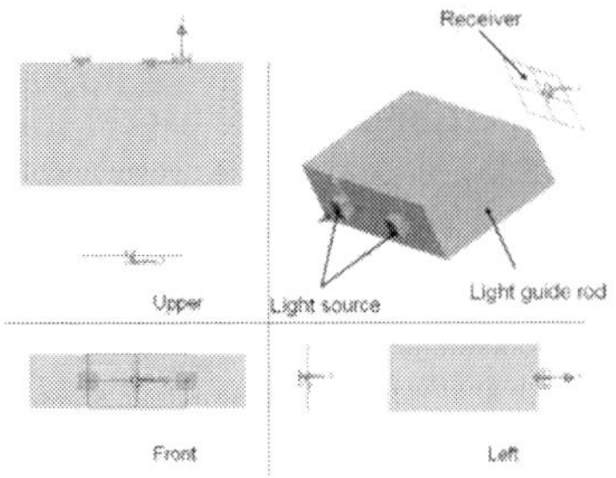

Figure 3. The standard LGR model.

2.2. Simulation process and material

In this paper, the goal is to perform the multi-objective optimization for the LGR automatically. In order to design experiments and analyze these data, the RGA method written in the MS Office Excel VBA software connecting with LightTools software is used to change parameters' setting, read and process illuminance simulations, perform optimization algorithm, and do the whole loop by default settings.

The LGR material is the MIE scattering material composed of the polycarbonate (PC) mixing with the μm size polymethylmethacrylate (PMMA) particles. Usually, the light cannot penetrate the inorganic molecules and its shape is often sphere. The refractive index of selected particles must be different and the added weight ratio is usually 1% to 5%. The MIE scattering particle's parameters will affect the optical result.

2.3. Simulation convergence analysis

In this paper, the convergence analysis is estimated by the average value (AV) and average difference (AD) of illuminance flux. The convergence criteria is taken as the fluctuation percentage of the last four measured AV and AD values, which are all required under one percentage. The measured AV and AD for the LGR are shown in Figure 4, where the numbers of illuminance simulations are from 10,000 to 30,000. It is found that when the light number is larger than 24,000 the fluctuation percentage of the last four measured AV and AD value are less than one percentage and the numerical simulations are convergent.

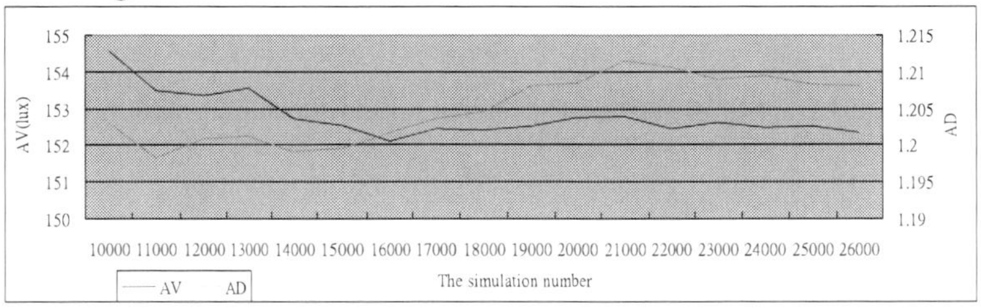

Figure 4. The simulations of the AV and AD value with different simulation numbers for LGR.

2.4. The real-coded genetic algorithms

The genetic algorithms which make use of the real-coded chromosomes are termed as the real-coded genetic algorithm (RGA). In a continuous searching space, the RGA is more natural than binary-coded GA to perform the optimization process [3-4]. The optimization searching RGA is a simulating evolution mechanism on a computer-based platform in conjunction with natural selection and genetic mechanism. The RGA continuously searches for the better chromosomes in this way until the convergent index is satisfied. The reproduction procedure adopts the roulette wheel selection, and the crossover and uniform mutation adopt the methods in [5-6].

The RGA process [7-8] is briefly described by the following nine steps: setting the constraint specification, determining the fitness function, generating the initial population, evaluating the fitness value, reproduction, crossover, mutation, evaluating the fitness value for offspring chromosomes, and constructing the new population.

Before executing the RGA process, the population size, maximum generation number, crossover probability, mutation probability, fitness function, range of each parameter must be assigned for the RGA by our expert knowledge and numerical experience. How to define the fitness function is the key point of the RGA, since the fitness function is a figure of merit, and is computed by using any domain knowledge. In this paper, we employ the GRD as the fitness function. This is the main difference from the traditional GA by using the specialized fitness function.

In this paper, there are three parameters including LED viewing angle, distance between LED, and scattering particles' density, and the population size is set 40. The probability of the jth chromosome into the mating pool can use the following equation:

$$fit\ ratio_j = \frac{fit\ value_j}{\sum_{i=1}^{k} fit\ value_i} \tag{1}$$

where k is the population size. Here, the arithmetic crossover operator is defined as follows:

$$x_{o1} = (1-\alpha)x_{p1} + \alpha x_{p2} \tag{2}$$

$$x_{o2} = \alpha x_{p1} + (1-\alpha)x_{p2} \tag{3}$$

where x_{p1} and x_{p2} are two genes in parent chromosomes, x_{o1} and x_{o2} are two children, and α is selected randomly between 0 and 1. The crossover probability is given between 0.8 and 1 by ordinary. In this paper, the crossover probability is set 0.8. Here, uniform mutation is used and is defined as follows:

$$x_{new} = LB + \beta(UB - LB) \tag{4}$$

where x_{new} is the gene after mutation, β is the mutation probability and is selected randomly between 0 and 1, *LB* and *UB* are the minimum and maximum values of the gene's ranges, respectively. In general, the mutation probability is often given a low value, and set 0.01 in this paper.

Finally, repeating steps to search for the optimal solution until the end of the maximum generation. In this paper, the maximum generation number is 30.

2.5. *Quality characteristic and factor combination*

The performance indicators of the LGRs are the average value (AV) of output illuminance and uniformity. The average value of illuminance is the larger the better (LTB), but the average difference (AD) of uniformity is the smaller the better (STB).

In this paper, we consider three factors: LED viewing angle, distance between LED, and scattering particles' density for the LGR. For these three factors of the LGR, one restricted level factor and two random level factors are shown in Table 1.

Table 1. The RGA factors and their levels for the RGA

	factor	Level A	Level B	Level C
A	LED viewing angle ($dgree$)	120	60	30
B	Distance between LED (mm)	0	–	17
C	Scattering particles' density ($/mm^3$)	4,000	–	5,000

2.6. *Grey relational analysis (GRA)*

Grey theory is the most used for the co-relational analysis and modeling in the incomplete information and its model is not clear. In this paper, the GRA is briefly described as follows. The pre-processing method is to normalize the data. Converting all parameters' values to the numbers between 0 and 1, and all targets of factors are adjusted to the larger-the-best (LTB), and all factors are then defined. This method has different equations according to various quality characteristics:

Larger-the-best:

$$x_i(k) = \frac{x_i^0(k) - \min[x_i^0(k)]}{\max[x_i^0(k)] - \min[x_i^0(k)]} \tag{5}$$

Smaller-the-best:

$$x_i(k) = \frac{\max[x_i^0(k)] - x_i^0(k)}{\max[x_i^0(k)] - \min[x_i^0(k)]} \tag{6}$$

where $\max[x_i^0(k)]$ and $\min[x_i^0(k)]$ are the maximum and minimum of sequence $x_i^0(k)$, respectively. It can avoid adjusting to a single quality characteristic and ignoring the other one.

In the grey relational space, there is a sequence $x_i = (x_i(1),\ x_i(2),\ ...,\ x_i(k)) \in x$, and $i = 0,\ 1,\ 2,\ ...,\ m$, $k = 0,\ 1,\ 2,\ ...,\ n \in N$, so that $x_0 = (x_0(1),\ x_0(2),\ ...,\ x_0(n))$, $x_1 = (x_1(1),\ x_1(2),\ ...,\ x_1(n)), ...,\ x_m = (x_m(1),\ x_m(2),\ ...,\ x_m(n))$. The GRC is defined by

$$\gamma(x_i(k),\ x_j(k)) = \frac{\Delta_{\min} + \zeta\Delta_{\max}}{\Delta_{ij}(k) + \zeta\Delta_{\max}} \tag{7}$$

where $i, j = 1,\ 2,\ 3,\ ...,\ m$, $k = 1,\ 2,\ 3,\ ...,\ n$. The $x_i(k)$ is the reference series, $x_j(k)$ is the specific series, where $\Delta_{ij}(k) = \|x_i(k) - x_j(k)\|$ is the absolute difference between $x_i(k)$ and $x_j(k)$. $\Delta_{\min} = \min\{\Delta_{ij}(k), i = 1, 2, ..., m; j = 1, 2, ..., m\}$ and $\Delta_{\max} = \max\{\Delta_{ij}(k),\ i,\ j = 1, 2, ..., m\}$ are the minimum and maximum absolute difference in the sequence, respectively. The ζ is the distinguishing coefficient, and $\zeta \in [0,1]$. In general, the distinguishing coefficient is taken as 0.5, and can also be adjusted according to the actual needs.

Since the GRC is obtained, we can calculate GRD by averaging the GRCs as

$$\Gamma(x_i,\ x_j) = \frac{1}{n}\sum_{k=1}^{n}\gamma(x_i(k),\ x_j(k)) \tag{8}$$

where n is the number of performance characteristics. In this study, two output responses of the AV and AD illuminance are applied to calculate the integration. The normalization is only used for the LTB type of equation (5). Next, the GRC calculation will proceed in equation (7).

For these multi-objective cases, we try to combine this flow and integrate the quality characteristics into an equation. Using equation (5) or (6) to normalize the AV and AD values of illuminance flux, but not their S/N values, we calculate the absolute difference as

$$\Delta(k) = \|1 - x_{AV}(k)\| + \|1 - x_{AD}(k)\| \tag{9}$$

where $x_{AV}(k)$ or $x_{AD}(k)$ are the AV and AD values of illuminance flux, respectively. The absolute differences $\Delta(k)$ is used to evaluate the illuminance effects with the STB, that means the two normalized results are better as closer to 1. Its merit is to avoid adjusting to a single quality characteristic and they are equally important. Comparing the proposed method (9) with the S/N ratio in Taguchi method, our method doesn't need so many expressions and can have the same function of multi-objective optimization.

2.7. Combining the RGA and GRA

In this paper, we combine the RGA and GRA for the multi-objective optimization design. The procedure of the RGA-Grey is shown in Figure 5. The GRA can deal with the complex relations and multi-objective cases. The RGA generates the factor combination, and needs the fitness function to find the optimal solution. In this paper, the combination idea is proposed to take the absolute difference (9) as the fitness function for the RGA.

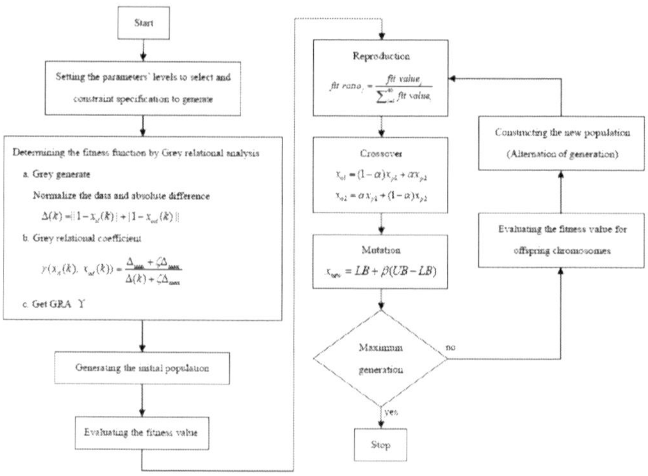

Figure 5. The flow chart of RGA-Grey.

In our proposed process, after the illuminance simulation for factor combination, the AV and AD values of optical effects are normalized into the range between 0 and 1, which is also called the Grey relational generating. The GRC is then calculated from the normalized optical effects to express the relationship between the estimated factor combination and the previous combination for each optical effect. The value is the LTB and the survival chromosome probability is proportional to this value. The re-produced chromosomes then undergo a re-combination that consists of crossovers and mutations. The whole process will do loop until achieving the stop condition or the optimal solution is obtained.

3. Results and discussions

The results of our proposed RGA-Grey multi-objective optimization design for the standard model is found that the best factor combination is: LED viewing angle is 120°, distance between LED is 11.7mm, and scattering particles' density is 4,687.132 $/mm^3$. By using this factor combination, the illuminance flux is 0.456(lm) and AD is 0.162. The GRC of standard method deals with quality characteristics individually, and then takes their average. In our improved method, the GRC can process quality characteristics by using equation (9) at the same time. By this way, we can integrate into a single performance index even though don't take their average. The results of the standard GRA are the best factor combination is: LED viewing angle is 120°, distance between LED is 11.9mm, and scattering particles' density is 4,510.648 $/mm^3$. By using this factor combination, the illuminance flux is 0.445(lm) and the AD is 0.166. As a result, the differences of illuminance flux and the AD are both about 2.5%, it is found that the results are very similar as shown in Table 2. However, in our proposed method, less equations are used and can reduce 58.2969 seconds for one generation of the RGA, which are performed in a computer with Windows XP Service Pack 3 operating system, Intel Pentium 3.4GHz processor and 1GB of RAM.

Table 2. The results of the standard method and improved method

	Factor sets	Illuminance (lm)	Average difference
Standard method	120° /11.9mm / 4,510.648 $/mm^3$	0.445	0.166
Improved method	120° /11.7mm / 4,687.132 $/mm^3$	0.456	0.162
Difference	3.9%	2.5%	2.4%

In the GRA method, the raw data is converted to the S/N ratio by a simplified calculation. The comparisons between the proposed method and the RGA method are shown in Figure 6. It is found that the GRCs are all required for the two methods, but the proposed one does not need the S/N calculation.

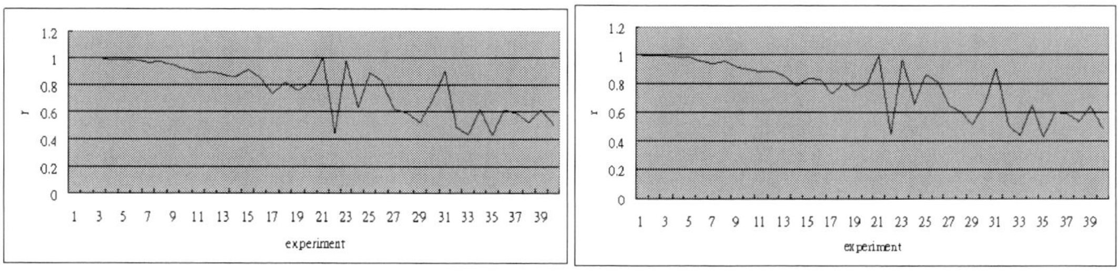

(a) RGA-Grey (b) GRA whose raw data is converted to the S/N ratio

Figure 6. The compared results of the RGA-Grey and the GRA whose raw data is converted to S/N ratio.

3.1. Comparison with Taguchi-Grey method

In order to compare with Taguchi-Grey method, where levels and the responses of the GRD are shown in Table 3. According to the responses of GRC which is shown in Figure 7, we select the larger value as the best factor set, which is a set of A1, B3, and C2. That is, the LED viewing angle is 120°, distance between LEDs is 12mm, scattering particles' density is 4,500 $/mm^3$.

Table 3. LGR Selected Taguchi factors and their levels

	A	B	C
factor	LED viewing angle ($dgree$)	Distance between LED (mm)	Scattering particles' density ($/mm^3$)
Level 1	120	17	5,000
Level 2	60	14.5	4,500
Level 3	30	12	4,000

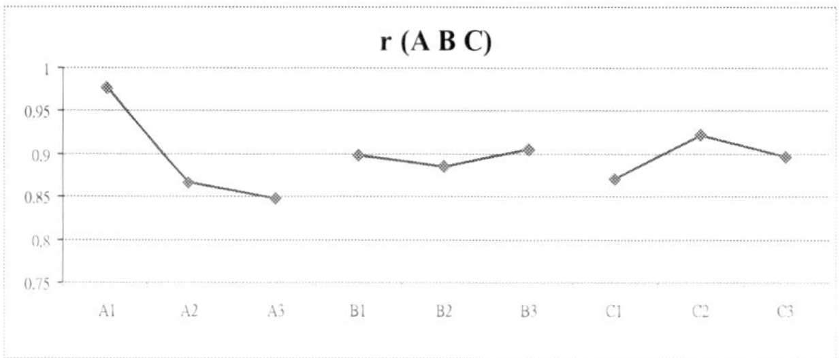

Figure 7. The response of Grey relational grade

From our analysis, it is known that the LED viewing angle and distance between LEDs are very similar in the RGA-Grey and Taguchi-Grey optimal solutions. However, the scattering particles' density in Taguchi's method is restricted in levels, but is a random value in the RGA method. Therefore, the result in the RGA method can be found a more precise value. The optimal solutions between the RGA-Grey and Taguchi-Grey methods are compared in Table 4.

Table 4. The gains of the RGA-Grey and Taguchi-Grey for LGR standard model

	Factor sets	Illuminance (lm)	Average difference
Taguchi-Grey	120° / 12mm / 4,500 $/mm^3$	0.434	0.168
RGA-Grey	120° / 11.7mm / 4,687.132 $/mm^3$	0.456	0.162
	Gain	5.07%	3.57%

3.2. Comparison with Pareto genetic algorithm

The Pareto optimal solutions' set can be provided for decision-makers' reference and evaluation. The conclusion illustrated Pareto genetic algorithm (PaGA) can search across the Pareto optimal solution set [9]. The difference of the PaGA and RGA-Grey is that the PaGA result is a set of solutions and that of the RGA-Grey is only a solution. To confirm the RGA-Grey method can make equal quality characteristics, we use one simple way to combine multi-objective functions into a scalar fitness function for the PaGA. The weighted sum approach [10] is

$$P(k) = w_1 * x_i(k) + w_2 * x_j(k) \tag{10}$$

where $x_i(k)$ or $x_j(k)$ is the normalized one of illuminance simulations' results, $i = 1, 2, 3, ..., m$, $j = 1, 2, 3, ..., m$, and w_1 or w_2 is a constant weight for $x_i(k)$ or $x_j(k)$. In this case, w_1 and w_2 are taken as 0.5 for the equal optical effects of the LGR. We take equation (10) in the place of $\Delta(k)$ and the PaGA result is shown in Figure 8. According to this result, we choose three sets (A, B, and C) to compare with that decided by the RGA-Grey method, and the compared result is shown in Table 5. It is found that the optical effects of PaGA A set is very similar to the RGA-Grey method, and its mean is same as the $\Delta(k)$.

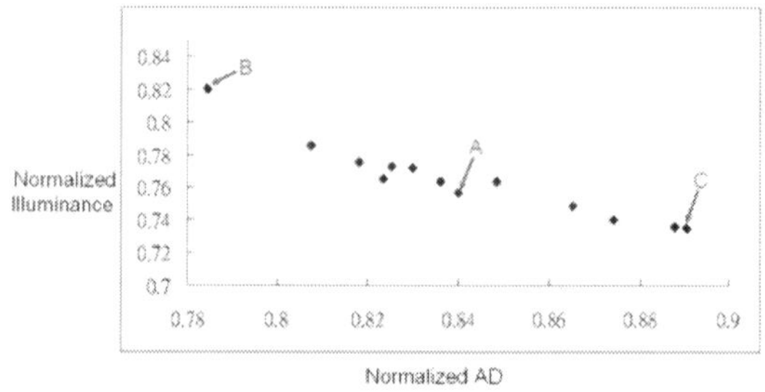

Figure 8. The result of PaGA.

Table 5. The gains of RGA-Grey and PaGA for LGR standard model

	Factor sets	Illuminance (lm)	Average difference
RGA-Grey	120° / 11.7mm / 4,687.132 $/mm^3$	0.456	0.162
PaGA A	120° / 10.9mm / 4,051.613 $/mm^3$	0.450	0.170
	Gain	-1.32%	-4.93%
PaGA B	120° / 11.8mm / 4,137.943 $/mm^3$	0.462	0.171
	Gain	1.32%	-5.56%
PaGA C	60° / 11.9mm / 4,051.613 $/mm^3$	0.344	0.134
	Gain	-24.56%	17.28%

In order to confirm optimal results of the standard model's are effective, we perform the same process with the same conditions for the origin model. The optimal result is: LED viewing angle is 120°, distance between LEDs is 11.9 mm, and scattering particles' density is 4,694.867$/mm^3$.

The illuminance simulations of the LGR by using the best set from the RGA-Grey method are compared with the initial set, and their results are shown in Table 6 and Figure 9.

Table 6. The gains of the optimized and initial factor sets for LGR origin model.

	Factor sets	Illuminance (lm)	Average difference
Initial	120° / 17mm / 4,000 / mm^3	22.543	1.684
Optimized	120° / 11.9mm / 4,694.867 / mm^3	22.931	1.195
	Gain	1.72%	29.04%

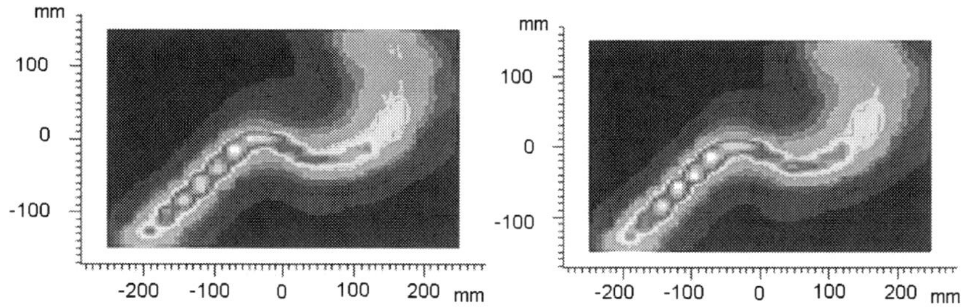

(a) RGA-Grey (b) initial

Figure 9. The compared results of the RGA-Grey and initial sets for LGR origin model.

Converting the lighting image of these real samples to 16 colors and grey values, the results are shown in Figure 10. It is seen that the chart line by the RGA-Grey method is the smoothest than the initial and Taguchi-Grey ones. It is known that the best set is found by the RGA-Grey method, because its interval of bright and dark is the smallest than the others. A comparison of the integrated performance index between the initial and optimal sets shows that the illuminance flux increases from 22.54(lm) to 22.93(lm) and the AD value decreases from 1.68 to 1.19. The positive gains for the total illuminance flux and the AD value, compared with the initial condition, are reported as 1.72% and 29.04%, respectively. These optical properties are obviously improved.

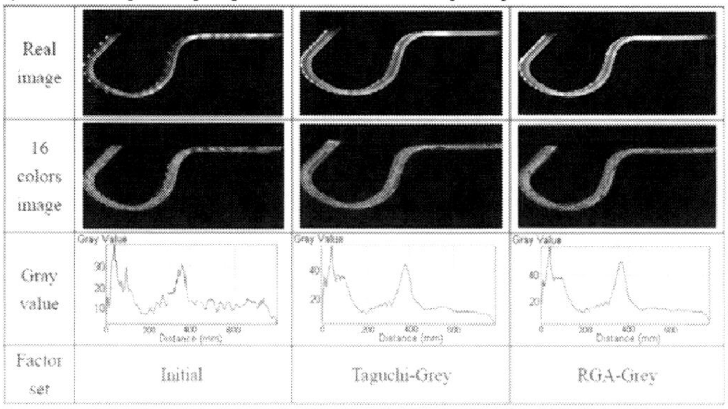

Figure 10. The compared results of real samples.

4. Conclusions

By the RGA method, the best factor combination is more ideal than Taguchi's method, in which factors are restricted in levels. In our proposed examples, not all factors are random, but some of them have several specifications and can't change freely. However, it is a real condition for product setting for manufacturers. The RGA-Grey method can obtain the same result as the process flow whose raw data is converted to the S/N ratio and save about 7.6% time for per generation of RGA. The GRA could combine with not only the RGA method but also Taguchi method to deal with the multi-objective optimization problem. For this case, the method can make quality characteristics of both are as equally important as possible and don't need to decide by decision-maker as Pareto. Our proposed method can also reduce the developing time and production cost from simulation samples to mass productions, and enhance the competitiveness for the industry manufacturer. Finally, the results of the optimized combination confirm the optical properties with AV and AD of illuminance flux are improved 3~5% of gain by using the proposed method in this paper.

5. References

[1] Yang, Y.S., Huang, W., 2012. A grey-fuzzy Taguchi approach for optimizing multi-objective properties of zirconium-containing diamond-like carbon coatings. *Expert. Syst. Appl.* **39**, 743-750.
[2] Pan, L.K., Wang, C.C., Wei, S.L., Sher, H.F., 2007. Optimizing multiple quality characteristics via Taguchi method-based Grey analysis. *J. Mater. Process. Technol.* **182**, 107-116.
[3] Örkcü, H.H., Bal, H., 2011. Comparing performances of backpropagation and genetic algorithms in the data classification. *Expert. Syst. Appl.* 38, 3703-09.
[4] Murata, T., Ishibuchi, H., 1996. MOGA: Multi-Objective Genetic Algorithms. *Computers ind. Engng.* **30**, 957-968.
[5] Kim, J.W., Kim, S.W., Park, P.G., Park, T.J., 2002. *On the similarities between binary-coded GA and real-coded GA in wide search space.* in: Proceedings of the 2002 Congress on IEEE Evolutionary Computation (CEC'02), 681–686.
[6] Lin, S.T., Huang J.N., 2002. Numerical integration of multibody mechanical systems using baumgarte's constraint stabilization method. *J. Chinese Inst. Eng.* **25**, 243–252.
[7] Konak, A., Coit, D.W., Smith, A.E., 2006. Multi-objective optimization using genetic algorithms: *A tutorial. Reliab. Eng. Syst. Safe.* **91**, 992-1007.
[8] Yang, Y.S., Shih, C.Y., Fung, R.F., 2012. Multi-objective optimization of the light guide rod by using the Taguchi method and Grey relational approach. *J. Intell. Manuf.* (In press).
[9] Che, Z.H., Chiang, C.J., 2010. A modified Pareto genetic algorithm for multi-objective build-to-order supply chain planning with product assembly. *Adv. Eng. Softw.* 41, 1011-22
[10] Tzeng, C.J., Lin, Y.H., Yang, Y.K., Jeng, M.C., 2009. Optimization of turning operations with multiple performance characteristics using the Taguchi method and Grey relational analysis. *J. Mater. Process. Technol.* **209**, 2753-59.

IOP Conf. Series: Materials Science and Engineering **46** (2013) 012034 doi:10.1088/1757-899X/46/1/012034

A Simplified Model of a Reinforced Square Hollow Section (SHS) T-Joint for Stress Evaluation in Bus Superstructures

T Vichiensamuth[1], **M Pimsarn**[2], **K Takahashi**[3] , **T Tantanawat**[4]
[1] International College, King Mongkut's Institute of Technology Ladkrabang, Bangkok, Thailand
[2] Faculty of Engineering, King Mongkut's Institute of Technology Ladkrabang, Bangkok, Thailand
[3] International Development Engineering, Graduate School of Science and Engineering, Tokyo Institute of Technology, Tokyo, Japan
[4] National Metal and Materials Technology Center. National Science and Technology Development Agency, Patumthani, Thailand

Abstract. This study aims to create a simplified model of a reinforced square hollow section (SHS) T-joint found in bus superstructures. The approach is to use a combination of one- and two-dimensional finite element models to represent a reference three-dimensional finite element (solid) model of the joint and determine stress concentration factors (SCFs) as functions of the geometrical variables of the joint. This approach requires the stiffness of the simplified model to be equivalent to the stiffness of the reference solid model. Trial models, therefore, must be proposed and their stiffnesses must be evaluated against the stiffness of the reference solid model. The best trial model is then selected based on the stiffness error function defined to represent the deviation of the simplified model's stiffness from the reference model's stiffness. After a trial model with minimum stiffness error is selected, its SCFs, relating the maximum stress in the simplified model to the maximum stress in the reference solid model, are determined. Since the maximum stress is assumed to be at the weld toe where structural discontinuity exists, the maximum stresses on both simplified model and reference solid model are evaluated based on a hot spot stress (HSS) method. In this study, three trial models, namely Model A, Model B, and Model C, were investigated. Model B, consisting of beam and shell elements with particular constraints on the joint-reinforcement geometry, was found to provide the minimum stiffness errors of 8.09%, 6.87%, and 6.44% for three different joint dimensions. The SCFs were then determined as a function of the thickness-to-width ratio of the joint under static in-plane bending load. The resulting simplified model allows the stress evaluation on the bus superstructures to be done more quickly compared to a solid model while maintaining the accuracy of the solutions. Consequently, the designs of bus superstructures can be explored more thoroughly, leading to a better design.

1. Introduction

Buses are one of the vehicle types that people use as a means for transportation in many countries. In Thailand, the size of buses, which is signified by the bus length, can be 8m, 10m or 12m. A number of

[5] E-mail:pooh_kmsj@hotmail.com

bus makers are located in different provinces in Thailand such as Chonburi, Ratchaburi, and Nakornrachasima. In the bus production, the design of the bus superstructure is one of processes that are vital for the safety of passengers. The strength of the superstructure has to be sufficient in order to ensure the safety of passengers while the weight of the superstructure has to be minimal in order for the bus makers to stay competitive in terms of product cost and fuel consumption. Various forms of structural steels, such as hollow square beam, I-beam, C-beam, and plates, are normally used in the bus superstructure. Different types of joints are used to assemble those structural steels. Welding is the jointing technique widely used in the fabrication of the bus superstructure. Although the welding process is perfectly controlled but the strength of the structure at welded joint areas are decreased because of peculiarities such as inhomogeneous material, welding residual stress, and weld geometry [1]. Those peculiarities cause difficultly in stress evaluation at welded joints. Moreover, most bus superstructures are designed by experiences of experts in a company. Those experiences, if without engineering knowledge, limit the continuous improvement of the design. Engineering knowledge, therefore, needs to be applied in order to increase the effectiveness of the bus superstructure design process. Computer Aided Design (CAD) and Computer Aided Engineering (CAE) are ones of the tools that can be used to evaluate stresses on the bus superstructure under several loading conditions.

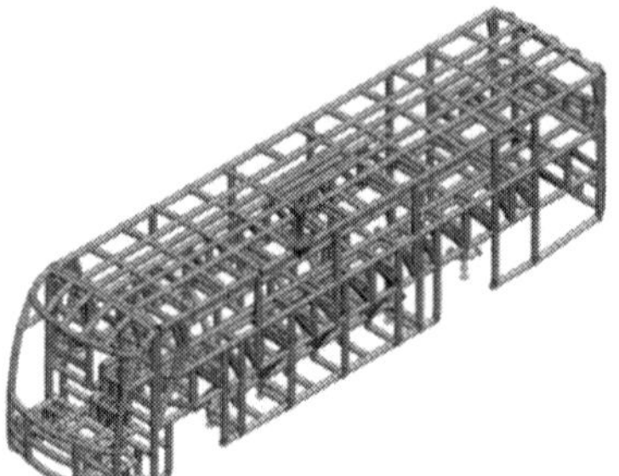

Figure 1. Bus superstructure wireframe

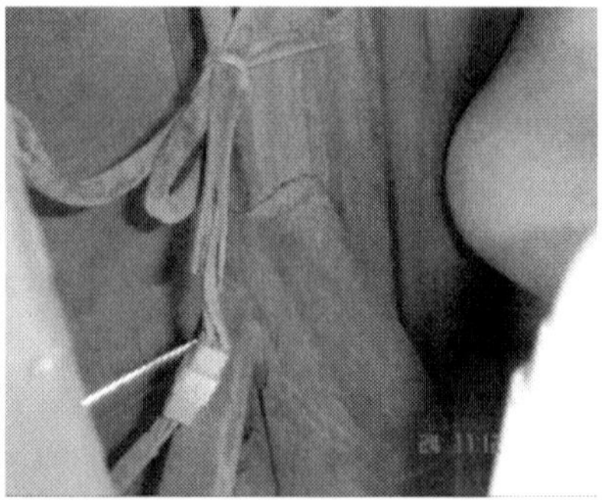

Figure 2. Reinforced Square Hollow Section T-joint

Manokruang S. [2] presented the bus structure design using finite element analysis (FEA) under static loading condition. It was found that a wireframe, which consists of one-dimensional finite elements as shown in figure 1, could be used to identify structural members that might initiate structural failure in bus superstructures. However, according to the results, the maximum stress normally occurred at welded joints where structural discontinuity existed. The absence of the details in joint geometry, therefore, limits the applications of the wireframe model. Stress calculation approaches for welded structure have been developed for many years. Local stress approaches namely Nominal Stress method, Hot Spot Stress method, and Notch Stress method were presented by Radaj D., Sonsino C.M., and Fricke W. [3] to evaluate stresses of welded structures. Advantages and disadvantages were also presented for each method. Useful recommendations of welded structures are also given as guidelines, e.g. International Institute of Welding (IIW) [4], Eurocode [5]. However, they are only applicable to thick structures having a thickness greater than 5 mm. In bus superstructures, weight has to be reduced as much as possible to lower the material cost and fuel consumption. Thin-walled hollow square beams are usually used in the bus superstructure design. Hot spot stress (HSS) method can be applied to various applications, particularly to an automotive field. Savaidis G. and Fokilidis G. [6],[7] studied various welded joints of bus structures by HSS method. There were good agreements between experiments based on strain gage measurement and FEA using shell elements. In addition, Brennan F.P. [8] studied stress concentration factor (SCFs) of welded T-joint and skewed T-joint under tension and pure bending by two-dimensional plane stress analysis. The closed-form solutions of weld geometry parameters were represented to evaluate the SCF model.

The strength of joints can be increased by reinforcements. Those reinforcements are applied to a number of joints in bus superstructures but they are usually ignored in a stress analysis. Joints in bus superstructures may be reinforced by various means. Some joints may be reinforced by steel plates. One of interesting means to reinforce the joints is to use a hollow square beam to reinforce square hollow section (SHS) T-joint as shown in figure 2. A hollow square beam is cut to a right angle and then assembled to the SHS T-joint as shown in figure 3 by welding process. The purpose of this study is to create a simplified finite element model of this reinforced SHS T-joint using beam and shell elements. The stiffness of the simplified model has to be equivalent to the stiffness of the reference model. In addition, the SCFs are calculated by determining the relationship between the maximum stress in the simplified model and the maximum stress in the reference solid model based on HSS method. The SCFs are presented as a function of joint geometries. The scope of this study is to focus on a reinforced SHS T-joint having thickness less than 5 mm under static in-plane bending.

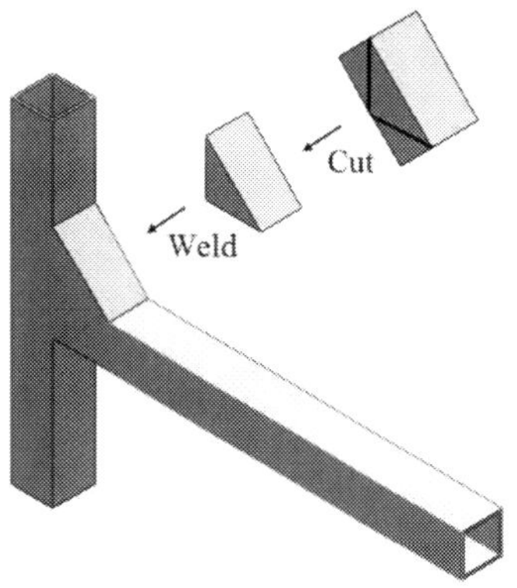

Figure 3. Reinforced tubular T-joint assemble

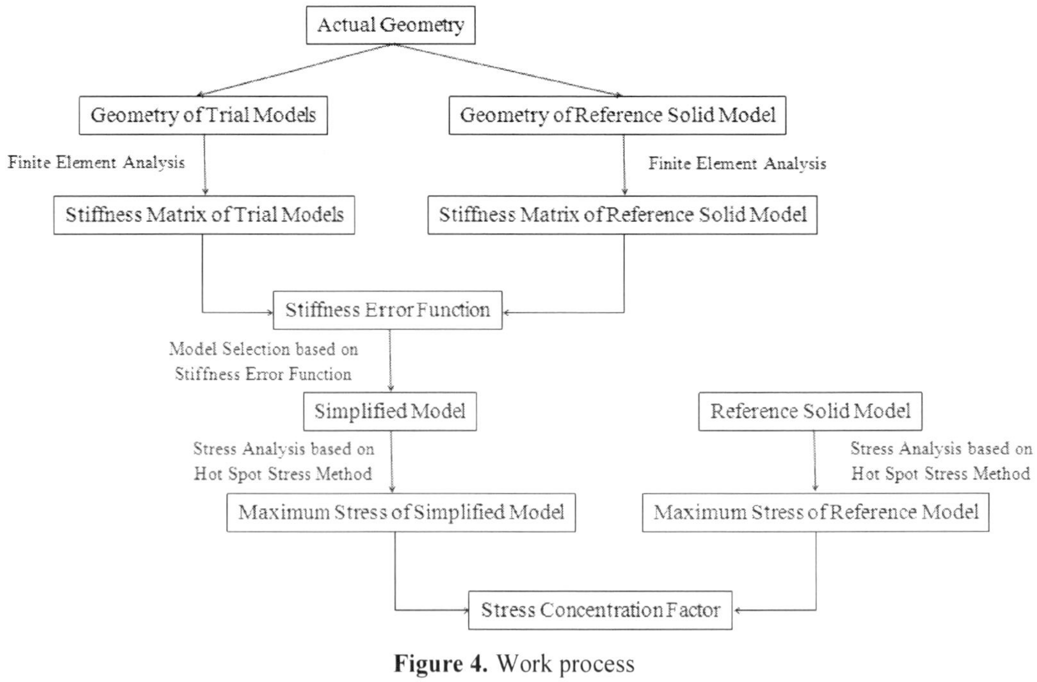

Figure 4. Work process

IOP Conf. Series: Materials Science and Engineering **46** (2013) 012034 doi:10.1088/1757-899X/46/1/012034

2. Methodology

Work process of this study is shown as a diagram in figure 4. After various joints in the bus superstructure had been reviewed, the reinforced square hollow section (SHS) T- joint was found to be an interesting geometry for the study. Geometry and dimension of the joint are shown in figure 5, where W is the width of hollow square beam and t is its thickness. The same cross-sectional dimensions of hollow square beams are used for both T-joint and the reinforcement. Width of 38-60 mm and thicknesses of 2.3-5.0 mm are dimensions of hollow square beams commonly sold in the market. Those dimensions are used to construct CAD models to be used for the analyses in this study.

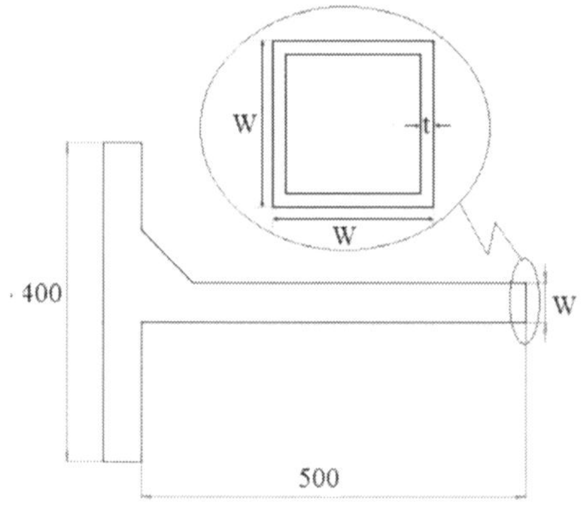

Figure 5. Geometry dimension

2.1. Computer Aided Design

Computer aided design (CAD) is the use of computer program to design the geometry for many applications, mostly for engineering. It makes engineers more convenient to design. The reference solid model and three trial models namely Model A, Model B, and Model C are analyzed in this study. The trial models are allowed to consist of beam and shell elements. For Model A, only beam elements are used to represent SHS T-joint. As shown in figure 6, the reinforcement is not modelled, which is a case usually found in the bus superstructure analysis. Model B and Model C consist of beam and shell elements with the same dimensional geometry. However, the difference between Model B and Model C is shell thickness, which is represented by parameter "T", specified to complete the definition of shell elements. As shown in figure 7 and figure 8, shell thickness of Model B equals to the width of the hollow square beam while shell thickness of Model C equals to two times of the hollow square beam thickness.

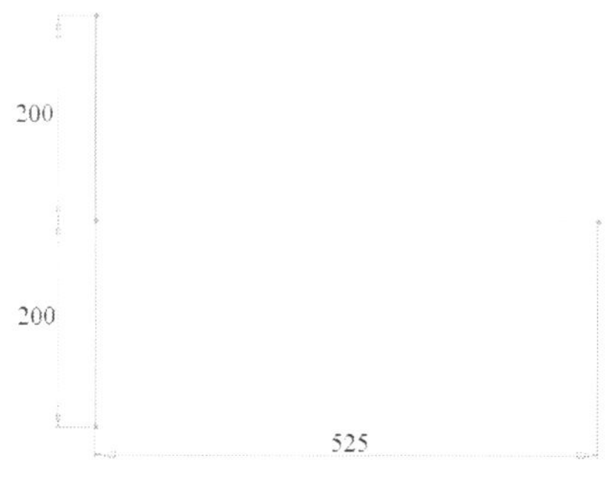

Figure 6. Trial Model A

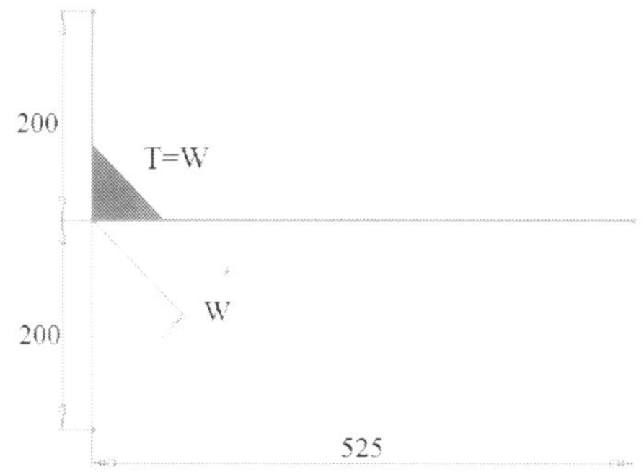

Figure 7. Trial Model B

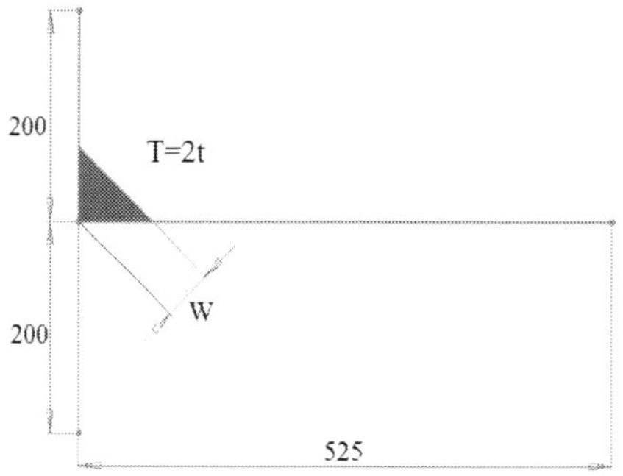

Figure 8. Trial Model C

2.2. Finite Element Analysis

Finite element analysis (FEA) is a numerical technique used to evaluate approximate solutions. FEA is used for structural analysis in various applications, particularly in automotive field.

The joint in bus superstructure has been analyzed using finite element analysis solver. CAD of reference solid model and trail models are transferred to CAE analysis. Loads and boundary conditions of the reference solid model and trial models are set up to determine the in-plane stiffness matrices. Furthermore, the reference solid model and simplified models, which is selected from trial models, are also analyzed to determine the stress concentration factors (SCFs) as a function of joint geometries to relate the maximum stresses in two models based on hot spot stress (HSS). Material properties are significant data for FEA. Table 1 shows the material properties of structural steel used in the FEA. Figure 9 shows bending load and boundary conditions of reinforced square hollow section (SHS) T-joint.

Reference solid model and trial models from CAD have been transferred to CAE model by meshing. 10-node tetrahedral elements with 3 DOFs at each node are used to mesh reference solid model. Trial models consist of 2 types of element in this study. The first type is 2-node beam element with 6 DOFs at each node. Cross-sectional properties of a hollow square beam have to be specified for the beam elements. The other type of element is 4-node shell element with 6 DOFs at each node. For this element, shell thickness has to be specified.

Table 1. Structural steel properties.

Property	
Elastic of modulus	2×10^5 MPa
Poisson's Ratio	0.3

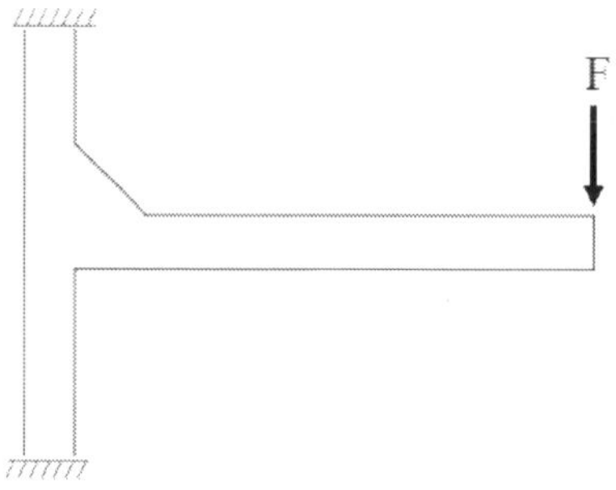

Figure 9. Bending in-plane loading

2.3. Stiffness

The structural behaviour of the simplified model must be equivalent to that of the reference model. This behaviour is represented as a stiffness matrix shown in equation (1). The stiffness matrices are calculated using equation (2) which is a basic equation of finite element method theory. The stiffness of each trial model calculated by using FEA is compared to the stiffness of reference model.

$$[K] = \begin{bmatrix} K_{11} & K_{12} & K_{13} & \cdots & K_{19} \\ K_{21} & & & & \\ K_{31} & & \ddots & & \vdots \\ \vdots & & & & \\ K_{91} & & \cdots & & K_{99} \end{bmatrix} \tag{1}$$

$$[F] = [K][U] \tag{2}$$

Where $[K]$ =Stiffness matrix, $[F]$= Load apply and $[U]$=Displacement

The stiffness error function in equation (4) is defined to represent the deviation of each trial model's stiffness from reference model's stiffness. The stiffness errors between reference solid model and trial models were calculated using equation (2), equation (3) and equation (4). The trial model which provides the minimum stiffness error is selected as the most suitable simplified model.

$$ERR_{ij} = \frac{K_{s,ij} - K_{ij}}{K_{s,ij}} \times 100\% \tag{3}$$

$$ERR = \frac{\left[\sum (ERR_{ij})^2\right]^{1/2}}{n} \tag{4}$$

Where: $K_{s,ij}$ = Stiffness matrix of reference model, K_{ij} = Stiffness matrix of trial model, ERR_{ij} = Stiffness error matrix of trail model, ERR = Stiffness error function and n = number of member of stiffness matrix

2.4. Hot Spot Stress Method

Hot spot stress, sometimes called Geometric stress, has been discussed in a former paper [9]. Hot spot is described as a point that high stress occurs. In welded joint, this point corresponds to the weld toe. Thus, HSS method is limited to the failure location occurring at weld toe in complicated structures. HSS method includes all stress raising effects of a structural detail excluding all stress concentrations due to the local weld profiles itself. HSS can be determined using reference points away from weld toe by extrapolation on surfaces in order to avoid stress singularity. A set of reference points in HSS method depends on thickness of weld components. Both linear extrapolation and quadratic extrapolation shown in figure 10 and figure 11 can be applied to evaluate the HSS. Surface stresses can be measured experimentally by strain gauges.

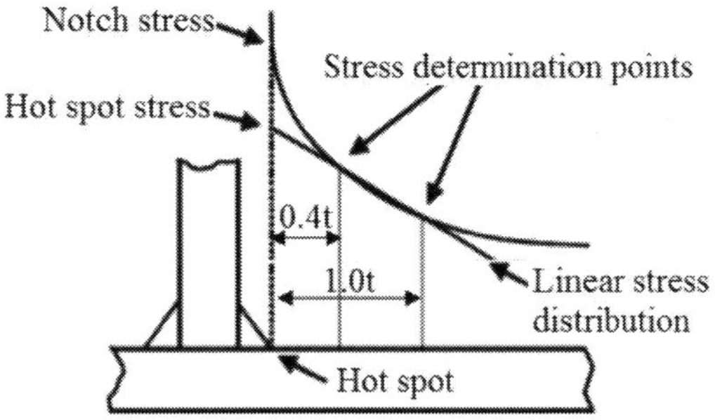

Figure 10. Linear extrapolation of Hot Spot Stress method

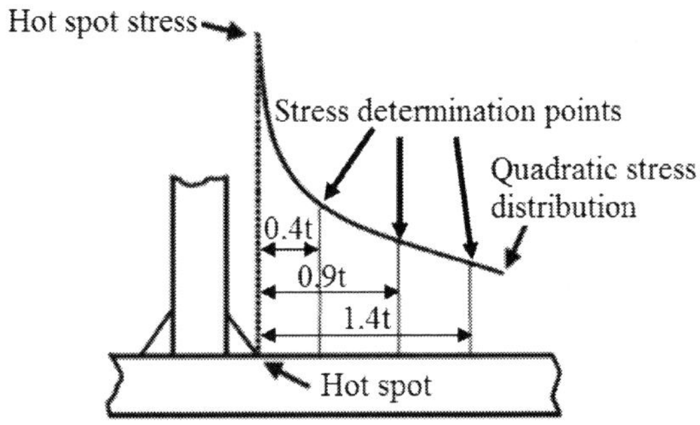

Figure 11. Quadratic extrapolation of Hot Spot Stress method

For a thin-walled structure, surface stresses are difficult to be determined by experiment because the distance between reference points are too small to place strain gauges. Instead of using experiment, FEA can be used to determine surface stresses without any difficulty. Based on a finite element model, it is possible to place nodes on lines where stress must be measured to evaluate HSS. Because local weld profiles are excluded in stress raising effects, weld geometry may be modeled or not modeled in FEA. In this study, weld geometry is not modeled. Reference points can be determined by distances away from structural discontinuity points. HSS of both reference solid model and simplified model are determined by extrapolating surface stresses located at distances from the edge of the reinforcement where the structural discontinuity exists as shown in figure 12 and figure 13. A quadratic extrapolation is recommended to be used for hollow square beams. According to IIW recommendations, fine mesh is required for FEA. The stress values at the reference points 0.4t, 0.9t and 1.4t away from structural discontinuity has been collected to calculate the quadratic HSS by equation (5).

$$\sigma_{HS,quadratic} = 2.52\sigma_{0.4t} - 2.24\sigma_{0.9t} + 0.72\sigma_{1.4t} \quad (5)$$

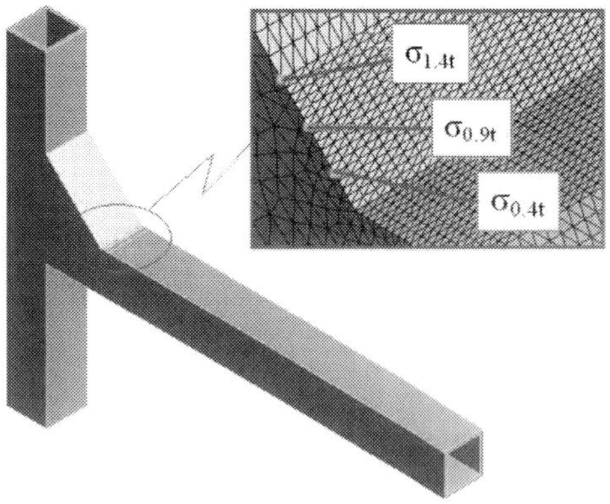

Figure 12. Finite element mesh detail of reference solid model

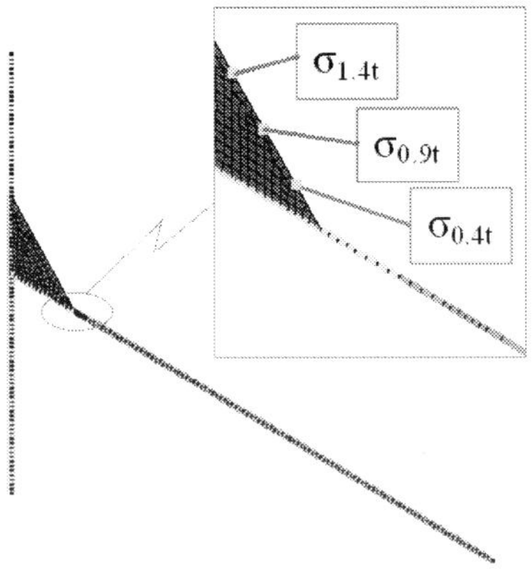

Figure 13. Finite element mesh detail of simplified model

2.5. Stress concentration factor

The maximum stress of reference solid model and simplified model were determined using HSS method on reinforced structures. SCFs are calculated based on maximum stresses of the two models. Each value of SCF is determined using equation (6) as a HSS of reference solid model to HSS of simplified model ratio.

$$SCF = \frac{\sigma_{HS,solid}}{\sigma_{HS,simplified}} \tag{6}$$

3. Result

The results of stiffness errors for Model A, Model B, and Model C are summarized as shown in table 2. Model B provides the minimum stiffness error, which represents the deviation of trial model's stiffness from the reference solid model's stiffness. This model is, therefore, the most suitable to be a simplified model whose stress concentration factors (SCFs) will be determined. This simplified model consists of beam elements and shell elements without weld geometry. Beam elements are positioned along neutral axes of hollow square beams. The thickness of shell elements is specified to be equal to the beam width.

Table 2. Stiffness error function

W × W × t	50 × 50 × 3.2	50 × 50 × 4.0	50 × 50 × 5.0
Model A	10.7424%	8.2063%	8.2013%
Model B	8.0852%	6.8732%	6.4382%
Model C	9.4627%	7.0188%	6.7590%

Maximum stress occurs at the discontinuity points of reference solid model and simplified model. Figure 14 presents the SCF of square hollow section (SHS) T-joint having thickness t of 3.2 mm with different beam width. The best fit equation of the beam width is equation (7). Figure 15 presents SCF of SHS T-joint having width W of 50 mm with different beam thickness. The best fit equation of beam thickness is a linear equation of (8).

$$SCF_{t=3.2mm.} = -0.0002W^2 + 0.031W + 0.3404 \tag{7}$$

$$SCF_{W=50mm.} = -0.2195t + 2.1324 \tag{8}$$

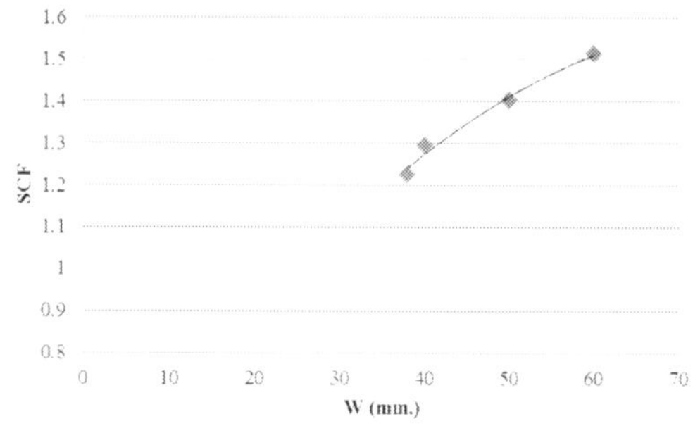

Figure 14. SCF static in-plane bending load for t=3.2mm.

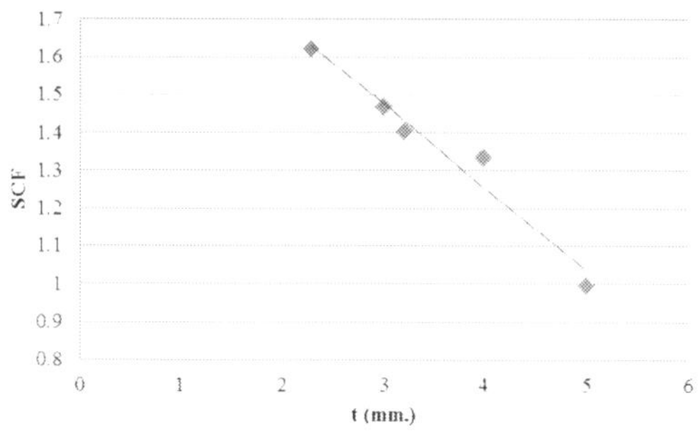

Figure 15. SCF static in-plane bending load for W=50mm.

Thicknesses to width Ratio of hollow square beam is presented as geometrical variable of SCF in figure 16. The best fit equation is equation (9)

$$SCF_{t/W.} = -62.188\left(\frac{t}{W}\right)^2 - 1.1101\left(\frac{t}{W}\right) + 01.7572 \tag{9}$$

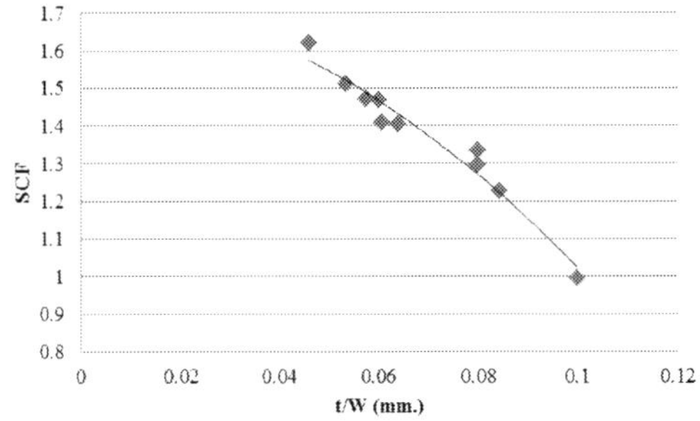

Figure 9. Bending in-plane loading

4. Conclusion

The reinforcement of square hollow section (SHS) T-joints, usually ignored in bus superstructure analysis, is considered carefully in this study. Trial models namely Model A, Model B, and Model C were investigated to evaluate the stiffness errors, which represent the deviations of trial model's stiffnesses from the reference solid model's stiffness. It was found that Model B, providing the minimum stiffness error, is most suitable to be used as a simplified model. Finite element analysis (FEA) was then used to evaluate hot spot stress (HSS) of both reference solid model and simplified model in order to determine the stress concentration factor (SCFs). The SCFs of reinforced SHS T-joint in bus superstructure without weld geometry were presented as a function of thickness to width ratio of hollow section beam. This simplified model allows the stress evaluation on the bus superstructure to be done more quickly compared to reference solid model while the accuracy of the solution is maintained. The SCFs are valid for SHS T-joints having thickness less than 5 mm. However, the simplified model, as well as its SCFs, determined in this study is limited to some values of geometric variables. The simplified model should be studied further for a more complete range in the values of geometric variables. Moreover, the simplified model can be improved by optimizing the geometrical parameters of the finite element model in order to represent more accurate stiffness of the joint. Nonetheless, only bending load is applied in this study. Other load types may occur in reality. Reinforced SHS T-joint has to be studied further on various load types to evaluate the accuracy of the stress results.

5. References

[1] Radaj D 1996 Review of fatigue strength assessment of nonwelded and welded structures based on local parameters Int J. Fatigue 18-3 pp 153–170
[2] Manokruang S 2010 *Design and Analysis of bus body structure for high decker bus (Standard No.1)* Master Thesis, King Mongkut's Institute of Technology Ladkrabang pp 1–93.
[3] Radaj D, Sonsino CM and FrickeW. 2009 *Recent developments in local concepts of fatigue assessment of welded joints* IntJ. Fatigue 31-1 pp 2–12
[4] Hobbacher A 2008 *Recommendations for fatigue design of welded joints and components* IIW doc XIII-2151r4-07/XV-1254r4-07 Int Inst Welding
[5] European Committee for Standardization 1993 *Eurocode 3, Design of steel structures. Part 1-1: General rules for buildings* ENV
[6] Savaidis G and Vormwald M 2000 *Hot-spot evaluation of fatigue in welded structural connections supported by finite element analysis* Int. J. Fatigue 22-2 pp 85–91

[7] Fokilidis A and Savaidis G 2007 *Experimental investigation of fatigue of thin-walled welded structures of commercial vehicle frames* Int. Conf. Diagnosis and Prediction in Mechanical Engineering Systems 2 pp 34–38
[8] Brennan FB, Peleties P and Hellier AK 2000 *Predicting weld toe stress concentration factors for T and skewed T-joint plate connections* Int. J. Fatigue 22-7 pp 85–91
[9] Niemi E, Fricke W and Maddox SJ *Fatigue analysis of welded components designer's guide to the structural hot-spot stress approach* IIW-1430-00 Int Inst Welding,

HTML Extraction Algorithm Based on Property and Data Cell

Detty Purnamasari[1] **I Wayan Simri Wicaksana**[2]
Suryadi Harmanto[3] **Lintang Yuniar Banowosari**[4]
[1,2,3] Information Technology [4]Information Management, Gunadarma University
Jl. Margonda Raya No. 100 Pondok Cina Depok-Indonesia

E-mail: [1]detty@staff.gunadarma.ac.id [2]iwayan@staff.gunadarma.ac.id
[3]suryadi@staff.gunadarma.ac.id [4]lintang@staff.gunadarma.ac.id

Abstract. The data available on the Internet is in various models and formats. One form of data representation is a table. Tables extraction is used in process more than one table on the Internet from different sources. Currently the effort is done by using copy-paste that is not automatic process. This article presents an approach to prepare the area, so tables in HTML format can be extracted and converted into a database that make easier to combine the data from many resources. This article was tested on the algorithm 1 used to determine the actual number of columns and rows of the table, as well as algorithm 2 are used to determine the boundary line of the property. Tests conducted at 100 tabular HTML format, and the test results provide the accuracy of the algorithm 1 is 99.9% and the accuracy of the algorithm 2 is 84%.

Keyword: extraction algorithm, html, table extraction, table property

Introduction

The Internet provides the data in various model and formats. The table is one form that is used to display the data. Likewise with the format used such as HTML and PDF formats. The data in the table on the Internet can be retrieved by manual means that copy-paste, but it is an obstacle if done on more than one table, so in this article discussed about development an approach in preparation to find an area, so that the HTML format table extraction into a database can be done. Extraction or called wrapper that is part of the application that makes the web resource become a resource that can be queried since it is in the form of a database, where the source is in the form of semi-structured or unstructured. [4]

Table considered as a representation of a structured data and related information in the form of 2 (two) dimensions. [6], the contents of the table are the data that are presented briefly. The table consists of the cell, which can contain labelling cell and data cell. [7].

Research on the extraction table been done by [7], which distinguishes the cell content by identifying the cell as property and the cell as an instance, and in this case the property is located in the first row of the table.

[8] Do the extraction of a table that presents biological data by 2 (two) ways: i). table detection (table identification of documents), and ii). Processing table (it extract data from tables). Detection of property has done by considering <hr> tag which is used to create a horizontal line in HTML.

Document Object Model (DOM) is also used by researchers to perform the extraction of HTML tables, some of them [5], and [2]. DOM is a base or a stand-alone language used to represent and make the connection between objects of different documents into HTML or XML web page, with the form of a tree structure as depiction. [3] Combine DOM with XY cut algorithm to find a table on the web page.

Search schema matching on the extraction of HTML tables made to merge the results of extraction, but the research on this area is found several problems including the table location in web pages, merged attribute, and word synonyms in the process of merging the data extracted. [1]

Approach in preparation for finding areas (arrays) on the extraction of HTML tables in this article with respect to the position of property that could be more than 1 (one) rows in the table. (Figure 1)

Name		**Phone Number**	
First	**Middle**		Property
Lussiana	Lee	8972898	
Finisha	Noor	3891810	
Lyre	Widi	1290109	Instance
Susan	Melany	4029049	
Juwita	Okta	3920421	

Figure 1 Examples of Table with Property More than 1 row

Figure 1 shows the table property in row positions, which are row 1 and row 2. Property could be more than 1 (one) row because of the merging of rows and columns, as in Figure 1 shows a merging the 1st and 2nd column "Name" and the merging of row 1 and 2 "Phone Number". Then it shows the content / data table (instance) are from row 3 to row 7.

This study develop two algorithms which are preparation to find areas (arrays) to perform the extraction of HTML tables and show the results of the accuracy testing of those algorithms. Algorithm 1 is used to calculate the number of columns and rows, and the second algorithm is used to determine the row boundary of the table property.

The article is divided into four parts: the first part as an introduction that contains the definition of the problem and refer to previous similar research papers. The second part discuss the approach taken to perform the extraction of HTML table that contains 2 (two) algorithms, and the third part contain testing of the algorithm, and the final part contains the conclusions and future work.

2. Table Extraction Approach

A table has property and data cell. Refer to property and data cell the data can be considered in some types. Actually, in general there are three types of table. The first type is a table that has property in top row of table and the most left of table. The second type is a table that has property in top row of table. The third type is a table that has no property.

Refer to type of table, extraction process need to identify the area of property and cell. The first step is to calculate number of row and column in a table. The second step is to consider which row or column as property cell.

There are 2 (two) algorithm as a preparation in finding the area (array) to perform the extraction of HTML tables. Algorithm 1, that is algorithm used to calculate the actual number of columns and row. This is done since if the table has a property of more than 1 (one) row, then there is a merging of rows and columns in the property cell of the tables, so we need an algorithm to calculate the actual number of columns and row of the table. The calculation of the number of columns and row are actually useful to know the size of the table (row x columns).

Algorithm1 Count for Actual Number of Row and Actual Number of Column

```
Read HTML
s = 0
If read <tr> then s = s + 1
    RsTotal = s
Jum <td> = count tag <td>...</td> in first tag <tr>...</tr>
CsTotal = 0
For i = 1 to jum<td>
    Read nilai cs (i)
    CsTotal = CsTotal + cs (i)
Next i ;
```

Notes:
- RsTotal : the total number of <tr> tag on tag <table>...</table>
- CsTotal : the number of *colspan* value
- Jum<td> : the number of <td> tag
- cs : *colspan* value
- i : <td> tag
- s : <tr> tag

Figure 2 is an illustration that conducted to run the algorithm 1. In Figure 2 looks <HTML> tag of the example table in Figure 1 with the actual number of rows = 7, which is calculated by summing the <tr> tag inside the <table>... </ table> tag. The actual number of columns = 3 is calculated by looking at the number of <td> tags inside the first <tr>... </ tr> tag and if there is *colspan* in it, then it calculate the number of *colspan*.

Algorithm 2 is used to find the value of the row boundary as property (rowmax_pro) by finding the largest value of *rowspan* (RsMax) present in each ith tag <td> ... </ td> on the sth tag <tr> ... </ tr> . If it is not found *rowspan* value > 1 any longer, then the row boundary of the property is found. Figure 3 is an illustration to run the algorithm 2.

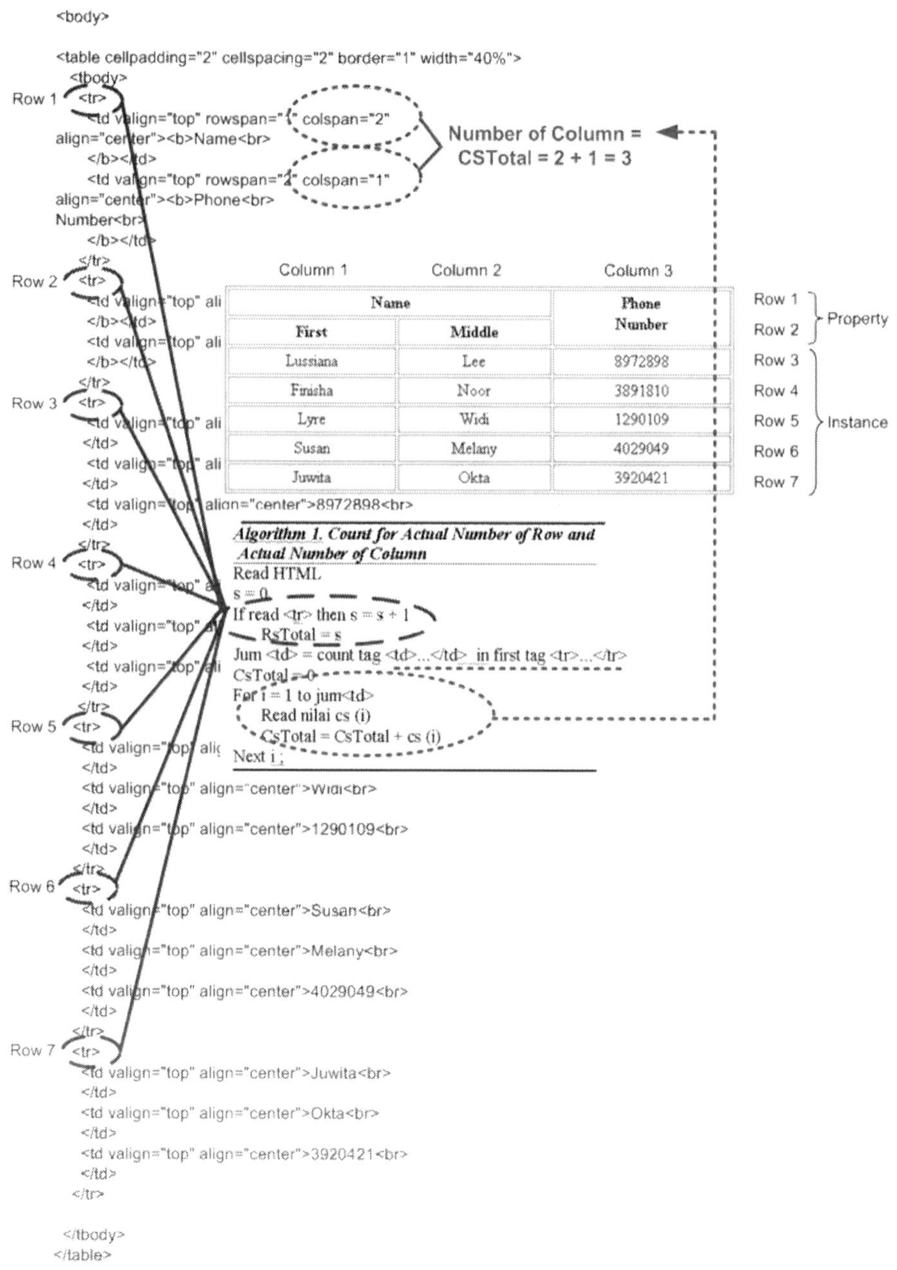

Figure 2. Illustration for Algorithm 1

Algorithm 2. Finding the Largest Rowspan Value, and Number of Row to be The Property Boundary

```
mBatas (0) = 1
while s = 1 do
    rsMax (0) = 1
    Count jum<td>
    For i = 1 to jum<td>
         If rs (i) > 1 then
           If rs (i) > = rsMax (i-1) then rsMax (i) = rs (i)
           else
         If rs (i) < rsMax (i-1) then rsMax (i) = rsMax (i-1)
     Next i ;
mBatas (s) = rsMax (i) + s – 1
if mBatas (s) < mBatas (s-1) then mBatas (s) = mBatas (s-1)
until mBatas (s) ;
rowmax_pro = mBatas (s) ;
```

Notes :

- RsMax : the highest value of *rowspan*
- mBatas : row value boundary as property
- rowmax_pro : row boundary which named as property
- i : <td> tag
- s : <tr> tag

Figure 3 shows an <HTML> tag from the example of table in Figure 1; they are the tag for row 1 up to row 3. <td> tag existing on the first <tr> ... </ tr> tags have the largest *rowspan* value which is 2, so the row boundary of the table property exist in a table on the 2nd row, because once doing a reading for a <td> tag that is inside the second <tr> ... </ tr> tags is no longer found any *rowspan*.

After obtaining the actual table size (row x columns) and the boundary row of the table property, then table extraction into a database is easy to do.

Illustration of implementation of the approach can be implemented in many areas. For instance in manufacturing area, the main issues is to find a raw material. Currently many suppliers provide the information in Internet by using table form. Automotive industry need to have spare parts, such as tire, engine audio system. There are many pro ducts of audio system that inform by table in Internet. To find the appropriate spare parts the manufacturing will copy many tables from many suppliers of audio system. The next step, the content of table will merge to be one table. This effort is acceptable for number of data and sources are limited. If number of data and sources are huge, it can be hard effort and difficult to avoid an error. By implementation the two algorithms, the automatic process data harvesting process in table of Internet can be performed.

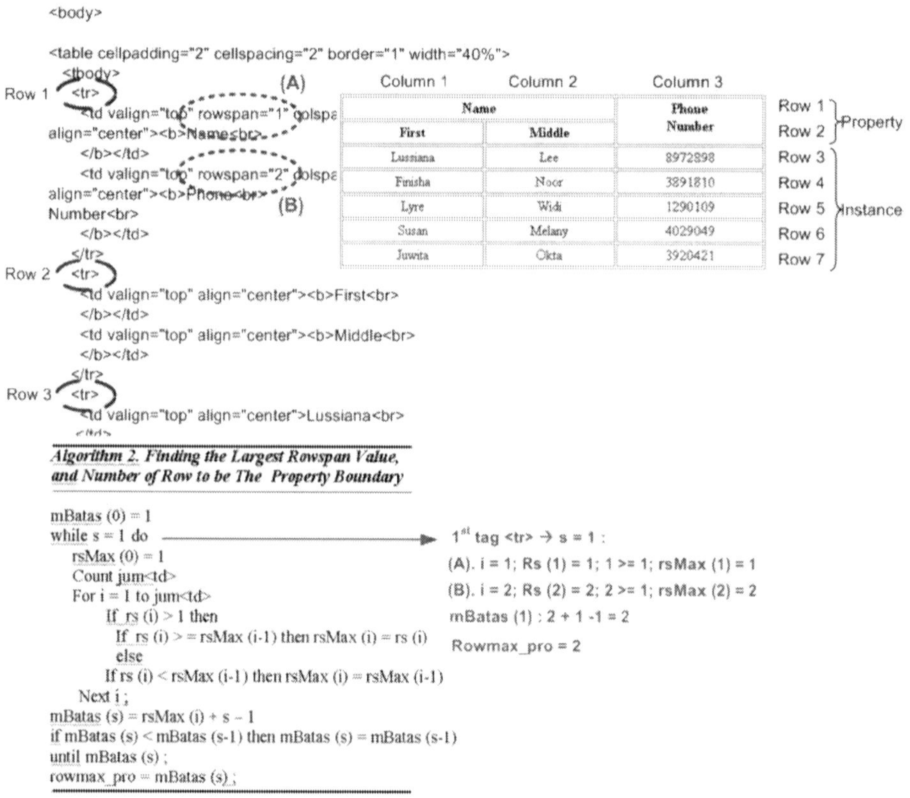

Figure 3 Illustration for Algorithm 2

3. Experiment

Both algorithms in this article are implemented using the PHP programming language. The test preparation is provide a table with various forms in HTML format totalling one hundred tables. The format of tables that used in experiment is as follow:

1. Basic form table consists of one or more row and columns, where the first row and the first column on the left of the property.
2. Basic form table consists of one or more row and columns, where 1 is the top line of the property, without any column as a property set.
3. Basic form table consists of one or more row and columns, where the tables are not property.
4. Basic form table consists of one or more row and columns, where the rows and columns of the property, and no merging row in the column property.
5. Basic form table consists of one or more row and columns, where the rows and columns of the property, and no merging columns on the property row.

6. Basic form table consists of one or more row and columns, where the rows and columns of the property, and no incorporation of rows and columns on the property row.
7. Basic form table consists of one or more row and columns, where the top line of the property, there is no merging row and columns on the property row.
8. Basic form table consists of one or more row and columns, where the table does not have a property but there was merging lines.
9. Basic form table consists of one or more row and columns, where the table does not have a property but there was merging columns.
10. Basic form table consists of one or more row and columns, where the table does not have a property but there was merging of rows and columns.
11. Basic form table consists of one or more row and columns, where the tables have property on the position of row and there is a merger of both row and column of the merger on the property or instance. (Complex tables)

In Figure 4 shows one form of tables that are used as test inputs. The table in Figure 4 is a complex table which has property in first until third row with some row consists of span row.

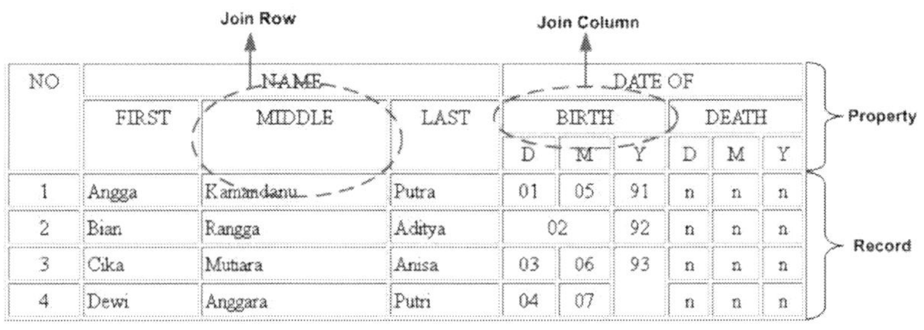

NO	NAME			DATE OF					
	FIRST	MIDDLE	LAST	BIRTH			DEATH		
				D	M	Y	D	M	Y
1	Angga	Kamandanu	Putra	01	05	91	n	n	n
2	Bian	Rangga	Aditya	02		92	n	n	n
3	Cika	Mutiara	Anisa	03	06	93	n	n	n
4	Dewi	Anggara	Putri	04	07		n	n	n

Figure 4 Examples of Complex Table Form for Trial

The purpose of testing the algorithm 1 is to determine the ability of the algorithm to count the number of rows and number of columns of the table, while the second algorithm testing aims to determine the ability of the algorithm to find the limits of the table properties ranging from row 1 to a particular row number in the table. Here is the test scenario:

1. Algorithm 1 and algorithm 2 is written in the PHP programming language.
2. Input is HTML tables with different forms.
3. Output measured is accuracy of the actual number of rows and columns of the table that will be extracted (from algorithm 1), and the accuracy of determining the row boundary of the table property (from algorithm 2)

In Table 1 shows a summary of the test results of algorithm 1 to determine the actual number of rows and columns.

Table 1 Testing Summary of Algorithm 1

Table Model	Manual		Program		Accuracy (%)
	Number of Row	Number of Column	Number of Row	Number of Column	
1	4	3	4	3	100
2	5	3	5	3	100
3	5	4	5	4	100
4	6	3	6	3	100
5	6	4	6	4	100
...	...	...	...	...	...
78	5	17	4	17	90
...	...	...	...	...	...
98	13	13	13	13	100
99	13	13	13	13	100
100	16	19	16	19	100

Table 1 shows the comparison of the number of columns and number of rows retrieved manually (human visual) and the number of rows and columns obtained from algorithm 1 with the program. The average percentage of the number of rows and columns accuracy of 100 table test is 99.9%.

Then Table 2 shows a summary trial in algorithm 2 to determine the accuracy of the table to determine the property boundaries on what row.

Table 2 Testing Summary of Algorithm 2

Table Model	Number of Row to be property Boundary		Accuracy (%)
	Manual	Program	
1	1	1	100
2	1	1	100
3	1	1	100
...	...	...	...
59	1	2	50
60	3	2	66,7
...	...	...	...
99	3	3	100
100	4	4	100

Table 2 shows there are a tables that has the accuracy is not 100% (e.g., in models of tables 59 and 60). The reason of the result because the tables are complex model. Refer to the result if neglected the tables achievemnt is 100%, however the achievemnt is 99% if include table 59 and 60.

Refer to the result of algorithm 2, the approach has limitation in simple table to provide the best result.

4. Conclusion

The experiment out for the algorithm 1 and algorithm 2, it was found that the percentage of accuracy algorithm 1 to get the actual number of rows and columns of the table is 99.9%, the result is better then Tangli [7] researh with achievement 91%.

Percentage accuracy of the algorithm 2 to determine the table properties row boundary is 84%. Result of algtorithm is enhance achievement compare to Wong [8] result (83%).

Looking at the percentage of accuracy of the two algorithms, it is said that the approach taken to prepare the area (array) as the beginning of the process of extracting HTML tables work well.

In a subsequent study, it will use the algorithm 1 and algorithm 2 to conduct the cell contents of the property, and making the contents of cell as an instance in the table extraction process.

References

[1] Embley D. W., Hurst M., Lopresti D., Nagy G. 2006 *Table-Processing Paradigms : a Research Survey* International Journal of Document Analysis, page 66-86

[2] Gultom R., Sari R. F., Budiardjo B. 2011 *Proposing the new Algorithm and Technique Development for Integrating Web Table Extraction and Building a Mashup* Journal of Computer Science 7 (2), page 129-142

[3] Krupl B., Herzog M., Gatterbauer W. 2005 *Using Visual Cues for Extraction of Tabular Data from Arbitrary HTML Documents* Proceeding WWW '05 Special interest tracks and posters of the 14th international conference on World Wide Web

[4] Lerman K., Knoblock C., Steven M. 2001 *Automatic Data Extraction from Lists and Tables in Web Sources* Proceedings of Automatic Text Extraction and Mining Workshop ({ATEM}-01)

[5] Lin J., Wong J., Nichols J., Cypher A., Lau T. A.2009 *End-User Programming of Mashups with Vegemite* Proceedings of The 13th International Conference on Intelligent User Interfaces, page 97-106

[6] Liu Y., Mitra P., Giles C.L. 2008 *A Fast Pre processing Method for Table Boundary Detection : Narrowing Down the Sparse Lines using Solely Coordinate Information* The Eighth IAPR International Workshop on Document Analysis Systems, page 431 – 438

[7] Tengli A., Yang Y., Ma N. L. 2004 *Learning Table Extraction from Examples* Proceedings of The 20th International Conference on Computational Linguistics

[8] Wong W., Martinez D., Cavedon L. 2009 *Extraction of Named Entities from Tables in Gene Mutation Literature* Proceedings of The Workshop on Current Trends in Biomedical Natural Language Processing, BioNLP '09

MSWT-01, flood disaster water treatment solution from common ideas

Gamawan Ananto[1], Albertus B Setiawan[2] and Darman MZ[3]
Manufacture Engineering Dept, Bandung State Polytechnic for Manufacturing, Jl. Kanayakan 21 Bandung 40135, INDONESIA

Email: gamawan@polman-bandung.ac.id

Abstract. *Indonesia has a lot of potential flood disaster places with clean water problems faced. Various solution programs always initiated by Government, companies CSR, and people sporadical actions to provide clean water; with their advantages and disadvantages respectively. One solution is easy to operate for instance, but didn't provide adequate capacity, whereas the other had ideal performance but more costly. This situation inspired to develop a water treatment machine that could be an alternative favor. There are many methods could be choosed; whether in simple, middle or high technology, depends on water source input and output result quality. MSWT, Mobile Surface Water Treatment, is an idea for raw water in flood area, basically made for 1m^3 per hour. This water treatment design adopted from combined existing technologies and related literatures. Using common ideas, the highlight is how to make such modular process put in compact design elegantly, and would be equipped with mobile feature due to make easier in operational. Through prototype level experiment trials, the machine is capable for producing clean water that suitable for sanitation and cooking/ drinking purposes although using contaminated water input source. From the investment point of view, such machine could be also treated as an asset that will be used from time to time when needed, instead of made for project approach only.*

Keywords: *water treatment, raw water source, capacity*

1. Introduction

Indonesia has a lot of places with potential disaster that flood might be happened frequently, and clean water lacking is one of problems faced in such places due to its water sources contaminating. Since water is a human basic needs, water procuring and managing became a very vital issue. For providing clean water, especially for places that had potential flood disaster in rainy season, various solutions always initiated by all related parties such as Government mitigation programs [9], CSR program from some companies, and survival actions of people around these unlucky places. These actions had their advantages and disadvantages respectively, or each strong points and constraints as well. One solution is easy to operate for instance, but didn't provide adequate capacity, whereas the other had ideal performance but could more costly [1].

For locations where no water source, clean water are supplied from other place using trucks for transportation and distribution, whereas in other areas that have improper water source condition people are treating their water sporadically in very small capacity level. For other cases, some companies CSR program are arranged by social foundations to organize the projects, collect the branded standard water machines and operate the distribution [3]. These actions are very appreciated and they are improved continuously for more efficient result from time to time [6]. Using big trucks may constrained by disaster location infrastructure, whereas provide imported equipments that available in the market, although are capable to produce better quality output or guaranteed result but has the costly consequence with very specific or limited implementation, and usually initiated by project based approach. For such case, new equipment should be re-procured for next event, as a new project.

From the requirement side, clean water and drinking water quality standard are defined, such as in SNI 01-3553-2006 Indonesian 'Badan Standardisasi Nasional' [4] as a reference in treating surface and ground water. In some cases there are still obstacles to reach such quality standard either caused by high turbidity or iron/ manganese (Fe/ Mn) contamination, or organic/ ammonium/ undissolved compound content. Some conditions show that people are even pushed in using their water for sanitation only, whereas for cooking and drinking are unsolved.

Above mentioned problems are inspiring to develop water treatment that could be an alternative favor. It innovates a water treatment prototype for appropriate needs, that might use available raw water such as river or flood water at disaster location, in order to provide clean water demand easier for the people. Some factors that should be considered in such development idea are technology that will be choosed, system capacity (in m^3 per hour) that possible to produce and mobile mechanism system in order to make easier in operation and moving to reach disaster sub locations or sub sectors where it's needed. Furthermore, energy or electric source for the system may also be settled.

2. Methodology

For water handling as people do for their need, there are many methods that could be choosed; whether in simple, middle or high technology, depends on water source feed input and water output result quality. Source water with low turbidity level for instance, might use rapid or slow sand filtration, the easy and simple methods that do not need any high skill [11] [12]. Other ways that more sophisticated are possible to conquer more complicated water source input. A few water treatment technologies in 'figure 1' [7] shows that conventional filtration process may overcome undissolved compound content up to around 1μm dimension, whereas the most extreme condition such as seawater or brackish with salts content has to use RO (reverse osmosis) membrane.

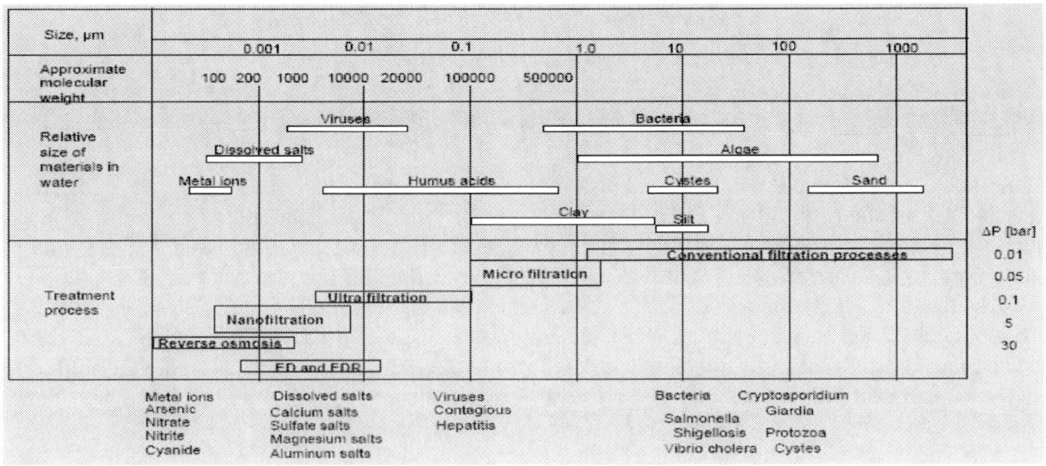

Figure 1. Water treatment process technology level [7] (*Heijman S.G.J, CT4471, Nanofiltration and Reverse Osmosis, 2007*)

Generally, water treatment design referred to raw water sample lab analysis, then its technology will be choosed conditionally [2]. Using existing methods like conventional filtration, micro filtration (MF) or ultra filtration (UF) will solve the problem with suitable solution. One situation may need conventional filtration only whereas the other could be added with MF for instance [10]. The smaller permeated particle size, the higher pump pressure needed, and possibly the smaller its recovery value as well. As known, conventional filtration process and MF do not have any rejection, thus all of input water goes into the system. The UF membrane, though had no rejection neither but a small precentage of input should be circulated back to initial route. Using RO catridges gave the best result but it needs higher pressure for pumping and some precentage will be eliminated as reject waste.

In a few cases these also combined with basic method such as coagulation & flocculation to improve rejection or make the filter lifetime longer. 'Figure 2' shows the simple and very general stages in water treatment technology selection as described. Considering output water quality result, it might also be added UV or other disinfection system to the system.

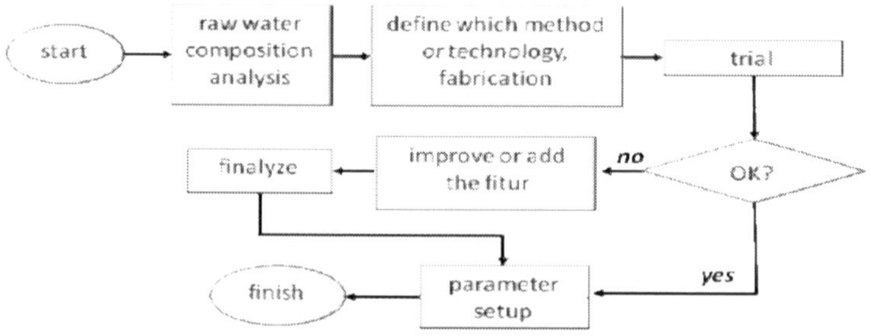

Figure 2. General stage in water treatment technology selection

There are three system examples from market that could be studied and considered for benchmarking. The first is a product from Japan, a Portable Water Treatment [5] (*KYUSUI Manual Book, 2011*), that used such technologies for disaster equipment which made for 150 liters (0.15m^3) per hour capacity. It is designed with MF and UF for filtration, followed with UV lamp for disinfection, thus without any chemical implemented which its block diagram showed in 'figure 3'. This kind of machine is very elegant, easy to operate and only approximately 22 kg weight, but in general its capacity seems rather too small for flood disaster.

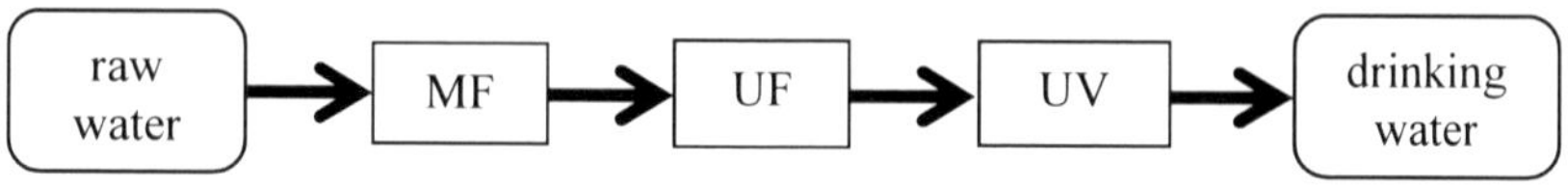

Figure 3. Portable Water Treatment example Block Diagram

Other technologies that could be viewed is WWT (well water treatment) and RWT (river water treatment) machine that manufactured by Czech Republic [8]. These products have the minimal capacity of WWT for 1m^3 per hour (WWT-01) whereas RWT for 5m^3 per hour (RWT-05). This RWT-05 is designed for surface water source such as river or lake, and could be adopted for flood water with some necessary improvement. The disadvantage is that physically it has big dimension (L-W-H of 2x3x1.8 in meter) that might give rise to obstacle in mobile operation due to its total weight and infrastructure access width. In the other hand, the WWT-01 has appropriated dimension (1x2x1.8 in meter) but its capability is only for well water sources.

The following pictures show the block diagram of WWT-01 (figure 4) and RWT-05 (figure 5). From such figures the following components in both products could be explained:

- Hydrocylone : removal of most solid suspended substances with part size larger than 0,2 mm.
- Static Mixer : intended for spontanious mixing different chemical solutions.
- Retention Tanks : precipitate iron and manganese through oxidation reaction.

- Sand Filter/ Active Carbon Filter Chamber : ensures water filtration from undissolved substances.
- Dosing Pump : dose chemical solution with automatic dosage measure.
- Sieve Filter, Automatic Self Cleaning: eliminates all undissolved substances > 100μ
- Turbo Mixer (used only for RWT-10 system, or more) : for effective mixing and homogenization of dosed flocculants with crude water
- Flocculations Chamber : for optimum drain of cores created during coagulation and condensation

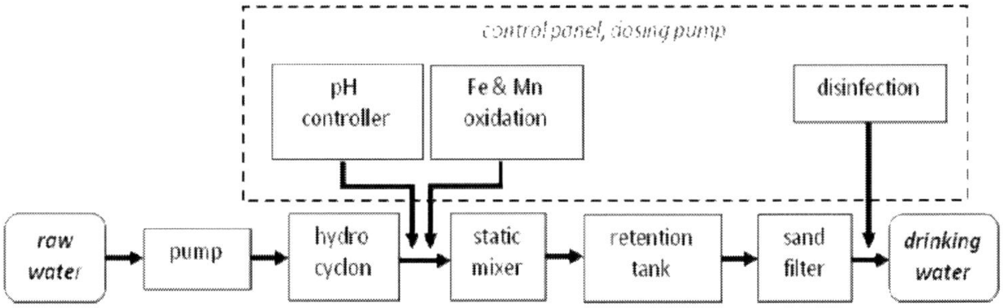

Figure 4. WWT-01 Block Diagram

WWT-01 has a more simple process than RWT-05 since ground water do not need any coagulant nor flocculants. In RWT-05, there is even an initial valve that has to be switched according to raw water quality. For certain condition, input water will be oxidized first before stored in Tank-A (sees figure 5) and flow to the main system. Otherwise, the input go directly to sieve filter as a surface water standard treatment. In this condition, Tank-A functioned paralelly with Tank-B.

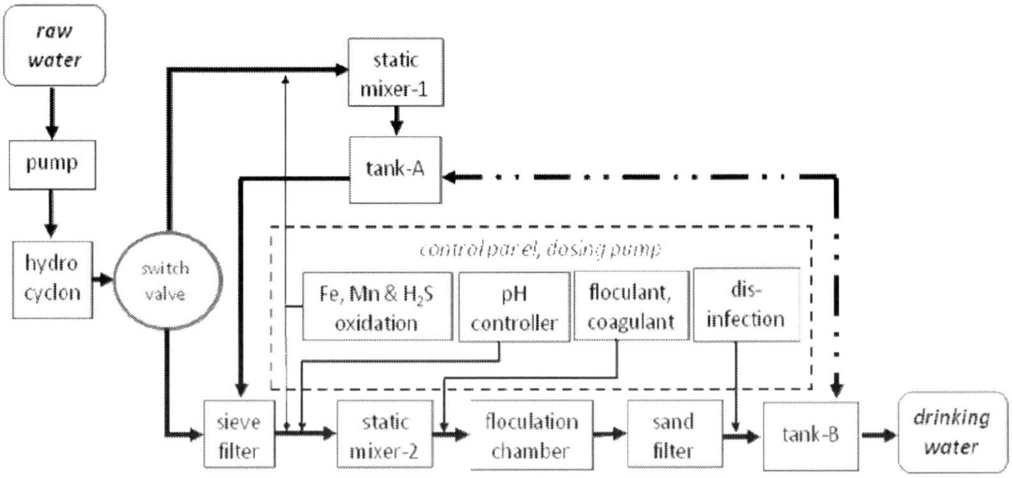

Figure 5. RWT-05 Block Diagram

The differences between these two kind of machines are that WWT-01 uses retention tank for oxidation reaction, whereas RWT-05 equipped with sieve filter due to potential raw water higher turbidity, turbo mixer and flocculation chamber that needed for flocculation and coagulation process that not necessary in WWT. These components are used for surface water source processing since it might has more than 10,0 NTU (Nephelometric Turbidity Units) of maximum 5 NTU that recommended by quality standard for clean water.

The study to develop water treatment (WT) equipment refers to previous researches and market products, which technology level will be used or combined. It is designed for river or raw water source in flood area, basically made for limited community, that is $1m^3$ per hour or 18-20 m^3 per day capacity, which is equal with approximately 100-150 man requirements. This number is considered for temporarily operation, it then has to be made easier in operation due to limited weight and easily move the unit to different places. Combining the mechanical and electrical parameters of Japan Portable Water Treatment, Czech RWT-05, WWT-01 and other references were inspiring for developing "new sub varian' prototype for disaster mitigation.

Mentioned reasons caused MSWT-01--Mobile Surface Water Treatment-- identity, choosed. 'Surface Water' tells the input which water source in flood area is included, whereas 01 indicates $1m^3$ per hour capacity. Combining several existing technologies, related literatures and necessary modification, the design should be also appropriated with local parts availability due to maintenance aspect. Above mentioned 'water treatment process technology level' (Figure-1) and exploring/ searching information resulted that the basic processes in common ways like coagulation, flocculation, and filtration will still be used. The technologies are possibly not special, thus the important thing is how to combine such methods and put these modular processes in a compact design elegantly.

3. Discussion

Due to the similarity, the RWT basic/ general system & components could be easily adopted into WWT (hydro cyclone, dosing pump, static mixer and sand/ carbon filter). The firm differences are flocculation chamber, sieve filter and turbo mixer in RWT that can't be found in WWT. The study just simplified its system that flocculation chamber and turbo mixer, which are relatively expensive, won't be used in MSWT-01, as a part of cost reduction aim.

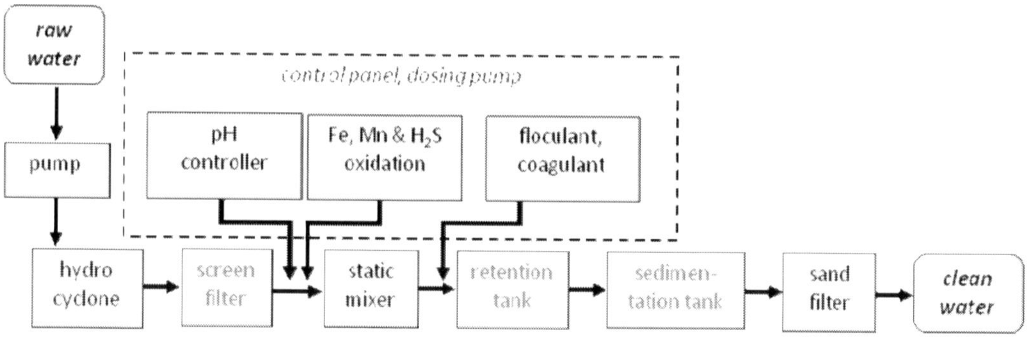

Figure 6. MSWT-01 for Disaster Mitigation Block Diagram

The study still has to prove experimentally that the only necessary additional component could be sieve filter, although, since its unavailability in local market, other type of screen filter will be used, without automatic self cleaning capability. Regarding the replacement of turbo mixer and flocculation chamber, implementing the sedimentation tank could be considered in order to make the sand filter performance longer, as shown in 'figure 6'.

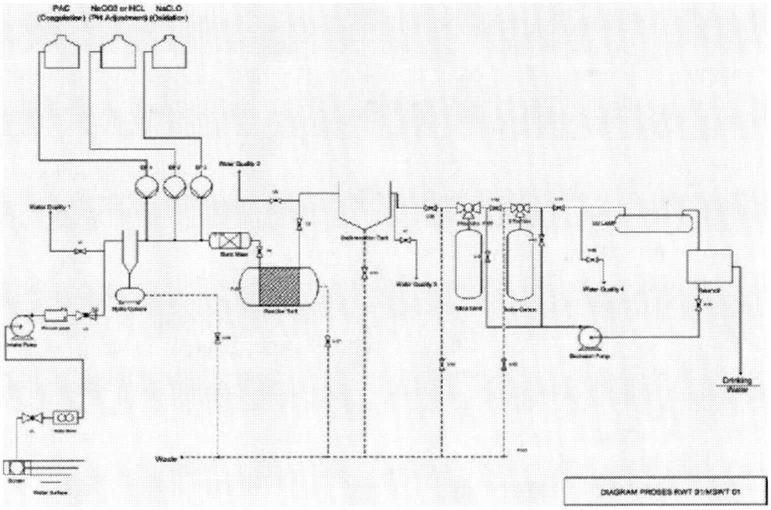

Figure 7. MSWT-01 Components with UV lamp disinfection system

Through prototype level experiment trials, the machine capability has to be proved for producing clean water that suitable for sanitation and cooking/ drinking purposes although using contaminated water input source. One of important thing is how to combine such methods, implementing for people needs, and developing network with disaster mitigation management stake holder for wider arrangement. From the investment point of view, such machine could be also treated as an asset that will be used from time to time when needed, instead of made for project approach only.

'Figure 7' shows all components for MSWT-01, but in the real prototype the sedimentation tank and UV lamp not implemented yet. Considering the poor quality of input raw water at flood disaster location, implementing the disinfection process is more elegant, although the output clean water as default will not be used for drinking purpose. UV lamp is shown on 'Figures 7', but as described, other alternatives for disinfection system could be choosed such as Ultra Filtration, adding NaClO etc, with their each advantage and disadvantage respectively.

Table 1. Sample results from 4 different places

Parameter	Unit	Std. max spec.	Sample-1		Sample-2		Sample-3		Sample-4	
			Source	Result	Source	Result	Source	Result	Source	Result
Color	TCU	15	< 5	0.5	< 5	0.65	6.9	2	14.6	6
Turbidity	NTU	5	0.28	0.2	3.17	0.35	37	1.2	57.7	2
Fe content	mg/L	0.3	< 0.1	0.1	0.69	0.23	0.15	0.1	0.49	0.2
Mn content	mg/L	0.4	< 0.03	0.03	0.02	0.02	< 0.01	0.01	0.65	0.3
Hardness)*	mg/L	500	64	60	32	32	46.5	45	114	111.7

)* *$CaCO_3$ content*

The MSWT-01 trial from 4 raw water samples mentioned that the samples from several places had different condition or content, and might be in extreme differences. 'Table 1' shows several essential parameters only of not so terrible samples. From each sample, it is analyzed either the source and result, where each result samples is taken after a period of machine operation, in order to get good representation and avoid from mixing up with previous sample remained in the machine. It showed that from the highest turbidity (sample 4 with 57.7 NTU) for instance, MSWT could reduce to the acceptable level. The experiment should be done for more samples that have extreme turbidity for instance, in order to attempt with the real condition in flood area. In necessary condition, a location that has flood water source with a high turbidity for instance, sedimentation tank could be added to the system.

4. Conclusion

MSWT-01 (Mobile Surface Water Treatment) Prototype for Disaster Mitigation is designed for river or surface raw water source in flood area, made for $1m^3$ per hour or 18-20 m^3 per day capacity, which is equal with 100-150 man requirements. Adopted and combined from RWT-05/ river water treatment with $5m^3$ per hour capacity, WWT-01/ well water treatment $1m^3$ per hour from Czech Republic product, several references and common existing technologies, is capable to produce clean water that suitable for sanitation and cooking/ drinking purposes although using contaminated water input that taken from improper or dirty sources. It is proved through number of experiment trials at prototype level. More samples that represent real disaster condition are needed to develop in next stages. For necessary condition, it could be added with sedimentation tank and disinfection system.

Comparing with other three example market available products, MSWT-01 could provide higher capacity than Japan Portable Water Treatment (150 liters or $0.15m^3$ per hour), higher capability than WWT-01 that made for ground water only but had the similar dimension, and more simple than RWT-05 which used Sieve Filter/ Automatic Self Cleaning and Flocculation Chamber.

Next step is that this MSWT will be equipped with mobile feature and electric generator in order to make easier in operational and moving to reach disaster sub locations or sub sectors where it's needed, thus clean water demand could be provided in nearly any condition. Moreover, since the technologies are possibly very common and not special, the important thing is then how to combine such methods, implementing for people needs, and developing network with disaster mitigation management stake holder for wider arrangement. From each stake holder entity or social organization point of view, the machine could be also treated as an investment that will be used from time to time when needed, instead of made for project approach only that new equipment should be re-procured for every event, as a new project.

5. References

[1] Ananto Gamawan 2010 *Smallest Unit Water Treatment opportunity from entrepreneurship point of view* Thesis (Bandung: MM Universitas Padjadjaran)

[2] Ananto Gamawan 2012 *MSWT-01 (Mobile Surface Water Treatment) Prototype, an Alternative For Disaster Mitigation* ISSN 1412-9612 (RAPI XI FT UMS)

[3] Badan Koordinasi Nasional Penanggulangan Bencana (BAKORNAS PB) *Pedoman Penanggulangan Bencana Banjir Tahun 2007/2008* (Pelaksana Harian BAKORNAS PB, Jakarta)

[4] Badan Standardisasi Nasional 2006 *Air Minum Dalam Kemasan* (SNI 01-3553-2006)

[5] Clear System Ltd and UBE Environment Research Ltd 2011 *KYUSUI Manual Book* (Yamaguchi)

[6] Daryono SSi MSi 2010 *Mitigasi Bencana Banjir* (Program Doktor Ilmu Geografi UGM)

[7] Heijman S G J 2007 *CT4471, Nanofiltration and Reverse Osmosis* (TU Delft) p2.

[8] IWET PT 2007 *Well Water Treatment/ WWT-01 & River Water Treatment/ RWT-05 Manual Book* (Tasikmalaya: IWET)

[9] Kementrian Pekerjaan Umum, Dirjen Cipta Karya 2010. <http://ciptakarya.pu.go.id>.

[10] Len Boselovic and Mc Clatchy 2009 *Carbon wins contract for Phoenix water treatment* ProQuest document ID1923254051 (Washington: Calgon - Tribune Business News Dec 17, 2009)

[11] Mohammad Arifur Rahman, Shamim Ahsan, Satoshi Kaneco, et.all 2005 *Waste water treatment with multilayer media of waste and natural indigenous materials* (Journal of Environmental Management 74) Issue 2:107-110

[12] Ziv-El, Michael C and Rittmann Bruce 2009 *Water quality assessment of ground-water treated with a membrane bio-film reactor* (AWWA Journal Article.cfm-3.htm) item number = 52401

Effect of Reflow Profile on Intermetallic Compound Formation

I. Siti Rabiatull Aisha[1*], A. Ourdjini[2], M. A. Azmah Hanim[3], and O. Saliza Azlina[4]

[1]Faculty of Mechanical Engineering, Universiti Malaysia Pahang, 26600 Pekan, Pahang, Malaysia
[2]Faculty of Mechanical Engineering, Universiti Teknologi Malaysia, 81300 Skudai, Johor, Malaysia,
[3]Faculty of Engineering, Universiti Putra Malaysia, 43400 Serdang, Selangor, Malaysia,
[4]Faculty of Mechanical and Manufacturing, Universiti Tun Hussein Onn Malaysia, 86400 Parit Raja, Batu Pahat, Johor, Malaysia

E-mail: rabiatull@ump.edu.my

Abstract. Reflow soldering in a nitrogen atmosphere is a common process consideration in surface mount technology assembly. This is because the use of nitrogen in reflow equipment may benefit the process as well as the quality of the end product, where it can increase the reliability of the solder joint. So far, many papers have reported effects of cooling speed, type of solder pastes and solder fluxes on the reliability of lead-free solder joints. While the effects of reflow conditions on intermetallic compound (IMC) formation at the solder joint such as the atmosphere during the reflow process are still unclear. The present study investigated thoroughly the effect of different reflow soldering atmosphere, which is air and nitrogen on IMC formation and growth. Several techniques of materials characterization including optical, image analysis, scanning electron microscopy and energy dispersive X-ray analysis will be used to characterise the intermetallics in terms of composition, thickness and morphology. In addition, the effects of cooling rate and isothermal aging were also studied for the solder alloy Sn–4Ag–0.5Cu on electroless nickel/immersion gold (ENIG) surface finish. From the study, it was found that reflowing under nitrogen atmosphere had better effect on IMC formation and growth compared to reflowing under air. Besides, the cooling rate of solder during reflow also appears to have a significant effect on the final structure of the solder joint, and controlling the growth behaviour of the IMC during subsequent isothermal aging.

1. Introduction

Flip chip has been widely used in electronic industry due to its low cost, small form factor, high I/O count, and high performance[1,2]. Another advantage of flip chip process is its compatibility with conventional surface mount technology (SMT) process, as the reflow is being utilized to attach the solder bumped flip chip to the substrate. Reflow is a controlled heating process for the solder to melt and wet on a pad in order to create interconnections between the die and substrate.

However, with downsizing and requirements for faster and cheaper electronic devices, a strong demand has arisen for high soldering reliabilities. A reliable soldered joint should perform both mechanical and electrical functions without failure during its service life. Manufacturers need to optimize many materials and processing factors to control the quality of their products. Solder joint reliability is influenced by various factors such as the type of solder paste, solder paste volume, pad design, aperture design, board finish, assembly arrangement, and reflow temperature profile/atmosphere[1,3]. However, the current study will only focused on the effect of reflow profile and reflow atmosphere onto intermetallic compound (IMC) formation.

Several parameters related to reflow profile may affect the solder joint formation including soak time, ramp rate, peak temperature, time above liquidus and ramp rate during cooling[2]. Previous research investigating lead-free reflow profile of solder bumps on FR4 substrates reported that the most significant factor in achieving a joint with a thin IMC layer and fine microstructure is the peak temperature[3]. In another study on the effect of the reflow peak temperature and time above liquidus (TAL) on both Sn-Pb and Sn-Ag-Cu lead free solder joint shear strength with four different chip resistors on FR4 substrate confirmed that both reflow peak temperature and TAL of reflow profile are the most critical factors in terms of package shear strength for Sn-Ag-Cu solder joints[4]. In recent years, greater attention has been drawn to understanding the formation and growth of the IMC layer and its effect on solder joint reliability in electronic packaging industries with Pb-free application needs[5, 6]. It is generally accepted that an increase in IMC layer thickness decreases the lifetime of solder joints. Typically, the IMC layer thickness strongly depends on reflow process conditions such as time above liquidus and peak temperature[7-9]. The correlation between reflow process and IMC layer has been investigated to determine an ideal reflow process condition that would achieve reliable solder joints during thermal cycling testing using statistical analysis[10].

2. Materials and Methods

First, the copper substrate was subjected to a pretreatment process to remove oxides and activate the surface before the desired finish layers are deposited. Then, an ENIG surface finish was deposited on the substrate having the dimensions (width x length x thickness) of 45 x 50 x 1 mm. Prior to soldering, a dry solder mask was laminated onto the plated substrates using a laminated machine. Then, the solder mask was exposed to an ultraviolet (UV) light through a patterned film. The exposure was to ensure an array of pads is made onto the solder mask upon subsequent development stage in the developing solution. The substrates were then populated with Sn-4Ag-0.5Cu (SAC405) solder spheres of 500μm. The solder spheres were arranged in several rows and bonding to form the solder joints was made by reflow in a furnace with the peak reflow temperature set at 250°C. Prior to soldering, all substrates were treated with a no clean flux to remove surface oxide. In order to reveal the morphology of IMC formed during the soldering process a useful method of selective chemical etching of the top surface was employed and examination of the IMC was made by means of scanning electron microscopy. Energy dispersive x-ray (EDX) was used to identify the type and composition of IMC formed.

3. Results and Discussion

3.1. Effect of Reflow Profile

It is an established fact that the final structure of solder joint depends on the reflow profile, and in particular the cooling rate. In order to determine the influence of cooling rate on the solder joint structure, two different reflow profiles were used as shown in Figure 1 and 2. The reflow profile shown in Figure 1 has slower cooling rate than that in Figure 2. In addition, the atmosphere within which reflow was conducted was also different. Reflow was conducted in air (Figure 1) and under nitrogen (Figure 2). In subsequent discussion, the reflow profiles will be labeled as reflow A for the slower post-reflow cooling rate and reflow B for the fast post-reflow cooling rate.

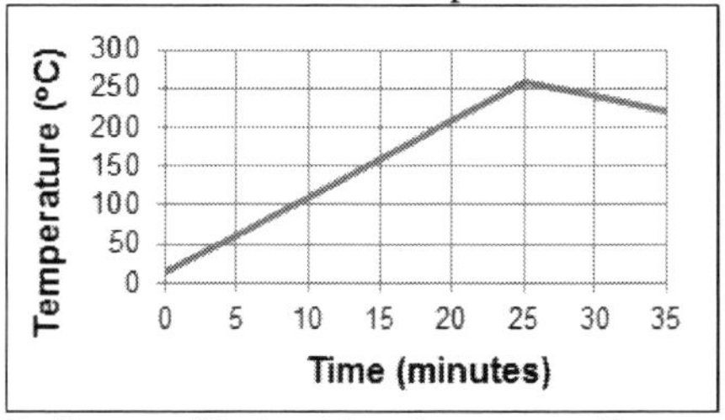

Figure 1. Reflow profile A with a slow post-reflow cooling rate.

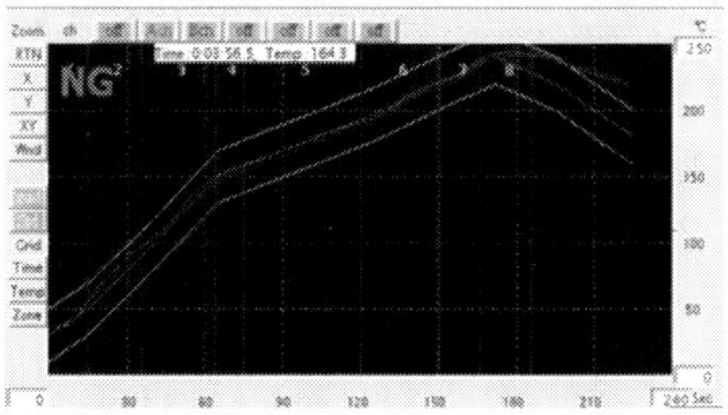

Figure 2. Reflow profile B with a fast post-reflow cooling rate

After reflow soldering, Sn and Cu (from solder alloy) and Ni and Cu (from the coating and substrate) diffuse towards each other producing IMC. However, the size and shape of this IMC varies depending on reflow condition. Figure 3 shows the top surface views of samples reflowed by the two different profiles. It can be seen that, there was small circle formed at the interface within which the size of IMC is different from that observed throughout the solder. However, the position of this circle is different between the two reflow profiles was used. This is probably due to the effect of different atmospheric conditions used, where the interfacial reaction between the solder alloy and substrate was much uniform when nitrogen was used since oxidation can be prevented. Therefore, the solder joint will be much stronger[11].

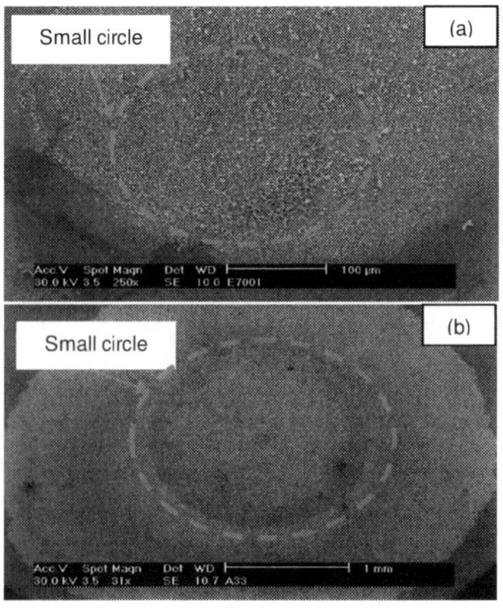

Figure 3. SEM top surface of whole solder after reflow soldering of Sn-4Ag-0.5Cu using different reflow profiles; (a) reflow profile B on ENIG, and (b) profile A on ENIG.

Furthermore, the post-reflow cooling rate also affects the IMC formation. Generally, the fast cooling rate favors nucleation but since the holding time at high temperature zone is shorter, the effective time for IMC growth is shorter. This results in the formation of a thin layer of fine grained IMCs as shown in Figure 4(a, b). The reason why these fine grains are in needle shape has something to do with the direct reaction between Sn in the solder and the decomposed Ni from the metastable Ni coating (ENIG). While for the slow cooling rate, nucleation rate is lower but the IMC grains have enough time to grow. As a result, chunk type IMC becomes dominant and the IMC layer thickness is larger (Figure 4(c, d)).

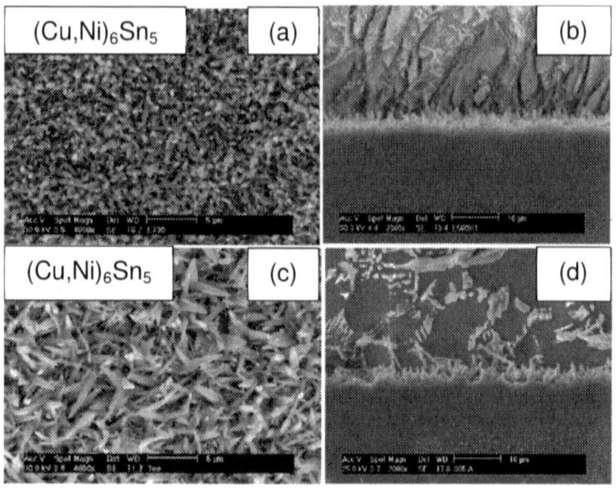

Figure 4. SEM top surface and cross sectional views of Sn-4Ag-0.5Cu after reflow soldering using different reflow profiles; (a, b) reflow profile B, and (c, d) profile A.

It was also found that the IMC growth increased upon aging and post-reflow cooling rate. This is due to more grain boundaries produced in smaller IMC grain size when reflow profile B was used. When more grain boundaries exist, the diffusion become much easier and faster compared to larger IMC grain size. As a result IMC grows to a higher thickness for the same aging condition in faster cooling rate samples (Figure 5).

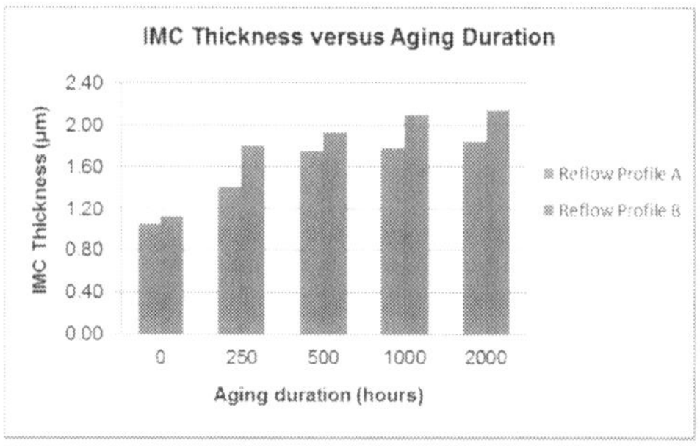

Figure 5. Effect of aging duration on IMC thickness using different reflow soldering profiles.

Apart from that, needle-type, boomerang-type and chunk-type (Cu, Ni)$_6$Sn$_5$ IMC grains were formed between Sn–4Ag–0.5Cu solder and ENIG at different post-reflow cooling rates as can be seen in

Figure 6 and Figure 7. When cooling rate decreases, the amount of needle-type grains also decreases, but the amount of chunk-type and boomerang-type grains increases due to grain growth and coalescence (Figure 7).

The formation of (Cu, Ni)$_6$Sn$_5$ and (Ni, Cu)$_3$Sn$_4$ IMCs are based on the Cu_6Sn_5 and Ni_3Sn_4 crystal structures, respectively. As can be seen in Figure 6 and Figure 7, Cu_6Sn_5-based IMC was the primary formed compound during reflow, although the Cu content (0.5 wt.%) in the solder was extremely low compared to Sn and Ni. The stable ternary IMC initially formed at the interface after reflow was the (Cu, Ni)$_6$Sn$_5$ due to diffusion of Cu from the solder alloy towards Ni substrate. During aging, interfacial reaction between Ni and Sn-Ag-Cu, Sn, Cu and Ni atoms must diffuse through the existing IMC layer to form interfacial reaction products such as (Cu, Ni)$_6$Sn$_5$. If there was no more Cu present in the solder material, the Ni_3Sn_4-based IMC could be stable. However, the solder used in the current work was able to provide enough Cu towards the substrate since even after 2000 hours aging, (Cu, Ni)$_6$Sn$_5$ IMC was still dominant. Nevertheless, Figure 7(e, f) shows that there was (Ni, Cu)$_3$Sn$_4$ formed after 2000 hours aging. This is due to the effect of different reflow atmospheric conditions used whereby reflow conducted in oxygen produced non-consistent IMCs' type throughout the interface since corrosion product prevented the Cu atoms to diffuse towards the substrate. Besides, isothermal aging renders the IMC rounder, bigger and more compact for both reflow profiles used.

Figure 6. SEM top surface of ENIG, as reflow and aged at 150°C, reflowed with Sn-4Ag-0.5Cu solder, mag 4000x using reflow profile B; (a) as reflow mag 2000x, (b) as reflow at small ring, (c) as reflow at near outside, (d) aging 500hours at small ring, (e) aging 500hours at near outside, (f) aging 2000hours at small ring, and (g) aging 2000hours at near outside.

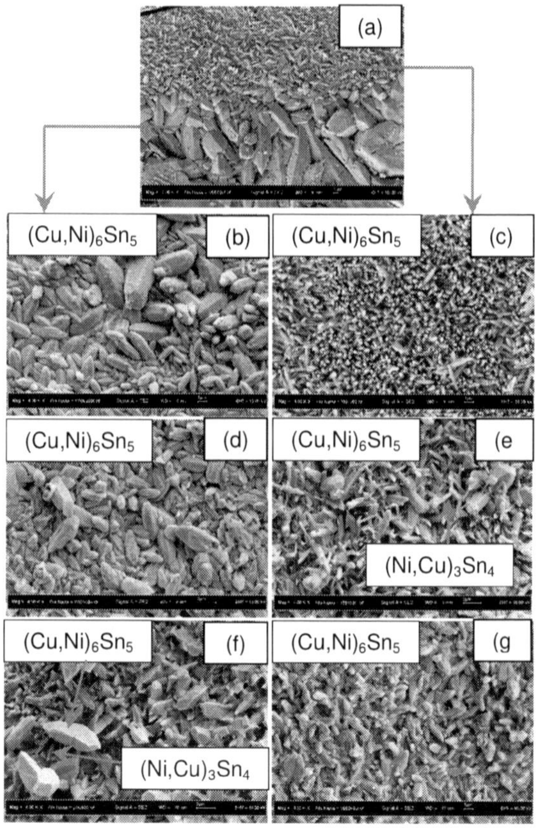

Figure 7. SEM top surface of ENIG, as reflow and aged at 150°C, reflowed with Sn-4Ag-0.5Cu solder, mag 4000x using reflow profile A; (a) as reflow mag 2000x, (b) as reflow at small ring, (c) as reflow at near outside, (d) aging 500hours at small ring, (e) aging 500hours at near outside, (f) aging 2000hours at small ring, and (g) aging 2000hours at near

4. Conclusion

Several conclusions can be made from the study:

i. Different reflow profiles produce different morphologies of IMC at the interface during reflow which may affect the growth behavior of the IMC after aging.
ii. The IMC's formed during reflow under nitrogen atmosphere were more consistent than those produced when reflow is conducted under air. However, faster cooling rate produced finer IMC grains which might be translated into a better solder joint strength.

iii. Isothermal aging resulted in increased IMC thickness for both reflow profiles used in air and under nitrogen.

Acknowledgments
The authors would like to acknowledge the financial support provided by Universiti Malaysia Pahang (vot 120359) for providing the research facilities.

References

[1] Amalu E. H., Ekere N. N. 2012 Journal of Materials Processing Technology, **212** 471
[2] Lau C. S., Abdullah M. Z., Ani F. C. 2012 Microelectronics Reliability,
[3] Tegehall, P. E. 2011. The ELFNET Book on Failure Mechanisms, Testing Methods, and Quality Issues of Lead-Free Solder Interconnects, pp. 179-195.
[4] Pan J. 2006 Soldering & Surface Mount Technology**18** 48
[5] Ladani L. J. 2008 IEEE Transactions on Components and Packaging, **31** 51
[6] Huang W., Loman J. M., and Sener B. 2002 Microelectronics Reliability, 42, 1229
[7] Tu P. L., Chan Y. C., and Lai J. K. L. 1997 IEEE transactions on Components and Packaging, B **87**
[8] Wang, L., Wang, J., Liu, X., and Liu, X. 2010. International Conference on Electronic Packaging Technology & High Density Packaging (ICEPT-HDP). pp. 181-184.
[9] Huang W., Loman J. M., and Sener B. 2002 Microelectronics Reliability **42** 1229
[10] Arenas M. F., He M., and Acoff V. L. 2006 Journal of Electronic Materials **35** 1536
[11] Law C. M. T., Wu C. M. L., Yu D. Q., Wang L., and Lai J. K. L. 2006 Journal of Electronic Materials **35** 89

Web-based Factors Affecting Online Purchasing Behaviour

Mohd Shoki Md Ariff[1], Ng Sze Yan[1], Norhayati Zakuan[1*], Ahamad Zaidi Bahari[1] and Ahmad Jusoh[1]

1 Department of Management, Faculty of Human Resource and Management, Universiti Teknologi Malaysia (UTM), Skudai, Johor, Malaysia

E-mail: norhayatimz@utm.my

Abstract. The growing use of internet and online purchasing among young consumers in Malaysia provides a huge prospect in e-commerce market, specifically for B2C segment. In this market, if E-marketers know the web-based factors affecting online buyers' behaviour, and the effect of these factors on behaviour of online consumers, then they can develop their marketing strategies to convert potential customers into active one, while retaining existing online customers. Review of previous studies related to the online purchasing behaviour in B2C market has point out that the conceptualization and empirical validation of the online purchasing behaviour of Information and Communication Technology (ICT) literate users, or ICT professional, in Malaysia has not been clearly addressed. This paper focuses on (i) web-based factors which online buyers (ICT professional) keep in mind while shopping online; and (ii) the effect of web-based factors on online purchasing behaviour. Based on the extensive literature review, a conceptual framework of 24 items of five factors was constructed to determine web-based factors affecting online purchasing behaviour of ICT professional. Analysis of data was performed based on the 310 questionnaires, which were collected using a stratified random sampling method, from ICT undergraduate students in a public university in Malaysia. The Exploratory factor analysis performed showed that five factors affecting online purchase behaviour are Information Quality, Fulfilment/Reliability/Customer Service, Website Design, Quick and Details, and Privacy/Security. The result of Multiple Regression Analysis indicated that Information Quality, Quick and Details, and Privacy/Security affect positively online purchase behaviour. The results provide a usable model for measuring web-based factors affecting buyers' online purchase behaviour in B2C market, as well as for online shopping companies to focus on the factors that will increase customers' online purchase.

Keywords: B2C, website quality, online buyers, online purchasing behaviour

1.0 Introduction

Electronic commerce (E-commerce) refers to the process of buying, selling, transferring, or exchanging products, services, or information via computer networks [1]. Based on Kenneth and Carol [2], nowadays, it is inextricably for e-commerce linked to the Internet. Business to Consumer (B2C) is one of the natures of transactions or interactions in e-commerce. According to Turban, et.al [1], B2C refers to e-commerce model in which businesses sell to individual shoppers. The volume of online B2C transactions is increasing annually. According to the latest MasterCard Worldwide online

Shopping Survey, consumers in Malaysia are shopping online more and have become more confident about online security. The number of Malaysian people shopping online has grown by 12% in year 2011 compared to the year before (55% in 2010) [3]. The growing number of B2C transactions provides a huge prospect of online buyers for E-marketers in Malaysia, and therefore, examining factors affecting online buyers' behaviour in B2C market is crucial. However, it was reported that e-commerce in Malaysia is not as popular as in western countries. Although almost every Internet user surveyed said in general they like the idea of shopping on the internet, in fact, only a small number of Malaysians actually buy online [4]. Thus, if E-marketers know the factors affecting online buyers' behaviour in B2C market, and the effect of these factors on behaviour of online consumers, then they can develop effective marketing strategies to (i) attract more Malaysian to get involve in online purchase; (ii) convert potential customers or less active online buyers into active one; and (iii) improve quality of online transactions by focusing on web-based factors that are perceived as important by online buyers.

According to Syed et al. [5], the dominant factor which influences consumer perceptions of online purchasing is the website design characteristics. With reference to website design/ quality, it was found that navigation and content were the most sought after factors of online purchasing [6]. On the other hand, Zeithaml et al.[7]suggest that entertainment criteria identified in the studies of more general websites are not relevant in the online purchase context. Shergill and Chen [8] found that website design quality was an important issue in customer satisfaction. When the customers satisfied with the website quality, this will increase their involvement in online purchasing. Based on Yam et al. [9], privacy and security are the examples of the factors which affect consumers' willingness to buy from online retailers. Grace and Chia [10] stated that consumers' attitudes and beliefs related to convenience and security concerns (shaped during the online purchase process) have significant effects on their intention to purchase online. Thus, as suggested by Nunes [11], if E-marketers intend to ignore the fundamental truths about online consumer behaviour and which web-based factors valued most by online buyers in different situations or countries, most of the promises of E-marketing in the B2C context will not be fulfilled. Thus, the effect of web-based factors on online purchasing behaviour in the Malaysia context should be determined.

The evolution in B2C studies has point out the importance of enhancing quality of web-based transactions to ensure sustainability of business in e-commerce. E-marketers need to address web-based factors which are significantly affecting buyers' online purchase behaviour, especially among young consumers. Young consumers are playing an important role in online shopping and this provides an emerging prospect for online retailers [5]. Particular to the behaviours of online consumers, skills in information and communication technology (ICT) and age has been found to be strongly related to adoption and use of information systems, including online retailing web-based systems [12]. In Malaysia, as reported by Haque and Khatib [13], most internet users are youths (15-20 years old; 50 per cent) and young adults (20 - 29 years old; 39 per cent). These groups of users consist of those between 15 – 29 years of age, in which most of them are students in secondary schools and universities. Thus, students in tertiary education, especially those who have knowledge in ICT and e-business, have significant influences to B2C market. ICT students are regarded as ICT literate; therefore studying their behaviour towards online purchasing provides a new landscape in understanding behaviour of online buyers. In addition, given that the ICT students are considerably 'expert' in information technology systems, revealing their online purchase behaviour in B2C market would be interesting to research. Hence, this study provides empirical evidence on online purchasing behaviour of Malaysian young customers, specifically future ICT professionals. Therefore, this study aims to answer to the following questions; (i) what are web-based factors affecting online purchasing behaviour of information technology literate consumers, i.e. ICT students?; and (ii) what is the effect of web-based factors on online behaviour of ICT students?

2.0 Literature Review

2.1 The buying process in an e-market

In consumer behavior research, it is very important for marketers to understand consumer decision making process. This will help them predicting consumers' purchase behavior, especially in the context of online purchasing. Current literature on consumer online purchasing decisions has mainly concentrated on identifying the factors which affect the willingness of consumers to engage in internet shopping. Based on Na Li and Ping Zhang [14], online purchasing behaviour refers to the process of purchasing products or services via the Internet. Consumers' attitude towards online shopping is a prominent factor affecting actual purchasing behaviour. The potential buyers recognize a need for some merchandise or service and they will go to the internet and search for related information before they make a buying decision. They will evaluate the alternatives and choose the best which they satisfied for it. Hence, the online transaction is conducted.

Customers purchase goods online in different modes. The most common mode is purchasing from catalogs at fixed prices. Sometimes prices are negotiated or discounted. Another mode is dynamic pricing, which refers to non-fixed prices, such as those in auctions or stock (commodity) markets. According to the buying process in an e-market, the process starts with logging into a seller's site, registering (if needed), and entering into an online catalog (E-catalog) or the buyer's account. E-catalog can be very large, so a search mechanism may be needed. Besides that, the online buyer needs to compare prices. If the buyer unsatisfied, the buyer may abandon the site. On the other hand, if satisfied, the buyer will select the item and place it in a shopping cart. The buyer might then return to the e-catalog to choose more items. When shopping is completed, the buyer goes to a check out page where a shipment option is selected from a menu. A payment option may be available. Payment method may be PayPal, credit card and others. After checking all the details for accuracy, the buyer submits the order.

2.2 Theories related to the assessing factors influencing online purchasing behaviour

The classic Theory of Reasoned Action [15], the Theory of Planned Behaviour [16], and Technology Acceptance Model [17] have been extensively adopted for explaining and predicting user behaviour in an online shopping environment. Theory of Reasoned Action (TRA), is a theory of attitude-behaviour relationships which links attitudes (considered as beliefs that a person accumulates over his lifetime), subjective norms, behavioural intention and behaviour in a fixed causal sequences. Attitudes are said arise from the beliefs about behavioural outcomes and the evaluations of those outcomes. A person's overall attitude toward an object is derived from his beliefs and feelings about various attributes of the object. According to TRA, a person's behaviour (i.e. online purchase, vote, etc.) is determined by his/her intention to perform this behaviour. Intention is the cognitive representation of a person's readiness to perform a given behaviour, and it is considered to be the immediate antecedent of behaviour. So, the stronger the intention of an individual to perform a particular behaviour, the greater the most likely the particular behaviour will be performed [16].

An expansion of the TRA is the Theory of Planned Behaviour (TPB) which is proposed by Ajzen [18]. In fact, TPB differs from the TRA with its addition of the perceived behavioural control (PBC). TPB suggests that a central factor in human behaviour is behavioural intention, which is affected by attitude toward behaviour, subjective norm, and perceived behavioural control [18, 16]. PBC reflects a person's perception of the ease or difficulty of implementing the behaviour in question. It concerns with the beliefs about the presence of control factors that may facilitate or hinder individuals from performing the behaviour. In TPB, behavioural control directly affects the intention to perform behaviour, such as to use internet-based transaction or online purchasing, and may directly affect behaviour in situations where the user intends to perform the behaviour, but is prevented from doing so [18, 16]. PBC relates to the extent to which the person believes that she/he has control over personal or external factors that may facilitate or constrain the behavioural performance [16]. It is assumed that PBC together with behavioural intention, can be used directly or indirectly to predict behaviour achievement.

In a further extension of TRA, TAM was introduced by Davis [17], which described an individuals' acceptance of information technology. In general, TAM is a theoretical model that evaluates the effects of things like system characteristics on user acceptance. In understanding acceptance of information technology systems, such as online retailing websites, TAM was used as the basis of the theoretical framework to build a model to determine intention of users to use a system or a website. TAM stressed on adoption degree of a person in using a technology and aspects that affect his or her acceptance or intention to use the technology. It suggests that users' adoption of information technology systems, in this case is online retailing websites, is determined by their intention to use the systems, which in turn is determined by their beliefs towards the systems. Users' beliefs about information technology systems are influenced by their perceived usefulness and perceived ease of use of the systems.

2.3 Instruments used to measure quality of web-based transactions

The electronic service quality (e-SQ) was highly concerned due to it highly related with the success or failure of an internet-based company or online purchasing or transaction. Several scales have been developed in measuring the website e-SQ. However, different types of services required different approaches to assessing the website e-SQ. Among the e-SQ measurement or assessment techniques, eTailQ, WEBQUAL, SITEQUAL and e-SERVQUAL are the most popular techniques used by researchers to determine e-SQ in different internet context. The eTailQ, a reliable and valid scale for the measurement of etail quality, was developed by Wolfinbarger and Gilly [19]. The authors suggested that four factors—website design, fulfillment/reliability, privacy/security and customer service—are strongly predictive of customer judgments of quality and satisfaction, customer loyalty and attitudes toward a website. ETailQ was one of the newest, most mature and scholarly models of website quality and it encompasses the entire life cycle associated with electronic transactions. Besides that, each of its factors, variables and items exhibit high levels of inter-item reliability and validity.

WEBQUAL was developed by Loiacono, Watson and Goodhue [20] in order to evaluate website quality focusing on the website interface. The WEBQUAL was used to predict the re-visit/re-use behaviour of web users based on their perceptions of overall website quality. There are four dimensions in WEBQUAL - usefulness, ease of use, entertainment and complimentary relationship (Loiacono, Watson and Goodhue, 2002). WEBQUAL 4 developed by Barnes & Vidgen [21] was the new generation for WEBQUAL and it composed of 22 items of three dimensions, namely, quality of information, quality of interactivity/confidence and empathy and usability of the site/usability and design. SITEQUAL was developed by Yoo & Donthu [22] with nine items distributed over four dimensions of ease of use, design, speed and security.

The e-SERVQUAL [23] is a method for measuring website e-SQ. E-SERVQUAL measures website e-SQ based on the customers perceptions and the scale contains the core scale (E-S-QUAL) and recovery scale (E-RecSQUAL). There are four dimensions in E-S-QUAL - efficiency, fulfilment, reliability and privacy – and for e-RecSQUAL, the three dimensions proposed are responsiveness, contact and compensation.

2.4 The influence of web-based factors on online purchasing behaviour

A customer's online buying experience consists of everything from information search, product evaluation, decision making, making the transaction, delivery, returns and customer service [19]. Web-based factors influence online purchasing behaviour either positively or negatively [5]. By using a sample of 214 online shoppers, Ranganathan and Ganapathy [24] found four key dimensions of B2C web sites - information content, web design, security and privacy – and although all these dimensions have an impact on the online purchase behavior, security and privacy were found to have greater effect on the purchase intent of online buyers. With reference to website quality, Ruchi et al.

[6] suggested that navigation and content are the most sought after factors consumers are looking for when dealing with a web-site. Categorization of information, colour usage, layout/space usage, graphics usage, establishing website's identity and presentation of information are also the important factors. The web site design describes the appeal of the user interface design that will increase the willingness of customers to visit a website more often and to stay longer with the site.

Vijayasarathy and Jones [25] stated that reliability is a measure of customers' perceptions on delivery of online purchased order as promised by e-retailers and it is closely associated with risk. According to Na Li and Ping Zhang [14], online consumers apparently want to receive the right quality and right quantity of items that they have ordered within the time frame promised by the retailers, and they expect to be billed accurately. Online service providers who deliver the promised services within promised time frame only will be considered as reliable [5].

Customer service's dimension of responsiveness, courtesy and understanding/knowing the customer are included in the original conceptualization of service quality [26]. In internet or web-based transaction, compare to the voice interaction, the response on email inquiries are more important [19]. This is why most companies used online customer service as a way to provide help and service to their online customers.

According to [27], privacy refers to the degree to which the online shopping web site is safe and protects the customers' information. Security refers to consumers' perceptions about the security of the online transactions as well as the protection of financial information from unauthorized access in an online retailing context [28]. Prior researches on online shopping context indicated that consumers' perceptions of privacy have a significant and positive effect on their trust in the online vendor [27]. Based on Ahuja et al., [29], security and privacy were the biggest barriers to online shopping and online buyers' purchasing behaviour was significantly affected by these factors.

3.0 Methodology

3.1 The Conceptual Framework of the Study

Based on review of TRA, TPB and TAM theories and e-SQ instruments related to the determination of consumers' online purchase behaviour, as discussed in Section Two of this study, the dependent variable was determined, i.e. online purchasing behaviour. The independent variable identified in this study was web-based factors. The selection of the factors was based on the following steps:

- Step 1: Review of literature and current trend in the usage of internet and online purchasing in Malaysia resulting in the need to examine web-based factors affecting behaviour of ICT professional;
- Step 2: Review of theories related to website quality and e-SQ instruments. This process had contributed to the identification of four factors in the eTailQ instrument – fulfilment/reliability, website design, privacy/security, and customer service – as web-based factors that may affect the behaviour of online buyers [19]. These factors have been widely validated as predictors of online purchasing behaviour [8, 19].
- Step 3: Review of previous studies related to the consumers' online purchasing behaviour revealed that information content was one of the factors which are affected online purchasing behaviour [21, 24] and this factor will either attracts or repels online customers. According to Szymanski and Hise [30], consumer satisfaction with regard to the online purchasing is largely determined by the quality of the site's product information. Web site information is often scanned, rather than read in detail and users always wanted useful and reliable information which they can access quickly. Hence, website information quality was included as one of the web-based factors.

- Step 4: The impact of those web-based factors must be examined in order to determine the extent to which they exert influence on attitude of online buyers. Therefore, attitude towards online purchasing behaviour was established as dependent variable of the study.

3.2 Research Instruments, Sampling Procedure and Data Collection

The four web-based factors - website design, fulfilment/ reliability, customer service and privacy/security - of eTailQ, developed by Wolfinbarger and Gilly [19] were used in this study. The scale was chosen as it has already been tested and validated as strong predictive of customer judgments of quality and satisfaction, customer loyalty and attitudes toward the website. 14 items of the four web-based factors originally proposed in e-TailQ [19] with three additional items (for Web Design factor) based on the work of Shergill and Chen [8] was employed.

Website information quality (with seven items), which is one of the dimensions in WEBQUAL4, proposed by Barnes and Vidgen [21], was additional factor included in the web-based factors of this study. 24 items of these five factors (Refer to Table 1) were developed in the questionnaire. Three questions (The idea of using this website to purchase a product or service is appealing, I like the idea of purchasing a product or service on this website, and Using this website to purchase a product or service at this store would be a good idea), which were adopted from Hans, Tibert and Marcel [31], were used to measure attitudes towards or online purchasing behaviour. The items were rated using five point Likert scales of *1=strongly disagree, 2=disagree, 3=somewhat agree, 4=agree and 5=strongly agree.*

All ICT undergraduate students of a public university served as a population of this study. In multivariate research, the sample size required should be 5 to 10 times of variables for 10% and 5% margin error [32]. The number of questions in the questionnaire was 27 and, therefore 270 questionnaires must be collected to ensure 5% of margin error. However, 320 questionnaires were distributed, and 310 of them were gathered and completed for further analysis. The Stratified Random Sampling technique (based on students' year of study and academic programs) was used in order to increase the accuracy of the data and the representativeness of the population of the study. In distributing and collecting the questionnaire, the students were first asked whether or not they have experience in online purchasing. Those who have the experience were asked to complete the questionnaires and then the researchers collected them back instantly.

4.0 Result and Analysis

4.1 Profile of the Respondents

The brief profile of respondents of this study indicated that:

- 47.4% are male respondents and the rest 52.6%.are female respondents.
- Most of the respondents (37.4%) are 1st year students. 29.6% are 2rd year students, followed by 3rd year students (22%) and 21% are the final year students.
- In term of academic program, 83 students (26.8%) are pursuing their degree in Computer Science (Software Engineering), which is the highest response rate among the six programs of ICT in this university. The least students (25 students or 8.1%) participated in this study are pursuing their degree in Bachelor of Computer Science (Industrial Computing).

4.2 Instrument assessment and determination of web-based factors affecting online purchasing behaviour

As indicated in Table 1, the KMO measure of sampling for the web-based factors was .919, which was greater than 0.6, supported by Bartlett's test of Sphericity of 0.00 that was less than 0.05, allowing the research to proceed with factor analysis. The results of Total Variance Explained of Exploratory Factor Analysis (EFA) indicated that five factors in the initial solution with eigenvalues greater than 1.0 have been extracted with the cumulative percentage of 64.83%. All the 24 items of these five web-based factors were retained based on the results of component matrix with factor loading ≥ 0.5. However, based on result of EFA, the original five web-based factors were reorganized and renaming as follows:

- Factor 1 - Information Quality. This factor was retained as in the proposed research framework. 7 items of Information Quality were accurate information, believable information, timely information, relevant information, information is easy to understand, information at right level of details and appropriate format of information.
- Factor 2 was a combination of items of Fulfilment/Reliability and Customer Service. Hence, it had been renaming as Fulfilment/ Reliability /Customer Service. The items were receiving the correct products, on-time delivery, and product represented accurately by the site, responding to customer needs, problem solving and prompt reply.
- Factor 3 consists of four items proposed in the original items in Website Design factor, and it was retained as Website Design. The items were good selection, competitive prices, understanding the needs of customers and comfort and ease of surfing.
- Factor 4 consists of four items proposed in the original Web Design factor. The items were in-depth information, time efficiency, ease in completing transaction and level of personalization. Therefore, it was labelled as Quick and Details.
- Factor 5 was Privacy/Security. Three items of originally proposed in the Privacy/Security factor – privacy protected, feel safe in transaction and adequate security – were retained.

Reliability coefficient assessing the consistency of the entire scale, in which *Cronbach's Alpha* is the most often used; and it is generally agreed that Cronbach's Alpha should exceed 0.70. In this study, the alpha value for the all five factors was 0.940. Specifically, the Cronbanch's Alpha values for Information Quality was .902, Fulfilment/Reliability/Customer Service (.842), Website Design (.828), Quick and Details (.814), and Privacy/Security (.865) respectively.

4.3 Examining web-based factors affecting online purchasing behaviour

As presented in Table 2, result of Multiple Regression Analysis indicated that Information Quality (β=.292, p-value<0.05), Quick and Details (β=.151, p-value<0.05) and Privacy/Security (β=.017, p-value<0.05) were positively and significantly affected online purchasing behavior. The effect of Fulfilment/Reliability/Customer service (Sig. .109, p-value<0.05) and Web Design (Sig. .569, p-value<0.05) of web-based factors on online purchasing behavior was insignificant.

5.0 Discussions

Based on the EFA analysis, web-based factors that are considered by ICT students when they do online purchase were Information quality, Web Design, Fulfilment/Reliability/Customer service, Quick and Details and Privacy/Security. The result of Multiple Regression indicated that Information quality, Quick and Details and Privacy/Security were positively and significantly affected the online purchasing behaviour of the students. Based on these findings, the major implication of this study can be discussed in the following three perspectives:

- **Issue of Web Design factor** – Four items of the originally proposed Web Design factor were found to have insignificant effect on online purchase behaviour. This is consistent with Syed, et.al [5] who found insignificant relationship between Web Design and online purchasing behaviour. Young consumers of ICT students' perceived characteristics of web site design as less important factor that would likely to influence their online buying behavior. These characteristics include elements of good selection, competitive prices, understanding the needs of customers and comfort and ease of surfing feature in the websites. In contrast, four items of Quick and Details (originally there were items of Web Design factor) were positively and

Table 1: Rotated Factor Matrices of Perceived Factors affecting Online Purchase by Online Buyers

Variables		Factors 1	2	3	4	5
IQ1	The website provides accurate information.	**.688**	.209	.140	.284	.286
IQ2	The website provides believable information.	**.766**	.105	.110	.120	.292
IQ3	The website provides timely information.	**.770**	.159	.225	.041	.213
IQ4	The website provides relevant information	**.726**	.240	.183	.199	.094
IQ5	Information provided was easy to understand	**.635**	.322	.317	.172	.100
IQ6	The website provides information at the right level of detail	**.536**	.435	.072	.379	.031
IQ7	The website presents the information in an appropriate format	**.555**	.390	.040	.401	.080
FR8	I received my product which I ordered from the site	.277	**.617**	.218	.179	.133
FR9	I received my product by the time promised by the company	.371	**.575**	.196	.092	.152
FR10	I received the product represented accurately by the site	.221	**.655**	.261	-.012	.177
CS11	I feel the company was willing and ready to respond to customer needs.	.174	**.724**	.106	.139	.150
CS12	The website showed a sincere interest in solving the problem when I face it.	.115	**.712**	.088	.190	.100
CS13	The inquiries were answered promptly	.089	**.623**	.118	.319	.285
WD14	The website has good selection	.041	.306	**.695**	.309	.157
WD15	The site has competitive prices	.075	.262	**.794**	.118	-.056
WD16	The website understands my needs	.348	.145	**.722**	.141	.123
WD17	I feel comfortable in surfing this site	.344	-.008	**.669**	.273	.073
WD18	The website provides in-depth information	.305	.219	.184	**.726**	.039
WD19	The site doesn't waste my time	.123	.154	.306	**.734**	.158
WD20	It is quick and easy to complete a transaction at this website	.260	.193	.266	**.554**	.167
WD21	The level of personalization at site is about right, not too much or too little	.181	.262	.486	**.503**	.293
PS22	I feel my privacy was being protected by this site	.112	.357	.084	.151	**.766**
PS23	The transaction of the website makes me feel safe	.237	.236	.064	.056	**.850**
PS24	The website provides adequate security features	.350	.092	.097	.182	**.783**
	Total variation explained by these factors					64.833%
	Cronbach's Alpha			0.940		
	Kaiser-Meyer-Olkin Measure of Sampling Adequacy			0.919		
	Barlett's Test of Sphericity:					
	Approc. Chi-Square			4410.471		
	D.F.			276		
	Significance			.000		

Table 2: Regression between Web-Based Factors and Online Purchasing Behaviour

Model	Unstandardized Coefficients		Standardized Coefficients	t	Sig.	VIF
	B	Std. Error	β			
1 (Constant)	1.069	.216		4.960	.000	
Information quality	.299	.072	.292	4.149	**.000****	2.302
Fulfilment/reliability/customer service	.117	.073	.108	1.608	.109	2.095
Website Design	.035	.062	.037	.570	.569	1.910
Quick and Details	.156	.074	.151	2.121	**.035***	2.363
Privacy/Security	.102	.051	.117	2.007	**.046***	1.574
Note: * p<0.05; ** p<0.01	F statistic			78.546		
	R^2			.346		

significantly affected consumers' online purchasing behaviour. This finding is consistent with the work of Wolfinbarger and Gilly [19] who was suggested that web design is a strong predictive of customer judgments of quality of and attitudes toward the website. The design characteristics of a website that exhibit much influence on young ICT students' online behaviour were in-depth information, time efficiency, ease in completing transaction and level of personalization. In this study, ICT professional valued time efficiency and ease in completing transactions, and this finding support the idea of Ruchi et al. [6] who were highlighted that navigation of a website are the most sought after factors consumers are looking for when dealing with the web-site. This suggests that efficiency in online transaction is important as being addressed in e-SERVQUAL scale [23]. In addition, this group of consumers valued rich product related information that stimulate information processing before they commit in any online purchase decision.

- **Information quality** was found to have significant effect on online purchase behaviour, which is consistent with the study of Ranganathan and Ganapathy [24], Zhang and Von Dran [33]. Considering the online market is highly competitive, and users experience and expectation from online services are fast changing, it was necessary for the online companies to provide quality information in the websites. Two types of information; non-value-added and value-added; should be used by search mechanisms in web-based stores, which can lead to shoppers gaining a feeling of increased convenience, and allow them to make quick and right online purchase decision.
- The finding of this study highlighted that **privacy/security** is important in an online buying situation, which is consistent with the work of Shergill and Chen [8], Wolfinbarger and Gilly [19], Ranganathan and Ganapathy [24], Ahuja, et.al [29], Chen and Tsao et al. [34]. Considering ICT students are information technology literate, issues related to transaction security, financial and identity theft, misuse of personal information are universal regardless of the extent to which online consumers possess knowledge in information or web-based systems. Consumers' perceived privacy/security of online purchasing has a critical effect on their decision making, and therefore, procedure and policy related to safeguard of individuals' personal information and the security of online transaction systems must be well addressed in the B2C market.

6.0 Conclusions

In this research, respondents who are information technology literate were chosen to examine the effect of web-based factors on their online purchase behavior. Information Quality (β=.292, p-value<0.05), Quick and Details (β=.151, p-value<0.05) and Privacy/Security (β=.017, p-value<0.05) of web-based factors were positively and significantly influence the behavior of online purchasing of this group of consumers. Despite meeting the research questions of the study, further study should be carried out to validate this finding by extending involvement of ICT students to cover all Malaysian public universities. Further, specific online shopping website was not determined in this study, and this make very difficult for E-marketers to focus on website improvement. Therefore, specific online shopping website (e.g. eBay) is suggested to cover.

References

[1] Turban E.King D., Lee, J. K. and Viehland D. (2009) Electronic Commerce 2009: A Managerial Perspective. Prentice Hall.
[2] Kenneth C. Laudon and Carol Guercio Traver (2010), 'E-Commerce: Business, Technology, and Society'. PearsonEducation, 6th edition, USA.

[3] The Star Online, More Malaysians Shopping Online, Retrieved May 18 2012, from http://thestar.com.my/metro/story.asp?file=/2012/5/18/metrobiz/11306418&sec=metrobiz
[4] John and Lim (2001), 'Drivers and Impediments to E-Commerce in Malaysia', Malaysian Journal of Library & Information Science, Vol. 6, no. 2, December 2001: 1-19
[5] Syed, Zaharah, Hishamuddin and Nilufar (2008)'Young consumers online shopping: an empirical study', Journal of Internet Business, Issue 5, pp. 81-98.
[6] Ruchi, Ashish and Gupta. (2010) 'To study the impact of website factors on consumer online buying behaviour in India', IJMMR, Volume1, Issue 1, pp. 1-16.
[7] Zeithaml, Parasuraman, and Malhotra, (2002) "A Conceptual Framework for Understanding e-Service Quality: Implications for Future Research and Managerial Practice," MSI Monograph, Report # 00-115.
[8] Shergill, G.S. & and Chen, Z. (2005). Web-based shopping: Consumers' attitudes towards online shopping in New Zealand, Journal of Electronic Commerce Research, VOL. 6, No.2.
[9] Yam B. Limbu, Marco Wolf and Dale L. Lunsford (2011), 'Consumers' perceptions of online ethics and its effects on satisfaction and loyalty', Journal of Research in Interactive Marketing, Vol. 5 No. 1, 2011, pp. 71-89.
[10] Grace and Chia (2009), 'Factors influencing satisfaction and loyalty in online shopping: and integrated model', Online Information Review, Vil. 33 Iss: 3 pp. 458-475.
[11] Nunes, P.F.(2011), "Marketing: Dazed and Confused", Outlook Journal, No. 1, pp. 41-47.
[12] Feller, B. (2003). Students Rule Net Population. Marketing News, 34 (24).
[13] Haque, A. and Khatibi, A., (2005). E-shopping: Current Practices and Future Opportunities towards Malaysian Customer Perspective. Journal of Social Sciences, vol. 1, Issue 1, pp. 41-46.
[14] Na Li and Ping Zhang (2002), 'Consumer Online Shopping Attitudes and Behaviour: Assessment of Research', English Americas Conference on Information Systems, pp.508-517.
[15] Ajzen, I. and Fishbein, M. (1980), Understanding Attitudes and Predicting Social Behaviour, Prentice-Hall, Englewood Cliffs, NJ.
[16] Ajzen, I. (1991) "The Theory of Planned behaviour", Organizational *Behaviour and Human decision Processes*, Vol. 50, 179 211.
[17] Davis, F. (1985). A technology acceptance model for empirically testing new end-user information systems: theory and results. Unpublished Doctoral dissertation, MIT Sloan School of Management, Cambridge, MA.
[18] Ajzen, I., (1985), from intentions to actions: A theory of planned behaviour. In J.Kuhl, &J.Beckmann (Eds.), Action control: From cognition to behaviour. New York: Springer-Verlag.
[19] Wolfinbarger, M. & Gilly, M. C.(2003). eTailQ: dimensionalizing, measuring and predicting etail quality, Journal of Retailing, Volume 79, Issue 3, Pages 183-198.
[20] Loiacono, E. T., Watson, R. T., & Goodhue, D. L. (2002). A 12-factor quality super-model from a user's perspective. *American Marketing Association: Winter Marketing Educators' Conference 13* (Austin, Texas: American Marketing Association, 2002): 432-438.
[21] Barnes, S. and Vidgen, R., 2003. Interactive e-government: evaluating the web site of the UK Inland Revenue. *Journal of Electronic Commerce in Organizations*, 2 (1).
[22] Yoo and Naveen Donthu (2001). Developing a Scale to Measure the Perceived Quality of an Internet Shopping Site (SITEQUAL). *Quarterly Journal of Electronic Commerce*, 2 (1), 31-47.
[23] Zeithaml, Parasuraman, and Malhotra, "A Conceptual Framework for Understanding e-Service Quality: Implications for Future Research and Managerial Practice," MSI Monograph, Report # 00-115, 2002.
[24] Ranganthan, C. and Ganapathy, S. (2002) 'Key dimensions of business-to-consumer Web sites', Information and Management, Vol. 39, pp. 457-465.
[25] Vijayasarathy, L. and Jones, J.M. (2000) 'Print and Internet catalog shopping: assessing attitudes and intentions', Internet Research: Electronic Networking Applications and Policy, Vol. 10, No. 3, pp. 191-202.
[26] Parsuraman, A., Zeithaml V. & Berry L. (1985), A conceptual model of service quality and its implications for future research, Journal of Marketing, Vol 49, pp 41-50.
[27] Chiu, C.M., Chang, C.C., Cheng, H.L., Fang, Y.H. (2009), "Determinants of customer repurchase intention in online shopping", *Online Information Review*, Vol. 33 No.4, pp.761-84.
[28] Roman, S. (2007), "The ethics of online retailing: a scale development and validation from the consumers' perspective", *Journal of Business Ethics*, Vol. 72 No.2, pp.131-48.
[29] Ahuja, M.K., Gupta, B., Raman, P. (2003), "An empirical investigation of online consumer purchasing behavior", *Communication of the ACM*, Vol. 46 No.12, pp.145-51.
[30] Szymanski, D.M., and Hise, R.T. "e-Satisfaction: An initial examination", Journal of Retailing (76:3) 2000, pp 309-322.
[31] Heijden, Hans v. d., Tibert Verhagen, and Marcel Creemers (2003). "Understanding online purchase intentions: contributions from technology and trust perspectives," *European Journal of Information Systems*, 2, 41-48.
[32] Hair, Joseph, F., Jr., Babin, Barry, Money, Arthur H., and Samouel, Philip. (2003). Essentials *of Business Research Methods*. Wiley: Malloy Inc., USA.
[33] Zhang, P., and von Dran, G. M. (2000). Satisfactor and dissatisfactors: A two-factor model for website design and evaluation. *Journal of the American Society for Information Science*, 51(4). pp 1253-1268.
[34] Chen, Tsao, Lin and Hsu (2008), 'A conjoint study of the relationship between website attributes and consumer purchase intentions'.PACIS 2008 Proceedings. Paper 224.

IOP Conf. Series: Materials Science and Engineering **46** (2013) 012039 doi:10.1088/1757-899X/46/1/012039

Design The Cervical Cancer Detector Use The Artificial Neural Network

Dwi Intan Af'idah, Eko Didik Widianto, Budi Setyawan
Computer System Engineering, Diponegoro University
Jalan Prof. Sudharto, Tembalang, Semarang, Indonesia
E-mail: dwiintanafidah@yahoo.com

Abstract. Cancer is one of the contagious diseases that become a public health issue, both in the world and in Indonesia. In the world, 12% of all deaths caused by cancer and is the second killer after cardiovascular disease. Early detection using the IVA is a practical and inexpensive (only requiring acetic acid). However, the accuracy of the method is quite low, as it can not detect the stage of the cancer. While other methods have a better sensitivity than the IVA method, is a method of PAP smear. However, this method is relatively expensive, and requires an experienced pathologist-cytologist. According to the case above, Considered important to make the cancer cervics detector that is used to detect the abnormality and cervical cancer stage and consists of a digital microscope, as well as a computer application based on artificial neural network. The use of cervical cancer detector software and hardware are integrated each other. After the specifications met, the steps to design the cervical cancer detection are: Modifying a conventional microscope by adding a lens, image recording, and the lights, Programming the tools, designing computer applications, Programming features abnormality detection and staging of cancer.

1. Intoduction

Cancer is one of the non-infectious disease which become a public health issue not only in Indonesia but also in the world. In the world, 12% of the deaths is caused by the cancer, and it has become the second position in cause of death after the cardiovascular disease. In Indonesia, the prevalence number of cancer and tumor is 4.3 of each 1000 people. (Riskesdas, 2007)

According to Prof. Tjandra Yoga, the Director General of P2PL, Ministry of Public, the cervical and breast cancer control sample project had been done in order to detect immediately the cervical cancer. This action used the Visual Inspection method using Acetic acid (IVA). This method was done by rubbing the cervical area with an acetic acid solution. After that, there would be a color change within which shown whether there was an abnormality in the cervical cell or not.

This such early detection method is really practical and cheap. But, it has a low detection accuracy for it can't detect the cancer stage. There's another method which has better detection rather than IVA method, called PAP Smear method. The disadvantage of this PAP Smear method is expensive and needs a well-experienced pathology-citologist, so this method can't be applied widely.

The information technology has developed more and more and it creates smarter and easier-to-use software for many purposes. Some software is used to detect the cervical cancer stage, as well as its abnormality. This software uses the artificial neural network technology. It is able to detect the cancer stage and abnormality through the microscopical image of cervical tissue sample.

The Food and Drug Administration (FDA) has acknowledged the software due to the procedure of the cervical cancer detection. Even more, some researchs show that the cancer stage and abnormality detection through this application software has better accruracy rather than the conventional PAP smear one. (WebMD2002, HTAC 2002)

According to the case above, considered important to make the cancer cervics detector that is used to detect the cervical cancer stage and abnormality and consists of a digital microscope, as well as a computer application based on artificial neural network.

The women's knowledge about this disease are still low. According to the survey which involved 5,423 Asian women within 9 countries--including Indonesia, it has been proven that only 2 percents of the women who have the knowledge about HPV infection. (Emilia, 2010)

The women's knowledge level about the cervical cancer also affects to the less women that take a cervical examination. In fact, it's necessary to determine whether the cervical cells is normal or not. If there's an abnormality, it can be detected earlier, so it becomes easier to both prevent and to cure. (Diananda, 2008)

Therefore, it needs a tool that is able to detect properly and practical. In other words, the women can easily check their cervical condition.

1. Material And Method

The cervical cancer detector uses software and hardware integrated each other. Below is the explanation about the specifications of the required hardware and software.

- Hardware
  - a. desktop computer/ notebook is used to make subsystem of computer application and to do programming when subsystem acquired image.
  - b. Microcontroller ATMega 32 as a controller in digital microscope.
  - c. Lens and image recording to caputre the cervics tissue
  - d. Microcontroller Digital Signal Processing (DSP) for image processing results of capturing a digital format (image acquisition subsystem)
  - e.Serial communication module to transmit the results of image capturing to the computer
- Software
  - a. AVRStudio 4, to do programming on the controller, DSP microcontroller and serial communication module
  - b. NetBeans, to make the application interface
  - c. Plug-in NeurophStudio, to make detection feature on an application made

After the specifications met, the followings are the step to design the cervical cancer detection:

a. Modifying a conventional microscope by adding a lens, image recording, and the lights
b. Programming the controller lens
c. Programming the serial communication modules between the microscope – computer
d. Designing computer applications
e. Programming features abnormality detection and staging of cancer
f. Do training to the application in order to identify abnormalities and the stage of cancer
g. Testing the tool

Below is the steps in flowchart :

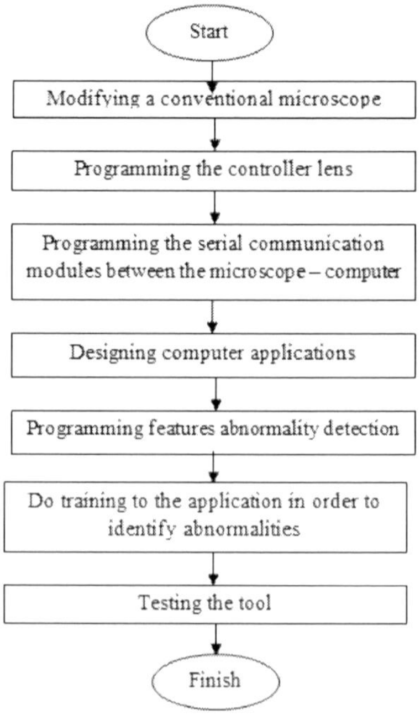

Figure 1. The flowchart to design system

Early step Modifying a conventional microscope by adding a lens, image recording, and the lights. This is done in order to obtain clear images and easy process. Additional lenses are used in order to correct the image acquisition. And image recording needed for a microscope capable of recording images captured. The lamp used to illuminate the object in order to be seen clearly.

Figure 2. Modifying a microscope

The next step is to control the lens. This used to wear can be adjusted as needed. After that, it needs special programming so that the serial communication modules between the microscope – computer can be connected. It is that image in the microscope can be directly sent to a computer and image processing will be done next.

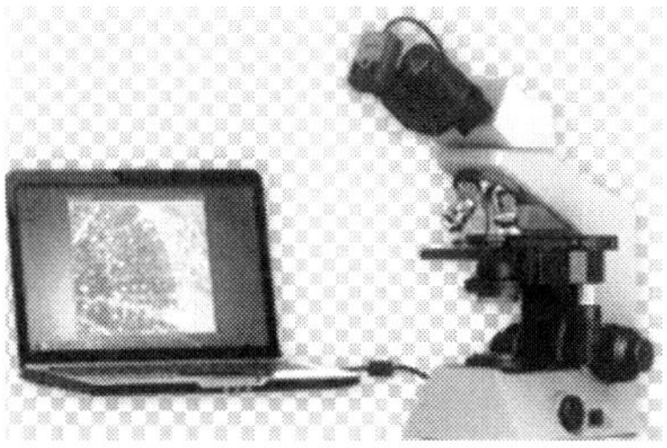

Figure 3. Connect a microscope and computer

Design of computer applications is done using the artificial neural networks. The artificial neural network is an information processor system which imitates the characteristics of biological network. This network is made by generelizing the mathematical function of some biological tissues, and has the following characteristics:

- The information processing occurs in many simple element, that is neuron
- The signal is being delivered between the neurons through the connectors
- The connector between neurons has a value which can strengthen or weaken the signals
- The given value can be done through a training or learning algorithm
- To determine the outputs, each neuron uses the activation function which imposed on the number of the received inputs. The outputs value is being compared with the treshold

(Siang, 2009)

From analysis or the artificial neural network can be in the program if the image is a regular cell or cancer cells. If you find any differences in cell (found cervical cancer), it can be programmed cancer cell is in what stage.

The next , do training to the application in order to identify abnormalities and the stage of cancer the artificial neural network is used in order to understand the pattern of input that result in the appropriate output.

Testing is done by collecting a tissue sample screening results. The results of screening are then processed in the image acquisition subsystem in order to become a digital format and can be read by a computer. Furthermore, the image will be processed by the software and are classified into the categories of normal, low-grade squamous intraepithelial lesion (LSIL), or high-grade squamous intraepithelial lesion (HSIL).

3. Conclusion

1. The use of such software in a cervical cancer detection procedure has better accuracy than the conventional Pap smears. The conventional pap smears method needs an experienced pathologist to detect cervical cancer. But, the number of them bit in order the detection of cervical cancer use pap smears make accurancy is not good.
2. Synchronization between hardware and software is necessary to obtain the optimal system function. It is caused tranfered image from the microscope to the software will not be done if the synchronization between the hardware (microscope) and the software is not good.
3. Neural network that works like the expert experts hail obtained optimal

References

[1] Diananda. 2009. Mengenal Seluk - Beluk Kanker. Jogjakarta: Katahati

[2] Hislop, T. G., Band, P. R., Deschamps, M., Clarke, H. F., Smith, J. M. & Ng, V. T. Y. (1994). Cervical Cancer Screening In Canadian Native Women: Adequacy of The Papanicolaou. The Journal of Clinical Cytology and Cytopathology. 38(1). 29-32.

[3] Mat-Isa, N. A., Mat-Sakim, H. A., Mashor, M. Y. & Othman, N. H. (2003). A Review on Application of Neural Networks In Cancer Disease : Case of Cervical and Breast Cancer. Proc. of The Int. Conf. on Robotics, Vision, Information and Signal Processing. 538-543.

[4] Samadi. 2010. Yes, I Know Everything About Cervical Cancer!. Jakarta: Tiga Kelana.

[5] Siang, JJ. 2009. Jaringan Syaraf Tiruan dan Pemrogramannya Menggunakan Matlab. Yogyakarta : Andi

[6] Suprapto. 2010. Penggunaan Pengolahan Citra Digital pada Pemeriksaan Pap Smear dalam Pendeteksi Kanker serviks. Jurnal Teknologi Informatika. 1(2): 149-168

5S PROGRAM TO REDUCE *CHANGE-OVER* TIME ON FORMING DEPARTMENT (CASE STUDY ON CV PIRANTI WORKS TEMANGGUNG)

Septika Rosiana Dewi[1], Budi Setiawan[2] and Susatyo Nugroho W P[3],
[1,2] Undergraduate Student of Industrial Engineering Department, Faculty of Engineering, Diponegoro University.
[3] Lecturer of Industrial Engineering Department, Faculty of Engineering, Diponegoro University.

E-mail: septika.rosiana.dewi@gmail.com

Abstract. *Productivity is one aspect that determines the success of a company in the competitive world of business. There are seven main types of activities that do not have value-added in manufacturing processes such as overproduction, waiting time, transportation, excess inventory, unnecessary motion and defects. The whole activity is a waste (waste) that can cause harm to the Company. Therefore, in production activities is important to pay attention so that the objectives of production productivity can be achieved. Problems experienced by CV Piranti Works is a production target is not achieved resulting in a lost sale raises the cost of which can cause harm to the Company. From the analysis conducted major known cause of the problem is the length of time required for changeover. This is supported by the high non-value added activity in the changeover activities. Lean Manufacturing is an approach to make system more efficient by reducing waste. This study refers to the book compiled by Takashi Osada (2004) and several other references. In this research used method 5S (Seiri, Seiton, Seiso, Seiketsu, and Shitsuke) for the of forming departement. The purpose of this research is to design a work environment using the 5S method (Seiri, Seiton, Seiso, Seiketsu, and Shitsuke) and make arrangement of equipment and working tool cabinet design with TRIZ methods. From these results, is expected to eliminate or reduce of non-value added activity and improved the changeover time so as to meet production targets completion of the company.*
Keywords: Production target, 5S, changeover, non value added activity, Lost sale

1. Introduction

Productivity is one of the aspects that determine the success of a company in the competitive business world that increasingly stringent. According to Liker (2006) Toyota has identified seven major types of activities that do not have an value added in both business and the manufacturing process, such as overproduction, waiting time, transportation, excess inventory, unnecessary motion and defects. All of these activities is a waste which can result in losses to the Company. CV.PIRANTI WORKS is one of the companies that produce by order (job order). The products include: wooden radio, wooden toys and stasionary / office stationery supplies. The final products should be maintained and even improved quality. To improve the quality of the product need a good working environment, clean and ease to reach equipment and raw materials in the production process. This company has five working groups in which each working group addressing a different job. Of the five working groups, Forming Department is a working group that has greatest percentage of the number and frequency of reject, compared with other working groups drawn from the radio product type Ikono data in July-December 2010, with percentage is 50.01% and the number of reject reject frequencies of 58.48%. So it can be said that the Forming Department is critically working group.

CV Piranti Works implement *make to order* system. During this time, most of the orders received from distributors abroad. This is because the company is more focused on overseas markets. So far, CV Piranti Works often making shipping products delays to several major distributors, because production quantities are not able to achieve the targets set. CV Piranti Works determined that the lead time to complete the order from the distributor is two months or 60 days (for the product with the type of radio). Based on the experience of the owners along that within one month of work to complete the radio product type "Ikono" by the number of 350 pieces. Ikono product is a product of the most booked for this. Here is an example case of delay problems faced by the Company, in which the order comes from the Main Distributor in Germany (Area ware.com) in 2010 order. During this time, in case of delays occurred, the distributor would receive compensation from the CV. Piranti Works. This case data shown in Table 1 below :

Tabel 1. Data of Delay Production Target , Product Type Radio "Ikono"

No	Order Time	Quantity	Finished Target	Actual Quantity	Number of delays
1	Maret	100	Mei	100	0
2	Mei	200	July	185	15
3	Agustus	104	Oktober	90	14

(Source : Bagian Produksi CV. Piranti Works tahun 2010)

From the observations that have been made, the part-forming department is critical working group, so that the object of this study focused on this group. On this working group, identification of cause problems made by Fishbone diagram. Having obtained the issues that often comes up then did the analysis of the effect of the failure to determine the priority of solving the root causes using FMEA (Failure Mode and Effects Analysis). Robin E. McDermott (2009) say FMEA is a technique used to locate, identify, and eliminate potential failures, errors, and known issues of the system, design, process, or service before they reach the customer. FMEA Process in this research is to detect the risks identified during the process.

By identifying the effects, the causes of waste and control methods used can be calculated weighting value to see the waste. The cause of the greatest weight is the cause of the waste of potential waste of an opportunity that has the most impact on the system and the potential for reduced. From FMEA result of analysis obtained the main potential failure, that was very excessive lead time in machining process on part-forming department. This was because a redundant *change-over time.* The amount of changeover time is due to the too manye non-value added activities such as searching tools, xcessive transportation necessary to get the equipment for placement anywhere. So to resolve the root cause of problems that occur, this study is focused to address the high change-over time. Figure 1 below illustrates the graph data summary results are expected to represent the activity changeover happens.

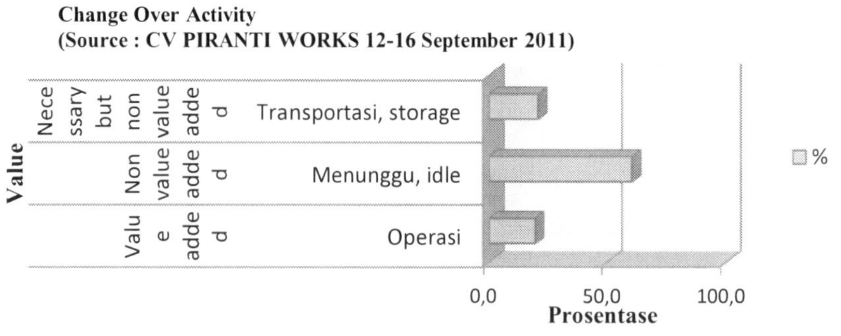

Figure 1. Change-over Performance

The purpose of this research is designing a work environment using the 5S method in part-forming department (working group), constructing equipment and facilities arrangements and designing a Tool-Rack, to eliminate non-value added as tool searching time or other equipment, and operators transportation, to reduce the limitations of the current system is the placement of the equipment.

2. RESEARCH METHODOLOGY

This study uses data collection using interview and direct observation in the field so after the design of the data obtained with the 5S method to do. The research methodology is shown in Figure 2.

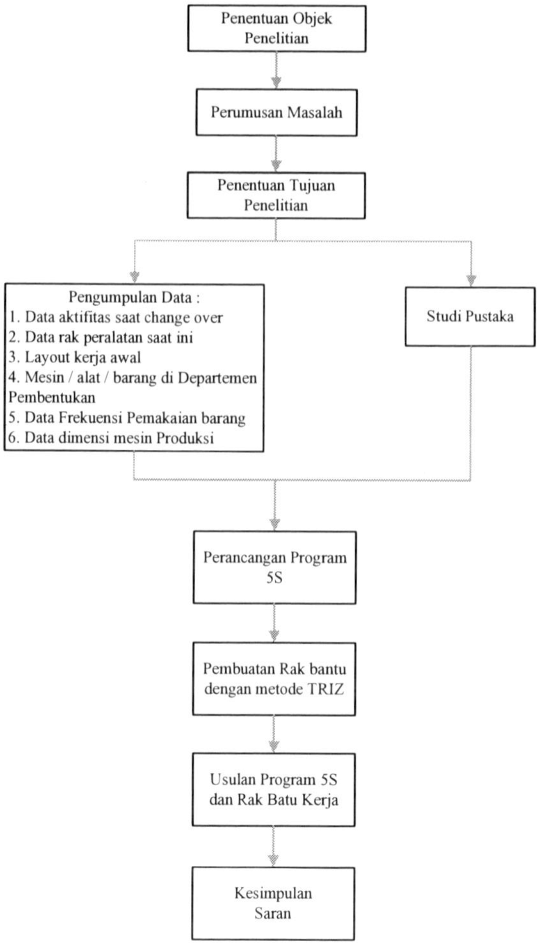

Gambar 2. Research Metodology

3. LITERATURE STUDY

3.1. Lean manufacturing

Lean manufacturing is a concept that aims to eliminate waste in all areas of production including product design, supplier networks, and factory management relationship with the customer. This concept comes from Japan. With the implementation of lean manufacturing, is expected to produce a quality product by reducing waste there. Definition of waste here is all the activities and resources that do not create or provide added value in the process. While the process of adding value (value-adding activities) is the process of conversion / transformation of raw materials, changes in the shape or quality: changes in raw material into a particular part or product.

Here are some techniques that can be used to apply the concept of lean manufacturing : Kanban / Pull Production, SMED (single minute exchange of die), 5S, Heijunka (Production Smoothing), Takt Time, Flow and Cellular Production.

3.2. 5S Program as Implementation of Kaizen Philosophy

Five S (5S) are five simple steps that are used to improve the working conditions that are named based on the initial letters of the five steps in Japanese (Masaaki Imai, 2001).

- ***Seiri :*** According to Takashi Osada (2004:23) that Seiri means organizing everything, sorting according to certain rules or principles. 5S means necessary to distinguish between what is needed, take a firm decision, and implement management stratification is not required to dispose of them. The emphasis here is the stratification and management seek to eliminate the causes that are not needed and eliminate the cause of a problem before occured.
- ***Seiton*** **:** According to Takashi Osada (2004:25) that Seiton means storing stuff in the right place or the right layout so that it can be used in a state of surprise. This is a way to eliminate the search process. The emphasis here is functional management and removal of the search process. If everything is kept in its place by the quality of security, it means you have a tidy workplace
- ***Seiso :*** According to Takashi Osada (2004: 27) that Seiso means cleaning stuff so it becomes clean. In terms of 5S, means taking out the trash, dirt and foreign matter and clean up everything. Cleaning is one form of the examination. Here the preferred cleanup as checks and creates a workplace that does not have a blemish.
- ***Seiketsu*** **:** According to Takashi Osada (2004: 28), in terms of 5S, Seiketsu or stabilization means constantly and repeatedly maintain sorting, structuring, and cleansing. Thus, include personal and environmental hygiene. In this matter, especially the stabilization of 5S and visual management. Innovation and unified visual management used to achieve and maintain conditions that strengthened so as to always act quickly and appropriately.
- ***Shitsuke*** **:** According to Takashi Osada (2004: 29), Shitsuke means of training and ability to do what you want to do even though it is difficult to do. In 5S terms, this means that the implant (or have) the ability to do things the right way. In this case, the emphasis is to create a workplace with good habits and behavior. By teaching each one what to do and ordered everyone to do it, then the bad habits will be wasted and will form good habits. People practice to make and comply with regulations.

4. RESULTS AND DISCUSSION

4.1. 5S Designing

The design is divided into categories by each S.

a. ***SEIRI***

Aiming to determine the items required and not required in Part-Forming Department Area.

- **Separate and determine the inter-goods groups.** In Table 2 below recapitulation sorting items in Part-Forming Department Area.

Table 2. Recapitulation Sorting Items

No	Tools/Machine	Category
1	Pallet komponen, gerinda, Dongkrak hidrolik, Tong sampah, Kardus alat kebersihan, Tempat duduk plastik, Isolasi, Mata Pahat, drill dan kertas amplas mesin amplas belt, Mistar, Jangka sorong, Penggaris siku, Kunci Inggris, Palu, kaca mata mesin, minyak pelomas dll, lem alteco, kertas amplas tangan, peralatan kerja pendukung, klem C.	Put in the shop floor area
2	*Table saw*, Ragum kecil, dan Lampu penerangan tambahan	Store in equipment and machinery warehouse.

IOP Conf. Series: Materials Science and Engineering **46** (2013) 012040 doi:10.1088/1757-899X/46/1/012040

b. ***SEITON***

Aiming to save and put things in place and the proper layout.

- **Designing equipment lay-out**

To design this made a bulletin board. Bulletin board here used to attach an explanation batch production running or to be run so as to eliminate idle activity and transport operators to find the head unit. The notice board is also used as a means of attachment to another announcement. The illustration shown in Figure 3 below.

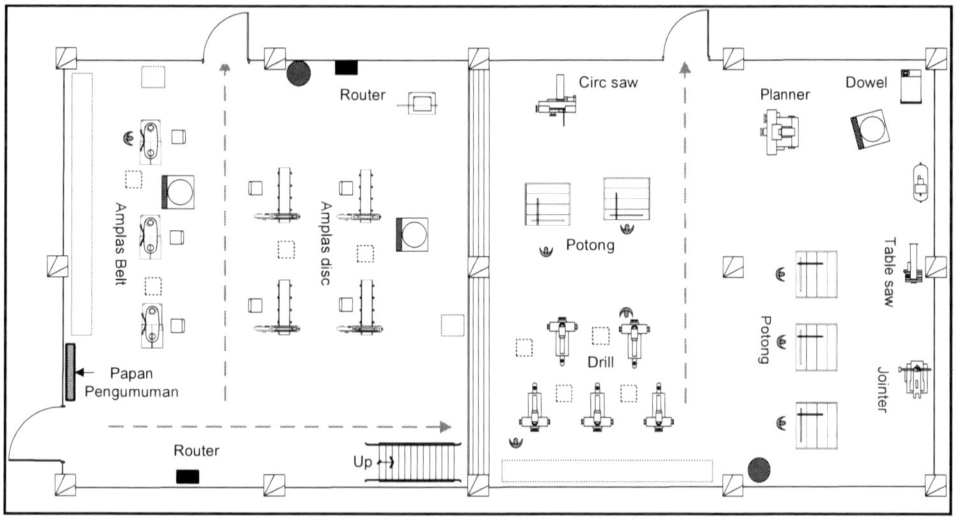

Figure 3. Position of Announcement Board

- **Labeling of jig equipment**

The labeling here is intended to provide a clear identification of the goods so as to facilitate the search. But the administration should ensure the name is obvious, not even confusing to workers who want to wear them. Here is an example of the label on the jig work equipment shown in Figure 4 below.

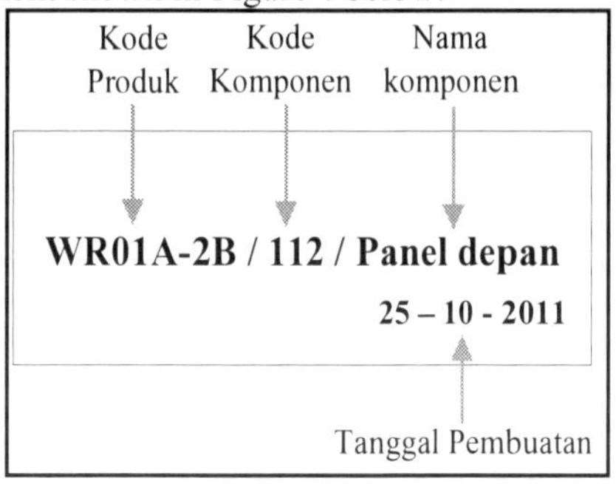

Figure 4. Jig Labeling

- **Production Area Marking**

Making the boundary line location is only done on a production machine and some tools like the WIP storage racks. Figure 5 below will explain about giving part-forming Area boundary.

IOP Conf. Series: Materials Science and Engineering **46** (2013) 012040 doi:10.1088/1757-899X/46/1/012040

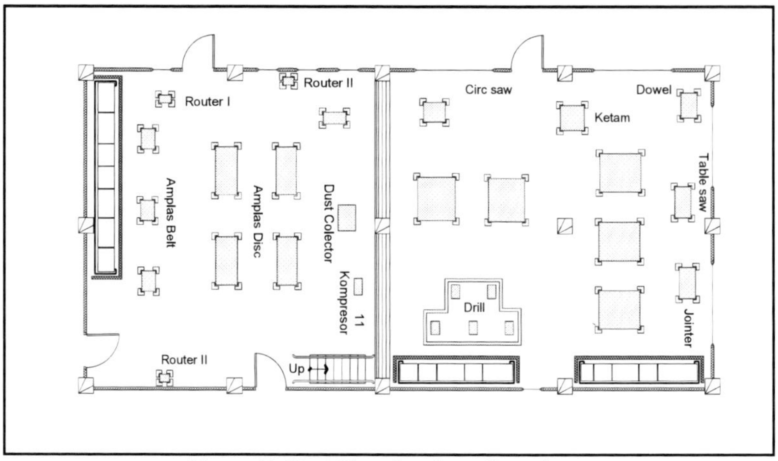

Figure 5. Part-Forming Area Boundary.

- **Work Equipment Rack with TRIZ methods**

Here are the results for extra equipment rack design work. The equipment rack design results can be seen in Figure 6 below.

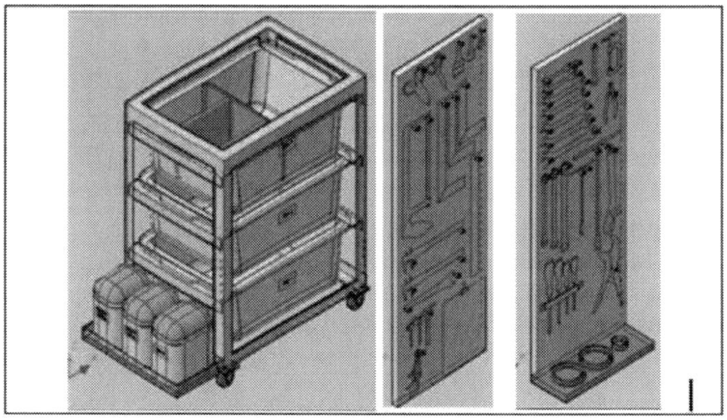

Figure 6. Work-Equipment Rack.

c. ***SEISO***

Aimed at maintaining the cleanliness of work areas

- **Checklist design of equipment and machines work cleaning.**

Checklist is used to directly monitor the performance of the application Seiso. Here is a sample checklist of cleaning equipment and machinery work. The results of the draft checklist for applications Seiso contained in Figure 7 below.

Kategori		**Keterangan**
Lantai	Tidak ada barang cacat (*reject*) berserakan	
	Garis batas tidak patah	
	Sudut - sudut ruangan bersih	
Mesin	Tidak terdapat serbuk / bubuk kayu	
Alat kebersihan	Tergantung rapi di tempat penyimpanan	
Meja kerja	Bersih dari serbuk / bubuk kayu	
	Komponen tidak berserakan	
Jig	Tertata pada tempatnya	
	Pemberian sesuai dengan ketentuan	
	Cara penyimpanan benar	
Rak	Bersih dari serbuk / bubuk kayu	
	Cara penyimpanan benar	
Peralatan	Tidak ada kotoran	
	Tidak ada label / stiker yang tidak diperlukan	
	Semua memiliki tempat dan disimpan di tempatnya	
	Cara penyimpanan benar	
Keterangan : V : oke X : tidak oke		

Figure 7. Cleaning Checklist Form

- **Making responsibilities lists of work area cleaning.**

  The initial step is to divide the area of production area into sections and allocating responsibility on the sub group. The division of the region can be seen in Figure 8 below.

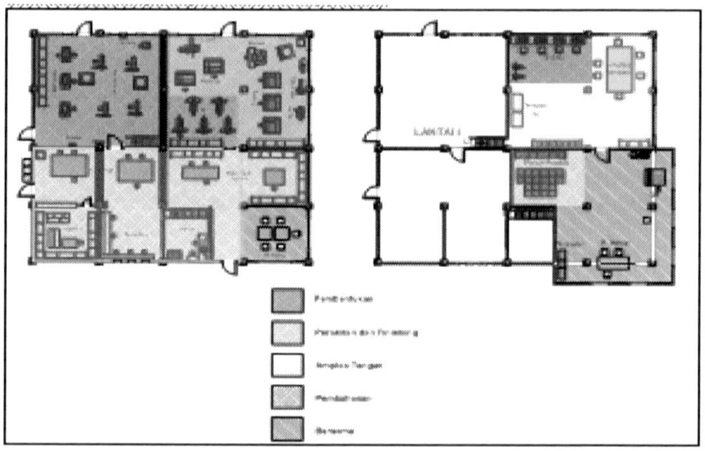

Figure 8. The Division of The Cleaning Region

d. ***SEIKETSU***

Aimed at maintaining the neatness and cleanliness in Part-Forming Area.

- **Determination of the discarded items place.**

  In this case, the supply box to be used as a garbage dump in the process. Garbage in this case is all the dirt including components derived from the rejects parts and non process. Non process the samples considered are not used anymore.

- **Announcement Board Example**

  It aims to make the room look neat with a bulletin board. Surely the correct way and neat attachment. So that the necessary procedures for the settlement procedures of the bulletin board. The way to do that is in the form of posters placed on one wall in the area of part-forming. The poster image can be seen in Figure 9 below

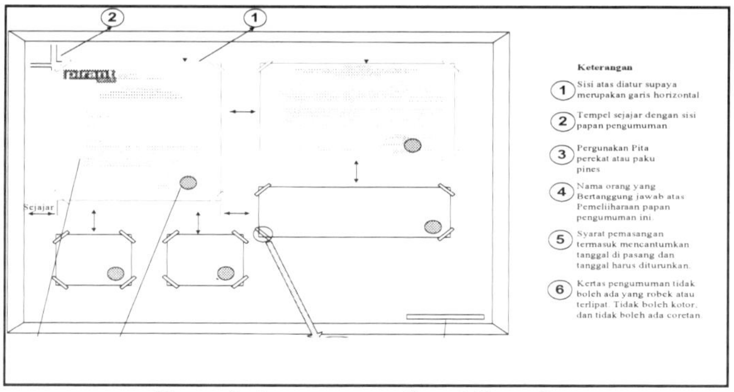

Figure 9. Anouncement Board Procedure

- **Preparation of Work Procedures For Equipment Order.**

  This procedure is associated with a new rack design that uses tools card (cards work equipment). This procedure uses tools such as instruction card and instructions for lending equipment work. The instruction works can be seen in Figure 10 as follows:

piranti works	CV. PIRANTI WORKS	DISUSUN : TANGGAL PEMBUATAN : NO. REVISI :
	INSTRUKSI KERJA	TANGGAL PELAKSANAAN : NO :
DISETUJUI OLEH DEPARTEMEN : MESIN :	PENGGUNAAN *TOOLS CARD*	
PENGERTIAN	*Tools Card* adalah kartu identitas peminjaman yang dimiliki oleh tiap mesin.. Yang berfungsi untuk melacak keberadaan peralatan yang tidak terdapat di tempat penyimpanan..	
PELAKSANA	Operator Mesin pada kelompok Pembentukan	
	A. Pengambilan *Tools Cards* 1. Periksa warna kartu, apakah warna yang diambil sudah Sesuai dengan ketentuan berikut ini.. ✓ Mesin Amplas belt : Merah ✓ Mesin amplas disc : Kuning ✓ Mesin drill : Hijau ✓ Mesin Router 1 : Biru ✓ Mesin Router 2 : Ungu ✓ Mesin Potong : Abu - Abu ✓ Mesin Bubut : Putih 2. Ambil *tools card* dari tempat penyimpanan *tools card* dengan jumlah peralatan yang akan diambil **B. Peletakan *Tools Cards*** 1. Setelah peralatan diambil, letakkan *tools card* pada pengait tempat peralatan tersebut disimpan, jika yang diambil adalah peralatan kerja 2. Letakkan *tools card* pada pegangan laci rak mata pahat tempat mata pahat disimpan jika yang diambil adalah mata pahat **C. Pengembalian *Tools Cards*** 1. Ambil *tools card* lalu kembalikan peralatan di tempat *tools card* tersebut digantungkan 2. Pastikan posisi penyimpanan alat sudah benar 3. Letakkan *tools card* sesuai poin 3 dan 4 jika akan melakukan peminjaman lagi 4. Letakkan kembali *tools card* pada penyimpanan awal jika tidak melakukan peminjaman lagi 5. Pastikan posisi kartu ditempat penyimpanan semula sesuai Dengan kelompok warnanya	
CATATAN	Mohon diperhatikan instruksi ini untuk mempermudah pelacakan keberadaan peralatan kerja	

Figure 10. ***Tools Card Procedure***

e. ***SHITSUKE***

Aims to keep the implementation Seiri, Seiton, Seiso, Seiketsu to run sustainable.

- **Daily checksheet for 5S implementation.**

Daily Checksheet is intended for workers in the 5S program so easy working environment that includes the production area can be maintained. The purpose of making a checklist is to assist the operator in performing activities of cleanliness that no tool/machine that has not been cleaned. Another aim is that production managers can easily monitor the condition of the real from the production area. In Figure 11 there Checksheet containing containing disciplinary categories 5S implementation,Aims to keep the implementation Seiri, Seiton, Seiso, Seiketsu to run sustainable.

Kategori Checksheet 5S			Tanggal						Keterangan	Tindakan
	No	Jenis	1	2	3	4	5	6		
Lantai	1	Membersihkan lantai dari kotoran / debu								
	2	Membersihkan sisa potongan material / part (geram)								
	3	Tidak ada barang cacat berserakan								
	4	Memastikan *aisle* dan garis batas mesin produksi								
	5	Memastikan lantai produksi bebas dari kotoran, oli dan minyak								
	6	Ada petunjuk yang jelas siapa yang bertanggung jawab								
Mesin	7	Tidak terdapat serbuk / bubuk kayu								
	8	Membersihkan sisa tumpahan oli dan minyak pelumas pada mesin produksi								
	9	Membersihkan sisa potongan material / part (geram)								
	10	Mesin dapat berjalan dengan baik								
Pallet komponen	11	Memebersihkan kotoran pada pallet								
Alat Kebersihan	12	Tertata pada tempatnya								
	13	Tdak ada yang rusak								
Meja kerja	14	Tidak terdapat serbuk / bubuk kayu								
	15	Komponen tidak berserakan di meja kerja								
Jig	16	Tertata pada tempatnya								
	17	Pemberian sesuai dengan ketentuan								
	18	Cara penyimpanan benar								
	19	Tidak ada yang rusak								

Figure 11. Daily Checksheet For 5S Programs Implementation

5. ANALYSIS

5.1. Analysis of 5S Designing

a. ***SEIRI*** : Designing Seiri begins by sampling the goods or equipment or machinery located in the production area and the area of the experiment. Of data items that are used to classify data obtained by function type of goods / equipment / machinery and the frequency of use of goods. From the classification of the data types of goods /

equipment / machinery is divided into types of items needed and unneeded items. For items that are necessary in the production area amounted to 30 kinds. As for the unnecessary items totaling 3 kinds of table saw, a small vise, and additional lighting lamps. Further action for 21 different types of work equipment / machine work after sorting placed determined in accordance with a predetermined (fixed establishment in the area) and for items not used as table saw, a small vise, and additional lighting lamps removed from the production area and warehouse equipment placed so that if necessary can be easily retrieved.

b. ***SEITON*** : Designing Seiton done in various stages: design placement of equipment on the part of the establishment, labeling the jig, the installation of the line marking in the production area as well as the design of work equipment rack. To design equipment laying on the formation is divided into two kinds of design for machine production and design a bulletin board. Based on the observations made for the placement of permanent production machine types are placed in its original position due to the limited extent of the area and did not interrupt the flow of material available today. The placement of a bulletin board attached explanation here is needed for batch production running or to be run so as to eliminate idle activity and transport operators to look for the head unit. And also that the paper work was not accumulate instructions on the production. Therefore given the notice board. In addition the bulletin board is a necessity for companies, because not all news can be disseminated orally, but some are in writing, hence why bulletin board is very important. Fitting a line to mark the placement of the goods made in all areas of production so that it can be seen that after the installation of the production area boundary markings look more presentable. This is because each area has had a limit so that production machinery can be more easily trimmed if at any time move. Jig labeling is intended to provide a clear identification of the goods so as to facilitate the search. But the administration should ensure the name is obvious, not even confusing to workers who want to wear them. The labeling here can be permanently affixed so stick to the jig. It may be easier to find the jig back if the need for this service which often turned to the R n D before using it and it will take work from the operator.

c. ***SEISO*** : As the concept of Seiso is how to maintain the cleanliness of the floor, machinery and equipment in the production area. The purpose of the design was to create a clean working area. And operators are expected to undertake regular cleaning in order goods or equipment or machinery used keep it clean and prevent damage to goods or equipment or machinery. Seiso is done in two steps, namely the design of list creation workspace cleanup responsibilities. The division here adapted to the working group that is no longer needed so that the grouping of cleanliness so that coordination can be directly done before. The second design are the checklist cleaning equipment and machinery work design. The checklist in this case is used as an observation when the application is run and the initial Seiso used only by the working group, designed Checklist is an example of the formation alone. At a later stage this checklist will be developed into a checksheet for use as a tool 5S monitor to keep running continuously.

d. ***SEIKETSU :*** The basic concept used in designing Seiketsu is how to maintain the production area is always in immaculate condition and clean. The design is done there are 3 stages, namely, the determination of the laying of the items to be discarded, making sample boards and manufacturing procedures to order/borrow equipment. Determination of placement of goods to be disposed done by providing a box that will be used as a garbage dump in the process. In the area of part-forming supplied two box placed at corners and edges of the wall so as not to interfere with the process of considering the labor of currently available work area is not too large. Making an example of bulletin boards are intended for some announcements obtained from the owner or the Production Manager, this announcement contains not only the only production but appeals from owners. Preparation of work procedures for borrowing equipment relates to the design of new equipment rack where the presence of this equipment lending procedures be known who are the borrowers and equipment to facilitate the monitoring equipment.

e. ***SHITSUKE*** **:** Shitsuke is a discipline that must be implemented in order to carry out the procedure Seiri, Seiton, Seiso, Seiketsu can be sustainable. The design is done on the design concept is just Shitsuke daily checksheet for the implementation of 5S, checksheet contains all aspects of the assessment of 5S. Checksheet only examples used in the part-forming department/working group and used it as a tool to evaluate 5S Program.

6. CONCLUSION

The conclusion that can be drawn from this study are:

a. In Seiri, steps taken are: to separate and determine the inter-goods groups in the area of part-forming.
b. In Seiton, steps taken are: designing groundbreaking equipment on the part-forming department, equipment jig labeling, giving the line marking in the area of production and design of equipment racks working with TRIZ methods.
c. In Seiso, steps taken are: list-making responsibility for cleaning the work area, and a design checklist cleaning equipment and machinery work.
d. In Seiketsu, steps taken are: the determination of the place of the items to be removed, making the example boards, and manufacturing procedures to order/borrow equipment.
e. In Shitsuke, steps taken to design daily checksheet for the implementation of 5S.

7. RECOMMENDATION

The following recommendation can be given for the future practical program :

a. Development of a 5S program is not just done one particular workgroup, preferably applied in all production areas.
b. In a preferred design utilizing equipment that can be reused so that costs can be reduced.

8. REFERENCES

[1]. Ahrens, Thorsten. (2006). *Lean production: Successful implementation of organisational change in operations instead of short term cost reduction efforts*, Germany : Lean Alliance.

[2]. Ernawati, Dira. *Pengembangan Model Pengukuran Kinerja Supply Chain di PT. Indomapan*. Jawa Timur : Teknik Industri, FTI-UPN Veteran.

[3]. E, McDermott. Robin, J. Mikulak Raymond, R. Beauregard. (2009). *The Basic of FMEA 2nd Edition*. New York USA, Taylor & Francis Group, LCC.

[4]. Kadarina, Nurlaila. Rizkya, Indah. Hartini, Sri. Saptadi, Singgih. (2009), *Analisis Pemborosan Perusahaan Mebel dengan Pendekatan Lean Manufacturing*. J@TI Undip, Vol IV No 2 Mei 2009

[5]. K, Liker. David Meier, Jeffrey. (2006). *The Toyota Way Field book*. Jakarta : Penerbit Erlangga, PT Gelora Aksara Pratama.

[6]. Osada, Takashi. (2004). *Sikap Kerja 5 S*. Jakarta : Lembaga Manajemen PPM dan Penerbit PPM.

[7]. Ridwan, Asep ; Ratna Ekawati. (2008). *Rancangan Sistem Proses Produksi Dengan Menggunakan Value Stream Analysis Tools (VALSAT)*. Lampung : Seminar Nasional sains dan Teknologi-II.

IOP Conf. Series: Materials Science and Engineering **46** (2013) 012041 doi:10.1088/1757-899X/46/1/012041

Current trend of robotics application in medical

OA Olanrewaju[1], AA Faieza [2] & K Syakirah[3]

[1,2,3]Department of Mechanical and Manufacturing Engineering, Faculty of Engineering, Universiti Putra Malaysia, 43400 UPM Serdang, Selangor, Malaysia

E-mail: [1]lafiolanre@yahoo.co.uk, [2]dr.faieza.upm@gmail.com, [3]syaki_k@yahoo.com

Abstract. The applications of robotics in recent years has emerged beyond the field of manufacturing or industrial robots itself. Robotics applications are now widely used in medical, transport, underwater, entertainment and military sector. In medical field, these applications should be emphasized in view of the increasing challenges due to the variety of findings in the field of medicine which requires new inventions to ease work process. The objective of this review paper is to study and presents the past and on-going research in medical robotics with emphasis on rehabilitation (assistive care) and surgery robotics which are certainly the two main practical fields where robots application are commonly used presently. The study found that, rehabilitation and surgery robotics applications grow extensively with the finding of new invention, as well as research that is being undertaken and to be undertaken. The importance of medical robot in medical industry is intended to offer positive outcomes to assist human business through a complicated task that involves a long period, accuracy, focus and other routines that cannot be accomplished by human ability alone.

1. Introduction

There has been a lot of research carried out in the field of roboticsin this day and age because of the growing demand in the usage of robots, the application of robotics can be seen in almost all sector of the economy e.g. Transport, Manufacturing, Medical, Defence, Entertainment. However the word 'Robot' originated from the Czech word 'Robota' meaning 'Slave Labourer'. In the early 20's Karel Capek a Czechoslovakian dramatist first used the word robot in his play titled 'Rossum's Universil Robots' where he illustrated a robot as a timeless worker, till date there has not been a clear definition of robots. A robot is a mechanically controlled, reprogrammable, multi –use automatic system with several degrees of freedom. This may be either fixed in place or mobile [1].

Robots can be classified using several criteria such as the application area, control techniques, actuators, geometrical configuration, intelligent robots and others. The merits of classifying robots are very huge because it makes it easier in identifying robots to be used for a particular task. In this review we will be discussing the current trends in medical application of robotics.

Medical and healthcare robots are designed for different terrain and tasks. Those that involve the development of wheel chairs, rehabilitation manipulators for assisting disabled and elderly persons can be categorized as Macro-robotics while those that involve the development of tools for surgery such as minimally invasive surgery, image-guided surgery, computer-integrated advanced orthopaedics, stereotactic guidance can be categorized as Micro-robotics. The last category is the Bio-robotics which involves the development of modelling and simulating biological systems for providing a better knowledge of human body and system [2]. Figure 1 shows the classification of robotics according to their application in the medical field.

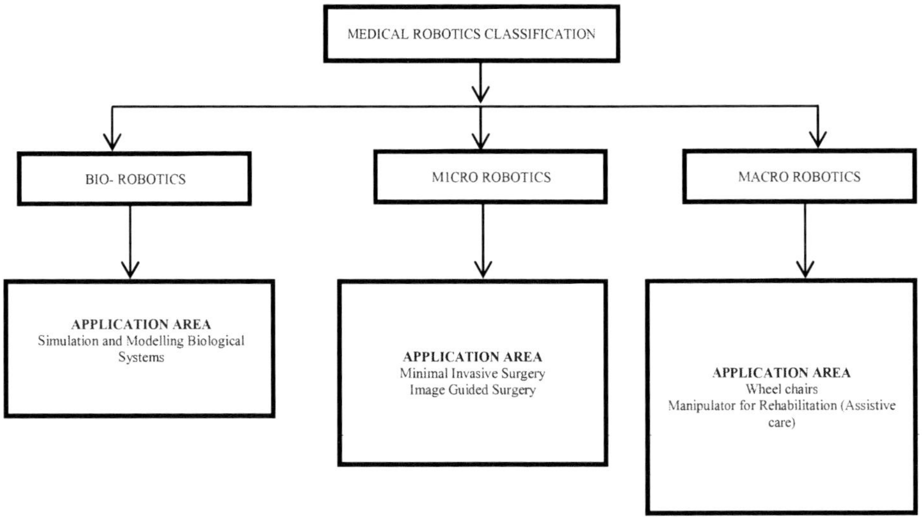

Figure 1. Shows the classification of robotics according to their application in medicine [2].

2. Rehabilitation robotics

Rehabilitation robotics involves the use of assistive devices for disabled and aged people.Examples of such devices are smart wheelchairs, artificial limbs, exoskeletons and dedicated interfaces which involve the replacement of impaired functions by means of advanced prostheses and ortheses to help them go about their day to day activities at home, office, and hospitals.The main motive of robotics in residential care which is also known as rehabilitation robotics is to completely or partially substitute the disabled or elderly user's manipulative function by placing a robot arm between the user and the environment [2].

There have been so many developments in recent years from neurorehabilitation for patients with neuromuscular injuries or diseases to sensory therapy which enables patient relearn how to move in case of chronic or cerebral stroke patients. An example of such robots is the Massachusetts Institute of Technology (MIT) - Manus rehabilitation robot which is now commercially available. It provides so many functions which include assisting, resisting and stretching which are all based on real time response. One of the main features of this robot is that it can acquire data to quantify the recovery of the patients [4]. Figure 2 shows the InMotion 2.0 Shoulder Robot developed by MIT for intensive rehabilitation of patients with stroke or brain injuries; and with limited strength or movement in the upper extremities.

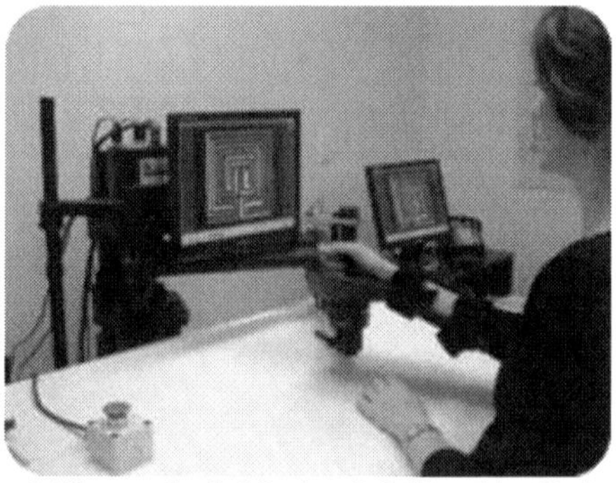

Figure 2. Shows the InMotion 2.0 Shoulder Robot[21].

Personal care for disabled population is another area in which rehabilitation robotics application can be seen. The growth rate of disabled populations which includes those with physical and social disorders which may be developed from birth or age related, demands a growing need for personal care to provide mobile aids ranging from simple self-stabilizing canes for the visually impaired to intelligent/smart wheel chairs for physically disabled to increase mobility and accessibility for them to go about their normal daily activities independently in their homes. A sophisticated device that enables severely disabled to perform tasks like personal toiletry and getting in/out of bed are still in their early development stages. To fulfil the various requirements of the disabled population the Politecnico di Milano designed an autonomous wheelchair control system called (LURCH) which means Let Unleashed Robots Crawl the House.

The LURCH provides options for the user to choose from different autonomous and interfaces ranging from a touch screen, a joystick, an electromyography (EMG) interface and a brain-computer interface (BCI) which transmits the user's intention by analysing the user's brain signals. Thus the LURCH system is not integrated with the wheelchair at the digital control bus stage but only relies on the simulation signals from the joystick in analogue domain [5]. In Prosthetics the National Health Interview Survey indicated that the people using Prosthetics are well over 3.5 million in the U.S.A and the amount is expected to double up to 7.3 million by 2020. Prosthetics are devices used in assisting, supporting a weak or non-functional joint, muscle, limb or substituting for a missing human body part. The use of Prosthetics dates back as early as ancient Egyptians time where a mummy's big toe had been amputated during his life time and replaced with a well-crafted wooden toe. From then till now massive development has taken place from the development of the Belgrade hand [16] and the UTAH arm [17] in theearly 60's and 70's to the C-leg and I-limb which is microprocessor controlled. A good illustration and summary of rehabilitation robotics from the late 60's to early part of 21st century can be found in the commentary by Hillman [15].

3. Surgical robotics

Surgical Robotics can be classified in two main areas: those based on "Image-guidance and minimally invasive". Most surgical robots today are being controlled directly by a surgeon mostly in a teleoperation mode in which human controls the input device and on the patient end robot follows the input. One of the main goal of that partnership between surgeons and robots exploit the capabilities of both performing the task better than either can perform alone [5].

Image-guided surgery includes orthopaedic surgery, spine surgery, neuro-surgery, reconstructive/plastic surgery and ORL surgery. The mode of operation behind image-guided surgery is the use of a robot workstation integrated into the surgical suite where some of the parts of the patient's body are fixed by means of suitable fittings. A very representative example of image guided surgery is the RoboDoc in knee and hip surgery developed by da Vinci systems.

Da Vinci is a teleoperated system in which the surgeon controls the surgical robot at a console and the robot arms follow those motions Da Vinci consists of three arms to hold two tools and an endoscope which is mounted to a single bedside cart. The grasper tool has 2 DOF inside the patient to be operated on while the EndoWrist shown in figure 3 below enhances the articulation and makes it go about complex manipulations with ease. The console incorporates separate video screen for each eye to display 3D video from the 3D endoscope. The operational end of the tools is mapped to the surgeon's hands to provide more detailed control [14].

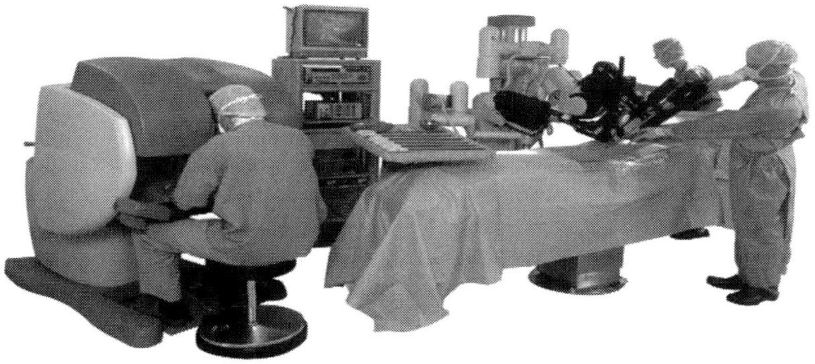

Figure 3. Shows the da Vinci Surgical System [22].

4. Current research and development in surgical robots

This section of the review paper focuses on the current research in surgical robots with the view that such research will lead to new capabilities of future commercial systems.

4.1. Neuro Arm and MrBot

One of such is NeuroArm and MrBot which are researching on improved MR-compactible robots. The MrBot developed by Johns Hopkins University is designed for (Magnetic Resonance Imagers) MRI-guided access of the prostate gland, it consists of a parallel linkage arm with a novel pneumatic stepper motor actuator to reduce the MR interference. while the NeuroArm robot which is being developed by University of Calgary in collaboration with MacDonald Dettwiler and Associates and IMRIS is an MRI-guided two- arm with 3 DoF neuro-surgical robot with a piezoelectric motor actuator [10, 11]. Figure 4 and figure 5 show the pictures of NeuroArm and MrBot.

Figure 4. NeuroArm [23].

Figure 5. MrBot [23].

4.2. HeartLander

HeartLander is a heart surgical robot; it is basically a minimally invasive robot which uses suction to move around the surface of the heart. It provides a better alternative to closed- chest, beating heart endoscopic surgery and endocardial approaches. It is used for intrapericardial drug delivery, cell transplants etc. [12].Table 1 shows how the robot works under the physician control and figure 6 shows the design of HeartLander.

Table 1. Robot working activity under physician control [24].

1	Enters the chest through an incision below the sternum
2	Adheres to the surface of the heart
3	Travels to the desired location
4	Administers the therapy

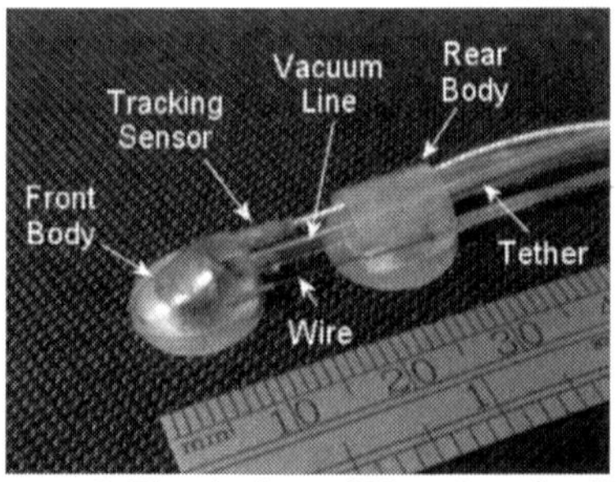

Figure 6. The design of HeartLander [24]

4.3. Amadeus

Titan Medical Inc. is currently developing Amadeus. Amadeus is a four-armed laparoscopic surgical robot system which uses a snakelike multiarticulating arm for improved manoeuvrability, and the system is being designed to accommodate teleoperation for long-distance surgery. The first human trial operation using Amadeus is planned for late 2013 [25]. The elements that make up the Amadeus Robotic Surgical System is shown in Table 2, while figure 7 shows the Amadeus Robotic Surgical System by Titan Medical Inc.

Table 2. Other elements that make up the Amadeus Robotic Surgical System [25].

Multi articulating arms	To enable obstacle avoidance in the body. Give multiple approach paths to a surgical target. Free the arms from fulcrum constraint at the body port site.
Communications	To allow the world's expert surgeons to be in the room with the patient. Increase access to specialized care. Save transportation cost.
Enhanced vision systems	Allow surgeons to see anatomy with better resolution and more viewing angle.
Force feedback	To let surgeons know how hard they are pulling on tissues, sutures or other structures for the first time.

Figure 7. Shows the Amadeus currently developing by Titan Medical Inc. [25].

4.4. Raven II
Raven II developed by the University Of Washington and University of California Santa Cruz is an academic surgical robot which is used to research into endoscopic telesurgery.It was designed to maximize surgical performance based clinical measurements [13]. The surgical robot consists of 7 degree of freedom each cable driven with two patient side arms. The arms are smaller, lighter and less expensive compared to the current robots for laparoscopy. Teleoperation experiments have been carried out with Raven II using data transmission through an unmanned aircraft. In early 2012, five surgical robotics where provided to different research labs to spur collaboration and further developmental efforts [13].Figure 8 shows the picture of Raven II Surgical Robots.

Figure 8. Raven II Surgical Robots [26]

5. Conclusion
From the introduction of this paper, it is obvious that there is a huge demand and advantages of using medical robots. The use of robots in medical, agricultural and construction field combined are still not on par when compared to the use of robots in manufacturing industries. In medical community for example, the reasons why robots did not gain immediate acceptance is both psychological in the sense that robots may be perceived as "competitors" by physicians and as potentially dangerous exotic machines by patients but presently the surgical robots are now rarely designed to replace a member of

the surgical team but to augment the medical teams by imparting superhuman capabilities such as accuracy that will otherwise by physically impossible with humans alone. Although there has been so much development from rehabilitation robotics to surgical robotics; where telemonitoring and telepresence surgery is been employed which provides global access to medical/health care but there is still so much to be done in terms of advancement and improvement for example in telepresence which allows a surgeon at one location to operate on a patient at a different location by employing an integrated surgical robotic system, the issue of speed of transfer of information from operator to the robot and the lag time from operator to execution time is a major issue. In addition economic issues such as cost-effectiveness of medical robotics and ethical issues such as patient privacy have to be addressed in the future.

References

[1] Haider A.F. Almurib, Haidar Fadhil Al-Qrimli and Nandha Kumar, A Review of Application Industrial Robotic Design, Ninth International Conference on ICT and Knowledge Engineering,pp.105 -112, (2011).

[2] Dario P., Guglielmelli E., Allota B. and Carrozza M., "Robotic for Medical Applications" (Editiorial), IEEE Robotics and Automation Magazine, vol. 5, pp.44-55(1996).

[3] Broekens J., Heerink M., and Rosendal H.,"Assistive social robots in elderly care review." Gerontechnology, vol.8 no.2, pp.94-103, 2009.

[4] Van der Loos H.F.M., Wegner J.J., Smaby N., Chang K., Madrigal O., Leifer L.J, and Khatib O.,"ProVAR assistive robot system architecture," in proc. IEEE int. conf. Robotics and Automation, Detroit,MI,1999,vol.1,pp.741-746.

[5] Mataric M. J., Eriksson J., Feil-Seifer D. and Winstein C., "Socially assistive robotics for post-stroke rehabilitation," Neuroeng J., Rehabil., vol.4 no.5, 2007.

[6] Flury P., Lopez P., Glauser D., Villotte N., Burckhardt C.W.: Minerva, a Robot dedicated to Neurosurgery Operations,Proc. of the 23rd ISIRS, Barcelona, Spain,1992,pp.729-733.

[7] Wolpaw J.R, Birbaumer N., McFarland D.J., Pfurtscheller G., and Vaughan T.M.," Brain-computer interfaces for communication and Control," Clin. Neurophysiology. vol.113, no.6, pp.767-791, 2002.

[8] Haider A.F. Almurib, Haidar Fadhil Al-Qrimli and Nandha Kumar, A Review of Application Industrial Robotic Design, Ninth International Conference on ICT and KnowledgeEngineering,pp.105 -112, (2011).

[9] Taylor R.H. and Stoianovici D., "Medical robotics in computer-integrated surgery" IEEE Trans. Robot. Automat., Vol.19, no.3, pp.765-781, 2003.

[10] Lang M.J., Greer A.D, and Sutherland G. R.," Intra-operative robotics: NeuroArm," Intraoperative Imaging, vol.109, pp.231-236, 2011.

[11] Sutherland G., McBeth P., and Louw D.," Neuroarm: an mr compactible robot for microsurgery," in international Congress series, vol.1256, pp.504-508, Elsevier, Amsterdam, The Netherland, 2003.

[12] Patronik N., Riviere C., El Qarra S. and Zenati M.," The heartlander: a novel epicardial crawling robot for myocardial injections," in International Congress Series, vol.1281, pp.735-739, Elseevier, Amsterdam, The Netherland, 2005.

[13] Lum M.J.H., Friedman D.C.W., Sankaranarayanan G. et al., "The RAVEN: Design and Validation of a telesurgery system," International Journal of Robotics Research, vol.28, no.9, pp.1183-1197, 2009.

[14] Sung G.T and Gill I. S. "Robotic laparoscopic surgery; a comparison of the da Vinci and Zeus systems," Urology, vol.58, no.6, pp.893-898, 2001.

[15] Hillman M., Rehabilitation Robotics from Past to Present. Proceeding of the Eighth International Conference on Rehabilitation Robotics, April 2003.

[16] Jacobsen S. C. et al.; Development of the UTAH Artificial Arm,IEEE Transactions on Biomedical Engineering, BME-29(40).pp,249-269.

[17] Jacobsen S.C. et al.; Design of the UTAH/MIT Dexterous Hand,Proc. of the 1980 IEEE Int. Conference on Robotics and Automation,San Francisco, CA, USA,pp.1520-1532.
[18] Kazerooni H., " Exoskeletons for human performance augmentation" In Springer Handbook of Robotics, B. Siciliano and O. Khatib,Eds.,Springer, New York, NY,USA,2008
[19] Bogue R.," Exoskeletons and robotic prosthetics: a review of recent developments," Industrial Robot, vol.36, no.5, pp.421-427, 2009.
[20] Bruno Siciliano, Oussama Khatib, Springer Handbook of Robotics, ISBN: 978-3-540-23957-4, (2008).
[21] Sister Kenny Rehabilitation Institute, "Webpage on AllinaHealth", http://www.allinahealth.org/ahs/ski.nsf/page/ar_robot
[22] Intuitive Surgical Incorporation, "Webpage on da Vinci clinical evidence," Jan 2013, http://www.intuitivesurgical.com/product/davinci_surgical_system/davinci_surgical_system_standard/
[23] Jason Stafford, Department of Imaging Physics, Physics of MRI Safety, http://www.aapm.org/meetings/amos2/pdf/59-17207-59975-979.pdf
[24] The Robotics Institute, "Webpage on HeartLander", http://www.cs.cmu.edu/~heartlander/index.html
[25] All About Robotic Surgery, The Official Medical Robotics News Center for AVRA Surgical Inc., Jan 2013, http://www.allaboutroboticsurgery.com/surgicalrobots/surgicalrobotspage2.html
[26] Raven II in the Media, "Webpage on University of Washington, Electrical Engineering", Jan 2013, http://www.ee.washington.edu/news/2012/ravenII_media.html

Advanced Manufacturing Systems in Food Processing and Packaging Industry

Mohd Shafie Sani[1] **and Faieza Abdul Aziz**[2]

[1]Department of Industrial Electronics, German-Malaysian Institute, Jalan Ilmiah, Taman Universiti, 43000 Kajang, Selangor, Malaysia

[1,2]Department of Mechanical and Manufacturing Engineering, Faculty of Engineering, Universiti Putra Malaysia, 43400 UPM Serdang, Selangor, Malaysia

E-mail: [1]shafie@gmi.edu.my, [2]dr.faieza.upm@gmail.com,

Abstract. In this paper, several advanced manufacturing systems in food processing and packaging industry are reviewed, including: biodegradable smart packaging and Nano composites, advanced automation control system consists of fieldbus technology, distributed control system and food safety inspection features. The main purpose of current technology in food processing and packaging industry is discussed due to major concern on efficiency of the plant process, productivity, quality, as well as safety. These application were chosen because they are robust, flexible, reconfigurable, preserve the quality of the food, and efficient.

1. Introduction

Food industry is one of the biggest industries in the world. In 2011, Plunkett Research reports that the total sale of food products in US is about 1508.5 billion US Dollar. This is comprised of poultry, meat, crops, vegetables, agricultural products and etc. Key components of food industry success are its food processing and packaging operation. It becomes necessary for companies to discover ways to improve their productivity in terms of maintaining safety, using sustainable materials in packaging, implementing flexible and standardized technology, and maintaining good quality of foods [1]. Reducing wastes is important, while shortening lead times is the goal of achievement. Recent advances in engineering such as Radio Frequency Identification (RFID), nanomaterials [2], and fieldbus technology bring greater opportunities to the processing and packaging industry.

2. Background

Food processing is the methods and techniques used to transform raw ingredients into food for human consumption. Food processing takes clean, harvested or slaughtered and butchered components and uses them to produce marketable food products [3]. There are several different ways in which food can be produced which are one off production, batch production, mass production or just-in-time production [4]. Food processing is any deliberate change in a food that occurs before it's available for us to eat. It can be as simple as freezing or drying food to preserve nutrients and freshness, or as complex as formulating a frozen meal with the right balance of nutrients and ingredients [5]. Processing also includes pre-processing and cleaning which sometimes is referred to as post harvesting processes [6].Traditional processing principles are based on thermal processing where the combination of temperature and time plays a significant role in eliminating the desired number of microorganisms from the food product without compromising its quality. Non-thermal processing methods such as PEF (Pulsed Electric Field), UV (Ultraviolet), and Ozone yield products with more 'fresh-like' flavour than those produced by traditional thermal processes due to fewer chemical and physical changes although they are not effective in reducing the activity of bacterial spores [7].

The principal roles of food packaging are to protect food products from outside influences and damage, to contain the food, and to provide consumers with ingredient and nutritional information [8]. Traceability, convenience, and tamper indication are secondary functions of increasing importance. The goal of food packaging is to contain food in a cost-effective way that satisfies industry requirements and consumer desires, maintains food safety, and minimizes environmental impact. A great deal of automation strategies are constantly being utilized in every phase of processing and packaging. Due to limitation of feasibility study and research in food processing, most of the studies focus on trends in food packaging and materials [6].

2.1 Materials in Food Packaging
Papers and clothes are flexible, lightweight, less waste-to-discard packaging materials. Glass and metals have been used for high-value products and are corrosion resistant and stronger, respectively. Polymers (plastics) exhibit many desirable features like transparency, softness, heat seal ability and good strength to weight ratio [9]. The most commonly used plastics in packaging industry are based on petro chemical products such as polyethylene terephthalate (PET), poly vinyl chloride (PVC), polyethylene (PE), polypropylene (PP), polystyrene (PS), and polyamide. However, they cause adverse effect (neither totally recyclable nor biodegradable) to the environment causing risk to human health or ecosystems [6]. There is an increasing demand for identifying biodegradable packaging materials and finding innovative methods to make plastic degradable. Biodegradation is the process by which carbon-containing chemical compounds are decomposed in the presence of enzymes secreted by living organisms. The use of bio plastic is to replicate the life cycle of biomass by conserving the fossil fuels, carbon dioxide and water production.

2.2 Smart Packaging
Smart packaging purpose is to add extra features to the food and also benefits the consumers. It deals with the mechanical, mechanical, chemical, electrical and electronically-driven function that enhances the usefulness of the food products in a something that can be measure. Various aspects of smart packaging are available for instances; time-temperature food quality labels, usage of self-heating or self-cooling containers integrated with electronic displays indicating important information on nutritional qualities and expiry dates. One of the good examples are the use of self-heating coffee container based on CaO exothermic reaction [10] and the introduction of RFID (Radio Frequency Identification) tag in an conventional package. The key functionality is electronic and the major beneficiaries are the stakeholders along the entire supply chain. There have been an increasing numbers of researches on application of nanocomposite materials which scale from 1-100 nm [11]. The advantage of using nanocomposite materials is mainly to improve the mechanical and oxidation stability of the foods. The famous nanocomposites used are (i) Polymer clay nanoclay (ii) Silica nanocomposites of nanosilver. The benefit of nanoclay in polymer is increased stiffness, strength and smaller cell size. Furthermore, researchers at University of Connecticut are developing nanoparticle films with embedded sensors in packaging in order to detect pathogens. This make the consumers possible to detect food contamination by monitoring colour change in packaging [12].

3. Manufacturing Systems in Food Processing and Packaging
Manufacturing systems today is beyond our imagination for few decades ago. The vast development and research in manufacturing systems brings benefit to the food processing and packaging industry. Advanced manufacturing systems include machineries and systems, automation, SCADA and DCS as well as its systems integration. Machines in food industry play major part in producing high return of investment without compromising quality in food. Moreover most of the machines are controlled by microprocessor, programmable logic controller or monitored remotely by the presence of SCADA. To

make it more efficient in producing massive amount of food products, the integration of machineries and systems must be easily adaptable to changeover with minimal time consume in tool changing. Most systems in food processing and packaging include cartooning machines, wrapping machines, labellers, shrink machines, cooling and drying machines, feeding and placing machines, inspection and detection machines as well as cleaning and sterilizing machine [6]. There have been strict regulations regarding food safety and hygiene due to numerous cases of food irradiation. The food processing industries have to take serious on the food irradiation issues by designing equipment and machine to cater the customers need.

3.1 Automation

Rising labor cost, health and safety issues lead to the application of robot and automation in food processing and packaging. The need to automate food processing and packaging activities is driven by several key requirements for competitive success and in some industries, viability of manufacturing plants. The advantages of automation are to improve productivity, product quality and profitability.

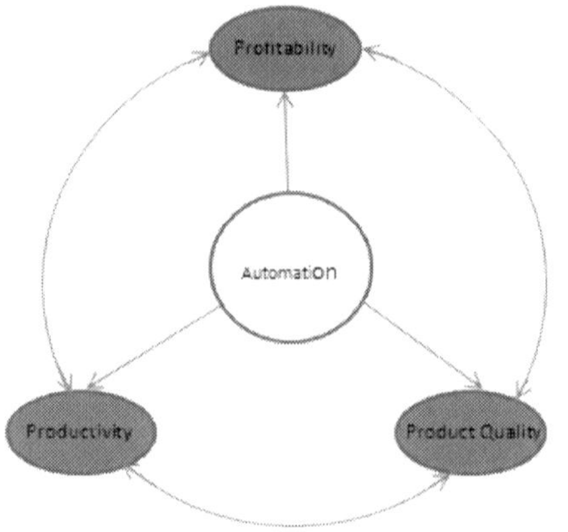

Figure 1. Plant Automation can improve profitability, product quality and productivity [13]

Productivity is directly related to how efficiently the input resources are utilized in translating them into marketable end products [13]. This is promising as a result of automation allows efficient scheduling of work and labor use. The food processing and packaging industry can preserve good records and data about past processes and significantly spot areas that can be targeted for more efficient allocation resources. According to research by University of Wisconsin-Madison, USA, an average of 30 percent increase in plant productivity by using three discrete microprocessor-based controllers designed to perform all continuous loops involving complex, integrated algorithms, valve interlocking, and some sequencing.

Quality assurance by far is the aim of any industry in the world. The capacity to process and package high quality foods consistently is the key for success in vastly competitive food industry [13]. Traditionally, the quality assurance activities in the industry are based on human visual inspection. This method is outdated and requires 100 percent involvement of human thus effect the total cycle time. As productivity increased vigorously, the quality supposed to be improved simultaneously. For example, inspection of the quality of potatoes and its sizes is achievable through the use of real time mathematical computer method using machine vision [14]. The overview of the system can be visualised in Figure 2 where the image acquisition process, pre-processing, thresholding, size measurement, defect detection and output for standard potates were taken place.

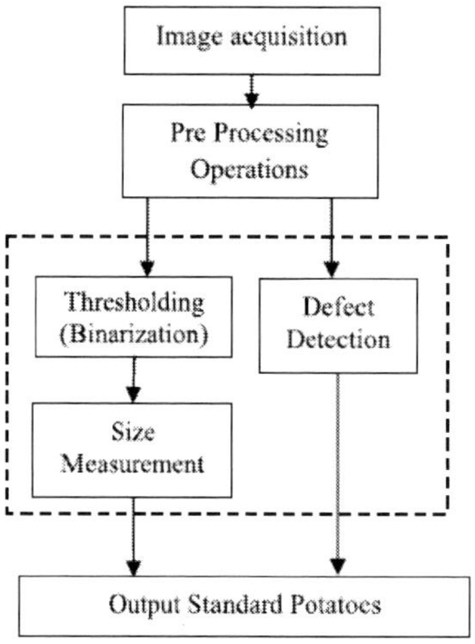

Figure 2. Potatoes machine vision inspection system's overview [14]

Although this system is still in research and developing stage, the benefit of the implementation of the system in large scale food processing and packaging industry can substantially improve quality of the end products.

The third aspect that contributes to the implementation of automation in food industry is the profit. Increased profit is the key element of business stability in eye of the management. With the improved in the productivity and quality, the profit margins will also be improved concurrently. The practice computer-controlled plant operations offer unlimited opportunities to maintain records of all events in the operations. For instance, generating usage reports helps in active inventory control. It can help to give quantitative picture of comparisons necessary for future planning [13]. This information can then readily be used to improve productivity, quality as well as profit.

3.2 Advanced Fieldbus Technology

Most of major industry in this world such as oil and gas, power generation, and waste management industries have adopted the advanced fieldbus technology into its automation and control systems since its existence. However, the implementation of this excellent technology into food processing and packaging industry as a whole is still in preliminary stage. Fieldbus is a control networking system, used for connecting field devices such as sensors, actuators, regulators, controllers, and man-machine interfaces with each other for factory automation [15]. It can be considered as sophisticated, compact and advanced digital communication method that reduce the wiring costs. Before the introduction of fieldbus technology to the automation world, most of food processing and packaging industry use conventional 4 to 20mA systems into its sensors and actuators network. The advantage is that the cost of installation is lower in low scale network due to fewer numbers of sensor and actuator. However, in mass production of food, the usage of conventional network system limits the performance thus affects the productivity. Table 1 describes the comparison between fieldbus system and conventional system [15]. When dealing with processing and packaging industries, there is a need for minimal production cycles with high performance and efficiency. By implementing fieldbus controls, an average company can save around 55% capital costs [15]. Some of the well-known fieldbus standards are ModBus, PROFIBUS DP, PROFIBUS PA, DNP3, CAN Open, P-NET, EIB, DeviceNet, LonWorks, FOUNDATION Fieldbus, WorldFIP, AS-I, SMART Distributed Systems, SERCOS, Interbus, etc [16]. Table 2 summarize the fieldbus technical data available in the market.

Table 1. Comparison of fieldbus with conventional system [15]

Category	Wiring	Information	Quantity of Information	Information Type	Control Functionality	Maintenance location	Maintenance Type
Conventional 4 to 20 mA system	Point-to-point	Unidirectional	Single data value	Measurement numerical value	Centered on a control system	Mainly in the field	Corrective maintenance
Field network	Multi-drop	Bidirectional	Multiple data values	Various data type	Distributed control	From a remote location	Predictive maintenance

Table 2. Fieldbus technical data [17]

Fieldbus Technology	**Technology Developer**	**Year Introduced**	**Physical Media**	**Max. Devices Nodes**	**Max. Distance**	**Primary Application**
Arcnet	Datapoint	1977	Coax, twisted pair, fibre	255	400-2000 feet	
AS-I	AS-I Consortium	1993	Two wire cable	31 slaves	100-300m	Assembly, packaging and materials handling machines
Bitbus	Intel		Twisted pair	32 without repeaters, 250 with repeaters	1.2km, 13.2 km	Intelligent I/O modules, Process control
CANOpen	CAN in Automation	1995	Twisted pair, optional signal and power	30	25-1000m	Sensors, actuators, automotive
ControlNet	Allen-Bradley	1996	Coax, fibre	99	250-1000m	Mission-critical, plant-wide networking of PCs, PLCs
Data Highway Plus (DH+)	Allen-Bradley		Twinaxial	64 per segment	3km	
DeviceNet	Allen-Bradley	1994	Twisted pair for signal and power	64	500m	Assembly, welding and materials handling machines
Filbus	Gespac		Twisted pair	32 without repeaters, 250 with repeaters	1.2km, 13.2 km	Remote I/O, data acquisition
Foundation Fieldbus H1	Fieldbus Foundation	1995	Twisted pair, fibre	240/segment, 65,000 segments	1900m	
Foundation Fieldbus HSE	Fieldbus Foundation	Current	Twisted pair, fibre	IP addressing, essentially unlimited	100-2000m	
IEC/ISA SP50 Fieldbus	ISA & Fieldbus Foundation	1992-1996	Twisted pair, fibre, and radio	IS: 3-7; non-IS 128	500-1700m	
Industrial Ethernet	DEC, Intel, Xerox	1976	Thin Coax, twisted pair, fibre, thick coax	1024, more via routers	185m (thin)	
Interbus-S	Phoenix Contact	1984	Twisted pair, fibre, slip ring	256	400m	Assembly, welding and materials handling machines
LONWorks	Echelon	1991	Twisted pair, fibre, power line	32,000 per domain	2000m	
Modbus Plus	Modicon		Twisted pair	32 per segment, 64 max	500m per segment	
Modbus RTU/ASCII	Modicon		Twisted pair	250 per segment	350m	
Profibus DP/PA	Siemens	DP: 1994 PA: 1995	Twisted pair or fibre	32 without repeaters, 127 with repeaters	200m, 800m	Inter-PLC communication, factory automation
Remote I/O	Allen-Bradley	1980	Twinaxial	32 per segment	6km	
SDS	Honeywell	1994	Twisted pair for signal and power	64 nodes, 126 addresses	500m	Assembly, materials handling, packaging, sortation
Seriplex	APC	1990	4-wire shielded cable	500+ devices	>500 feet	

WorldFIP	WorldFIP	1988	Twisted pair, fibre	64 without repeaters, 256 with repeaters	2 km, >10 km	Real-time control, process/machine

Fieldbus Technology	Technology Developer	Year Introduced	Physical Media	Max. Devices Nodes	Max. Distance	Primary Application
Arcnet	Datapoint	1977	Coax, twisted pair, fibre	255	400-2000 feet	
AS-I	AS-I Consortium	1993	Two wire cable	31 slaves	100-300m	Assembly, packaging and materials handling machines
Bitbus	Intel		Twisted pair	32 without repeaters, 250 with repeaters	1.2km, 13.2 km	Intelligent I/O modules, Process control
CANOpen	CAN in Automation	1995	Twisted pair, optional signal and power	30	25-1000m	Sensors, actuators, automotive
ControlNet	Allen-Bradley	1996	Coax, fibre	99	250-1000m	Mission-critical, plant-wide networking of PCs, PLCs
Data Highway Plus (DH+)	Allen-Bradley		Twinaxial	64 per segment	3km	
DeviceNet	Allen-Bradley	1994	Twisted pair for signal and power	64	500m	Assembly, welding and materials handling machines
Filbus	Gespac		Twisted pair	32 without repeaters, 250 with repeaters	1.2km, 13.2 km	Remote I/O, data acquisition
Foundation Fieldbus H1	Fieldbus Foundation	1995	Twisted pair, fibre	240/segment, 65,000 segments	1900m	
Foundation Fieldbus HSE	Fieldbus Foundation	Current	Twisted pair, fibre	IP addressing, essentially unlimited	100-2000m	
IEC/ISA SP50 Fieldbus	ISA & Fieldbus Foundation	1992-1996	Twisted pair, fibre, and radio	IS: 3-7; non-IS 128	500-1700m	
Industrial Ethernet	DEC, Intel, Xerox	1976	Thin Coax, twisted pair, fibre, thick coax	1024, more via routers	185m (thin)	
Interbus-S	Phoenix Contact	1984	Twisted pair, fibre, slip ring	256	400m	Assembly, welding and materials handling machines
LONWorks	Echelon	1991	Twisted pair, fibre, power line	32,000 per domain	2000m	
Modbus Plus	Modicon		Twisted pair	32 per segment, 64 max	500m per segment	
Modbus RTU/ASCII	Modicon		Twisted pair	250 per segment	350m	
Profibus DP/PA	Siemens	DP: 1994 PA: 1995	Twisted pair or fibre	32 without repeaters, 127 with repeaters	200m, 800m	Inter-PLC communication, factory automation
Remote I/O	Allen-Bradley	1980	Twinaxial	32 per segment	6km	
SDS	Honeywell	1994	Twisted pair for signal and power	64 nodes, 126 addresses	500m	Assembly, materials handling, packaging, sortation
Seriplex	APC	1990	4-wire shielded cable	500+ devices	>500 feet	
WorldFIP	WorldFIP	1988	Twisted pair, fibre	64 without repeaters, 256 with repeaters	2 km, >10 km	Real-time control, process/machine

3.3 Distributed Control System

Most of food processing and packaging industry mainly apply centralized control system on its automation. The outcome of adopting both decentralized control and open method will enhance the capability of functionality of the system from fixed to reconfigurable system[18]. The available fieldbus in the market has provided essential features in order to implement both DCS (Distributed Control Systems) SCADA (Supervisory Control and Data Acquisition) and architectures [19]. A distributed control system (DCS) refers to a control system usually of a manufacturing system, process or any kind of dynamic system, in which the controller elements are not central in location (like the brain) but are distributed throughout the system with each component sub-system controlled by one or more controllers [20]. In modern industrial automation, the concept of DCS has been upgraded to the next level where the application of software inside the DCS system becomes a vital component. SIEMENS for example, has introduced flexible, scalable, powerful and innovative distributed control system which is SIMATIC PCS 7. Figure 3 shows example of SIMATIC PCS 7 components integrated in field bus plant network. In this network, there are several components that have been introduced which are automation station, operation station, engineering station, terminal bus (Ethernet – TCP/IP protocol), plant bus (PROFIBUS), and field bus (PROFIBUS DP/PA) [21]. Devices connected to field bus are 'smart'. They deliver information about their health as well as the quality of the measured value [21]. The information of conventional 4-20mA analog communication is limited to control information only with insufficient management information. Nestlé Company for example, had adapted this powerful technology into their food processing automation system. The factory produces baby and toddler food for all of Europe. Up to 1000 jars and bottles are filled per minute on four production lines, which must conform to the highest product quality and safety standards. In order to meet these requirements and to ensure optimum efficiency of resources and maximum availability in the future, Nestlé had preferred SIMATIC PCS 7 systems into their food automation. The results were first class product quality, optimum efficiency of resources, and maximum availability [21].

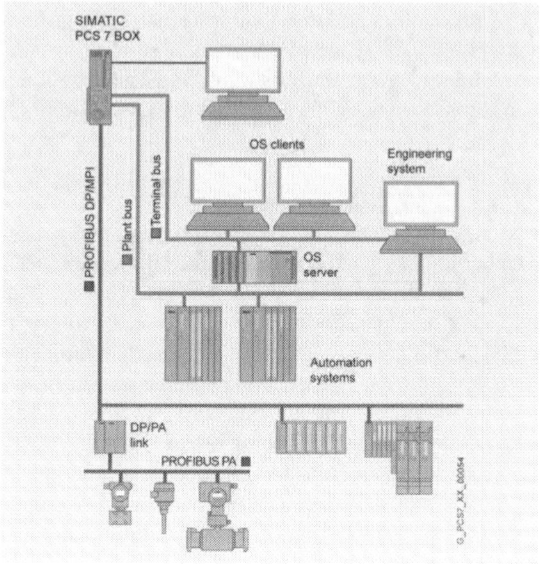

Figure 3. SIMATIC PCS 7 integration in plant network [21]

4.0 Food Safety Applications

The demand for safe and high quality food have led to innovation and novel approaches in food processing and packaging industry [22]. In order to comply with certain standard, for instance Food Safety and Quality Division, Ministry of Health Malaysia (FSQD) has imposed specific requirements on the food product to ensure the safety of food for Malaysian citizen.

Since food safety has become vigorously important in the industry, it has been compulsory to have implemented the automated in-line food inspection system [6]. The presence of contaminants such as metal, glass, bone, rubber etc must be inspected and eliminated. To accomplish this, non-destructive methods are chosen. The capability of non-destructive testing (NDT) to evaluate the food safety without compromising the quality becomes key element of its implementation. There are analysis of the effects of the drying process on the quality of food and propose the model to predict the changes in the biological and physical properties of the product through X-ray inspection [23]. Furthermore, X-ray technology also embeds artificial neural network methods and tools that provide sophisticated shape recognition functionality. The benefits of online X-ray system are the compatibility to changes in conveyor speeds, HMI (Human Machine Interface) compatibility, and system diagnostics. Besides, hyperspectral imaging is another excellent example of in-line inspection method. It is believed to increases the quality and the safety of the food products as well as improves productivity and throughput. Hyperspectral imaging offer the additional benefit of analysing the chemical composition of products hence increase production yields. By adopting this method into food processing and packaging industry, food products can be examined for disease conditions, ripeness, tenderness, grading, or contamination [24]. At the same time, machine vision application is become popular in the food industry. In this respect, the dual-band spectral imaging systems take place. The system has two-port camera system that comprise of two identical monochrome CCD cameras, and optical system and two narrow bandpass filters. The system is good in inspecting and testing off-the-shelf food products.

Traceability of food implies the ability to trace and follow a food, feed, or a food-producing animal or substance intended to be, or expected to be, incorporated into a food or feed, through all stages of production, processing, and distribution. The importance of traceability has grown due to the consumers increasing attention to food safety and food quality, and due to the increasing complexity of food supply chains [25]. A food processing and packaging industry must keep track of products, including the families of lots of production and distribution. This is crucial in order to react quickly to recall food products when a problem is reported. A track-and-trace system includes a data input system, software that includes a historian and analysis tools, and visualization and reporting [26]. Data input ranges from manual to such automatic tools as bar code and radio frequency identification (RFID).

5.0 Conclusion

As the food processing and packaging industry continues to remain competitive in an ever expanding global market, the need for technological advances that lead to increasing productivity, better product quality with enhanced safety assurance, and all at lower and lower cost, advances in automation and intelligent on-line control will inevitably continue at a rapid pace. In future, this industry is likely to transform conventional method to fully integrated and advanced manufacturing systems to further increase productivity without compromise the safety and hygienic issues. In this paper, the author have looked into the state of the art in food processing and packaging industry in terms of (i) packaging and materials, (ii) automation, control and fieldbus technology, and (iii) food safety technology. There is a need for a comprehensive research on overall food processing and packaging systems without neglecting the core element of any industry which is production management principle. Thus, food processing and packaging industry can benefit from the current findings in this area.

References

[1] Cheruvu, P., Kapa, S., & Mahalik, N. P. (2008). Recent advances in food processing and packaging technology. International Journal of Automation and Control, Inderscience, 2(4).
[2] Nambiar, A.N. (2009). Mass customization: where do we go from here? Accepted at *World Congress of Engineering -ICMEEM*'09, April, 2009, London, UK.
[3] Heldman DR, Hartel RW (1997). Principles of Food Processing. New York: Chapman and Hall.
[4] Food Engineering (2009). http://www.foodengineeringmag.com
[5] FDA (2009). U.S. Food and Drug Administration. http://www.fda.gov. Accessed March 2009.
[6] N.P.Mahalik, A.N.Nambiar (2010), *Trends in food packaging and manufacturing systems and technology*, Agricultural Sciences and Technology, California State University, USA.
[7] Dunne, C. P., & Kluter, R. A. (2001). Emerging non-thermal processing technologies: criteria

for success. *Australian Journal of Dairy Technology*, 56(2).
[8] Coles R. (2003). Introduction. In: Coles R,McDowell D, Kirwan MJ, editors. *Food packaging technology*. London, U.K.: Blackwell Publishing, CRC Press. p 1–31.
[9] Bohlmann, G. (2006). Biodegradable polymers. Sri consulting. http://www.sriconsulting.com/CEH/Public/Reports/580.0280/. Accessed 16.11.08.
[10] Goddard, N. D. R., Kemp, R. M., & Lane, R. (1997). An overview of smart technology. *Packaging Science and Technology*, 10, 129-143.
[11] Tonnie, A.O. (2007). A reference searching related to nanomaterials. *Food Packaging and Sustainability*.
[12] Selke, S. E. (2008). Nanotechnology and agrifood packaging: applications and issues.http://ifas.msu.edu/downloads/selke_AAAS_Nanotechnology_and_Agrifood_Packaging.pdf. Accessed 10.12.08.
[13] Gunasekaran. S (2011). Automation of Food Processing. *Food Engineering Vol. IV*. Univeristy of Winconsin-Madison, USA.
[14] Navid.R, B.S Mousavi, F.Soleymani (2011). A real- time mathematical computer method potato inspection using machine vision. *Computer and Mathematics with Applications* 63(2012) 268-279
[15] Matthew.Y, N.P.Mahalik, (2008). Extending fieldbus standards to food processing and packaging industry: A review. *Computer Standards and Interfaces* 31(2009) 586-598.
[16] I.Verhappen, (2006). Fieldbus Safety: Calculating your Safety Integrity Level, *Manufacturing Automation,* CLB MEDIA INC, Nov./Dec. 2006, pp. 1–3.
[17] David Green (2009), weighing-systems.com, www.weighingsystems.com/TechnologyCentre/fieldbus1.html)
[18] N.P. Mahalik, S.K. Lee. (2003). Flexible distributed control of production line with the Local Operating Network (LON) fieldbus technology: a laboratory study, *Internaional Journal of Integrated Manufacturing Systems*, 14(3), (MCB Press), pp. 268–277.
[19] Emerson Process Plant Web, (2012). http://plantweb.emersonprocess.com/customer/food_beverage_includes/technical_article/FOUNDATION_fieldbus.pdf.
[20] Stout, T. M. and Williams, T. J. (1995). "Pioneering Work in the Field of Computer Process Control". *IEEE Annals of the History of Computing* **17** (1).
[21] Siemens Simatic PCS 7, (2012) http://www.automation.siemens.com/mcms/process-control-systems/en/distributed-control-system-simatic-pcs-7/pages/distributed-control-system-simatic-pcs-7.aspx
[22] Pradeep,P., J,Jung., K,Sanghoon, (2012). Carbon dioxide sensors for intelligent food packaging applications. *Food Control* 25(2012) 328-333.
[23] Chen, X. D., & Patel, K. C. (2008). Manufacturing better quality food powders from spray drying and subsequent treatments. *Drying Technology*, 26(11), 1313-1318.
[24] Headwall Photonics (2009). Available from http://www.headwallphotonics.com/hyperspectral-applications-hyperspectralimaging-applicatoins.asp.
[25] Esther,K., Marjolein,V.D.S.,Theo,P.,Vicky,M., and Maria,G.(2012). Chapter 14 – Traceability. *Chemical Analysis of Food: Technique and Applications*. 465-498.
[26] Gary,M. (2008). Information Nuggets in the Track and Trace Data Stream. Automation World. http://www.automationworld.com/asset-management/information-nuggets-track-and-trace-data-stream

Microstructure and Nanoindentation Characterization of Low Temperature Hybrid Treated layer on Austenitic Stainless Steel

Askar Triwiyanto[1], Patthi Hussain[2], Mokhtar Che Ismail[3]

Department of Mechanical Engineering, Universiti Teknologi PETRONAS, Malaysia.

Email : asmsceng@gmail.com

Abstract. In this work, the hybrid treated layer on austenitic AISI 316L stainless steels were characterized to investigate the improvement on its surface properties. Characterization of this resulting layer was performed by FESEM (Field Emission Scanning Electron Microscope), USPM (Universal Scanning Probe Microscope) and nanoindentation. By using these methods, changes in the mechanical properties due to the diffusion of carbon and nitrogen at low temperature treatments have been traced. This hybrid treated sample has confirmed a considerable increase in hardness and a small rise in the elastic modulus compared to the untreated sample. It is found that all treated samples have enhance E/H ratio which exhibited the decreasing tendency to plastic deformation and reduced the mismatch of properties, while keeping deformation within the elastic range.

1. INTRODUCTION

Austenitic stainless steel (ASS) is used widely owing to its excellent corrosion resistance. However, the application of this material is severely limited by poor wear and friction behaviour. Stainless steel has very good corrosion resistance due to the presence Cr as alloying element to form stable thin passive layer Cr_2O_3 that protects the steel. Due to its inherent austenitic structure, this material has relatively low hardness as well as poor wear resistance which hinders a wider applicability of the material and may cause problems in existing applications. Consequently, surface engineering treatments for austenitic stainless steel are alternative ways to increase the surface hardness and improving the wear resistance. The gaseous thermochemical treatments to improve surface properties of material are typically carried out in carbon and/or nitrogen bearing gases and usually associated with temperature above 500 °C [1].

On the other hand, there was a phenomena where sensitization occurs which refers to the breakdown in corrosion resistance as a result of exposing the stainless steels in the temperature range from 550 °C to 850 °C. This problem in austenitic steel where precipitation of chromium carbides ($Cr_{23}C_6$) occurs at the grain boundaries at elevated temperatures, typically between 450 to 850°C;

[1] To whom any correspondence should be addressed.

diffusional reaction in forming chromium nitride/carbide leads to the depletion of Cr from its matrix in austenitic solid solution and consequently unable to produce Cr_2O_3 passive layer to make stainless feature. As a result, it reduces the corrosion resistance property of the stainless steel. This phenomenon causes reduction in ductility, toughness and aqueous corrosion resistance [2].

Hardening of the nitrided layer and the carburised layer is due to the incorporation of nitrogen and carbon respectively in the austenite lattice, forming a structure termed expanded austenite, which is supersaturated with nitrogen and carbon respectively [3,4]. More recently, a hybrid process has also been developed, which combines the nitriding and carburizing actions in a single process cycle by introducing nitrogen and carbon simultaneously into the austenite lattice to form a hardened zone comprising a nitrogen expanded austenite layer on top of a carbon expanded austenite layer [5-7].

Recent work has shown that low temperature hybrid process of nitriding-carburizing of austenitic stainless steel is possible in conventional tube furnace [8]. The low temperature thermochemical treatments can be performed at temperatures up to 450°C to diffuse a large amount of nitrogen and/or carbon in the treated layer of up to 20 μm thick, to form an expanded austenite (γ_x) (X=N,C) layer without formation precipitation of nitride/carbide which related to excellent corrosion resistance as well as wear performances [9,10].

Meanwhile, the surface and near-surface mechanical properties of thin films or coatings can be related to the final performance of materials. As a result, the need of better understanding in the field of depth-sensing nanoindentation would provides a quantitative method for mapping the mechanical properties, such as hardness and elastic modulus of the near-surface region [11, 12]. In order to investigate the high precision measurement of mechanical properties for thin layer Nanomechanical characterizations were performed on all treated and substrate materials. Nanoindentation was carried out to measure the mechanical properties of the implanted surface layers using Nano System 600, manufactured by Micromaterials UK, the schematic diagram for Nano System is shown in Figure 1.

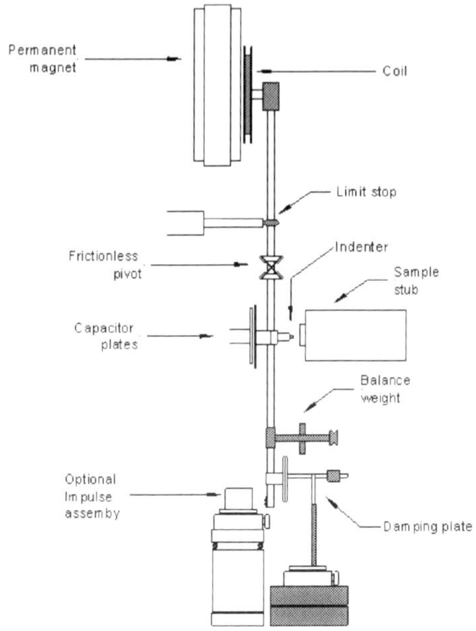

Figure 1 Schematic diagram of nano test pendulum system [20]

The aims of this paper is to characterize the expanded austenite layer on austenitic stainless steel substrates which representing a harden layer on a softer substrate respectively using the nanoindentation and morphological characterization. The hardness of expanded austenite is initially determined by the Oliver and Pharr method. In order to achieve proper measurement the use of very high applied load is avoided to minimize substrate effect where higher penetration of indenter which probably overtook the diffusion zone and entering the substrate bulk region [13].

2. EXPERIMENTAL

The specimens of 20 mm x 70 mm size rectangular coupon were cut from 2 mm thick hot-rolled plate with standard metallographical preparation for AISI 316L type austenitic stainless steel of the following chemical compositions (in wt.%): 17.018 Cr, 10.045 Ni, 2.000 Mo, 1.530 Mn, 0.030 C, 0.048 Si, 0.084 P, 0.030 S and balance Fe. The sample surface was ground on 320, 600, 800, 1000, 1200 grit SiC papers, and then polished using 1 μm Al_2O_3 pastes to the mirror finished.

Hybrid treatments were performed at 400 and 450°C in a horizontal tube furnace which involving reactive gases both NH_3 (for nitriding) and CH_4 (for carburizing) for a total duration of 8 h. The flow of gases were controlled by Aalborg flowmeter and the linear flow rate of gas mixture through the alumina retort were conditioned by gas mixing tank. In this system the sample were be placed in the quartz boat of about 12 cm in the center of isothermal zone of electric resistance tube furnaces.

Prior to treating, the specimens were soaked in concentrated HCl (2 M) solution for 15 minutes duration with the purpose to remove the native oxide film that commonly forms on stainless steel and protects the metal matrix from corrosion. This oxide layer is believed to act as a barrier for diffusional nitrogen transport [14,15]. After thermochemical treatments, the specimens were quenched in water. The cross sectioned treated specimens were first characterized by metallographic examination. To reveal the microstructure, the polished surface were etched in Marble's solution (4g $CuSO_4$ + 20 ml HCl + 20 ml distilled water). The structure and morphology of specimens were characterized using Field Emission Scanning Electron Microscope-FESEM (Zeiss Supra 55VP) and Universal Scanning Electron Microscope-USPM (NanoNavi: E-sweep) to reveal 3D surface topographical profile at higher resolutions.

Table 1 Treatment conditions and their corresponding layer thicknesses.

Treatment	Time (h)	Temp (°C)	Gas (%)			Layer Thickness
			CH_4	NH_3	N_2	(μm)
Hybrid process 316L	8	400	5	20	75	3.421
Hybrid process 316L	8	450	5	20	75	6.425

Table 2 Instrumented hardness settings

Variable	Value
Type of experiment	Depth vs. Load
Type of indentor	Berkovich pyramid
Maximum load	300 mN
Maximum depth	200nm
Dwelling time at maximum load	5 s
Loading rate	3.99 mN/s

Nanotest equipment which located at AMREC SIRIM Malaysia is made by Micro Materials Ltd. from UK and it is a nanomechanical property testing system with a Berkovich tip. The sample support was attached to a mechanism which permitted: (1) the sample to be brought into contact with the test probe of a NanoTest, (2) the applied normal force and the resulting depth of penetration to be monitored continuously. In the NanoTest unit as shown in the Figure 1, forces are generated by means of a coil and magnet system located at the top of a pendulum arrangement and displacements of the probe into the surface are monitored with a sensitive capacitor plate arrangement. In the recent measurement, the nanoindentation techniques were set as described in Table 2

3. RESULTS AND DISCUSSION

3.1 Layer morphology and surface topography analysis

Hardened layers with different morphologies were observed as a result of the various treatment conditions and the thicknesses of the layers produced in different specimens are shown in Table 1. According to the micrograph in Figure 2, expanded austenite layer is recognized as a featureless surface layer. For a similar treatment duration, the Plasma process is reported [16] to produce about 18 μm thick layer which is much higher compared to that of the present conventional hybrid treatment in horizontal tube furnace. In plasma process, the native oxide layer was removed mostly by bombardment of the plasma gas which is completely absent in conventional process. This is one of the reasons why conventional horizontal tube furnace produced small layer thickness compared to the corresponding plasma nitriding. Previous investigation revealed that nitriding at 450°C became effective after treatment for 6 h where a continuous treated layer was produced [17]. The two specimens processed under hybrid treatment conditions as shown in Figure. 2, actually produced duplex layers although not clearly revealed in the present micrograph, this separation of dual structures are observable under SEM.

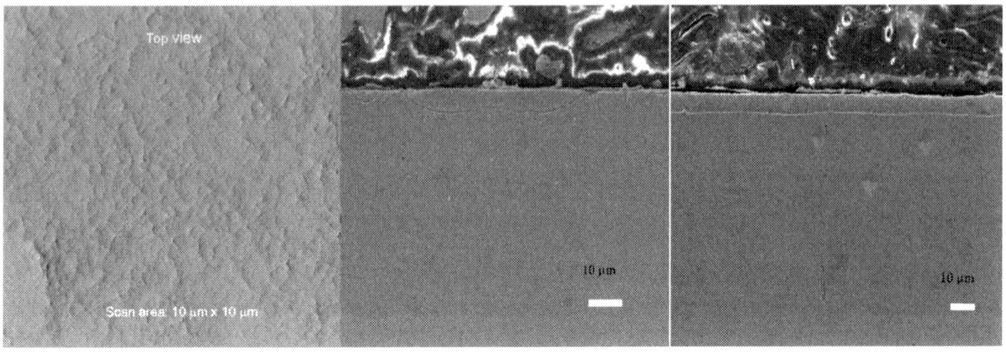

Figure 1 (a) Top view image from USPM, Cross-sectional optical micrograph; (b) hybrid 400 °C and (b) hybrid 450 °C sample.

3.2 Nanoindentation measurement profile

The instrumented hardness tests conducted on AISI 316L untreated and hybrid treated samples has confirmed a considerable increase in hardness 4 to 5 order and a small rise in the elastic modulus of the material after the hybrid thermochemical treatments as describe in Table 3. The curves in Figure 3 illustrate the applied load against the displacement of the indentor.

Previous investigation had explained that, the resilience and toughness of the material are of significant importance, and the ratio of hardness to elastic modulus has been reported to be a more appropriate index than the mere hardness, to rank the wear resistance [18,19]. Meanwhile, the best tribological performances have been reported for combinations of high surface hardness and relatively low elastic modulus, to reduce the tendency to plastic deformation and reduce the mismatch of properties, while keeping deformation within the elastic range which is an agreement with recent findings as describes in Table 3.

Table 3 Roughness, indentation hardness and elastic modulus for each hybrid treatment.

Treatment (°C)	Roughness, Ra (μm)	Hardness (GPa)	Elastic Modulus (GPa)	E/H
Untreated	0.12	2 ± 0.2	210 ± 2	97.22
Hybrid 400 °C	0.22	7.892 ± 0.7	167.518 ± 4.2	21.226
Hybrid 450 °C	0.28	8.471 ± 0.5	173.64 ± 3.1	20.50

In relation with surface roughness after thermochemical treatments as shown in Figure 1(a), Table 3 presents the general characteristics of the hybrid treated layers, due to the high nitrogen and/or carbon contents in the layers resulting from deposition process as describes elsewhere [5], a characteristics relief is usually observed which results in an increase in the surface roughness.

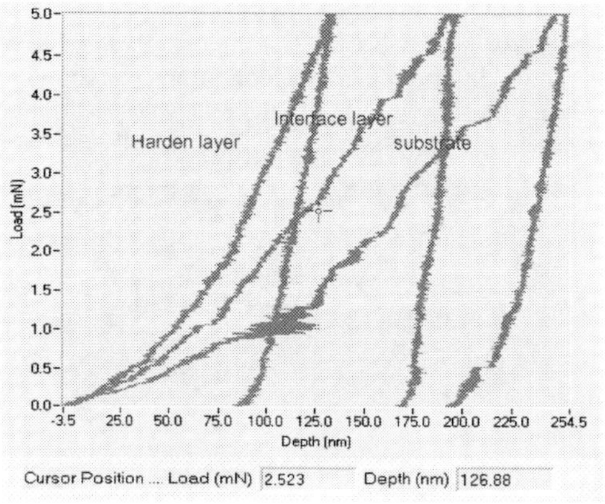

a)

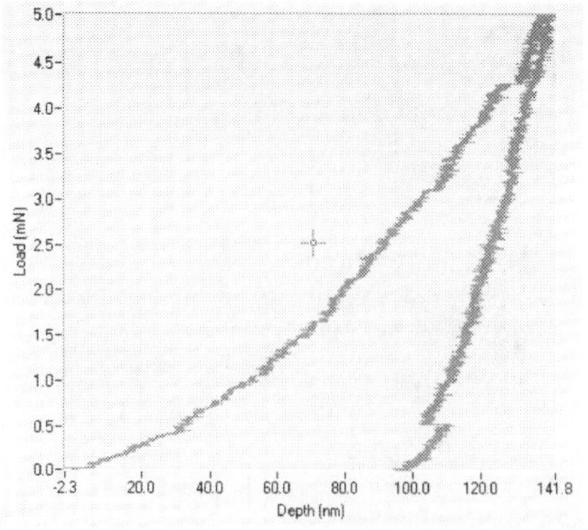

b)

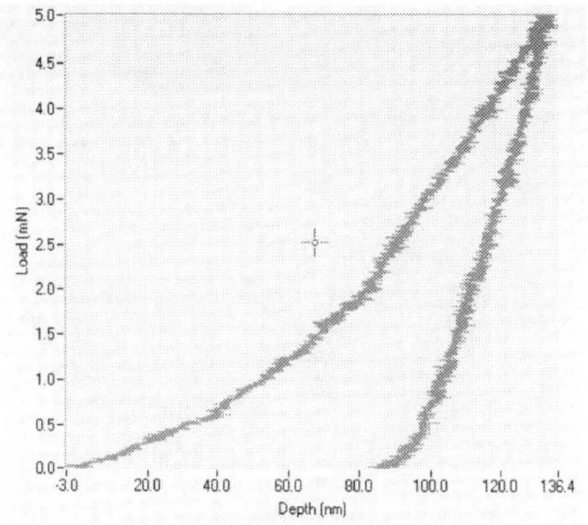

c)

Figure 2 Depth-sensing indentations performed on a) hybrid treated layer, b) hybrid 400 °C and c) hybrid 450 °C .

4. Conclusions

The thermochemical treatments of AISI 316L stainless steel in a horizontal tube process at 400 and 450°C have been demonstrate the possibilities to produce hard layer of an expanded austenite phase without precipitation of chromium carbide/nitride as confirmed by FESEM images. The layer produced in horizontal tube process is not uniform in thickness under the same treatment conditions. The layer thickness of hybrid treated at 400 °C is 6.425 μm while at 400 °C gave much smaller thicknesses for the same processing conditions. USPM observation of all treated surfaces shows a higher surface roughness after treatments. Nanoindentation tests reveal a higher elastic modulus and hardness for all treated samples compared to the untreated. It is found that all treated samples have enhance E/H ratio exhibit the decresing tendency to plastic deformation and reduce the mismatch of properties, while keeping deformation within the elastic range.

Acknowledgment

The authors would like to express their gratitude to Universiti Teknologi PETRONAS for supporting this research under STRIRF Grant No. 49/2011.

References

[1] A. Triwiyanto et al, 2012, in "Corrosion Resistance", book edited by Hong Shih, ISBN 978-953-51-0467-4.
[2] Clark. D.S. and Varney. W.R., 1962, Physical Metallurgy for Engineers, Litton educational publishers.
[3] D. B. Lewis, A. Leyland, P. R. Stevenson, J.Cawley and A. Matthews, 1993, Metallurgical study of low-temperature plasma carbon diffusion treatments for stainless steels, *Surf. Coat. Tech.* 60 416-423.
[4] Thaiwatthana, S., Li, X. Y., Dong, H. and Bell, T, 2002.: *Comparison studies on properties of nitrogen and carbon S phase on low temperature plasma alloyed AISI 316 stainless steel*, Surface Engineering, 18, *6,* 433-437
[5] M. Tsujikawa, D. Yoshida, N. Yamauchi, N. Ueda, T. Sone, S. Tanaka, 2005, Surface material design of 316 stainless steel by combination of low temperature carburizing and nitriding, *Surf. Coat. Tech.* 200, 507-511.
[6] Y. Sun and E. Haruman, 2008, Influence of processing conditions on structural characteristics of hybrid plasma surface alloyed austenitic stainless steel, *Surf. Coat. Tech.* 202, 4069-4075.
[7] X.Y. Li, J. Buhagiar, H. Dong, 2010, Characterisation of dual S phase layer on plasma carbonitrided biomedical austenitic stainless steels, *Surf. Eng.* 26, 67-73.
[8] A. Triwiyanto et al., 2011, Behavior of Carbon and Nitrogen after Low Temperature Thermochemical Treatment on Austenitic and Duplex Stainless Steel, J. *Applied Mechanics and Materials*, 110-116, 621.
[9] S. Thaiwatthana, X. Y. Li, H. Dong, T. Bell, 2002, Mechanical and chemical properties of low temperature plasma surface alloyed 316 austenitic stainless steel., *Surface Engineering.*, Vol.18 pp. 140-144.
[10] T. Christiansen, M.A.J. Somers, 2006, Characterization of low temperature surface hardened stainless steel, *Struers Journal of Materialography.*, Vol. 9, pp. 2-17.
[11] H.M. Pollock, 1992, "*Nanoindentation*", *ASM* Handbook, Friction, Lubrication, and Wear Technology, 18, pp. 419 – 429.
[12] Doerner, M. F. and Nix, W. D, 1986.: A method for interpreting the data from depth-sensing indentation instruments, Journal of Materials Research, **1**, *4,* 601-609
[13] Oliver, Pharr 1992: An improved technique for determining hardness and elastic modulus using load and displacement sensing indentation experiments.J. Mater Res.; 7: pp. 1564-1583.
[14] Yamauchi, N., Ueda, N., Demizu, K., Okamoto, A., Sone, T., Oku, K., Kouda, T., Ichii, K., Akamatsu, K., 2001, *Stainless Steel 2000*, Proc. Int. Current status seminar on thermochemical Surf. Eng., Maney Publishing, Osaka, p. 247.
[15] ASM Committee on Gas Carburizing, Carbonitriding, and Nitriding, 'Gas Nitriding', in *ASM Handbook-Heat Treatment*, Vol. 4, American Society for Metals, Materials Park, OH, pp. 199- 221, 1991.

[16] X.Y. Li, J. Buhagiar, H. Dong, 2010, Characterisation of dual S phase layer on plasma carbonitrided biomedical austenitic stainless steels, *Surf. Eng.* 26, 67-73.
[17] Y. Sun, E. Haruman, , H. Malik, A.G.E. Sutjipto, S. Mridha, K. Widi, 2006, Low Temperature Horizontal tube Nitriding of Austenitic Stainless Steel, *Solid State Phenomena.*, Vol. 118, pp. 125-130.
[18] Leyland, A. and Matthews, A, 2000.: *On the significance of the H/E ratio in wear control: a nanocomposite coating approach to optimised tribological behaviour*, Wear, **246**, *1-2,* 1-11
[19] Leyland, A. and Matthews, A, 2004.: *Design criteria for wear-resistant nanostructured and glassy metal coatings*, Surface and Coatings Technology, **177-178**, 317-324.
[20] Z.Y. Zhang, H.N. Dhakal, S.N. Surip, I. Popov, N. Bennett, 2011 "Characterisation of roof tile coating degradation using nano-indentation test and surface profilometry". *Polymer Degradation and Stability, Vol. 96, Issue 5, pp.833-838.*

Optical Properties of Nano-Porous Structure ZnO Prepared by Catalytic-Immersion Method

A Azlinda[1,2], K A Eswar[1,2], F.S Husairi[1,2], Z Khusaimi[1,2] and M Rusop[1,3]

[1]NANO-SciTech Centre (NST), Institute of Science, Universiti Teknologi MARA (UiTM), 40450 Shah Alam, Selangor, Malaysia
[2]Faculty of Applied Sciences NANO-SciTech Centre (NST), Institute of Science, Universiti Teknologi MARA (UiTM), 40450 Shah Alam, Selangor, Malaysia
[3]NANO-ElecTronic Centre (NET), Faculty of Electrical Engineering, Universiti Teknologi MARA (UiTM), 40450 Shah Alam, Selangor, Malaysia
E-mail: azlindazz@gmail.com

Abstract. This paper presents an efficient method to prepare ZnO nano-porous structure via immersion of Si substrate with catalyst assistance (gold) in the mixture of zinc nitrate hexahydrate ($Zn(NO_3)_2.6H_2O$) and urea (CH_4N_2O). Photoluminescence (PL) spectra of nano-porous ZnO revealed it had strong UV emission with low oxygen defect. Optical properties are found to be significantly affected by varying the Zn^{2+}: urea ratio but from surface morphology observation, the porous structure was stabile and seemingly unchanged.

1. Introduction

Nanosized ZnO, due to its unique properties such as chemical, structural, electrical, thermal and surface characteristics can be used in a wide range of applications depending on its shape-size controlled and material-dependent properties [1-4]. Being one of the dominant material for nanotechnology, ZnO versatility rise the range of application due to their structure, wide direct band gap (3.37 eV), large exciton binding energy (60 meV), size and thermal stability which make ZnO suitable for a wide range of devices [5, 6]. These unique properties of ZnO nanostructures attributed from its nanoscale shapes and size.

ZnO nanostructures is a versatile material that rich with various configurations such as nanorods, nanowires, nanotubes, nanoflower, nanosheet, etc [5-9] which have been successfully fabricated by variety of growth condition. Due to all these properties, ZnO is one of the most promising materials for electronic applications such as light-emitting diodes (LEDs), UV photoconductive sensor, chemical sensors and solar cells [10-13]. Reported that, ZnO nanorods can be successfully achieved by various deposition techniques such as spray pyrolysis, chemical vapor deposition (CVD), thermal treatment, sol-gel spin coating process, immersion, etc. [14-18] that have been used to generate a structure of ZnO nanostructures. Among the aforementioned techniques, immersion technique is preferred due to its simplicity, faster operation, less power consumption, and lower cost than the other techniques. It is well known that the properties of ZnO produced by immersion method are dependent on its preparation parameter such as material concentration, deposition temperature and time. Concentration plays an important role in controlling the morphologies and size of a ZnO nanostructure.

In this work, we study the influence of stabilizer molarity (urea) on the growth of ZnO using wet chemical approach on gold-seeded Si substrate and their properties were investigated. However, controlled synthesis of ZnO nanostructures using urea as a stabilizer still needs further investigation. But it is unfortunate that the papers on the synthesization of ZnO nanostructures are rarely reported to date.

2. Experimental details

2.1. Substrate preparation

One-side polished P-type silicon (Si) wafer (100) used as a substrate in this work was cut into 2 x 2 cm dimension or less for the growth of ZnO. It was ultrasonically cleaned with acetone, methanol, HF:water (1:10 ratio) and deionised (DI) water to remove organic and inorganic contamination that will affect the quality of ZnO. It was throughly dried with nitrogen gas before sputter-coated with gold for 60 s using gold target in argon plasma, which are similar to that described previously [18]. Then it was annealed at 500 °C for 30 min to ensure good adherence and improved crystallographic quality.

2.2. Synthesis of samples

The solution were prepared using a mixture of zinc nitrate hexahydrate ($Zn(NO_3)_2.6H_2O$) (purify 99.9 %), and urea (CH_4N_2O) that previously dissolved separately in DI water. The molar ratio of Zn^{2+}: urea was varied from 2:1, 1:1, 1:2, 1:4, and 1:6. Gold-seeded Si substrates were then immersed in Zn^{2+} aqueous solution at above condition in pyrex boiling tubes for 4 h at 90 °C. The deposited films on the substrates were annealed at 500 °C and the sample was subsequently taken out after cooling it down to room temperature. The chart of synthesization process is shown in figure 1.

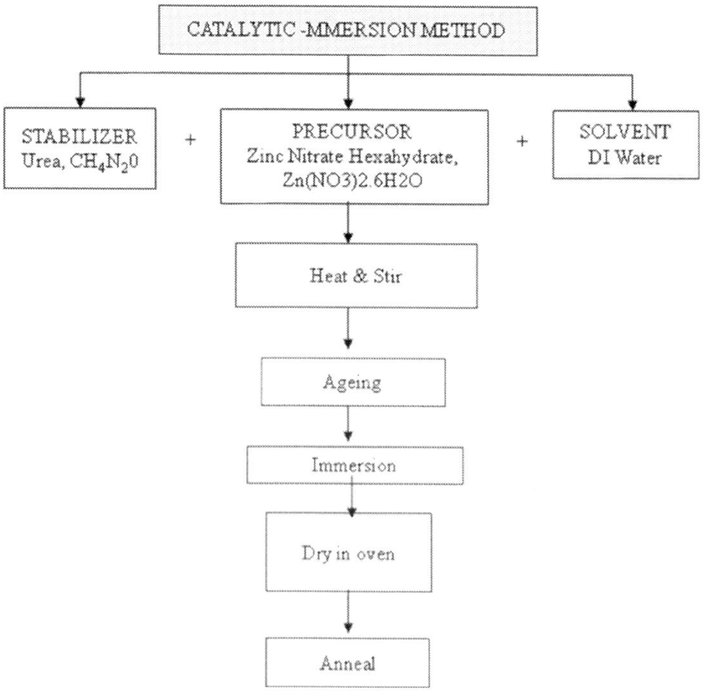

Figure 1. Chart of the research methodology

2.3. Characterisation of samples

The obtained products were characterized with a field-emission scanning electron microscope (FESEM) using JEOL JSM-7600F, to determine the surface morphology of the samples. The optical properties were analyzed using photoluminescence spectroscopy (PL), PL-Raman Horiba Jobin Yvon HR800 using He-Cd laser source and UV-VIS-NIR spectrometer (Varian Cary 5000) using diffuse reflectance accessory (DRA).

3. Results and discussion

3.1. Surface morphology

From FESEM micrograph in figure 2(a), it is obviously noted that the samples consist of hierarchical nano-sheets with network pores structure (ZnO nano-porous) grow in all direction of the template surface. The irregular pores structure of ZnO thin films grow on gold-seeded Si substrate remained similar with the synthesized parameters whether in size or structure, similar with reported by Khusaimi et.al [19]. The diameter of ZnO pores is about 50 – 75 nm in all solutions even at higher molar ratios (higher concentration of urea). EDX spectrum show the composition of the sample consists of Zn and O element which show the purity of the samples produce. This composition consistent with PL spectra in figure 2 that show nano- ZnO porous ZnO thin films have low oxygen defect.

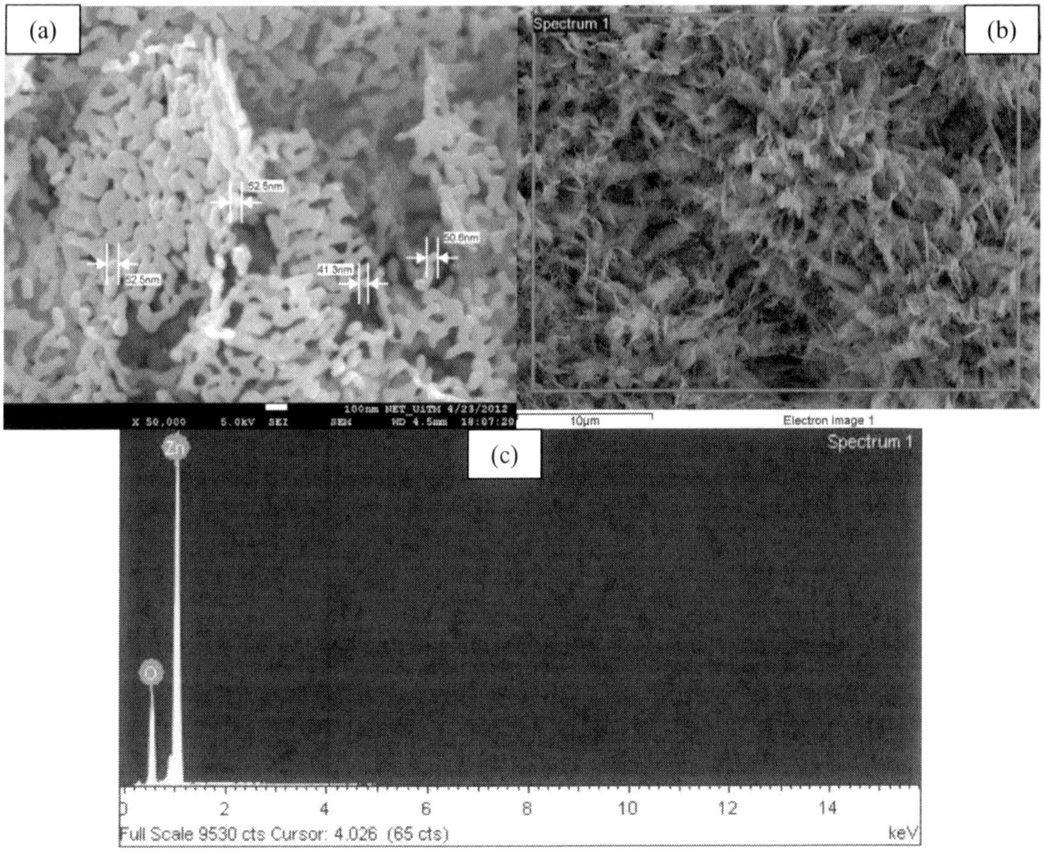

Figure 2. FESEM Micrograph of nano-porous structure ZnO (a) high magnification, (b) low magnification and (c) EDX spectrum at magnification (b).

3.2. Optical properties

PL spectra of ZnO thin films growth at varying stabilizers ratio are presented in figure 3. The measurement was conducted at room temperature. PL spectrum for as deposited sample shows the existence of 2 peaks centered at UV emission is about 402 nm (3.09 eV) attributed to near band gap exciton emission and the broad emission in the blue-green region of 500 – 700 nm might be due to oxygen deficiency or due to zinc vacancy in ZnO films[20]. According to the PL spectra grown at higher concentration of precursor ($Zn(NO_3)_2.H_2O$) solution at 0.10 M with 0.05 M of urea (2:1), a weak violet emission with the lowest intensity of visible emission can be observed. As ZnO nanostructures grown at equal molar ratio concentration (1:1), a strongest PL emission at violet region and relatively weak visible emission were detected. As the concentration of stabilizer increase (1:2, 1:4 and 1:6), the violet emission intensity gradually decreased. It can be seen that the violet emission has much sharper peak which might be due to the crystallization perfection [20]. Figure 4 shows the intensity of UV over visible emission at 402 and 600 nm respectively. The result revealed that 1:1 ratio is the optimum parameter to prepare Zn^{2+} solution to grow ZnO nano-porous due to highest intensity in UV/Vis intensity ratio graph that can led for further investigation to be promising material for optoelectronic devices.

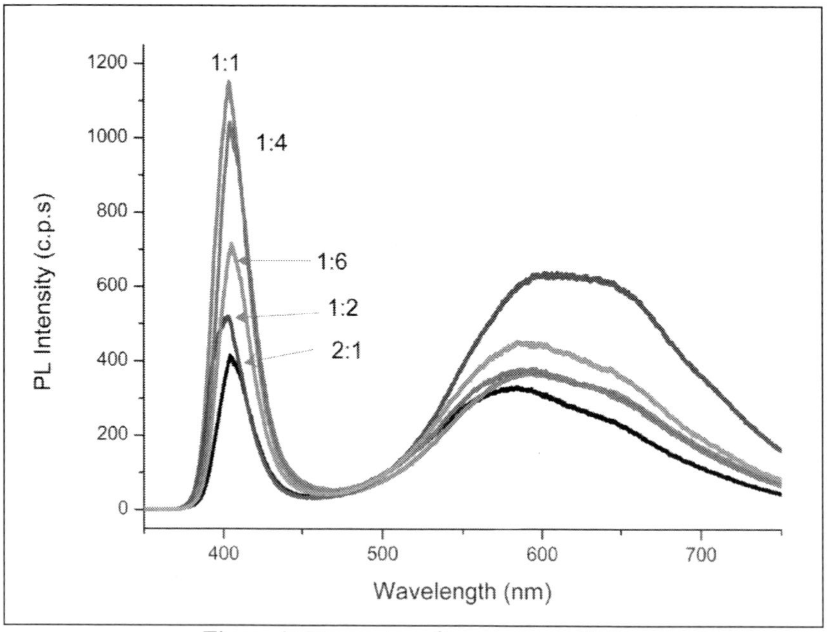

Figure 3. PL spectra of nano-porous ZnO

Figure 5 shows optical and absorbance properties of nano-porous ZnO on gold-seeded Si substrate measured using external diffuse reflectance accessory (DRA). The transmittance spectra of ZnO nano-porous structure with 2:1 molar ratio solution exhibited highest transparency in the visible region (66%) followed by 1:1 (53%) and the lowest transmittance is 2:1 (50%). This spectra shows that the transmittance tend to increase with the increasing of stabilizer concentration. Since ZnO thin films in this work were deposited with different stabilizer solution concentration, the thickness and the

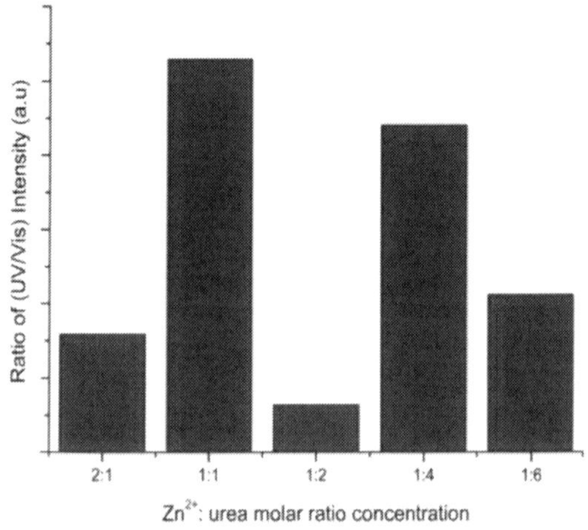

Figure 4. UV/Vis intensity ratio of nano-porous ZnO thin films

imperfect alignment of hierarchical ZnO nano-sheet with network pores structure of the films may attributed to the reduction of optical transmittance due to random scattering of incident light [21, 22]. However, the absorptivity of ZnO thin films 1:6 shows an improvement in absorption properties in UV region (below 400 nm) compared with the other molar ratio concentration. All these results indicated the optical properties of ZnO were highly dependent on the Zn^{2+}: urea solution concentration [23].

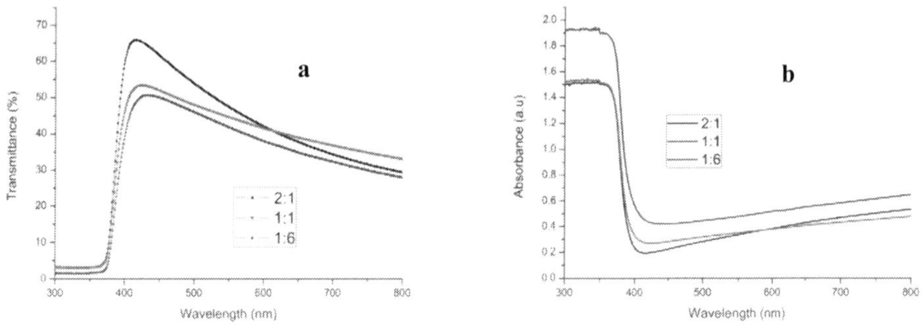

Figure 5. (a) Transmittance and (b) absorption spectra of ZnO at varying stabiliser molarity.

4. Conclusions

We have successfully grown ZnO nano-porous structure on Si substrate (100) with the assistance of gold act as a catalyst via immersion technique. FESEM image showed the morphology of nano-porous ZnO is stabile and was not affected by variation of solution concentration. The diameter of pores is about 50 – 75 nm. The PL spectra show that 1:1 ratio concentration has a relatively stronger UV emission than the other samples. These nanostructures indicate that Zn^{2+}: urea solution plays an important role to determine the optical quality of the nanostructures. UV-Vis NIR spectra reveal that the highest optical transmittance achieved at higher solution concentration ratio.

5. References

[1] D. Japic, J. Antonio Paramo, M. Marinsek, Y. M. Strzhemechny, and Z. Crnjak Orel. (2012) "Growth-morphology-luminescence correlation in ZnO-containing nanostructures synthesized in different media", *Journal of Luminescence,* Vol. 132,pp.1589-1596.
[2] H. Zhang, B. Chen, H. Jiang, C. Wang, H. Wang, and X. Wang. (2011) "A strategy for ZnO nanorod mediated multi-mode cancer treatment", *Biomaterials,* Vol. 32,pp.1906-1914.
[3] S. J. Pearton, D. P. Norton, K. Ip, Y. W. Heo, and T. Steiner. (2005) "Recent progress in processing and properties of ZnO", *Progress in Materials Science,* Vol. 50,pp.293-340.
[4] L. Li, T. Zhai, Y. Bando, and D. Golberg. (2012) "Recent progress of one-dimensional ZnO nanostructured solar cells", *Nano Energy,* Vol. 1,pp.91-106.
[5] Z. Khusaimi, S. Amizam, H. A. Rafaie, M. H. Mamat, M. Z. Sahdan, N. Abdullah, and M. Rusop, "Growth pattern of zinc oxide nanorods on gold coated silicon surfaces," in *AIP Conference Proceedings*, 2009, pp. 867-871.
[6] G. S. Wua, Y. L. Zhuang, Z. Q. Lin, X. Y. Yuan, T. Xie, and L. D. Zhang. (2006) "Synthesis and photoluminescence of Dy-doped ZnO nanowires", *Physica E,* Vol. 31,pp.5-8.
[7] G. S. Wu, T. Xie, X. Y. Yuan, Y. Li, L. Yang, Y. H. Xiao, and L. D. Zhang. (2005) "Controlled synthesis of ZnO nanowires or nanotubes via sol–gel template process", *Solid State Communications,* Vol. 134,pp.485–489.
[8] Y. Cao, B. Liu, R. Huang, Z. Xia, and S. Ge. (2011) "Flash synthesis of flower-like ZnO nanostructures by microwave-induced combustion process", *Materials Letters,* Vol. 65,pp.160-163.
[9] K. Kakiuchi, E. Hosono, T. Kimura, H. Imai, and S. Fujihara. (2006) "Fabrication of mesoporous ZnO nanosheets from precursor templates grown in aqueous solutions", *Journal of Sol-Gel Science and Technology,* Vol. 39,pp.63-72.
[10] S. Kishwar, K. ul Hasan, N. H. Alvi, P. Klason, O. Nur, and M. Willander. (2011) "A comparative study of the electrodeposition and the aqueous chemical growth techniques for the utilization of ZnO nanorods on p-GaN for white light emitting diodes", *Superlattices and Microstructures,* Vol. 49,pp.32-42.
[11] M. H. Mamat, M. Z. Sahdan, Z. Khusaimi, A. Z. Ahmed, S. Abdullah, and M. Rusop. (2010) "Influence of doping concentrations on the aluminum doped zinc oxide thin films properties for ultraviolet photoconductive sensor applications", *Optical Materials,* Vol. 32,pp.696-699.
[12] D. Barreca, D. Bekermann, E. Comini, A. Devi, R. A. Fischer, A. Gasparotto, C. Maccato, G. Sberveglieri, and E. Tondello. (2010) "1D ZnO nano-assemblies by Plasma-CVD as chemical sensors for flammable and toxic gases", *Sensors and Actuators B: Chemical,* Vol. 149,pp.1-7.
[13] L. Lu, R. Li, K. Fan, and T. Peng. (2010) "Effects of annealing conditions on the photoelectrochemical properties of dye-sensitized solar cells made with ZnO nanoparticles", *Solar Energy,* Vol. 84,pp.844-853.
[14] T. Tani, L. Madler, and S. E. Pratsinis. (2002) "Homogeneous ZnO nanoparticles by flame spray pyrolysis", *Journal of Nanoparticle Research,* Vol. 4,pp.337–343.
[15] S. Music, D. Dragcevic, M. Maljkovic, and S. Popovic. (2003) "Influence of chemical synthesis on the crystallization and properties of zinc oxide", *Materials Chemistry and Physics,* Vol. 77,pp.521-530.
[16] P. Singh, A. Kumar, A. Kaushal, D. Kaur, A. Pandey, and R. N. Goyal. (2008) "In situ high temperature XRD studies of ZnO nanopowder prepared via cost effective ultrasonic mist chemical vapour deposition", *Bull. Mater. Sci,* Vol. 31,pp.573–577.
[17] M. H. Mamat, M. Z. Sahdan, S. Amizam, H. A. Rafaie, Z. Khusaimi, and M. Rusop, "Al doped zno thin film based ultraviolet photo-conductive sensor prepared by sol-gel spin-coating method," in *AIP Conference Proceedings*, 2009, pp. 591-595.
[18] A. Azlinda, Z. Khusaimi, S. Abdullah, and M. Rusop. (2012) "Characterization of Urea versus HMTA in the Preparation of Zinc Oxide Nanostructures by Solution-Immersion Method Grown on Gold-Seeded Silicon Substrate", *Advanced Materials Research,* Vol. 364,pp.45-49.

[19] Z. Khusaimi, M. Hafiz Mamat, M. Z. Sahdan, N. Abdullah, and M. Rusop. (2011) "The Effect of Stabiliser's Molarity to the Growth of ZnO Nanorods", *Defect and Diffusion Forum,* Vol. 312-315,pp.99-103.
[20] S. C. Navale, S. W. Gosavi, and I. S. Mulla. (2008) "Controlled synthesis of ZnO from nanospheres to micro-rods and its gas sensing studies", *Talanta,* Vol. 75,pp.1315-1319.
[21] M. H. Mamat, Z. Khusaimi, M. Z. Musa, M. F. Malek, and M. Rusop. (2011) "Fabrication of ultraviolet photoconductive sensor using a novel aluminium-doped zinc oxide nanorod-nanoflake network thin film prepared via ultrasonic-assisted sol-gel and immersion methods", *Sensors and Actuators A: Physical,* Vol. 171,pp.241-247.
[22] E. Bacaksiz, S. Yılmaz, M. Parlak, A. Varilci, and M. Altunbas. (2009) "Effects of annealing temperature on the structural and optical properties of ZnO hexagonal pyramids", *Journal of Alloys and Compounds,* Vol. 478,pp.367–370.
[23] I. M. Joni, A. Purwanto, F. Iskandar, M. Hazata, and K. Okuyama. (2009) "Intense UV-light absorption of ZnO nanoparticles prepared using a pulse combustion-spray pyrolysis method", *Chemical Engineering Journal,* Vol. 155,pp.433-441.

Acknowledgments
Authors would like to acknowledge Universiti Teknologi MARA (UiTM) (Research Management Institute, 600-RM|/DANA 5/3/RIF (666/2012)) for their financial support.

Thermal Degradation of Nanocomposited PMMA/TiO_2 Nanocomposites

Nik Noor Hafizah[1], Mohamad Hafiz Mamat[2], Che Mohamad Som Said[3], Mohd Hanafiah Abidin[4] and Mohamad Rusop[5]

[1,5]NANO-SciTech Centre (NST), Institute of Science (IOS), Universiti Teknologi MARA (UiTM), 40450 Shah Alam, Selangor, Malaysia

[1,3]Faculty of Applied Sciences, Universiti Teknologi MARA (UiTM), 40450 Shah Alam, Selangor, Malaysia

[4]Faculty of Sports Science & Recreation, , Universiti Teknologi MARA (UiTM), 40450 Shah Alam, Selangor, Malaysia

[2,5]NANO-ElecTronic Centre (NET), Faculty of Electrical Engineering, Universiti Teknologi MARA (UiTM), 40450 Shah Alam, Selangor, Malaysia

Email: n.noor_hafizah@yahoo.com

Abstract. The polymer nanocomposite is a new choice to conventionally filled polymers. The lack of proper binding between the filler and the polymer can lead the decrease of the thermal and other properties of the nanocomposites. In this study, the nanocomposited PMMA/TiO_2 nanocomposites were prepared using sonication and solution casting method at different weight percent TiO_2. The aims of adding TiO_2 in the PMMA is to study the effects of TiO_2 nanofiller on the thermal properties nanocomposites. FESEM results show the higher amounts of TiO_2 in PMMA increase the rough surface morphology of the samples. Further, the Raman results reveal that the TiO_2 nanofiller were successfully intercalated into the PMMA matrix. In addition, the thermal properties of nanocomposited PMMA/TiO_2 nanocomposites were increased with the addition of TiO_2 in the PMMA.

1. Introduction

In recent years, there has been a strong prominence on the development of the polymeric nanocomposites and the interest in the production of them has comes from the large variety of possible applications including this type of material. The features such as higher resistances towards high temperature, UV radiation, high gas impermeability, higher mechanical strength, low expansibility and processing flexibility can be achieved by the production of the polymer nanocomposites. The particulate composites reinforced with the micro or nanosized particles of various materials are the most commonly utilized composites in everyday materials. Particles are normally added to increase the matrix elastic modulus and also yield strength. Therefore, by scaling the particle size down to the nanometer scale, it has been proven that the novel material properties could be obtained. Further, the polymer filled with the nano-sized fillers exhibit a dramatic improvements in the mechanical, thermal and also flammability properties [1]. Many previous studies have been reported on the properties of the nanocomposites. A. Laachachi et al has mix the poly (methyl methacrylate) with the TiO_2 by molten the PMMA pellets and the filers using a rheometer at 225 °C and 50 rpm. The resulting composites then were pelletized using rotary cutter mill and moulded by compression. From the results, they found that the nanoparticles of TiO_2 are well distributed in the material with some tendency to aggregation and the TiO_2 nanoparticles significantly enhance the thermal stability of PMMA which attributed to the restriction of mobility of polymers chain confirmed by the increase of the thermal properties with the TiO_2 amount [2]. Further, the researcher has prepared the nanocomposite by coating method. In the study, they found the problem in coating PMMA and TiO_2 which is lack of proper binding sites on the surface that can leads in the bad adhesion and hence, can cause the weak properties of the sample [3]. In the literature, the presence of inorganic nanoparticles can help to an increase of thermal properties of the materials [4-7]. Therefore, in this study, the polymer PMMA was added with the TiO_2 nanopowder to study the thermal properties and thermal degradations of the nanocomposited PMMA/TiO_2 nanocomposites at different weight percent of TiO_2. The samples were prepared using the sonication and solution casting method and the samples were measured using field emission scanning electron microscopy (FESEM), Raman spectroscopy and thermogravimetry analysis(TGA).

2. Experimental Procedure

There are several steps involve in the preparation of PMMA/TiO_2 nanocomposites. First, the preparation of the PMMA and TiO_2 solution where beaker A contain of 0.6 g poly(methyl methacrylate) (PMMA) (by ChemAR) with 10 ml toluene, and beaker B contain of 1 wt% of titanium dioxide (TiO_2) nanopowder (by ALDRICH Chemicals) with 10 ml toluene. Both of the beaker A and B were sonicated for 30 min. Then, the solution in the beaker A and B were mixed together and was sonicated again for 1 hr. After that, the mixture of the solution was poured into the Petri dish and was dried in the fume hood for 24 hrs to produce the PMMA/TiO_2 nanocomposites in free standing form. The steps are illustrated in the figure 1 below. All the steps were repeated to prepare PMMA/TiO_2 nanocomposites with different weight percent of TiO_2. The amount of TiO_2 were varies from 1 to 20 wt%. The samples were then characterized using field emission scanning electron microscopy (FESEM), Raman spectroscopy and thermo gravimetric analysis (TGA). All the results will be discussed.

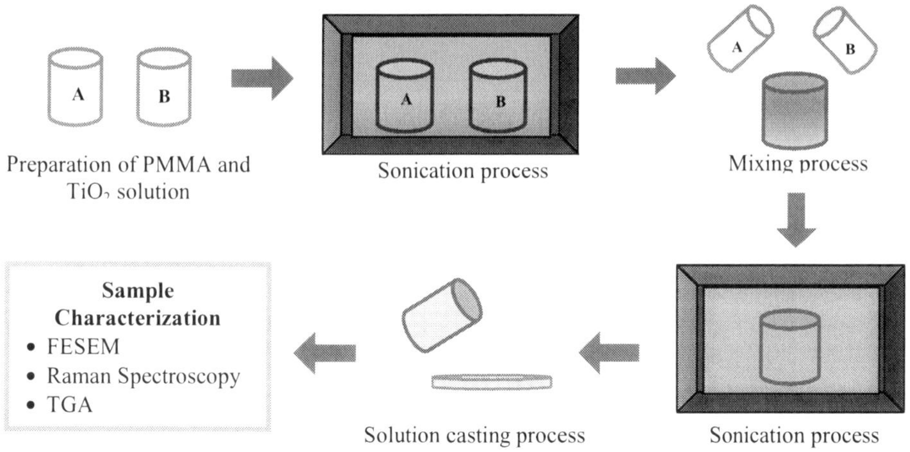

Figure 1: Flow process for the preparation of PMMA/TiO_2 nanocomposites

3. Results and Discussions

The FESEM images show the surface morphology of nanocomposited PMMA/TiO_2 nanocomposites samples at different TiO_2 content. TiO_2 nanoparticles were found to be loaded into the PMMA phase, as indicated by the white regions that were shown in the figure 2(b-f). The different amounts of TiO_2 nanoparticles give the different morphology of the nanocomposites samples. Sample 0 w% TiO_2 nanofiller as shown in the figure 2(a) that contains only PMMA shows the porous structure with smooth film surface. In contrast, the surface morphology of the samples that contain TiO_2 nanoparticles are coarse and small crack were observed as shown in the figure 2(b-f). When TiO_2 nanoparticles were added into the PMMA, the porous structures disappear.

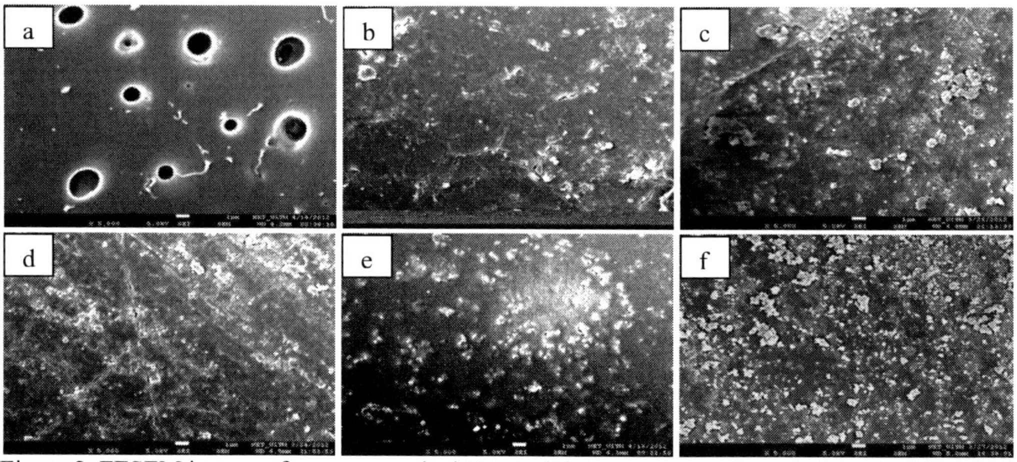

Figure 2: FESEM images of nanocomposited PMMA/TiO_2 nanocomposites of (a) 0 wt% TiO_2, (b) 1 wt% TiO_2, (c) 5 wt% TiO_2, (d) 10 wt% TiO_2, (e) 15 wt% TiO_2 and (f) 20 wt% TiO_2.

The different amounts of TiO_2 nanoparticles in the nanocomposites samples affect the structural properties of the nanocomposites samples where the cracks of the films morphology also increases as TiO_2 amount increase up to 10 wt% as shown in the figure 2(d). The dispersion of TiO_2 nanofiller in the polymer matrix plays important roles in identifying the final properties of the nanocomposites. Based on the previous study, the researcher reported that the good dispersion of TiO_2 filler in the PMMA will contribute to the increase in the glass transition temperature (Tg) value and also increase in mechanical properties of the nanocomposites samples [8]. In addition, other researchers found that the well dispersed of the metal oxides in the PMMA will improve its thermal stability and activation energy of the thermal degradation of the composites [9-12].

Further, Raman spectroscopy is used to study the phase content of the nanocomposited PMMA/TiO_2 nanocomposites. The TiO_2 nanoparticles are found to be encapsulated into the PMMA matrix. This can be seen in the figure 3(a-e), where the peaks of anatase were detected in the nanocomposites samples. The frequency of the Raman band for anatase TiO_2 peaks are identified at around: 147, 630, 513, and 519 cm^{-1} [13]. Liu et al also reported that the peaks detected are same with the peaks detected in this study [14, 15].

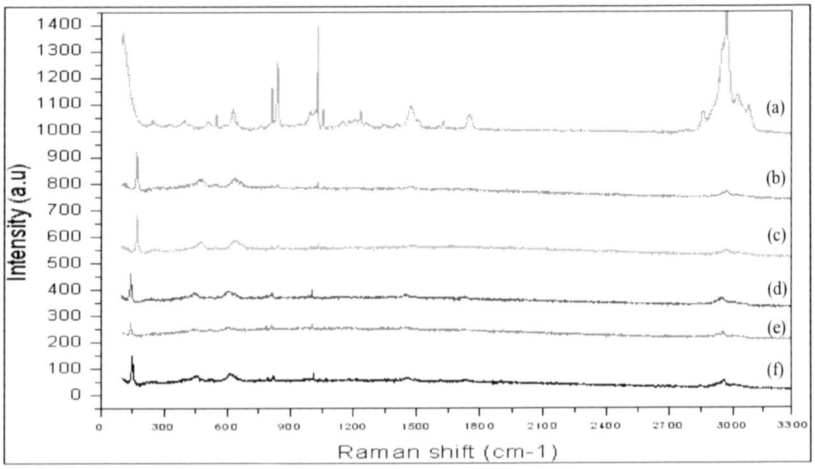

Figure 3: Raman Spectrum of nanocomposited PMMA/TiO_2 nanocomposites of (a) 0 wt% TiO_2, (b) 1 wt% TiO_2, (c) 5 wt% TiO_2, (d) 10 wt% TiO_2, (e) 15 wt% TiO_2 and (f) 20 wt% TiO_2

Then, TGA is used to measure the thermal degradation of the nanocomposited PMMA/TiO_2 nanocomposites samples. The thermograms show the sample PMMA has low thermal degradation compared to the samples of nanocomposited PMMA/TiO_2 nanocomposites.

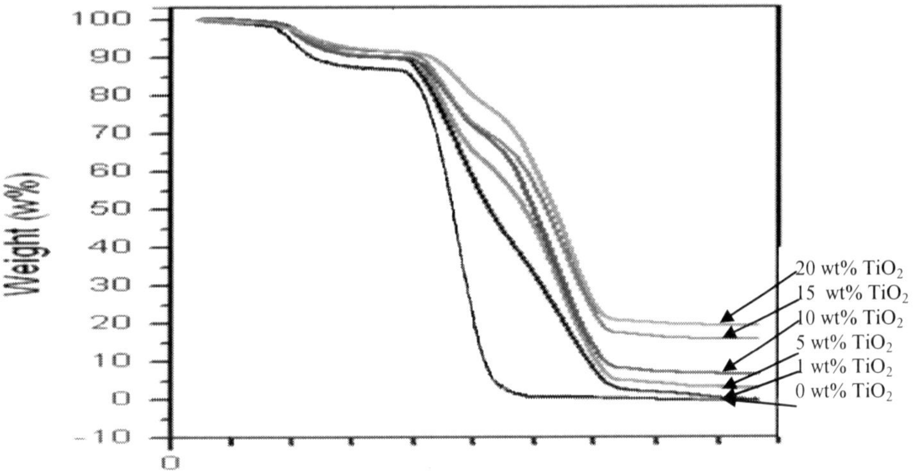

Figure 4: TGA analysis of nanocomposited PMMA/TiO_2 nanocomposites of (a) 0 wt% TiO_2, (b) 1 wt% TiO_2, (c) 5 wt% TiO_2, (d) 10 wt% TiO_2, (e) 15 wt% TiO_2 and (f) 20 wt% TiO_2

For the sample PMMA, the first stage of the thermal degradation occurs at around 120 °C which is same with the sample PMMA/TiO_2 nanocomposites. The second stage thermal degradation was observed at the temperature around 220 °C for the PMMA sample and the degradation temperature increase starts from about 240 °C for the PMMA/TiO_2 nanocomposites sample. The third stage observed at the temperature about 330 °C for the samples PMMA/TiO_2 nanocomposites and small stage observed for the sample PMMA. The thermal degradation temperature increases with the increase amount of TiO_2. From the thermograms, the degradation temperature that occurred below 150 °C can be attributed to the evaporation of incorporated water and decomposition of the peroxo groups that contain in the samples [15]. The degradation temperature that detected around 220 °C might be due to the decomposition of the small amount of head-to-head linkages in PMMA of unsaturated end groups of C=C bonds which are unstable [16, 17].

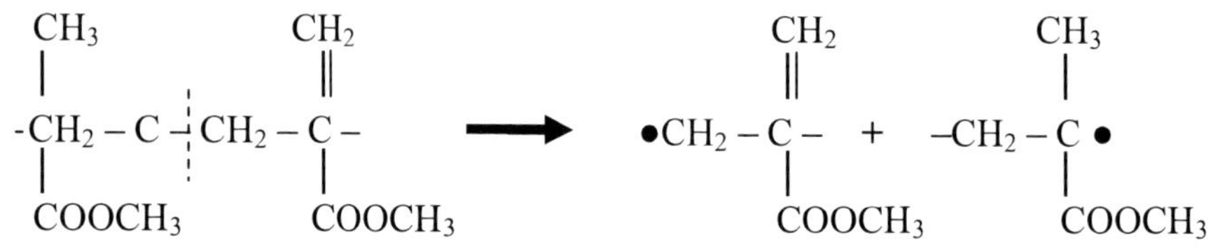

(1)

The next degradation stage is initiated by the random C-C scission. It is predominant at higher temperature and occurs at about 300 °C. Some researcher reported that at around this temperature, the weak of head-to-head linkages of the main chain were ruptured. In addition, the decomposition temperature was detected at this temperature also might due to the terminal vinyl group decomposition [18, 19]. In the literature, Garzella et al reported that the most intense peak is observed at around this stage is due to depolymerization that initiated by the random chain scission [20].

Meanwhile, the activation energy observed is higher than the first reaction. At high temperature about 300 °C, the dominant mechanism occurred because it decompose PMMA into methyl methacrylate (MMA) as the major product as shown in the reaction 2 [17, 21-23].

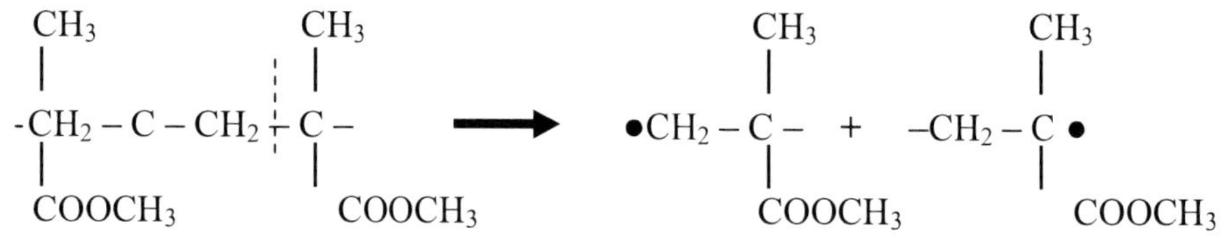

(2)

The intermediate products that further decompose produce the monomer. The reaction is shown below;

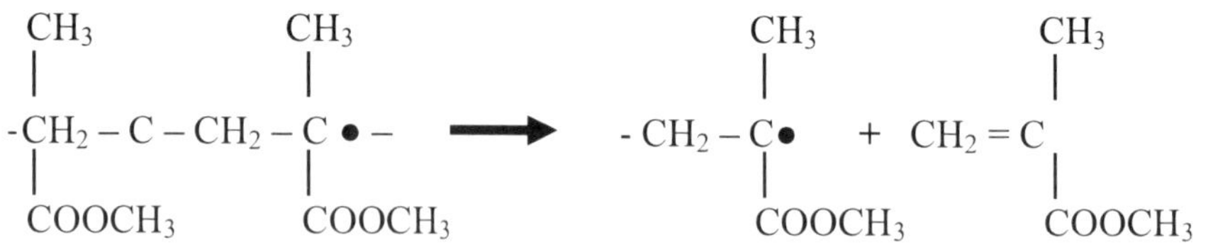

(3)

The degradation of the PMMA using TGA was done in the presence of oxygen. In the presence of oxygen, PMMA undergoes thermal oxidative decomposition due to the reaction of macro radicals with oxygen to form hydro peroxides which are unstable and will break down rapidly to give more free radicals. The main product of decomposition is monomer MMA. MMA will then further decomposed to produce smaller gaseous that are combustible. As described by previous studies, the monomer of MMA will break down to give small molecules by completing primarily two scission paths as shown in the equation below;

(4)

(5)

According to the authors, the equation 4 is the major reaction. Meanwhile equation 5 is the minor scission path [24].

From the thermograms, it is clearly indicates that the addition of the TiO_2 nanofiller in the PMMA shifted the degradation temperature to a higher temperature. The results are in accordance with the surface morphology that have been observed using FESEM where the researcher declared that the homogenously distributed TiO_2 nanofiller in the PMMA matrix will slowing the thermal degradation of the nanocomposites and hence, lead to the increase in thermal properties of the nanocomposites compare to the sample that contain of only PMMA. All the samples PMMA/TiO_2 nanocomposites show the samples decomposed completely at about 400 °C, same with the sample PMMA. The amorphous structure of TiO_2 is converted to anatase phase as the temperature increases from 370 °C. The results were confirming with the Raman Spectroscopy results in the previous section. There is no weight loss on TGA curve after 400 °C for the PMMA/TiO_2 nanocomposites samples, indicates that the decomposition of TiO_2 precursor has complete and can start as a first annealing temperature.

The thermal stability observed is higher in the presence of TiO_2 filler. The reasons of this to occur is because of the basis restriction of mobility of the polymer chains which resulting from some factors such as the bonds which due to the adsorption of the polymer on the oxide surface via a methoxycarbonyl group, and also due to the steric hindrance which is due to the presence of the solid particles of the additives [25]. For the TiO_2 as filler, it can be assumed that the restriction of the mobility by oxide particles is also responsible for the increase of thermal stability, most likely by inhibiting a key step of the reaction mechanism involved that is key step occurring during the two first weight losses. The increase of the degradation temperature were generally considered as an indication of enhancement on the thermal stability of the PMMA/TiO_2 nanocomposites [26, 27].

4. Conclusion

The nanocomposited PMMA/TiO_2 nanocomposites were successfully produced using the sonication and solution casting method. From the results, it is found that the addition of nanosized TiO_2 filler can modify and increase the thermal properties of the nanocomposited samples. The samples that contain TiO_2 nanofiller have higher thermal degradation compared to the pure PMMA itself. The increase of the thermal degradation also reveals that the TiO_2 nanofiller was successfully dispersed in the PMMA as shown in the FESEM images where the successfully dispersed the filler in the polymer leads to increase in the thermal properties of the nanocomposites. Further studies have to be done on to study on the mechanical properties of these nanocomposited PMMA/TiO_2 nanocomposites samples.

5. Acknowledgement

The authors would like to shows the million thanks and deep appreciations to the Research Management Institute (RMI), UiTM for the Research Intensive Faculty (RIF) (600-RMI/DANA 5/3/RIF(665/2012)) for the financial supports.

References

[1] S. S. Ray and M. Okamoto, *Prog. Polym. Sci.*, vol. 28, pp. 1539-1641, 2003.
[2] A. Laachachi, M. Ferriol, M. Cochez, D. Ruch, and J.-M. Lopez-Cuesta, *Polymer Degradation and Stability*, vol. 93, pp. 1131–1137, 2008.
[3] W. Su, S. Wang, X. Wang, X. Fu, and J. Weng, *Surface and Coatings Technology*, vol. 205, pp. 465-469, 2010.
[4] M. Yang and Y. Dan, *Colloid Polymer Science*, vol. 284, pp. 243-250, 2005.
[5] C.C. Weng and K.H. Wei, *Chemistry of Materials*, vol. 15, pp. 2936-2941, 2003.
[6] A. Laachachi, M. Cochez, M. Ferriol, J.M. Lopez-Cuesta, and E. Leroy, *Materials Letters*, vol. 59, pp. 36–39, January 2005.
[7] M. Marinovic-Cincovic, Z. V. Saponjic, V. Djokovic, S. K. Milonjic, and J.M. Nedeljkovic, *Polymer Degradation and Stability*, vol. 91, pp. 313-316, February 2006.
[8] M. Dixit, S. Gupta, V. Mathur, K. S. Rathore, K. Sharma, and N. S. Saxena, *Chalcogenide Letters*, vol. Vol. 6, pp. 131 - 136, March 2009.
[9] E. Džunuzović, M. Marinović-Cinović, K. Jeremić, and J. Nedeljković, *Polymer Degradation and Stability*, vol. 97, pp. 701-704, 2009.
[10] S. Majoni, S. Su, and J. M. Hossenlopp, *Polymer Degradation and Stability*, vol. 95, pp. 1593-1604, 2010.
[11] A. Laachachi, M. Ferriol, M. Cochez, J.-M. L. Cuesta, and D. Ruch, *Polymer Degradation Stability*, vol. 94, pp. 1373-1378, 2004.
[12] T. Orhan, N. A. Isitman, J. Hacaloglu, and C. Kaynak, *Polymer Degradation Stability*, vol. 96, pp. 1780-1787, 2011.
[13] M. Plodinec, I. Friščić, D. Iveković, N. Tomašić, D. S. Su, J. Zhang, and A. Gajović, *Journal of Alloys and Compounds*, vol. 499, pp. 113–120, 2010.
[14] Y. H. Liu, M.-C. Chang, P.-I. Liu, L.-C. Chung, H. Shao, M.-S. Huang, R.-Y. Horng, and A. C.-M. Yang, *Jornal of Material Science*, vol. 45, pp. 6212–6217, 2010.
[15] L. Ge, M. Xu, M. Sun, and H. Fang, *Materials Research Bulletin*, vol. 41, pp. 1596-1603, 2006.
[16] Miriam A. Freedman, James S. Becker, and S. J. Sibener, *The Journal of Physical Chemistry B*, vol. 112, pp. 16090-16096, 11 October 2008.
[17] T.-L. Tsai, C.-C. Lin, G.-L. Guo, and T.-C. Chu, *Industrial & Engineering Chemistry Research*, vol. 47, pp. 2554-2560, 2008/04/01 2008.
[18] J. M. Hwu, G. J. Jiang, Z. M. Gao, W. Xie, and W. P. Pan, *Journal of Applied Polymer Science*, vol. 83, pp. 1702–1708, 2002.
[19] A. Laachachi, D. Ruch, F. Addiego, M. Ferriol, M. Cochez, and J.-M. Lopez Cuesta, *Polymer Degradation and Stability*, vol. 94, pp. 670-678, 2009.
[20] C. Garzella, E. Bontempi, L.E. Depero, A. Vomiero, G. Della Mea, and G. Sberveglieri, *Sensors and Actuators B*, vol. 93, pp. 495–502, 2003.
[21] C. Cao, X. Zhang, Z. Zhang, Z. Liu, and H. Zhang, *IEEE Proceeding*, pp. 5829-5832, 2011.
[22] E. Kandare, G. Chigwada, D. Wang, C.A. Wilkie, and J.M. Hossenlopp, *Polymer Degradation and Stability*, vol. 91, pp. 1781-1790, 2006.
[23] P. R. Westmoreland, T. Inguilizian, and K. Rotem, *Thermochimica Acta*, vol. 367-368, pp. 401-405, 2001.
[24] W. R. Zeng, S. F. Li, and W. K. Chow, *Journal of Fire Sciences*, vol. 20, pp. 401-433, 2002.
[25] S. Zulfigar and K. Masud, *Polymer Degradation and Stability*, vol. 78, pp. 305–313, 2002.
[26] Y. L. Liu, C. Y. Hsu, and K. Ying, *Polymer*, vol. 46, pp. 1851-1856, 2005.
[27] T. Kashiwagi, A. B. Morgan, J. M. Antonucci, M. R. VanLandingham, R. H. Harris, W. H. Awad, and J. R. Shields, *Journal of Applied Polymer Science*, vol. 89, pp. 2072-2078, 2003.

AUTHOR INDEX

AUTHOR INDEX

Institute of Physics
Dirac House, Temple Back
Bristol BS1 6BE UK